47th Deep Foundations Institute Annual Conference 2022 (DFI47)

Leading with Purpose

National Harbor, Maryland, USA
4-7 October 2022

ISBN: 978-1-7138-7065-4

Printed from e-media with permission by:

Curran Associates, Inc.
57 Morehouse Lane
Red Hook, NY 12571

Some format issues inherent in the e-media version may also appear in this print version.

Printed with permission by Curran Associates, Inc. (2023)

For permission requests, please contact Deep Foundations Institute at the address below.

Deep Foundations Institute
326 Lafayette Avenue
Hawthorne NJ 0750

Phone: 973-423-4030
Fax: 973-423-4031

www.dfi.org

Additional copies of this publication are available from:

Curran Associates, Inc.
57 Morehouse Lane
Red Hook, NY 12571 USA
Phone: 845-758-0400
Fax: 845-758-2633
Email: curran@proceedings.com
Web: www.proceedings.com

DEEP FOUNDATIONS INSTITUTE

DFI

DFI47

LEADING WITH PURPOSE.

47th Annual Conference on Deep Foundations

October 4-7, 2022 | National Harbor, Maryland

2022 PROCEEDINGS

Keynote

Adapting Deep Foundation Techniques

Designing for Complex and Local Geology

Groundbreaking, Mega, and Local Projects

High Value Subsurface Characterization

Infrastructure, Policy, and Legislation

Load testing, Smart Monitoring, and Data Management

Local and National/International Case Histories

The Kitchen Sink

FOUNDATION CHALLENGES AND INNOVATIVE SOLUTIONS: THE WHARF PHASE 2, WASHINGTON, DC

Aaron L. Sacks, Mueser Rutledge Consulting Engineers PLLC, Washington, DC, USA, 202-554-0770, asacks@mrce.com
Joseph K. Cavey, Keller North America, Inc., Odenton, Maryland, USA, 410-551-1980, jkcavey@keller-na.com
David R. Good, Mueser Rutledge Consulting Engineers PLLC, New York NY, 917-339-9327, dgood@mrce.com
Srinivas Yenamandra, Mueser Rutledge Consulting Engineers PLLC, New York NY, 917-339-9399, syenamandra@mrce.com

ABSTRACT

Construction of Parcels 6 and 7, the second phase of The Wharf, began in 2019 and geotechnical construction was completed by December 2020. The overall project consisted of horizontal site, parks, and office buildings. The Wharf's Phase 2 development includes approximately 1.2 million ft^2 (110,000 m^2) of above-grade construction including five multi-story, mixed-use buildings constructed over up to three levels of underground parking. The site is located between the WMATA Green Line Tunnels to the north and the Washington Channel to the south; it is bisected by a 108-in. (2740-mm) diameter DC Water outfall pipe. This arrangement, along with very difficult ground conditions, created challenges in the geotechnical design and construction. The support of excavation for this complex project required several different methodologies, including sheet piles, soldier piles, displacement piles, tiebacks, deep foundation anchors, jet grouting, and internal bracing with multi-tier rakers. Ground improvement was also required for support of spread footings in soft soils over the garage footprint. Underpinning of the active 108-in. (2740-mm) diameter DC Water outfall pipe running on a north/south alignment through the center of the site between the two garage structures was also performed. Since portions of the support of excavation are below the groundwater table, an extensive pre-excavation dewatering program was necessary and was separate from the foundation package. On the north side of the project, support of excavation (SOE) was required on a curved alignment, offset approximately 10 ft (3 m) from the outbound Green Line tunnel. Due to the proximity of the tunnel, a stiff, near-zero displacement wall was required. A portion of the buildings extend over the tunnels, imparting load to the tunnels to approximately restore original overburden loads excavated for the basement space. The building foundation is a 3 to 4-ft (0.9 to 1.2-m) thick mat slab anchored into the Potomac clay formation. Continuous movement monitoring of the tunnels was performed by an automated system. This paper describes the design and construction of the SOE and foundations for this high-profile project, which required collaboration between the Owner, Engineer, Construction Manager, and Specialty Geotechnical Contractor to solve many challenges.

Keywords: Rigid Inclusions, bracing, jet grouting, rigid anchors, tie-downs, DC Mega Project, Keller, Displacement Piles, Auger Cast Piles, dewatering.

INTRODUCTION

The Wharf Phase 2 is a major development on Washington, DC's Southwest Waterfront. The overall project consists of the construction of two below-grade parking garages over the entire 7-acre site with 1.2 million ft^2 (110,000 m^2) of above-grade construction, including five multi-story, mixed-use buildings.

The project foundation walls are located only about 10 ft (3 m) from the Washington Metropolitan Area Transit Authority (WMATA) Green Line Tunnels which run in a curve along the north perimeter. The Washington Channel defines the south perimeter with the project bisected by a 108-in. (2740 -mm) diameter DC Water storm sewer between the two garages. To add to the complexity, plans called for removing soil from above the tunnels to within as little as approximately 10 ft (3 m) of the tunnel crown. These adjacencies (Fig. 1), with difficult ground conditions, required the innovative application of a variety of geotechnical design and construction techniques to deliver the foundations while also protecting adjacent critical infrastructure.

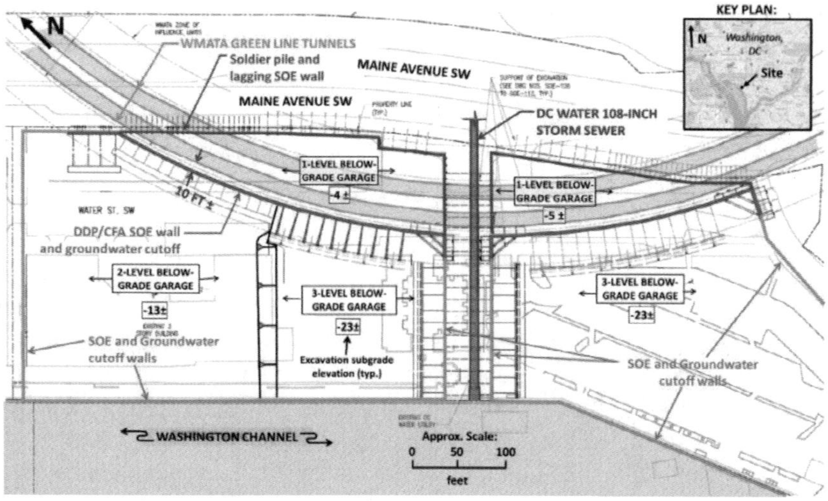

Fig. 1. The Wharf Phase 2 site plan showing foundation constraints and SOE elements

SUBSURFACE GEOLOGIC CONDITIONS

Site subsurface geologic conditions consist of 10 to 20 ft (3 to 6 m) of manmade fill overlying a relatively soft, geologically recent layer of alluvial silt and clay that thickens toward the Washington Channel. The alluvial deposits are underlain by the local "Terrace" soils, interlayered Pleistocene deposits of sand, gravel, and clay varying from loose to very compact. Hard clay and compact clayey sand of the Cretaceous-aged Potomac formation are present 40 to 50 ft (12 to 15 m) below grade. A typical soil profile, with the WMATA tunnels and excavation profile, is provided in Fig. 2.

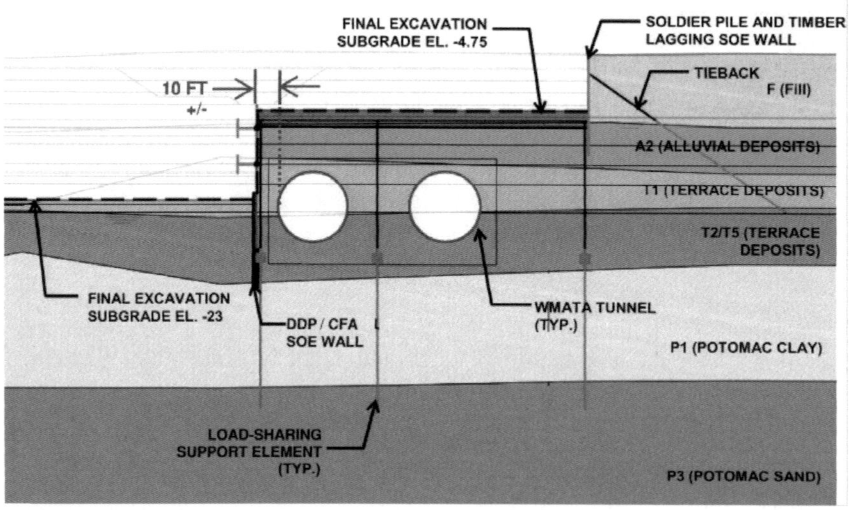

Fig. 2. Typical cross-section through tunnels and foundation showing soil profile

SITE CHALLENGES

WMATA Tunnels

WMATA's Green Line subway tunnels are twin, 17.5-ft (5.34-m) diameter tunnels with tunnel crown set 27 to 31 ft (8.2 to 9.5 m) below the pre-construction ground surface. The tunnels were constructed in the 1980s by earth mining with a shield and lined with a segmental steel liner. Liner segments were fabricated from 7/16-in. (11-mm) thick steel plates and include flange and stiffener plates. The liner plates were bolted together at the joints with lead caulk waterproofing.

Development plans called for one below-grade parking level above the tunnels and two or three below-grade parking levels in the area beside the tunnels (Fig. 1). Constructing this below-grade space required excavating asymmetrically in close proximity, both horizontally and vertically, to the tunnels.

DC Water Outfall Pipe

In addition to constructing the SOE around the WMATA Green Line on the north side of the site, it was necessary to provide SOE and underpinning of a 108-in (2740-mm) diameter DC Water outfall pipe running on a north/south alignment through the center of the site between the two garage structures (see Fig. 1).

Soft Soils on the South Side of the Site

Interior building columns for most of the site were supported on the upper Terrace and Potomac levels. However, on the south side of the site soil conditions below Garages 2A and 2B consisted of loose sand and medium clays.

SUPPORT OF EXCAVATION

With the excavation limits so close to the active Metro tunnels, maintaining the small allowable movement tolerance required an extremely stiff support of excavation (SOE) wall, carefully controlled sequence of excavation, and strategic use of dewatering (Fig. 3). Even with an extremely stiff SOE wall, construction effects such as loss-of-ground, and excessive relaxation of soil pressures commonly experienced during drilling piles had to be controlled by both "best practices" and engineered solutions.

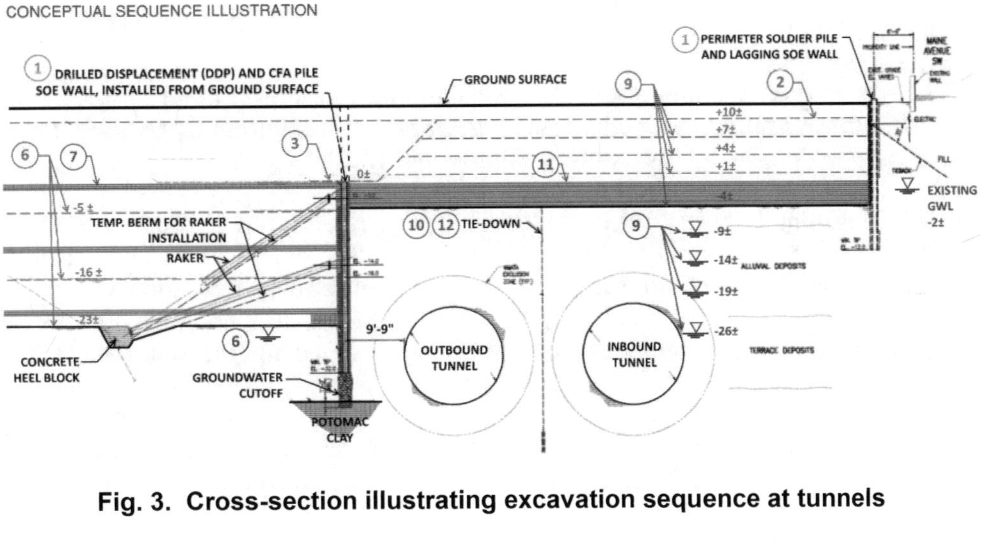

Fig. 3. Cross-section illustrating excavation sequence at tunnels

Key aspects of design features to meet the strict imposed movement threshold criteria, included:

Use of Dewatering to Counteract upward Tunnel Movement due to Excavation Above Tunnels

The excavation above the tunnel was calculated to cause upward movement of the tunnels exceeding the defined limits. As one measure to counter this, a system of deep wells was installed along the northeast side of the site. The wells were outfitted with variable speed pumps and pressure transducer controls and operated to lower the groundwater level around the tunnels in sequence with excavation of overburden.

Lowering groundwater levels decreases the buoyancy of the air-filled tunnels and increases the effective soil weight above the tunnels, restraining upward tunnel movement. In a similar "reverse" manner, the groundwater level was allowed to recover in sequence with the application of building load. The amount of dewatering by construction sequence was determined by a finite element soil-structure interaction (PLAXIS 2D) model, calibrated at each stage based on monitoring data recorded.

Hybrid Drilled Displacement Pile (DDP) and Auger Cast (CFA) Pile SOE Wall Adjacent to Tunnel

The threat of tunnel movement toward the excavation due to the construction effects of potential loss-of-ground and relaxation of local lateral earth pressures due to drilling of nearby SOE soldier piles cannot be calculated. Therefore, a system to preclude those concerns was engineered. The basic SOE design consisted of 18-in. (457-mm) diameter piles installed on a 24-in. (610-mm) [near-tangent] spacing, with a steel core beam in every pile. To provide a seal against groundwater and prevent soil loss through gaps between piles, 8-in. (203-mm) diameter holes filled with cement-bentonite grout were drilled between piles (see Fig. 4). The key to preclude tunnel movement toward the SOE was how those SOE piles were installed.

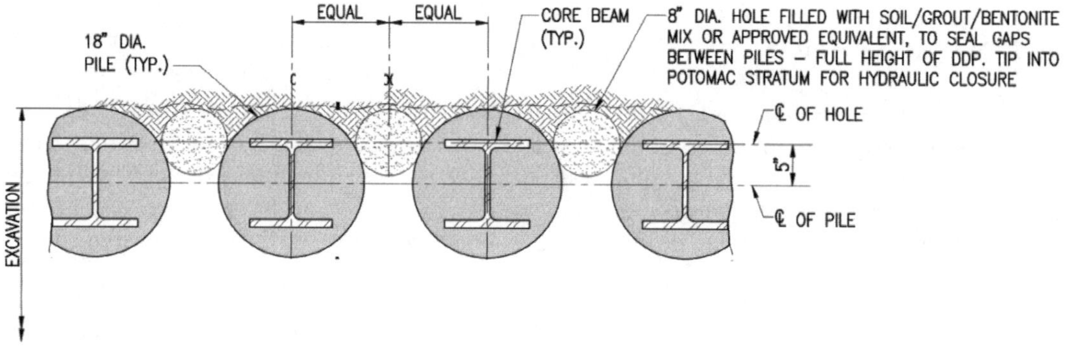

Fig. 4. Design of SOE wall adjacent to outbound tunnel (source: *Wharf Horizontal & Garage Phase 2, Support of Excavation* by Mueser Rutledge Consulting Engineers PLLC)

Mueser Rutledge Consulting Engineers PLLC (MRCE), designers of the SOE within the WMATA Zone of Influence, designed full-displacement drilling methods for the SOE wall. Rather than potentially losing or relaxing ground, installing drilled displacement piles densifies soil without vibrations, increasing horizontal earth pressure around the piles. The intent was to increase lateral stresses against the tunnel which would then relax slightly as the excavation proceeded and the SOE deflected toward the excavation.

To quantify the effect of displacement drilling, an off-site test had been conducted months earlier in similar soils to evaluate the concern of excess lateral pressure applied to the tunnel prior to excavation.

Because a small, but excessive by the project standards, lateral deformation of a portion of the tunnel did occur due to the initial displacement pile installation, a second on-site field test was conducted to include interlacing drilled displacement piles (DDP) with conventionally drilled continuous flight auger (CFA) piles. This combination of DDP and CFA provides the desirable "push-pull" capability, providing a system for relative precise control of tunnel movement.

The on-site test consisted of sequenced installation of a pattern of alternating DDP and CFA with monitoring by inclinometers at various distances. Based on the results of the test, a pattern of alternating DDP and CFA piles was selected for the SOE along the tunnels. The pattern was adjusted during construction based on tunnel deformation monitoring data, with an optimal DDP coverage ratio of approximately 25% (one DDP alternating with three CFA piles). This pattern resulted in a small, but measurable, horizontal compression of the adjacent tunnel liner during wall installation and allowed the tunnel to return to near-zero displacement when excavation was performed beside the wall. Installation of DDP and CFA piles is shown in Fig. 5.

Fig. 5. Installation of DDP and CFA piles

The DDP piles were installed by Keller using a bored pile drill rig to advance the holes into the Potomac Formation. CFA piles were drilled with an 18-in. (457.2) continuous flight auger advanced into the Potomac. A sand-cement 4,000 psi (27.6 MPa) grout was placed as the displacement tool (or auger) was extracted from the hole. After placing grout and extracting the drill tools, W10x88 beams were set into the hole. Setting the beams was difficult in some areas due to surface obstructions, cobbles at the top of the Potomac stratum, and the proximity of the piles to one another. In some areas, the beams could not be placed to their intended position within the hole and the bracing had to be adjusted.

At the conclusion of the DDP and CFA pile installation, Keller drilled and grouted 8-in. (203-mm) diameter holes between the piles and placed cement-bentonite grout to provide water cut-off. This grouting was originally performed as jet grouting. However, initial jet grouting resulted in measurable tunnel movement due to the higher pressures used in the jet grouting process. Therefore, small diameter holes, with low pressure cement slurry grouting, were installed to close the gaps between the DDP piles.

SOE Bracing

To minimize deflection of the SOE wall and consequent tunnel movement during excavation, MRCE designed a stiff bracing system consisting of steel wales, rakers, struts, and concrete heel blocks. The design included pre-loading every raker and strut using hydraulic jacks (Fig. 6). The wales, rakers and struts were located to avoid permanent structural elements (columns and basement wall construction joints). To minimize the excavation depth adjacent to the tunnels, raker heel blocks were integrated with column footings (Fig. 7). This required close coordination between the SOE design and structural foundation design and collaboration between the Foundation and SOE contractors during construction.

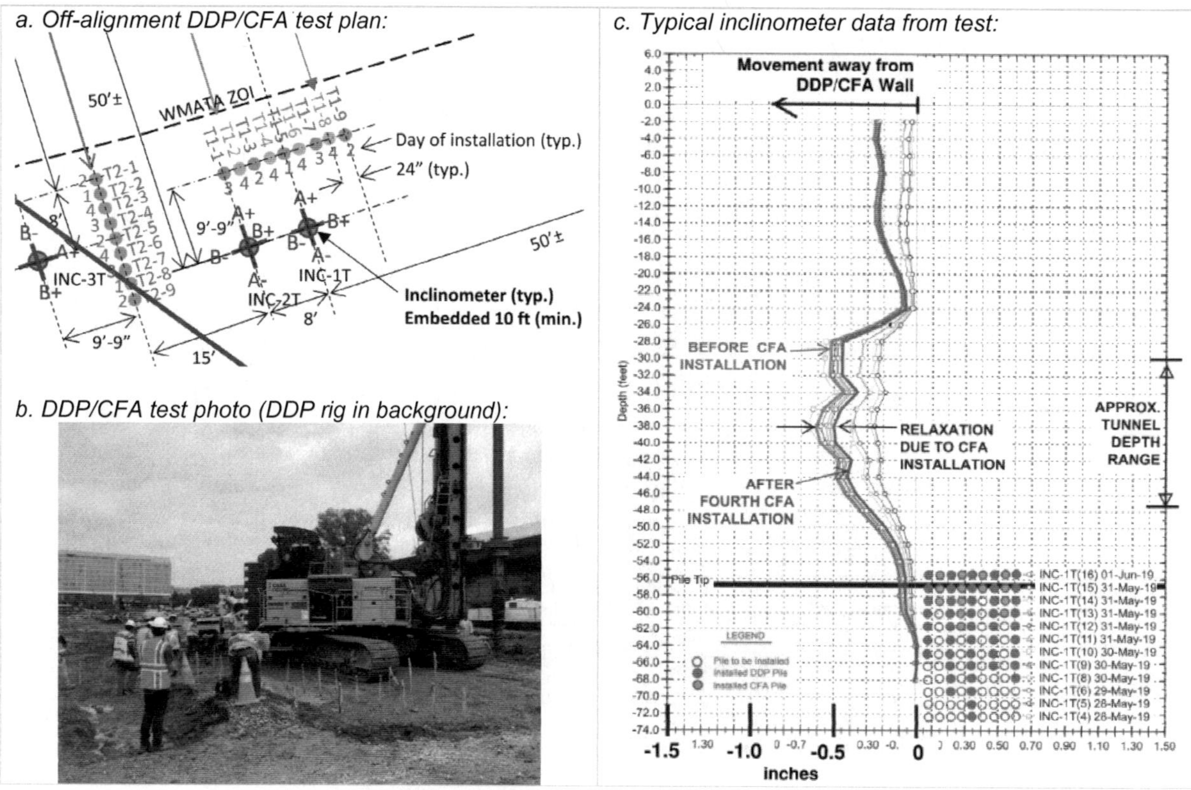

Fig. 6. Off-alignment DDP/CFA pattern testing, showing:

- **a.) Layout of DDP's (red) and CFA's (green) relative to inclinometers (blue)**
- **b.) DDP rig and tooling**
- **c.) Typical inclinometer data from test shows ground movement due to DDP and CFA installation**

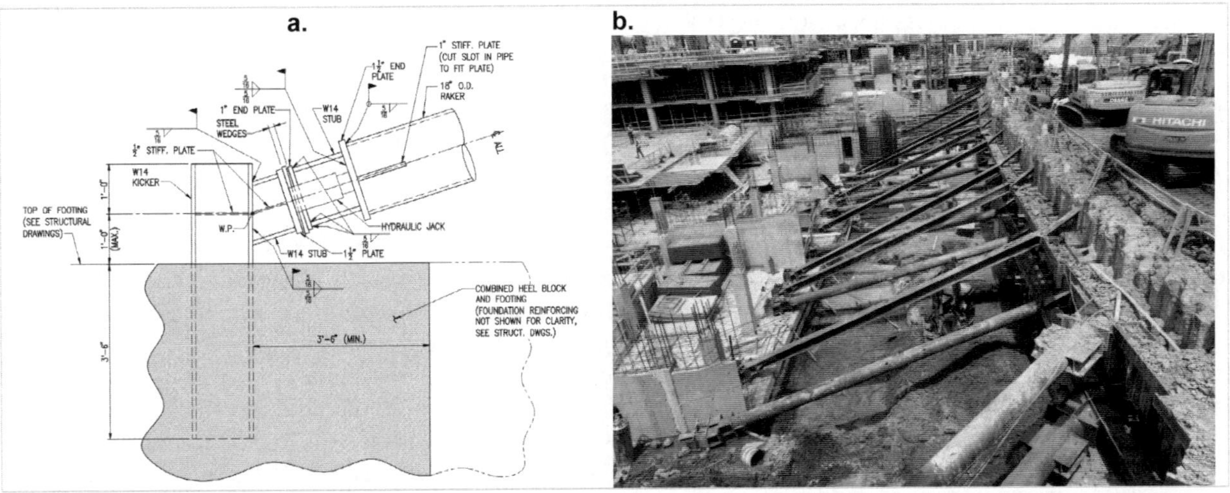

Fig. 7. SOE bracing: a. design detail; b. construction photograph showing DDP/CFA wall, raker braces, and heel blocks integrated with footings.

Underpinning of DC Water Outfall

Underpinning of an active 108-in. (2740-mm) diameter DC Water outfall pipe running through the center of the site between the two garage structures was performed using jet grouting (Fig. 8). The outfall pipe was first underpinned with 6-ft (1.8-m) diameter jet grout columns installed beneath the cradles that supported the pipe. During jetting operations, the outfall pipe remained in service, which required ongoing monitoring for movement.

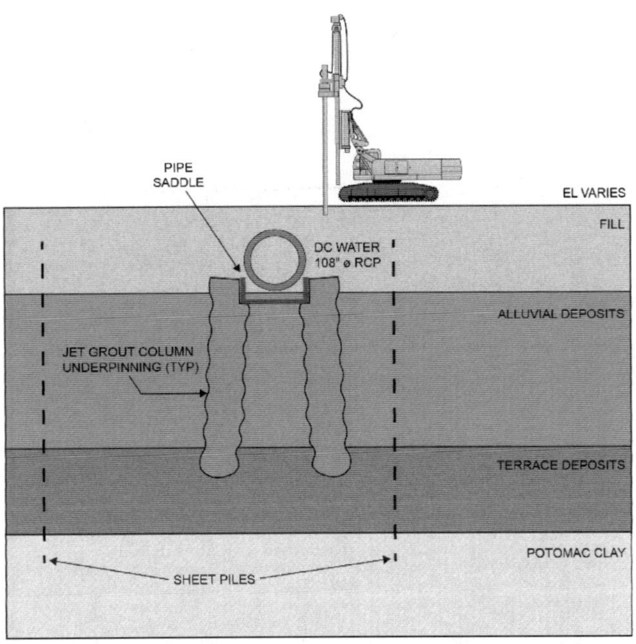

Fig. 8. Jet grout underpinning of 108-in. (2740-mm) RCP pipe prior to driving sheet piling

Steel sheet piling was installed on both the west and east sides of the pipe and tie rods were installed between the sheets for lateral support. An all-thread steel bar was used for each of the tie rods, connecting the west and east sheet piles above the pipe. These tie rods were installed by excavating trenches (and setting the bar) between the rows of sheet piles.

FOUNDATION CHALLENGES AND SOLUTIONS

Ground Improvement by Rigid Inclusions

Rigid inclusions (Fig. 9) were installed for ground improvement within a portion of Garage 2A to support the floor slabs at level P2 as well as to support the spread footings at level P2 to an average allowable bearing pressure of 8,000 psf (383 kPa). Post-construction settlement of no more than 1 in. (25 mm) was specified. Prior to production work, load tests on two individual RIs were performed to verify the design capacity. Sixteen-inch diameter RIs were used to support column foundations and 12-in. (305 mm) diameter RIs were used for slab support. Due to the depth of the soft alluvium and the soft terrace silt in the south area of the site, the RI design was based on two profiles. Profile 1 covered the northern portion of the site (approximately 75% of the treatment zone), and the RIs were seated in the upper sand Terrace deposits, terminating at Elevation -30.0 ft (-9.1 m) +/-. In Profile 2, the RIs were seated at Elevation -47.0 ft (-14.3 m) +/-, or approximately 3-ft. (1-m) into the Potomac Formation, and this zone applies to the southern portion of Garage 2A.

The first phase of the installation of RIs was performed from a berm at the north side of the garage prior to the installation of the wale and rakers along the DDP line. This sequence was used because the equipment to install the RIs would not be able to access the area once the wales and rakers were set. The remainder of the RIs for the slab support and foundation support were installed using two drill rigs from the subgrade elevation. A load transfer platform (LTP) that was 12 to 18 in (305 to 457 mm) in thickness was constructed beneath each foundation location after the RIs were installed.

Fig. 9. Rigid Inclusion Installation

Foundation Mat with Load-Sharing Support Elements Above Tunnels

Building foundations were designed by SK&A and Thornton Tomasetti to restore original overburden loads above the tunnels, using building weight. This was accomplished by a 3 to 4-ft (0.9 to 1.2-m) thick reinforced concrete mat specially designed to distribute building load above the tunnels. In some areas (Fig. 10), the applied mat bearing pressure exceeded original overburden loads, requiring the addition of load-sharing support elements (special deep foundation elements). The deep elements carry the excess building load and transfer them to soil below the tunnels.

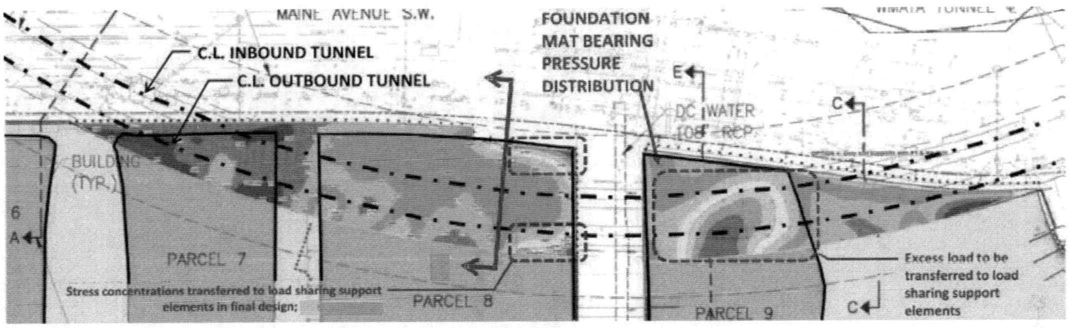

Fig. 10 Foundation bearing pressure map showing areas (red and orange shaded) where load-sharing support elements were required

The load-sharing support elements, termed "rigid anchors" (RA) for this project, consisted of 8-in. (203-mm) and 12-in. (305-mm) nominal diameter grouted hollow bars installed between and on both sides of the inbound and outbound tunnels, but only in areas requiring additional support (Fig. 11). The load-sharing design was achieved by an iterative "soil-spring" analysis, wherein estimated RA and soil subgrade stiffnesses were input into the foundation mat model and the rigid anchor layout adjusted until the desired bearing pressures above the tunnels were obtained. The RA design was further optimized based on field load tests, which resulted in shortening the design bond length of some anchors to obtain the desired stiffness and load sharing behavior. One key concern was that the rigid anchor elements could take too much load, resulting in lower confining pressures (on the tunnels) than desired. This concern was addressed by a detailed interpretation of load test results to assess the relationship between pile stiffness and bond length.

The upper portion of each rigid anchor was designed with a steel isolation casing coated with anti-friction coating to prevent unacceptable load transfer to the tunnels. The rigid anchors were also designed to provide uplift resistance in the flood load case. Upon structure top-out to allow elastic settlement to be nearly complete, the center bars were tensioned and locked off and the sleeves were filled with grout.

To minimize ground disturbance, RAs were drilled using low-strength grout as the drilling fluid and structural grout was pumped through each hollow bar after it reached final depth. Several different pre-drilling methods were used to control grout pressure build-up and the resulting pressure exerted on the tunnels. Most of the holes were pre-drilled with a recirculated slurry which allowed the Keller drilling teams to use a larger drill stem; the larger stem provided an increased up-hole fluid velocity than the hollow bars and therefore the drill cuttings were exhausted from the borehole without plugging. After pre-drilling, the larger drill rods were extracted from the hole and the hollow bars were drilled as a second (reaming) pass. Pre-drilling was largely successful in minimizing pressure buildup by clearing most of the plastic clay cuttings that otherwise would collect in a "collar" that plugged the annulus of the hollow bars. The 8-in. (203-mm) and 12-in. (305-mm) nominal diameter RAs were pre-augered with 6-in. (152-

mm) and 8-in. (203-mm) augers, respectively. Even with pre-drilling, some tunnel movement was detected during hollow bar installation. Therefore, in addition to pre-drilling, slotted drainpipes were installed to the top of the granular T2/T5 Terrace layer prior to installing RAs. These drains were installed to evacuate pressure buildup within the granular layer of soil adjacent to the tunnels that was believed to be transmitting hydraulic pressure to the tunnels.

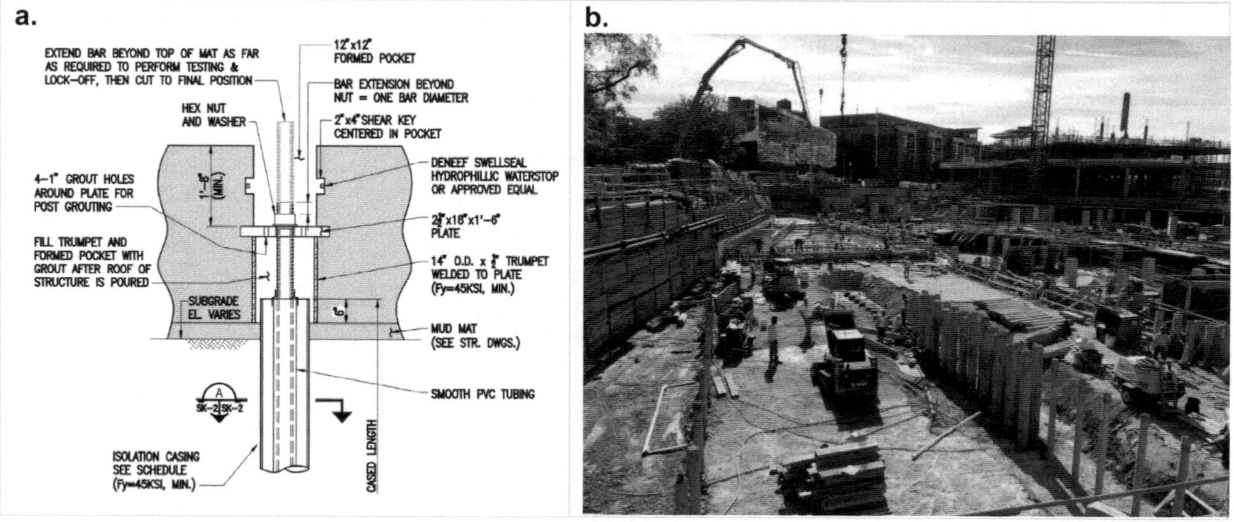

Fig. 11. Hollow Bar Rigid Anchor: a. head detail; b. construction photo showing installation of rigid anchors between and adjacent to tunnels. Photo was taken before cutting anchors to final length and pouring foundation mat

MONITORING

Continuous movement monitoring of the tunnels was performed by an automated system. Optical monitoring of SOE was performed using monitoring points on SOE walls and raker heel blocks.

CONCLUSION

The design and construction of the SOE, underpinning and foundations at the Wharf was challenging due to soft soils, a 108-in. (2740-mm) diameter storm drain bisecting the site, and sensitive WMATA tunnels and construction activity within the WMATA Zone of Influence. The work required detailed coordination and collaboration between the Owner, Structural Engineers, Geotechnical Engineer, Construction Manager and Specialty Geotechnical Contractor. Proximity of the WMATA tunnels and DC Water outfall provided challenges that were met with innovative design and construction methods. The successful completion of the project required the innovative application of numerous design and construction techniques in combination.

ACKNOWLEDGEMENTS

The authors would like to recognize Balfour Beatty, the Construction Manager on the project, and over 100 Keller construction personnel who safely performed this work in a very congested work environment under an accelerated schedule. The project was implemented in close coordination and technical collaboration with WMATA's Joint Development and Adjacent Construction (JDAC) engineering team.

ANCHORED HOLLOW BAR TANGENT PILE SYSTEM FOR UNDERPINNING

By
Kenneth A. Pidgeon, PE, Engineers Construction, Inc., Williston, VT, USA, (802) 863-6389, kpidgeon@ecivt.com
Benjamin H. MacKinney, PE, Engineers Construction, Inc., Williston, VT, USA, (802) 264-2286, bmackinney@ecivt.com
Kenneth R. Allard, EIT, Engineers Construction, Inc., Williston, VT, USA, (802) 779-4780 kallard@ecivt.com
Adriane G. Boscardin, PhD, PE, Boscardin Consulting Engineers, Inc., Amherst, MA, USA, (413) 549-3804, adriane@boscardinconsulting.com
Marco D. Boscardin, PhD, PE, DGE Boscardin Consulting Engineers, Inc., Amherst, MA, USA, (413) 237-4852, marco@boscardinconsulting.com - (413) 237-4852

ABSTRACT

This paper presents a case history for a unique application of a system of vertical micropiles with horizontal soil nail bracing and shotcrete facing for underpinning. The system was used to support and protect an existing 2-story hospital building during demolition and renovation. The goal was to limit building movements and have the building remain in use during construction of an abutting addition. Soils at the site consisted of clean, loose to dense, sand and gravel. The original underpinning design was revised after the initial excavation revealed unsuitable soil conditions for traditional underpinning pits. An alternative system consisting of vertical micropiles (hollow bars) through the footing, combined with shotcrete and soil nails was proposed, accepted by the owner, and successfully constructed. Excavation depths were up to 15 feet (4.6 m) below the existing footings. Wall movements were monitored at regular, horizontal and vertical intervals along the hospital wall, and less than 0.6 inches (15 mm) of settlement and 0.1 inches (3 mm) of horizontal displacement throughout underpinning and construction were reported. The challenges with the original proposed support of excavation design and other alternative approaches considered will also be discussed.

Keywords: underpinning, grouting, permeation grouting, grouted column, micropile, soil nail wall, sand, gravel, shotcrete, hollow bar

INTRODUCTION

This paper describes the support of foundation work for the renovation of an addition to the Brattleboro Memorial Hospital's Surgical Center, located in Brattleboro, Vermont. The work included supporting the foundation of the existing hospital building during the construction of a new surgery addition. The addition consisted of a 4-story steel framed structure located immediately adjacent to, but structurally independent of the existing 2-story hospital structure. The work included demolition and renovation of portions of the existing hospital, as well as new construction in an adjacent landscaped area.

The design of the new addition required excavation up to 15 feet (4.6 m) below the existing hospital's footings. The initial design consisted of vertically sequential pit underpinning where underpinning pits are constructed in 5-foot (1.5m) vertical units. The intent of underpinning is to safely transfer the existing building's foundation loads to a deeper foundation bearing stratum unaffected by the excavation of the new foundation, and to limit the existing building's displacement and related response to tolerable levels.

Dry running sands and the presence of coarse gravel layers at the site made pit underpinning as designed not reasonably constructible. These soils typically cannot maintain a vertical face greater than 6 inches in depth. Anticipated settlements associated with the vertically sequential pit underpinning were also of concern. Therefore, an alternative method combining grouted hollow bar micropiles and soil nails was used to address these issues. This method had the added benefits of expediting construction compared to the original pit underpinning scheme at a cost acceptable to the project.

PROJECT BACKGROUND AND SITE DESCRIPTION

The site of the Brattleboro Memorial Hospital is located in Brattleboro, Vermont, a small community in southeast corner of the state along the Connecticut River. The construction area was enclosed on three sides by the existing hospital, a two-story structure, with a landscaped area and parking lot present along the northern edge of the project area. The development included excavation below the existing footing for the construction of the new addition as well as demolition of a portion of the existing hospital. Our role on the project was to redesign and construct the underpinning wall.

The goal of the underpinning and SOE was to temporarily support the existing building foundation adjacent to the excavation so that deeper spread footings, basement walls, and grade beams for the new addition could be constructed. Excavations for the new foundations ranged from approximately 5 to 15 feet (1.5 - 4.6 m) deep adjacent to the existing building. Lengths of walls underpinned ranged from approximately 11 to 74 feet (3.4 – 22.6 m). The total area of the excavation was approximately 1,500 square feet (139 m^2). Figure 1 presents a plan view of the site.

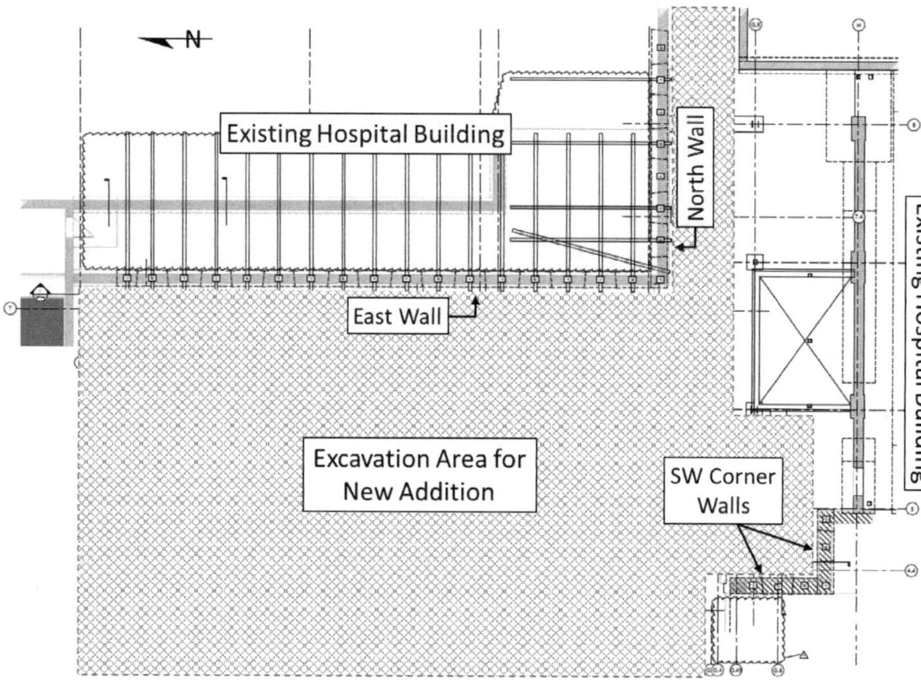

Fig. 1. Plan view of site

SUBSURFACE CONDITIONS

Soils at the site ranged from relatively clean medium to fine sands and gravel to very clean coarse sand, and gravel. Groundwater levels were well below the deepest proposed excavation depths. Seams and layers of gravel were observed in excavations at the site during the initial phases of the construction. The findings from the subsurface investigation's 2018 boring data were consistent with the geologic literature. The geologic literature for the area, Surficial Geologic Map of Vermont (C. Doll Ed., 1970), and USGS

Surficial Geology of the Brattleboro, VT Quadrangle (Stewart, 1956-1966), indicate that the soils in the project area consist of glaciolacustrine sands and gravels associated with historic glacial Lake Hitchcock.

Gradation test data from the subsurface investigation and subsequent test data collected during the start of construction indicated that the soils were clean, cohesionless sands and gravels with the fines (#200 sieve fraction) content of the soils tested from the underpinning area and elevations ranging from 5% to 11% for the subsurface investigation samples, and 1% to 3% for the later samples collected directly from the underpinning area at the start of construction. The lower percentage of fines found in the location of the underpinning work indicated that the in-situ soils were even cleaner than expected. This makes maintaining an unsupported, vertical face during sequential shallow underpinning pit excavations cuts more challenging. Figure 2 is a photograph of the soils encountered during excavation.

Fig. 2. An example of dry running sands and gravel located at the site, and micropile grouted columns after excavation.

UNDERPINNING DESIGN

The support of foundation design, in the owners bid documents, specified pit or pier underpinning. Pit or pier underpinning is a common form of underpinning that involves constructing individual, small vertical shafts to a deeper bearing elevation under the wall to be underpinned in an alternating sequence. This method of underpinning relies on the intermittent construction of alternating primary set of vertical piers using the existing building's ability to span small unsupported distances during pier construction (typically about 4 feet, 1.2 m) and then constructing secondary fill-in piers between the primary piers until a new, deeper wall is ultimately constructed to transfer the building loads to the new, deeper bearing elevation. In addition, this method relies on the soil having the ability to maintain a short vertical face so that the hand-excavation can be made and support of the excavation face in the form of horizontal timber support ("ring lagging") around the entire perimeter of the pit can be installed. Typically, the pits are excavated in 6-to-12-inch-deep (152 – 305 mm) increments, with no more than 12 inches (305 mm) of unsupported soil exposed in the side walls of the pit.

It should be noted that underpinning involves a load transfer process, which in itself usually results in some building displacement. Displacements in the range of 1/4 up to about 5/8 of an inch (6 – 16 mm) are typically expected in well controlled underpinning operations. Such displacements are typically associated with minor cracking in existing masonry walls (Boscardin and Cording, 1989) such as those on this project.

The bid-document underpinning pits were designed as approximately 2 to 4 feet (0.6 – 1.2 m) wide piers with the maximum unsupported length of wall specified as 4 feet (1.2 m). Pier heights ranged from approximately 5 to 13.5 feet (1.5 – 4.1 m), constructed in 5-foot (1.5 m) vertical units, and pier thickness of about 2 to 2.5 feet (0.6 – 0.8 m) or to the width of the existing footing.

Along the east wall (~74 feet, 22.6 m) with the deepest excavation depths (up to 15 feet, 4.6 m), one to three tiebacks were to be installed in each pier, with one tieback for pier heights less than about 10 feet, and three tiebacks for pier heights greater than about 10.5 feet (3.2 m). The north wall (~32 feet, 9.8 m) running east of the east wall has pier heights of approximately 7 to 9.5 feet (2.1 – 2.9 m), included one tieback in every other pier. No tiebacks were specified for the two shortest walls (~10.5-13 feet long, 3.2 – 4 m) located in the southwest corner of the site with pier heights of about 5 to 9.5 feet (1.5 – 2.9 m).

INITIAL CONSTRUCTION CHALLENGES

The underpinning work began as designed in the owner's bid documents, and immediately the excavation contractor encountered difficulty controlling ground loss of the cohesionless soils into the underpinning excavations during underpinning construction. The excavation contractor expressed concern regarding the ability to control the ground with the pit or pier underpinning scheme, and therefore the increased risk of potentially damaging displacements to the existing structure.

After the pit underpinning construction conditions and difficulties were encountered and evaluated, the general contractor requested assistance in identifying alternative approaches to complete the underpinning work.

RE-DESIGN CONCEPTS CONSIDERED

I. Soil Improvement and Field Verification Testing for Pit Underpinning

Additional engineers and contractors were brought in to augment the project team to identify alternative approaches to supporting the existing hospital foundation during excavation and foundation construction work. The team met in mid-October 2020 to observe the ground and site conditions exposed to date, and to discuss potential alternatives to complete the underpinning work. The initial discussions focused on how to proceed forward to allow for the construction of the underpinning piers as shown on the drawings. It was determined that ground improvement was needed to safely excavate and support the pits per the original design. Permeation grouting was identified as a possible alternative, using environmentally safe, rapidly setting chemical grouts.

A grouting program was developed, including preliminary costs, considering chemical grouting, micro-fine cement grouting, and ordinary Portland cement (OPC) grouting alternatives to permit construction of the pit underpinning in the dry, cohesionless soils encountered at the project site. Three additional samples of soils were collected and tested for gradation, specifically from the area of the proposed underpinning. The gradation tests showed that these additional samples were coarser (up to 2-inch, 51-mm, particles vs. 1-inch, 25-mm, particles) and cleaner (1% to 3% fines vs 5% to 11% fines [minus #200 sieve]) than the samples tested during the initial subsurface investigation. Therefore, a test grouting plan was developed to verify the applicability of permeation grouting at the site to enable the construction of the underpinning piers, along with a preliminary grout design for the actual grouting for the underpinning.

After preliminary costs for the chemical grouting were submitted, it was judged that the costs (specifically the material cost of the chemical grout) were not economically justifiable for the project. A grouting plan was finalized based on the coarseness and lack of fines in the three samples from the underpinning area which made the use of cementitious grout: Microfine cement grout, and ordinary Type I/II Portland cement grout feasible.

A test grouting program was performed to verify the grouting plan. The two types of cementitious grouts, OPC and Microfine cement grouts, were tested at a proposed grouting pattern using a more economical single phase, tip injection grout pipes. Generally, one would employ a grouted in-place, perforated sleeve pipe (tube-a-manchette) with multi-phases of targeted zone grout injections using packers, however this is a more expensive system. At that time, it was also suggested to use a soil nailed, shotcrete wall to support the grouted soil as a composite underpinning system. This system would be quicker to install than the pit underpinning and help gain schedule lost by the delays in implementing the original pit underpinning construction. Preliminary permeation grouting plans were developed that would work with both pit underpinning and the soil nailed, shotcrete wall systems.

The result of the permeation grouting tests, conducted mid November 2020, showed that there existed at depth even coarser seams of gravelly soils that took a large amount of grout. To effectively grout these soils and create the support required, successive cycles of grouting would be needed: (a) first to fill the coarsest voids in the gravels with a thicker, lower mobility grout, and then (b) on the order of at least two additional cycles of grouting with thinner, more mobile grout to fill and stabilize progressively finer voids in the soil.

After reviewing the results of the grouting test section, we recommended a composite shotcrete/soil nail wall with closely space drilling in vertical micropiles through the wall footings, instead of permeation grouting and pit underpinning.

II. Composite Soil Nail and Micropile Wall Underpinning

The grouted test section revealed large grout intakes in coarser seams of gravel at depth and limited grout penetration in finer zones (about less than 1 inch, 25 mm). Therefore, the design was changed to a composite shotcrete/soil nail wall with closely space, grouted vertical micropiles drilled through the wall footings (see Fig. 2 and 3). The composite shotcrete/soil nail and micropile wall system was designed and accepted so that installation could begin in early December 2020.

Slots for underpinning construction were drilled and cut in the footing at a horizontal spacing of approximately 10 inches (254 mm) on-center along the east wall and 14 inches (356 mm) on-center along the north wall using a 3.5-inch (89 mm) carbide button bit to accommodate hollow bar micropiles installed through these slots. Six-inch-(152-mm-diameter, carbide-tipped, cross-cut drill bits were attached to the hollow bars just below the bottom of the footing using small local excavations (hand dug) after a slot was cut and the hollow bar inserted through the slot. In the center of each micropile grout column along the 15-foot-high (4.6 m) soil nail wall sections, 52 mm hollow bars (52/26 bar, 1252 mm^2

of steel, ASTM 706 Grade 80) were used. The wall sections shorter than 15 feet (4.6 m) used 40 mm hollow bars (40/20 bar, 729 mm^2 of steel, ASTM 706 Grade 80). The 6-inch-(152-mm)-diameter drill bits produced approximately 8-inch-diameter (203-mm) micropile grout columns in these soils. Three-and-one-half-inch-diameter drill bits were used to produce 4-inch-diameter grout bodies for the soil nails. The grout mix was a neat cement grout using Portland Type I/II cement, and a target water-cement ratio of 0.45. The minimum cure time was 36 hours or a minimum unconfined compressive strength of 1500 psi (10.3 MPa) if a shorter cure time was used.

Micropiles extended from about 8 to 18 feet deep (2.4 to 5.5 m), with an embedment of about 3 feet (0.9 m) below the soil nail wall, and were installed with an inclination angle between 0° and 2° under the building footing. Micropiles were installed on an alternating schedule along the east wall, where micropile spacing was 10 inches (254 mm), by installing every other micropile (20 inches, 508 mm, on center) on Day 1, and the intermediary micropiles on Day 2. Along the other three walls where the wall heights were shorter and a micropile pile spacing of 14 inches (356 mm) was used, every third micropile was installed on Day 1 (28 inches, 711 mm, on center), the next 1/3 of the columns on Day 2, and the last 1/3 of the columns on Day 3.

Fig. 3. Micropiles drilled through slots in foundation footing

The micropiles were given at least one day for curing before building the soil nail wall. The soil nail wall was built by excavating in 5-foot lifts (1.5 m). If needed, any exposed grouted column was chipped back before applying a 4-inch thick (102 mm) layer of shotcrete (fc'= 6000 psi, 41 MPa) with 4-inch x 4-inch x W2.9 x 2.9 wwf (welded wire fabric), and 4, and 2 feet (0.6 m) #4 bars (square pattern). At each soil nail location, fully grouted, 40 mm hollow bar (40/20 bar, ASTM 706 Grade 80) soil nails were installed at 5 feet (1.5 m) horizontal and 5 feet (1.5 m) vertical spacings, starting at 2.5 feet (0.8 m) from the top and edge of the wall, with 8-inch x 8-inch x 1-inch (203 mm x 203 mm x 25 mm) steel (ASTM A36) bearing plate. Soil nails were 12 feet (3.7 m) long with a 10 kips (41 kN) design load. The same neat cement grout as for the micropiles was used. Two performance verification tests were performed on sacrificial soil nails and proof tests on 5% of production nails. Figure 4 shows the construction of the soil nail wall in progress.

Fig. 4. Construction of the shotcrete/soil nail wall in lifts.

After the new foundations, and related grade beams and pilasters were constructed, the excavation was backfilled to support a slab on grade with a top-of-floor at Elevation -12.0 (feet, -3.7 m). The top of the floor of the adjacent existing building is Elevation -9.5 (-2.9 m) and the bottom of footing support for the walls at these locations is at approximately Elevation -10.7 (-3.26 m) or lower.

Since the excavation area was planned to be backfilled with a slab-on-grade at Elevation -12.0 (-3.7 m), temporary lateral support was needed and provided by the soil nail walls. Corrosion protection of the soil nails was not required because only temporary support was needed. Final support of the wall is provided by the new addition that was constructed.

Under the east wall (the longest and deepest wall), the goal was to construct the interior wall with its back flush with the outside edge of existing footing. This required removal of the top 1.5 to 2 feet (0.5 – 0.6 m), to about Elevation -12.7 (-3.7 m), of shotcrete/soil nail wall and 4 inches of soil under the edge of the existing footing (from elevation -12.7 to the bottom of the footing) and replace it with concrete doweled into the footing above.

By the end of December 2020, the east wall, the longest wall with the greatest wall depths (up to 15 feet, 4.6 m), was completed before stopping for winter. The site was remobilized around mid-February 2021, after demolition was completed, and the remaining underpinning was performed.

CHALLENGES DURING CONSTRUCTION

The project site was very tight, surrounded on three sides by existing buildings to remain. A one-lane road in and out of the work area was the only access for material and equipment. Access to the bottom of the excavation for equipment was impractical, especially as its depth increased. An excavator-mounted hollow bar drill with a drifter/top hammer capable of installing 10-foot (3-m) long hollow bars was utilized. This drill was mounted on a 23 metric ton excavator capable of reaching the work from the top of the excavation (Fig. 5).

Drilling and grouting can be challenging during Vermont's winter, and this project was no exception. Underpinning work was performed on several days with low temperatures in the single digits Fahrenheit. The contractor used temporary structures, insulating blankets, forced hot air, and ground thaw units to keep the work area, mixing water, Portland cement, and finished shotcrete placements at necessary temperatures.

Fig. 5. Excavator-mounted hollow bar drill installing soil nails

The Portland cement grout was mixed with a high shear colloidal mixer and continuously pumped through the drill bit. Drilling with continuous grout flush allowed for greater permeation compared to drilling with water and adding grout after reaching depth. The sacrificial bit and hollow bar added additional strength to the micropiles compared to permeation grouting alone. The hollow bar was grouted up into the existing footing to maximize resistance to settlement.

Vibrations due to the drilling operations were of concern. Drilling was scheduled around operations in the surgical unit. Most vibrations were transferred through the existing footing during the drilling which took place relatively quickly. While drilling in the sand, vibrations were minimal and shutdowns were less frequent.

CONSTRUCTION AND MONITORING

The monitoring program consisted of horizontally and vertically spaced displacement monitoring points installed at 25-foot (7.6 m) intervals along the existing hospital walls adjacent to the excavation. Monitoring points were surveyed using optical and string line surveys generally at the beginning and end of each work day throughout the underpinning wall construction. There was generally less than 0.6 inches (15 mm) of vertical displacement and less than 0.13 inches (3 mm) of horizontal displacement measured during the installation of the underpinning (Fig. 6).

Five soil nail/micropile load test (2 verification tests and 3 performance tests) were performed to verify that the load carrying capacity and displacement behavior of the hollow bar elements conformed to project requirements. In these soils, the tests indicated that pull out tests on the hollow bar elements achieved at least 5 times required capacity with little or no creep behavior.

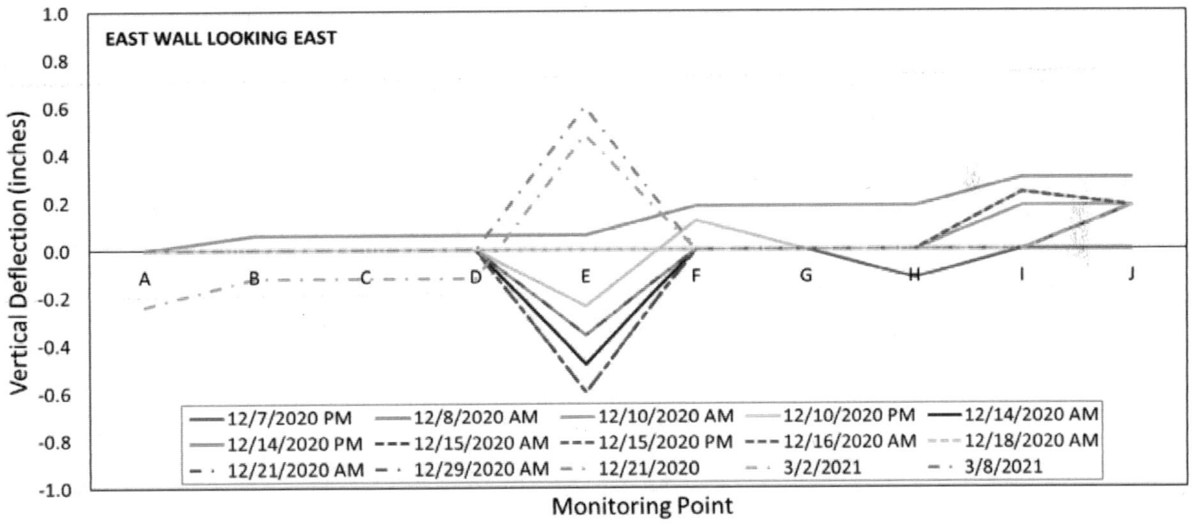

Fig. 6. Vertical displacements of East Wall of building monitored throughout underpinning wall construction

CONCLUSIONS

The shotcrete/soil nail/micropile underpinning/SOE system accomplished the goal of the original pit underpinning and soil anchor system with less risk and faster construction than the original system. In addition, the shotcrete/soil nail/micropile underpinning/SOE system is ultimately a simpler, less risky system for the conditions at this site than the alternative shotcrete/soil nail/permeation grout underpinning/SOE system, and showed little in the way of building movements.

Using a continuous grout flush during the installation of the micropiles resulted in greater permeation of the grout mixture into the soil than drilling with water to depth before adding the grout. Due to the space constraints of the site, an excavator-mounted hollow bar drill made installation of the soil nails simpler at deeper depths.

We recommend that on similar projects the specialty underpinning subcontractor be prequalified and that final underpinning system details be developed by and/or in coordination with the specialty contractor.

REFERENCES

Boscardin M.D. and Cording E.J. 1989. Building response to excavation-induced settlement. American Society of Civil Engineers, ASCE Journal of Geotechnical Engineering, 115(1).

Doll, C.G. 1970. Surficial Geologic Map of Vermont [map]. 1:250,000. State of Vermont, U.S. Geological Survey.

Stewart, D.P. 1956-1966. Brattleboro VT-NH 15 Minute Quadrangle Surficial Geologic Map [map], Original Draft for the Surficial Geologic Map of Vermont, 1970 Ed, by C. Doll. 1:62,500. State of Vermont, U.S. Geological Survey.

USING PILE REINFORCEMENT TO MITIGATE MULTIMODAL SLOPE FAILURES: A 3D LEM CASE STUDY

Ma, Cami, Javankhoshdel, Yacoub

ABSTRACT:

Piles are often used to reinforce slopes as they increase the available restoring force when calculating the factor of safety against sliding. In this paper, a 3D model of a slope containing multiple modes of failure is reinforced using pile elements. The results of the 3D Limit Equilibrium analysis containing piles are obtained using a newly introduced coupled approach which combines the results of finite element analysis (to obtain the displacement of soil along the pile) with the traditional limit equilibrium iterative formulation. The multiple modes of failure are obtained using the novel Multi-Modal Optimization (MMO) metaheuristic method which adapts the Particle Swarm Optimization (PSO) method to search for multiple local minima in a single slope model.

Key Words: Particle Swarm Optimization, Pile, Slope Stability, Multi Modal Optimization

1 INTRODUCTION

Traditional limit equilibrium (LEM) slope stability analysis methods seek to locate the slip surface with the lowest factor of safety (FS), known as the critical failure surface. LEM remains highly popular for both unreinforced and reinforced slopes due to its relative simplicity and ability to be combined with metaheuristic approaches to search for the most critical factor of safety (Cheng et al. 2007; Cheng et al. 2012; Zhao et al. 2008; Tian et al. 2009; Li et al. 2009; Chen and Chen 2011; Kalatehjari et al. 2012, 2014; Fisher and Goodale 2021; Javankhoshdel et al. 2021; Su and Shao 2021). These metaheuristic approaches employ naturally inspired processes to search for an optimal solution given a set of variable parameters. Among the various metaheuristic approaches, Particle Swarm Optimization technique (PSO) (Kennedy and Eberhart 1995) is one of the most popular. However, focusing the search effort on locating the single critical failure surface has its drawbacks. As pointed out by Reale et al. (2015) little research has been completed on slopes which could develop several critical slip surfaces with similar FS values. There are cases where the global minimum is of little practical importance, e.g. when the critical slip surface is too shallow to have any severe consequences, or when a slope is susceptible to multiple failure mechanisms, e.g. slopes with multiple benches and/or layers (Javankhoshdel et al. 2021). For slopes with multiple failure mechanisms, failure to detect some of the failure surfaces could lead to unsafe design. For instance, Javankhoshdel et al. (2021) has provided an example of an open pit mine which gives two failure surfaces with identical factors of safety on the north and south sides of the pit.

Recently, a method has been developed that has the ability of finding multiple minimum failure surfaces and their corresponding FS values - multimodal optimization (MMO) (Li et al. 2020). In this approach. a multimodal particle swarm algorithm, LIPS-R (Li et al. 2020), is used to find multiple local minima, or critical failure surfaces.

Piles are often used to reinforce slopes as they increase the available restoring force when calculating the factor of safety against sliding. In LEM, a restoring force is applied on the slipping surface where it intersects a pile element. Slopes that do not meet minimum safety factor requirements for overall stability stipulated in applicable codes and standards can be reinforced with piles to increase the factor of safety. However, if multiple failure modes exist in a slope, it is important to ensure that all the potential failure modes satisfy the minimum requirements instead of just the critical slip surface. The use of MMO is therefore useful to more efficiently identify and evaluate the factor of safety in multiple regions of a slope in a single analysis.

Recently, a coupled approach has been introduced which combines the results of finite element analysis of the displacement of soil along the pile with the traditional limit equilibrium iterative formulation. It combines the formulations for axial (Rocscience 2018) and lateral (Reese & Van Impe 2011) displacements to the piles to determine the restoring reaction of the pile onto the slip surfaces, depending on direction and the location and of the slip surface where it intersects the pile. This paper showcases the extension of the coupled analysis method for piles in LEM into 3D via the Slide3 and RSPile (Rocscience 2021) software.

In this study, a slope is reinforced with piles, and then the Uni-Modal Optimization (UMO) and MMO factor of safety analyses are conducted with considering the coupled LEM slope stability analysis and FEM pile analysis.

2. OVERVIEW OF THE MMO LEM METHOD

The PSO method has been proven to be efficient at solving unimodal problems in a wide array of disciplines. For many real-world problems it is often desirable to find several local optima. As such, the PSO method, among other population-based niching methods, has been extended to solve multi-modal optimization problems (Li et al. 2017).

Qu et al. (2013) proposed a distance-based locally informed particle swarm (LIPS) optimizer where instead of using the personal global best positions, the particles are updated using information from their neighbouring particles. This allows LIPS to eliminate the need to introduce additional niching parameters while obtaining the ability to form stable niches (Qu et al. 2013).

The LIPS-R algorithm used in the study by Li et al. (2020) inherits the key concepts from LIPS with an additional radius filter. This filter allows LIPS-R to return optima (niches) that are at least some distance away from each other, instead of many slightly different minima in the same region. At the end of the iterations, LIPS-R sorts all particles by its fitness value. It then filters the results by some percentage of the model space, r. In this study r=10% was found to perform well – this suggests the optima will be at least 10% of the search space away from the others. LIPS-R then returns all the optima that are at least r apart from each other.

3. LEM SLOPE STABILITY COUPLED WITH FEM PILE ANALYSIS

In LEM, support elements can be represented by using single line segments, which provide restoring forces to the slip mass at the location of the intersection of the slip surface with the support element. The magnitudes of these forces are typically either specified as constant values or calculated from advanced methods based on the properties of the support element. Similarly, the directions of these forces may typically be assumed to be tangent to the slipping direction, perpendicular to the pile, or somewhere in between. However, for advanced methods, the calculations of the support force and its direction can be complicated and rely on assumptions related to the mechanical deformation of the support and the surrounding soil.

For this study, the piles used to reinforce the slope are analyzed using the finite element method (FEM). A coupled procedure is required in which the displacement of the surrounding soil is assumed based on the slipping direction and intersection location of the slipping surface with the pile and applied as an initial condition to a pile-and-soil interaction model in FEM. Although it is extremely difficult to predict the realistic behaviour of the soil surrounding the pile during mobilization of the soil mass, the displacement magnitude and direction of the soil surrounding the pile is commonly assumed to be uniform above the point of intersection of the slip surface with the pile.

The deformation of the pile because of the soil displacement partitioned along each of its three principal axes (one axial direction and the major and minor lateral axes) is obtained via FEM assuming that the pile is a series of discretized deformable elements in an elastic foundation comprising of the surrounding soil. Reese & Van Impe (2011) proposed a formulation for lateral deformations of the pile, which is effectively summarized by Rocscience (2022), respectively. Rocscience (2018) proposed an analogous finite element method for determining the axial deformations of an axially loaded pile. The governing differential equations for the axial and lateral deformations of the pile are given in Eq. 1 (Rocscience 2018) and Eq. 2 (Reese & Van Impe 2011), respectively.

$$E_p A_p \frac{d^2u}{dx^2} + \tau \cdot C_p = 0 \qquad [1]$$

$$E_p I_p \frac{d^4y}{dx^4} + P_x \frac{d^2y}{dx^2} + E_{py} y - W = 0 \qquad [2]$$

Where u is the axial displacement and y is the lateral displacement along distance x along the pile. E_p is the elastic modulus of the pile, A_p is the cross-sectional area of the pile, I_p is the moment of inertia of the pile corresponding to the axis for which the lateral analysis is being performed, and C_p is the circumference of the pile segment at depth z. Along some distance x along the pile, τ is the soil unit skin friction, and E_{py} is the soil reaction modulus based on the magnitude of the distributed load W. Finally, P_x is the axial load on the pile head.

Nonlinear axial stress-strain (*t-z* curves) are based on recommendations from the American Petroleum Institute (2002) and lateral stress-strain (*p-y* curves) relationships for the soil are established based on Reese & Van Impe (2011). Note that by this method, none of the analyses along the three principal axes are assumed to interact – that is, the analyses of pile and soil displacement in each direction are conducted separately based on the partition of the soil displacement in each of the three directions.

After the deformations are obtained via numerical convergence in FEM, the internal shear forces of the pile at its intersection with the slipping surface are taken as the restoring reactions of the pile onto the slipping mass in a direction perpendicular to the length of the pile. Similarly, the internal axial force in the pile at the slipping location is taken as a restoring force in the longitudinal direction of the pile. Depending on the whether the pile is in tension or compression, the restoring force will be directed into the ground below the slipping surface or in the opposite direction, respectively. Note that in some cases, including both the axial and lateral components of the reaction of the pile may be too conservative. Either of these reactions may be omitted (i.e. taken to be zero) from the analysis if the pile is assumed to be axial- or lateral-only.

Effective Ground Angle

In some cases, the slip surface intersection with the pile will be shallow. The lateral component of the FEM analysis in the case of shallow failures for some clays and sands can affected by the effective ground slope. The effective ground slope was originally proposed by Reese (1958) and involves applying adjustment factors to the *p-y* curves in a conical wedge above the pile based on the geometry of the ground topography above the pile, as well as the batter angle and direction of the pile, according to Eq. 3 (Reese & Van Impe 2011). Deep failures are not affected by the effective ground slope. *p-y* curves which are not affected by the effective ground slope can also be specified. The effect of ground slope and batter angle can also be ignored by the practitioner altogether. The original methodology (Reese 1958) applies to 2D analyses of the pile along a vertical section and is extended in 3D for this paper.

$$\theta_i = \alpha_i - \beta_i \qquad [3]$$

Where θ_i is the effective ground angle for the lateral pile analysis in the ith lateral axis, and α_i and β_i are the ground angle and batter angle in the direction of the ith lateral axis, respectively.

In the proposed methodology, a vertical plane is taken in each of the principal axis directions of the pile. Solely for the purpose of obtaining the effective ground slope, β_i in each lateral direction is based on the batter angle of the pile projected onto the corresponding vertical plane, and α_i is the ground slope along the vertical plane. The determination of g_i and b_i is illustrated in Figure 1, which shows the pile starting from ground surface *G* intersecting the slip surface *S*.

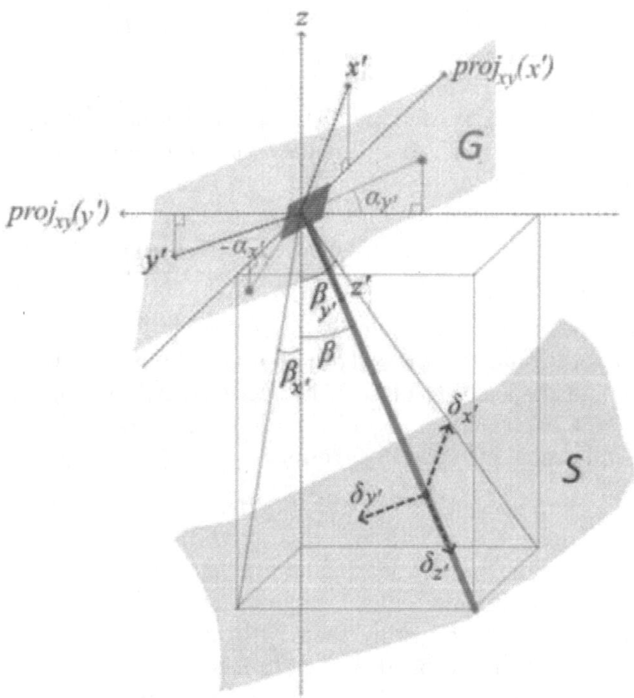

Figure 1. Determination of effective ground angle in the two principal lateral directions

The principal lateral axes of the pile are denoted *x'* and *y'*, while the axial direction of the pile is *z'* (note that this establishes a left-handed local coordinate system). The projection of *x'* onto the horizontal plane *xy* is denoted $proj_{xy}(x')$, and the projection of y' onto the horizontal plane is denoted $proj_{xy}(y')$. A vertical section is taken along x', for which α_x is taken as the angle between the horizontal $proj_{xy}(x')$ and the ground surface and β_x is the batter angle if the pile were projected onto this vertical section. The same is done to determine α_y and β_y in the y' direction. Note that the ground slope and batter angle are taken positive clockwise from the positive axis direction. β is the true batter angle of the pile, whose magnitude is equal or greater than the projected values β_x and β_y. Finally, the soil displacement vector δ is partitioned into the three principal components and applied uniformly from the top of the pile to the intersection of the pile with *S*.

4 NUMERICAL EXAMPLE

For the first part of the example, an unreinforced model is considered in this study which is shown in Figure 2. The slope consists of three layers of cohesive-frictional materials, each modeled using Mohr-Coulomb strength.

Name	Colour	Unit Weight (lbs/ft3)	Failure Criterion	Cohesion (psf)	Phi (°)
Stratum I (Tn)		115	Mohr Coulomb	155	30
Stratum II (Tm)		115	Mohr Coulomb	160	30
Stratum IIIa (Ta)		115	Mohr Coulomb	130	18

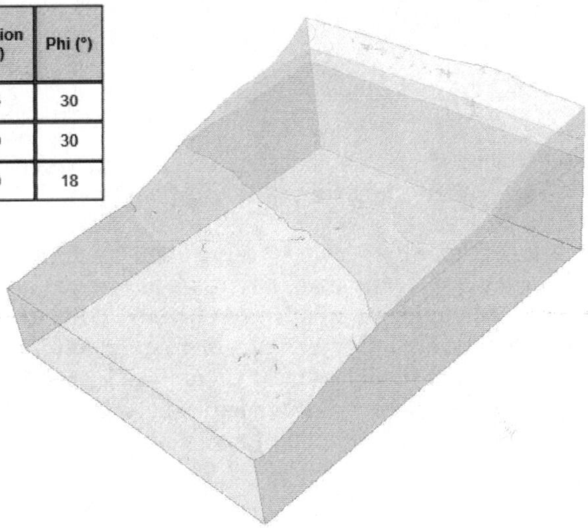

Figure 2. The geometry and parameters of the unreinforced slope model considered in this study.

For the unreinforced case, only MMO analysis was carried out to investigate the critical regions in the model. Limit Equilibrium analysis is carried out using Spencer approach. Figure 3 shows the safety map of the results of the unreinforced analysis. The critical regions in the model are coloured in red. It can be seen in Figure 4 that there is a small red critical region on the top region of the model in addition to the big red region on the center. These results show the necessity of looking at the multi-modal failure mechanism instead of unimodal failure.

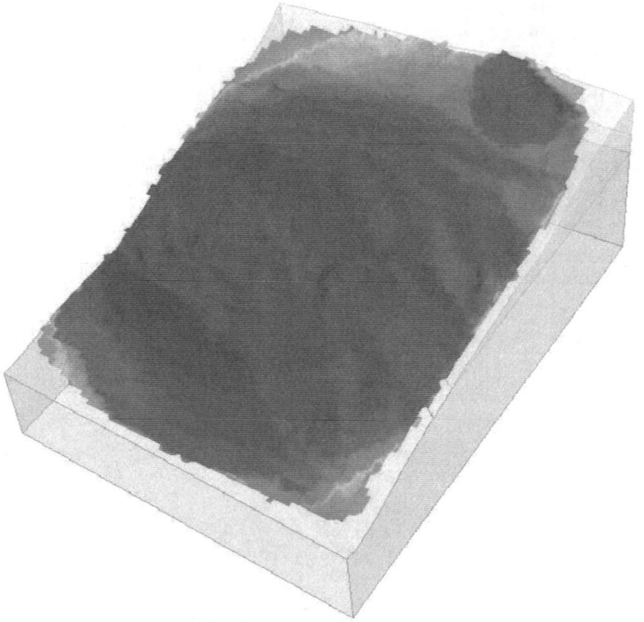

Figure 3. Safety map of the results for the unreinforced model.

Based on the results of the unreinforced model, two sets of piles have been added to the top and center of the model. For both top and center area, five rows of projected vertical piles with the horizontal and vertical spacing of 25 ft and the length of 75 ft have been added to the model (Figure 4). As mentioned in the previous section, the piles are analysed using a coupled approach with FEM which depends on the depth of the critical slip surface at the location of each pile The resultant of the shear strength and axial force of the pile will be used for the slope stability analysis.

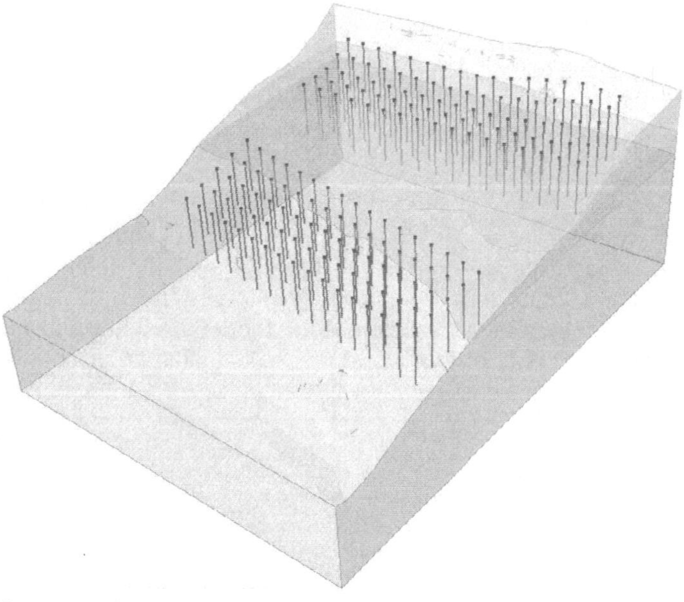

Figure 4. The pile reinforced model used in the study.

Figure 5 shows the results of the slope stability analysis of the model using UMO approach which gives the FS value of 1.44 with the critical slip surface located only at the center of the model. Even the safety map in Figure 6 shows the critical failure location only on the center of the model.

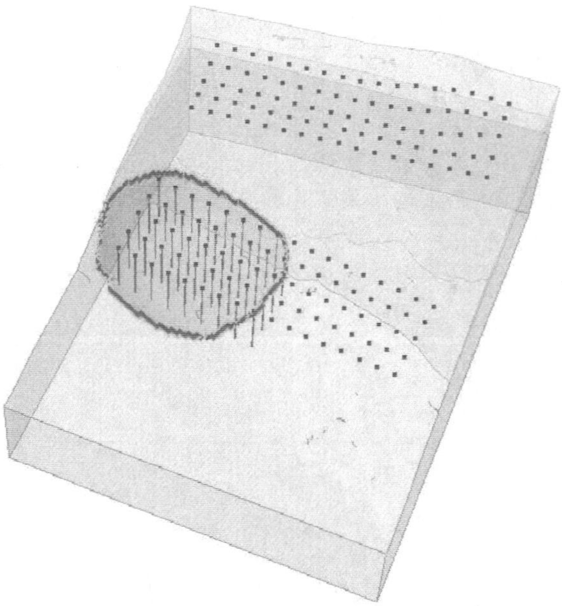

Figure 5. The critical slip surface found using the UMO approach corresponding to the FS = 1.44.

Results of the same model run using MMO method are shown in Figure 7. Results show two distinct failure modes with factor of safety of about 1.44. This emphasizes the importance of considering multi-modal optimizations and looking at all the possible failure regions instead of only the most critical one.

Figure 6. Safety map for the UMO analysis of the pile-reinforced model.

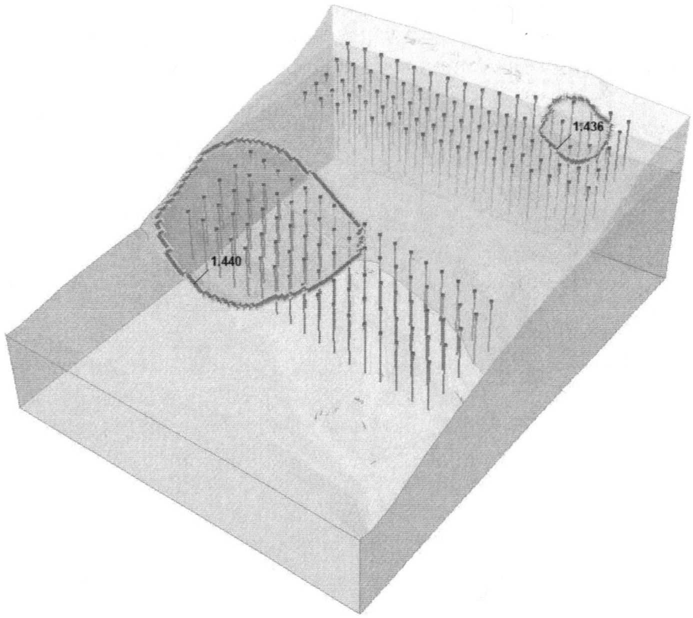

Figure 7. Results of the MMO analysis for the pile-reinforced slope.

The advantage of running the FEM pile analysis is that this method effectively provides a non-linear force diagram be used in the LEM analysis, which is also unique to the direction of slipping, as opposed to a force diagram that is assumed to be constant or independent of the slipping direction.

Figure 8 shows the support resultant forces of the piles which intersect each failure surface. In this figure, the values of the resultant forces are different at different locations of the critical slip surface because of the change of the depth of the slip surface that intersects the piles.

Figure 8. Support reaction forces for both critical slip surfaces.

5 CONCLUSION

Multi-modal Optimization (MMO) and Uni-Modal Optimization (UMO) for the case of a pile-reinforced slope have been used to find the critical slip surfaces and their corresponding factors of safety. The MMO approach provides two different regions for the failure surfaces with the same factors of safety which emphasizes the importance of using MMO approach and the limitations of the traditional LEM approach which only generates one critical factor of safety. A coupled FEM pile analysis and 3D LEM analysis was used to calculate the reaction forces of the piles against the critical slip surface.

6 REFERENCES

American Petroleum Institute. 2002. *API Recommended Practice 2A-WSD – Planning, Designing, and Constructing Fixed Offshore Platforms – Working Stress Design* (21st edition).

Chen, W., and Chen, P. 2011. "PSOslope: a stand-alone windows application for graphic analysis of slope stability." In *Advances in Swarm Optimization, Tan, Y., Shi, Y., Chai Y., and Wang, G., Eds., vol. 6728 of Lecture Notes in Computer Science*, pp. 56–63.

Cheng, Y.M., Li, L. and Chi, S.C., 2007. Performance studies on six heuristic global optimization methods in the location of critical slip surface. *Computers and Geotechnics*, 34(6), pp.462-484.

Cheng, Y.M., Li, L., Sun, Y.J., and Au, S.K. 2012. "A coupled particle swarm and harmony search optimization algorithm for difficult geotechnical problems," *Struct Multidiscipl Optim*, 45(4): 489–501.

Fisher, B., and Goodale, J. 2021. Global stability assessment of open pit slopes using LEM and FEM: A comparison between the factor of safety and strength reduction factor. *The Evolution of Geotech*, 436.

Javankhoshdel, S., Cami, B., Yacoub, T., Ma, T., and Abolfazlzadeh, Y. 2021. Multi Modal failure mechanism in open pit mines using LEM and FEM approaches. In *55th US Rock Mechanics/Geomechanics Symposium*.

Kalatehjari, R., A Rashid, A. S., Ali, N., and Hajihassani, M. 2014. The contribution of particle swarm optimization to three-dimensional slope stability analysis. *The Scientific World Journal.*

Kalatehjari, R., Ali, N., Hajihassani, M., and Fard M.K., 2012. The application of particle swarm optimization in slope stability analysis of homogeneous soil slopes. *International Review on Modelling and Simulations,* 5(1): 458–65.

Kennedy, J. and Eberhart, R. 1995. Particle swarm optimization, Proceedings of *ICNN'95 - International Conference on Neural Networks*, IEEE, Perth, WA, Australia, 4: 1942-1948

Kumar, U., Jayadeva, S.S. 2016. Benchmarking NLopt and state-of-art algorithms for Continuous Global Optimization via Hybrid IACOR. *Swarm and Evolutionary Computation*. 27: 116-131.

Li, L., Yu, G., Chu, X., and Lu, S. 2009. "The harmony search algorithm in combination with particle swarm optimization and its application in the slope stability analysis." In *Proceedings of the International Conference on Computational Intelligence and Security (CIS '09),* vol. 2, pp. 133–136, Beijing, China, December 2009.

Li, S., Cami, B., Javankhoshdel, S., Corkum, B., & Yacoub, T. 2020. Considering Multiple Failure Modes in Limit Equilibrium Slope Stability Analysis: Two Methods. *GeoVirtual 2020*. The Canadian Geotechnical Society.

Li, X. Epitropakis, G. M. Deb, K. and Engelbrecht, A. 2017. Seeking Multiple Solutions: An Updated Survey on Niching Methods and Their Applications, *IEEE Transactions on Evolutionary Computation*, 22(4): 518-538

Qu B. Y., Suganthan P. N. and Das S. 2013. Distance-Based Locally Informed Particle Swarm Model for Multimodal Optimization, *IEEE Transactions on Evolutionary Computation*, 17(3): 387-402

Reale, C., Xue, J., Pan, Z. and Gavin, K., 2015. Deterministic and probabilistic multi-modal analysis of slope stability. *Computers and Geotechnics*, 66, pp.172-179.

Reese, L.C. 1958. Discussion of "Soil modulus of laterally loaded piles." By Bramlette McClelland & John A. Fotch, Jr. *Proceedings,* ASCE (1081), *Transactions*, ASCE 123: 1071-1074.

Reese, L.C. and Van Impe, W. 2011. *Single Piles and Pile Groups Under Lateral Loading (2nd Edition)*. CRC Press, Boca Raton, FL, USA.

Rocscience. 2018. Axially Loaded Piles. Accessed online: https://static.rocscience.cloud/assets/verification-and-theory/RSPile/RSPile-Axially-Loaded-Piles-Theory.pdf (April 21, 2022).

Rocscience. 2022. Laterally Loaded Piles. Accessed online: https://static.rocscience.cloud/assets/verification-and-theory/RSPile/RSPile-Laterally-Loaded-Pile-Theory.pdf (April 21, 2022).

Su, Z., & Shao, L. 2021. A three-dimensional slope stability analysis method based on finite element method stress analysis. *Engineering Geology*, 280, 105910.

Sun, C., Chai, J., Luo, T., Xu, Z., Chen, X., Qin, Y., & Ma, B. 2021. Nonlinear shear-strength reduction technique for stability analysis of uniform cohesive slopes with a general nonlinear failure criterion. *International Journal of Geomechanics,* 21(1), 06020033.

Tian, D., Xu, L., and Wang, S. 2009. "The application of particle swarm optimization on the search of critical slip surface." In *Proceedings of the International Conference on Information Engineering and Computer Science* (ICIECS '09), pp. 1–4, Wuhan, China, December 2009.

Zhao, H.B., Zou, Z.S., and Ru, Z.L. 2008. "Chaotic particle swarm optimization for non-circular critical slip surface identification in slope stability analysis." In *Proceedings of the International Young Scholars' Symposium on Rock Mechanics: Boundaries of Rock Mechanics Recent Advances and Challenges for the 21st Century,* pp. 585–588, Beijing, China.

EFFECTS OF SELF-WEIGHT ON THERMOMECHANICAL RESPONSE OF ENERGY PILES

Arash Saeidi Rashk Olia, Department of Civil Engineering, Kansas State University, Manhattan, KS, Phone: 785-532-5862, e-mail: saeidi@ksu.edu
Dunja Perić, Department of Civil Engineering, Kansas State University, Manhattan, KS, Phone: 785-532-5862, e-mail: peric@ksu.edu

ABSTRACT

The increasing demand for renewable energy has attracted attention to energy geostructures as means of extracting clean geothermal energy from the ground. To further advance the understanding of soil-structure interaction in energy piles analytical solutions considering the self-weight of the pile were derived. These solutions were then incorporated in recently-derived thermomechanical weightless solutions to provide comprehensive analytical solutions for axial displacement, strain and stress in single energy piles. This was accomplished by comparing responses of energy pile under different loading scenarios that reflect presence and absence of the self-weight. Results of mechanics-based analysis revealed that the self-weight of energy piles embedded in stiff soils has no significant contribution to displacement, strain and stress that develop under thermo-mechanical loads. Overall, the significance of the self-weight contribution depends on the magnitude of thermal load as well as the soil-pile interface stiffness. The stiffness of end restraints imposed to the pile head and tip were also found to play an important role in contribution of self-weight in induced thermomechanical stresses in energy piles.

Keywords: Energy foundations, Soil-structure interaction, Analytical solutions, Thermo-mechanics

INTRODUCTION

Energy geo-structures such as piles, walls, and tunnels couple their primary role as structural supports to the role of exchanging heat with shallow geothermal resources. This relatively new feature contributes to the sustainability of human constructions through the establishment of low-carbon footprint buildings and infrastructure. Recently, energy piles have garnered attention worldwide due to their environmental and economic potentials.

The free thermal deformation of energy piles is restrained by surrounding soil and rock layers, thus inducing thermal stress in the pile. Consequently, design of energy piles requires the knowledge of the thermally induced stresses and strains in the pile as well as of interface shear and normal stresses. To accomplish this goal researchers have used physical and numerical models. Computational models have been historically used to address thermomechanical behavior of energy piles subjected to mechanical and thermal loads. However, these capabilities are often unavailable in common engineering circumstances. This fact, coupled with the significant time and expertise that are required to address through computer simulations the multi-physical phenomena involved with energy piles, often makes the numerical modeling daunting for the engineering practice. To this end, analytical solutions are powerful tools for both engineering practice and research purposes to address thermomechanical analysis of these structures (Rotta Loria et al. 2020).

Olia and Perić 2021 presented and validated analytical solutions for energy piles embedded in homogeneous soil layer were validated against geotechnical centrifuge test. These solutions also successfully predicted axial thermomechanical stress and strain resulted from a coupled thermo-hydromechanical finite element analysis (Olia and Perić 2020 a & b, Cossel 2019,). While most of the previous studies, and numerical models neglected the self-weight of the pile in thermomechanical analysis (Knellwolf et al. 2011, Perić et al. 2017& 2020) in this study analytical solution that accounts for the self-weight is derived to complement the previous solutions presented by Olia and Perić 2021. Furthermore, these solutions were implemented to investigate the extend of validity of this assumption by comparison of weighted and weigh-less solutions prediction against each other and experimental results.

ANALYTICAL SOLUTIONS

Olia and Perić 2021 presented solutions for a weight-less energy pile under thermal and mechanical load. Those solutions are extended to incorporate the self weight of the pile. To this end,Fig. 1 depicts Free Body Diagram (FBD) of an energy pile subjected to its own weight. Thus, the self-weight of the pile (W) must be in equilibrium with the resultant (Q_s) of the shear stress that is acting at the pile-soil interface, and the normal reaction (N_b) at the pile tip.

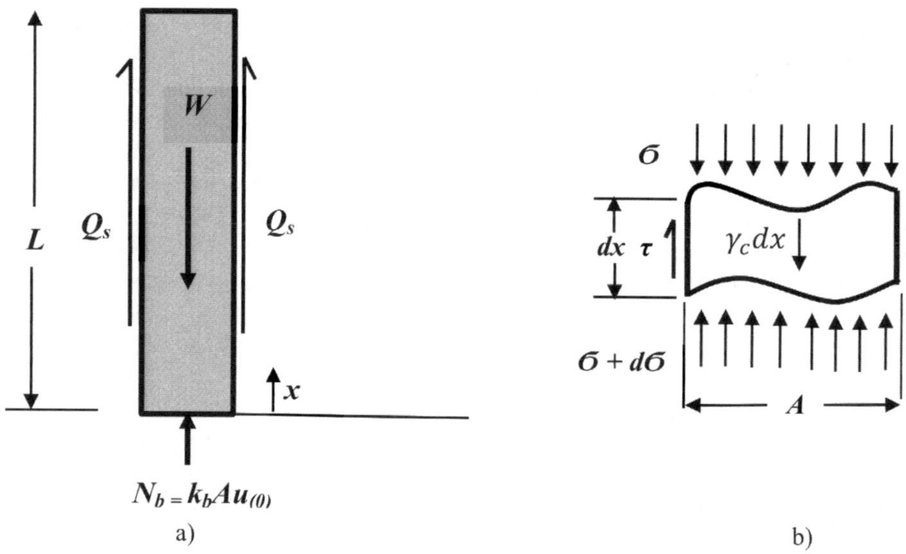

Fig. 1. FBDs of the energy pile subjected to its weight (a) the entire pile (b) an infinitesimal segment of the pile

As indicated in Fig.1 the positive x-axis is directed upward. The kinematic relationship is given by

$$\varepsilon = \frac{du}{dx} \quad [1]$$

where $u = u(x)$ is the axial (or vertical) displacement, and $\varepsilon = \varepsilon(x)$ is the axial strain. The thermo-elastic constitutive law is given by

$$\sigma = E(\varepsilon - \alpha\Delta T) \quad [2]$$

where $\sigma = \sigma(x)$ is the axial stress. In 1D formulation used herein u, ε and σ are simply referred to as displacement, strain and stress. The elastic modulus and the coefficient of thermal expansion of the pile are denoted by E and α respectively while ΔT is a temperature difference of the pile relative to the surrounding soil. The soil pile interface along the pile shaft is modeled with a continuous linear elastic shear spring whereby the magnitude of the interface shear stress, $\tau = \tau(x)$ is given by

$$|\tau| = k_S|u| \qquad [3]$$

Enforcing the equilibrium in the vertical direction of the infinitesimal pile segment (Figure 1.b) and combining it with Eqs. 1, 2, and 3 results in the following non-homogeneous second order ordinary differential equation (ODE).

$$\frac{d^2u}{dx^2} - \psi^2 u = \frac{\gamma_c}{E} \qquad [4]$$

where γ_c is a unit weight of the pile while ψ is a constant given by

$$\psi^2 = \left(\frac{p}{A}\right)\left(\frac{k_s}{E}\right) = \xi_g \xi_s \qquad [5]$$

where A and p represent cross sectional area and perimeter of the pile respectively, and k_s represents the stiffness of a continuous shear spring. Solving for the unknown function $u(x)$ from Eq. 4 results in:

$$u(x) = C_1 e^{\psi x} + C_2 e^{-\psi x} - \frac{\gamma_c}{E\psi^2} \qquad [6]$$

The boundary conditions at pile head and tip for a semi-floating pile under its own weight are given by

$$k_b u(0) = \sigma(0) \qquad and \qquad \sigma(L) = 0 \qquad [7]$$

where L represents the length of the pile, and k_b is the stiffness of the normal spring at the pile tip. Imposing the boundary conditions in Eq. 7 results in:

$$u(x) = \frac{\gamma_c}{E\psi^2}\left\{\frac{-k_b(e^{\psi x} + e^{\psi(2L-x)})}{[E\psi(1 - e^{2\psi L}) - k_b(1 + e^{2\psi L})]} - 1\right\} \qquad [8]$$

Axial strain induced by self-weight of the pile is obtained by combining Eqs. 1 and 8 as

$$\varepsilon(x) = \frac{\gamma_c}{E\psi}\left\{\frac{-k_b(e^{\psi x} - e^{\psi(2L-x)})}{[E\psi(1 - e^{2\psi L}) - k_b(1 + e^{2\psi L})]}\right\} \qquad [9]$$

Axial stress induced by self-weight along the pile shaft is obtained by combining Eqs. 2 and 9 as

$$\sigma(x) = \frac{\gamma_c}{\psi}\left\{\frac{-k_b(e^{\psi x} - e^{\psi(2L-x)})}{[E\psi(1 - e^{2\psi L}) - k_b(1 + e^{2\psi L})]}\right\} \qquad [10]$$

Analytical solutions for a weight-less semi-floating energy pile subjected to thermal and mechanical loads were previously presented by Cossel (2019), and Saeidi Rashk Olia and Perić (2021). Therefore, by combining the previous solutions with the ones presented herein the complete analytical solutions for a semi-floating energy pile subjected to thermal, mechanical and self-weight loads are given by:

$$u(x) = \frac{\alpha\Delta T sinh[\psi(x-x_0)]}{\psi cosh[\psi(L-x_0)]} + \frac{F[E\psi cosh(\psi x)+k_b sinh(\psi x)]}{AE\psi[E\psi sinh(\psi L)+k_b \cosh(\psi L)]} + \frac{\gamma_c}{E\psi^2}\left\{\frac{-k_b(e^{\psi x}+e^{\psi(2L-x)})}{[E\psi(1-e^{2\psi L})-k_b(1+e^{2\psi L})]} - 1\right\} \quad [11]$$

$$\varepsilon(x) = \frac{\alpha\Delta T cosh[\psi(x-x_0)]}{cosh[\psi(L-x_0)]} + \frac{F[E\psi sinh(\psi x)+k_b cosh(\psi x)]}{AE[E\psi sinh(\psi L)+k_b cosh(\psi L)]} + \frac{\gamma_c}{E\psi}\left\{\frac{-k_b(e^{\psi x}-e^{\psi(2L-x)})}{[E\psi(1-e^{2\psi L})-k_b(1+e^{2\psi L})]}\right\} \quad [12]$$

$$\sigma(x) = \underbrace{E\alpha\Delta T\left[\frac{cosh[\psi(x-x_0)]}{cosh[\psi(L-x_0)]} - 1\right]}_{\text{Thermal}} + \underbrace{\frac{F[E\psi sinh(\psi x)+k_b cosh(\psi x)]}{A[E\psi sinh(\psi L)+k_b cosh(\psi L)]}}_{\text{Mechanical}} + \underbrace{\frac{\gamma_c}{\psi}\left\{\frac{-k_b(e^{\psi x}-e^{\psi(2L-x)})}{[E\psi(1-e^{2\psi L})-k_b(1+e^{2\psi L})]}\right\}}_{\text{Self-weight}} \quad [13]$$

Positive displacement is directed upward, and tensile strain and stress are positive. Positive temperature difference (*ΔT*) between the pile and the surrounding soil corresponds to pile heating. Mechanical load in the form of an axial force *F* is positive in tension and negative in compression. The location of the thermal zero-displacement point, known as null point was obtained from the boundary condition at the pile tip, and is given by

$$x_0 = \frac{1}{\psi} tanh^{-1}\left[\frac{cosh(\psi L)-1}{\sinh(\psi L)+\frac{k_b}{E\psi}}\right] \quad [14]$$

It is noted that the solutions for displacement, strain and stress in end bearing energy pile are obtained from Eqs. (11) through (13) by setting $x_0 = 0$ and $k_b \rightarrow \infty$, thus resulting in

$$u(x) = \frac{\alpha\Delta T sinh(\psi x)}{\psi\, cosh(\psi L)} + \frac{F sinh(\psi x)}{AE\psi cosh(\psi L)} + \frac{\gamma_c}{E\psi^2}\left\{\frac{cosh[\psi(l-x)]}{cosh(\psi L)} - 1\right\} \quad [15]$$

$$\varepsilon(x) = \frac{\alpha\Delta T cosh(\psi x)}{cosh(\psi L)} + \frac{F cosh(\psi x)}{AE cosh(\psi L)} + \frac{\gamma_c}{E\psi}\left\{\frac{sinh[\psi(l-x)]}{cosh(\psi L)}\right\} \quad [16]$$

$$\sigma(x) = \underbrace{E\alpha\Delta T\left[\frac{cosh(\psi x)}{cosh(\psi L)} - 1\right]}_{\text{Thermal}} + \underbrace{\frac{F cosh(\psi x)}{A cosh(\psi L)}}_{\text{Mechanical}} + \underbrace{\frac{\gamma_c}{\psi}\left\{\frac{sinh[\psi(l-x)]}{cosh(\psi L)}\right\}}_{\text{Self-weight}} \quad [17]$$

VALIDATION OF THE ANALYTICAL SOLUTION FOR END BEARING PILE

A single end bearing energy pile subjected to a wide range of heating was tested in a geotechnical centrifuge at University of Colorado Boulder by Stewart and McCartney (2013). The geotechnical centrifuge with a g-level of 24, capable of modeling models for prototypes with 24 times the model's dimensions. In this context, a model energy pile with a diameter of 50.8 mm and a length of 533.4 mm, was used to model a prototype energy pile with 12.8 m of length, a diameter of 1.22 m. The soil pile stiffness was calibrated by Saeidi Rashk Olia and Perić (2021) to 55 MPa/m respectively. No mechanical load was present during the tests. Therefore, self-weight and thermal loads were the only contributors in induced strain and stress along the piles shaft. The energy pile with modulus of elasticity of 7.17 GPa, and unit weight of 24 kN/m^3 was embedded in a single layer of unsaturated Bonny silt with the coefficient of thermal expansion (α) of 7.5 × 10^{-6} 1/°C. Fig. 2 illustrates the effect of self-weight in predictions of thermal response of energy pile for 5°C, 10°C, 15°C, 19°C of heating. As shown, the effect of self-weight on thermal response of energy pile in more dominant for lower temperature magnitudes and diminishing as the temperature increases. To this end, at 5°C the the maximum displacement decreases by almost 45% when the self-weight is taken into account. However, the contribution of the self-weight in maximum displacement of the pile drops to 11% as the pile is heated to 19°C. The contribution of the self-weight to the maximum magnitude of vertical stress also drops from 43% at 5°C to 16% at 19°C.

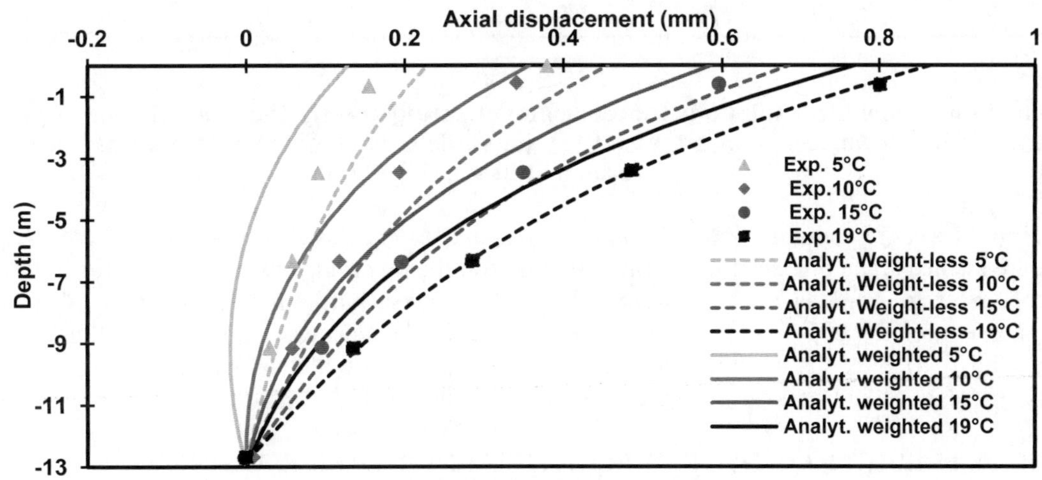

a)

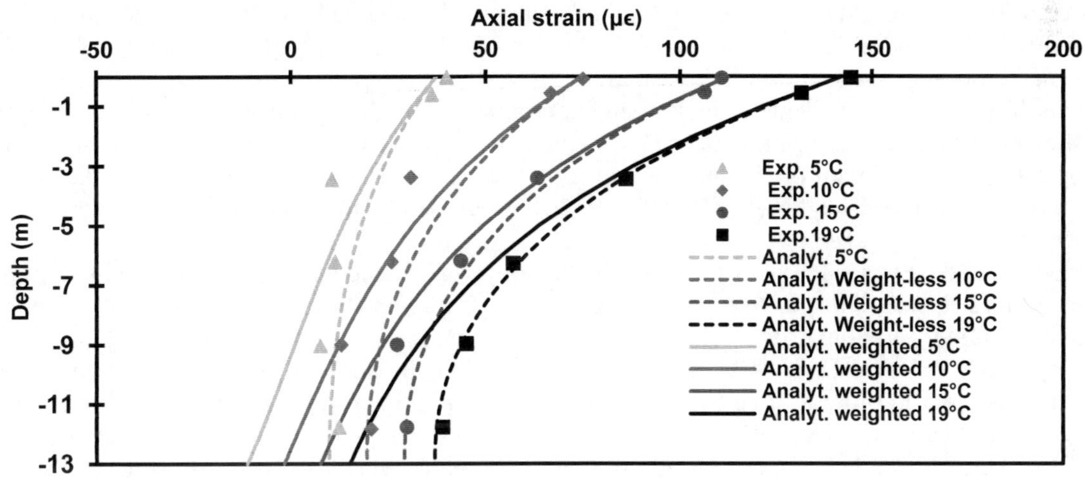

b)

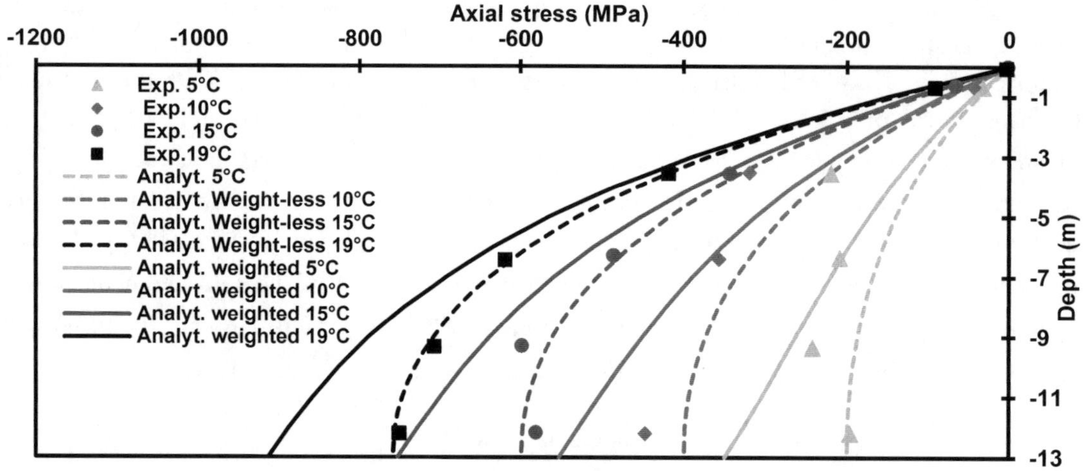

c)

Fig. 2. Validation of analytical solutions for ideally end bearing energy pile against centrifuge test results obtained by Stewart and McCartney (2013): (a) vertical displacement; (b) vertical strain; (c) vertical stress

It is noted that the aforementioned contributions correspond to the maximum values. These extreme values that are located at the pile head for displacement and at the pile tip in case of axial strain and stress. Therefore, the average differences along the pile length are lower. Furthermore, the observed trend of responses is as expected, i.e. in case of heating the upward displacement is larger in case that the self-weight is excluded. This results in larger compressive strain in the case that self-weight is excluded, and also larger compressive axial stress in case that self-weight is included.

ANALYTICAL PREDICTIONS OF SEMI-FLOATING ENERGY PILE RESPONSE

A series of full-scale pile tests on a single energy pile subjected to thermal and thermo-mechanical loads was conducted at the Swiss Federal Institute of Technology (Laloui et al. 2006; Laloui. et al. 1999). Out of a total of 97 piles at this site only one was equipped with heat pump and loops and performed as an energy pile. It has an average diameter of 1 m and approximate length of 26 m, with an elastic modulus and coefficient of thermal expansion equal to 29.2 GPa and 1×10^{-5}/°C, respectively. Test 7, which was conducted after the completion of the building construction, includes a simultaneous application of the axial compressive force of 1000 kN (Knellwolf et al. 2011) imposed by the building and a thermal load.

The energy pile at the Swiss Federal Institute of Technology was embedded in a four-layered soil profile that Perić et al. (2017) classified as low plasticity clays (CL) based on the laboratory test results reported by Laloui et al. (1999). Knellwolf et al. (2011) reported the stiffness of the elastic shear springs (k_s) to be equal to 16.7, 10.8, 18.2 and 121.4 MPa/m for soil layers A1, A2, B and C respectively based on the full-scale in situ pile tests. In order to investigate the effect of soil pile interface stiffness on the contribution of self-weight for thermo-mechanical response of a semi-floating energy pile two synthetic homogeneous soil profiles were assembled corresponding to the soil pile interface stiffness of soils A1 and C respectively. In addition, the normal spring with stiffness (k_b) of 1335.5 MPa/m (Knellwolf et al. 2011) representing the contact of the pile with the bedrock was selected. While the energy pile is subjected to 1000 kN of mechanical load, the selected history of the thermal loading selected herein is shown in Table 1 for heating and cooling scenarios.

Table 1. Temperature histories for the two loading scenarios

Time (days)	0	10	25	30
ΔT (°C)	0	10	20	3
ΔT (°C)	0	-10	-20	-3

Fig. 3 shows the prediction of head displacement for the two loading scenarios from Table 1 while accounting and not accounting for the self-weight of the pile. It is noted that both cases result in a negative head displacement prior to the commencement of thermal load. Compressive mechanical load induced these displacements. Both solutions predict similar trends of the head displacement whereby the contribution of the self-weight is more significant in case of soil A1 with k_s =16.7 MPa/m. In case of soil C, which has k_s = 121.4 MPa/m, the difference between including and excluding the self-weight of the pile is insignificant. Therefore, the stiffer soil pile interface decreases the significance of self-weight in analysis.

Vertical displacement under thermomechanical load including a compressive force of 1000 kN, and heating and cooling of 20°C is depicted in Fig.4 for soils A1 and C. As shown in case of a stiff soil (soil C) two analyses predicted almost the same results for ±20°C, thus rendering the effect of the self-weight insignificant. In the case of the smaller stiffness of the soil-pile interface (soil A1), the difference between the responses including and excluding the self-weight of the pile become noticeable. As expected, including the self-weight of the pile results in a smaller displacement magnitude in heating and larger displacement magnitude in cooling above the zero-displacement point known as the null point. Below the null point the trend is opposite, heating combined with the self-weight results in a larger displacement than cooling combined with the self-weight. displacement magnitude of displacement. This is due to different directions of thermal displacements above and below the null point. Also, the presence of the self-weight moves the location of the zero-displacement point under combined thermomechanical downward in case of cooling and upward in case of heating. The shift is negligible in a very stiff soil. Combined null point is the null point of axial displacement in the case of combined thermal and mechanical loads.

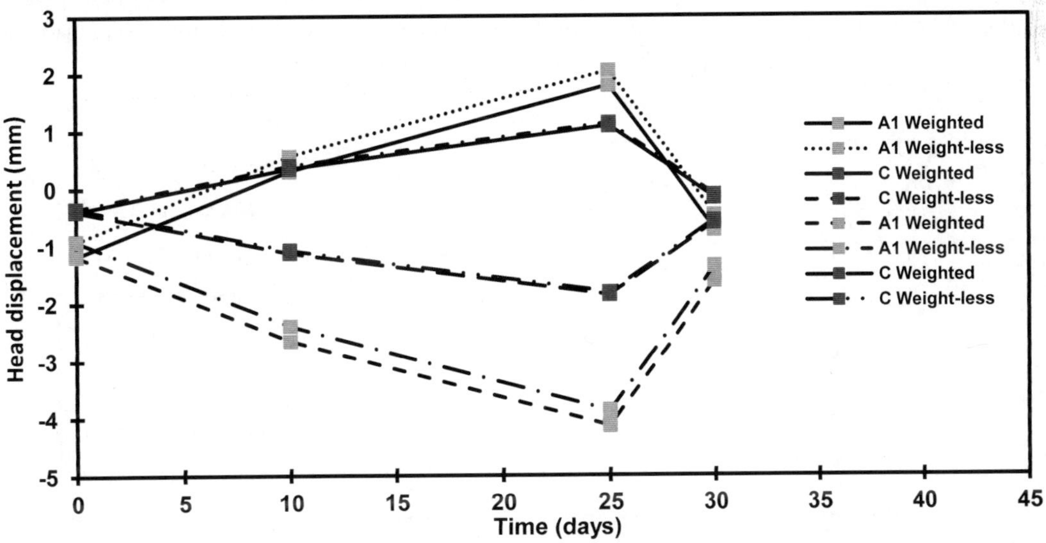

Fig. 3. Analytical prediction of thermomechanical head displacement versus time in presence and absence of the self-weight of the energy pile

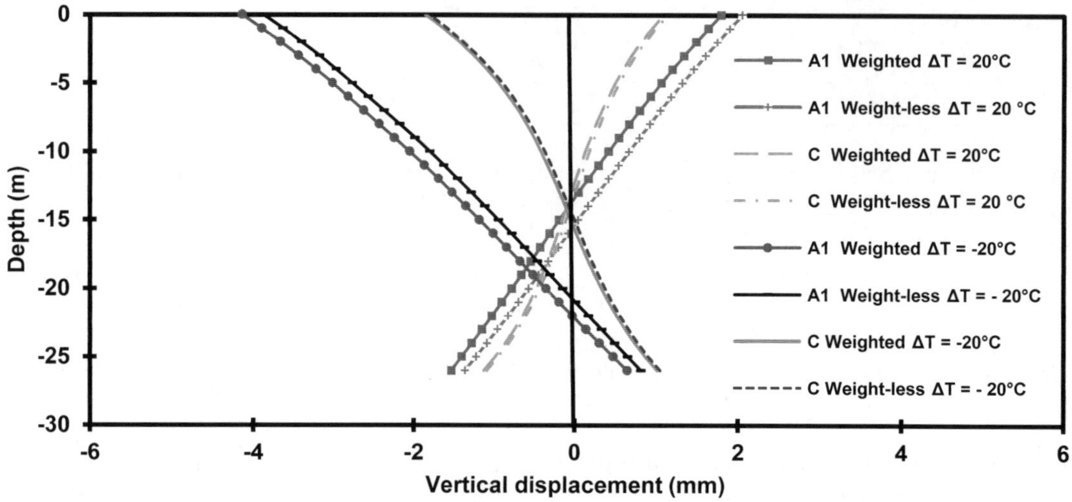

Fig. 4. Analytical prediction of thermomechanical vertical displacement in presence and absence of the self-weight of the energy pile

Figs 5 and 6 show vertical strains and stresses induced by the thermo-mechanical load whereby the axial compressive force of 1000 kN was applied at the pile head and pile was heated/cooled by ±20°C. Similarly, to displacement and strain response, the effect of the self-weight of the pile was insignificant for soil C while it became noticeable in soil A1. Even for the soil A1 the effect of the self-weight becomes less significant with increasing temperature, i.e. inclusion of thermal load results in the response that becomes closer to the one without including the self-weight as thermal load increases. The soil pile interface stiffness has a significant effect on the induced stress in the pile, i.e., the magnitude of the maximum compressive stress in case of compression and heating doubled, and the magnitude of the maximum tensile stress in case of compression and cooling almost quadrupled in soil C as compared to the same loading scenarios in soil A1, which is due to almost seven times higher stiffness of soil pile interface in soil C.

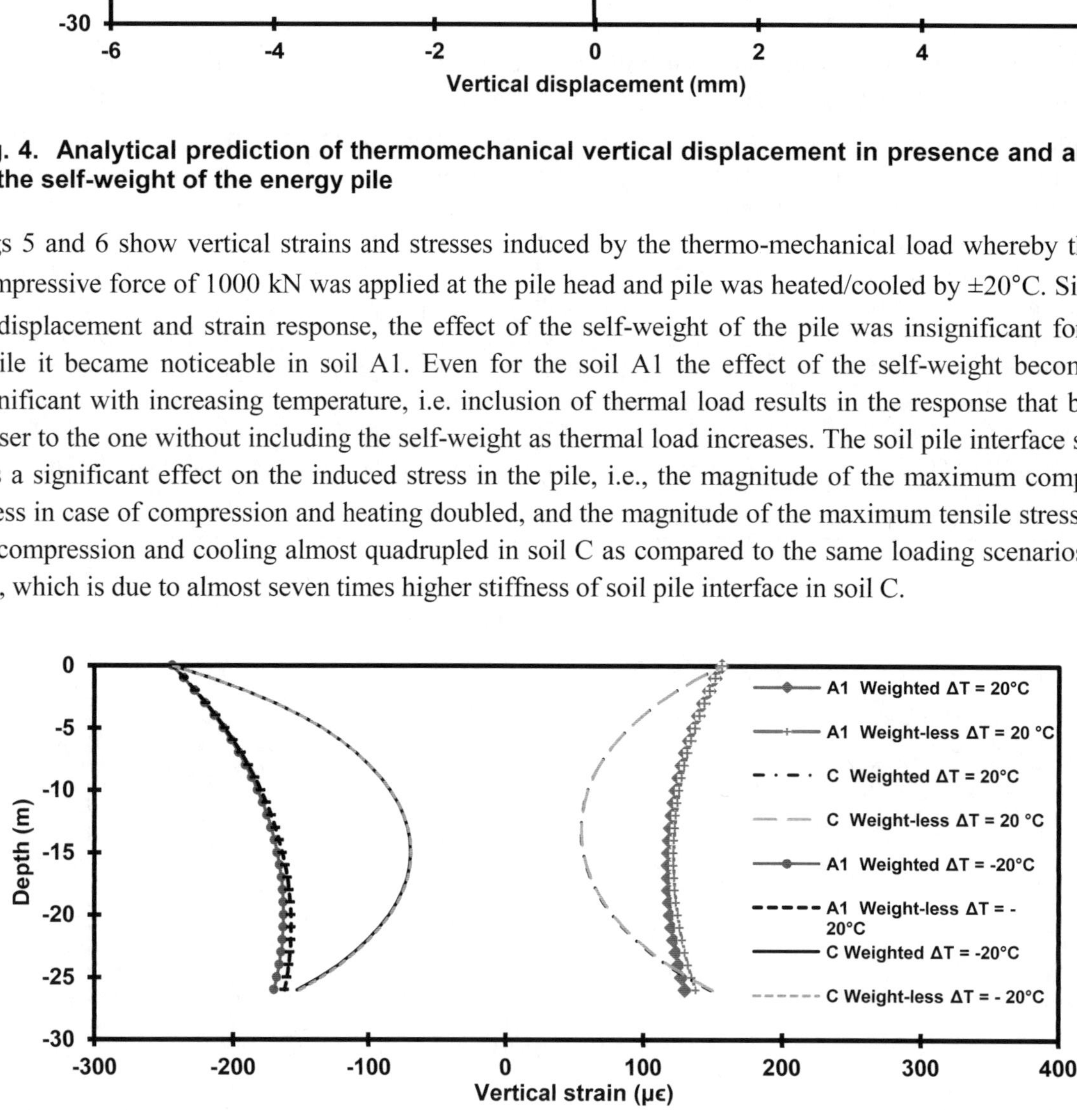

Fig. 5. Analytical prediction of thermomechanical vertical strain in presence and absence of self-weight of energy pile

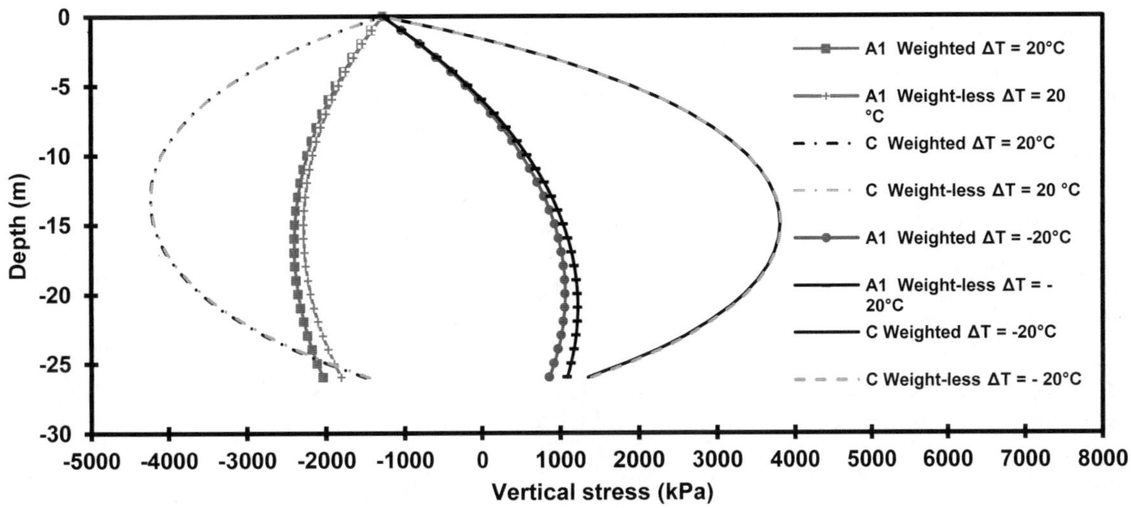

Fig. 6. Analytical prediction of thermomechanical vertical stress in presence and absence of self-weight of energy pile

CONCLUSIONS

Recently derived analytical solutions for response of a single energy pile embedded into a homogeneous soil profile and subjected to thermal, mechanical and self-weight loads were implemented to investigate the effects of self-weight on the soil structure interaction in energy piles. To this end, predictions for end bearing and a semi-floating energy pile subjected to thermal load have been validated against the centrifuge test results while including and excluding the self-weight. Further comparisons between responses of a semi-floating energy pile subjected to thermo-mechanical load with and without the self-weight were conducted to assess the significance of the contribution of the self-weight of the pile on vertical displacements, strain and stresses.
Prediction of displacement, strain and stress in end bearing energy pile subjected to thermal load showed that contribution of self-weight decreases in higher temperatures. While the self-weight induced displacements, stresses and strains are constant, and as an initial condition of the pile, upward displacements due to heating soon balances the self-weight settlements and under service thermal loads decreases the self-weight contribution significantly.

The main factor that determines the contribution of the self-weight on response of energy piles was found to be the stiffness of the soil-pile interface. To this end, the soil layer that exhibits the stiffest soil-pile interface at Swiss Federal Institute of Technology soil profile (soil C) demonstrated very little difference between the response obtained from weighted and weight-less analysis. Nevertheless, soil-pile interfaces of smaller stiffness result in different pile responses in case that the self-weight is included or excluded. Higher thermal loads decreases the impact of self-weight, thus makes this difference less pronounced.

REFERENCE LIST

Cossel, A. 2019. Analytical solutions for thermo-mechanical soil structure interaction in energy piles. M.S. thesis, Dept. of Civil Engineering, Kansas State University.

Knellwolf, C., H. Peron, and L. Laloui. 2011. Geotechnical analysis of heat exchanger piles. J. Geotech. Geoenviron. Eng. 137 (12): 890–902. https://doi.org/10.1061/(ASCE)GT.1943-5606.0000513.

Laloui, L., Moreni, M., Steinmann, G., Vulliet, L., Fromentin, A., and Pahud, D. 1999. Test en conditions réelles du comportement statique d'un pieu soumis à des solicitations thermos-mécaniques." Report of the Swiss Federal Office of Energy 1999.

Laloui, L., M. Nuth, and L. Vulliet. 2006. Experimental and numerical investigations of the behaviour of a heat exchanger pile. Int. J. Numer. Anal. Meth. Geomech. 30 (8): 763–781. https://doi.org/10.1002/nag.499.

Perić, D., Cossel, A. E., & Sarna, S. A. 2020. Analytical Solutions for Thermomechanical Soil Structure Interaction in End-Bearing Energy Piles. J. Geotech. Geoenviron., 146(7), 04020047. https://doi.org/10.1061/(ASCE)GT.1943-5606.0002269

Perić, D., T. V. Tran, and M. Miletić. 2017. "Effects of soil anisotropy on a soil structure interaction in a heat exchanger pile." Comput. Geotech. 86 (Jun): 193–202. https://doi.org/10.1016/j.compgeo.2017.01.005.

Saeidi Rashk Olia, A., and Perić, D. 2021. Thermomechanical Soil Structure Interaction in Single Energy Piles Exhibiting Reversible Interface Behavior. Int. J. of Geomech. 21(5), 04021065. https://doi.org/10.1061/(ASCE)GM.1943-5622.0002014

Saeidi Rashk Olia, A., and Perić, D. 2020a. Cross validation of analytical solutions against the computational model predictions of the response of end bearing energy pile. In E3S Web of Conferences 205, p. 05019). EDP Sciences https://doi.org/10.1051/e3sconf/202020505019

Saeidi Rashk Olia, A., and Perić, D. 2020b. Cross validation of analytical solutions against the computational model predictions of the response of semi floating energy pile. In E3S Web of Conferences (Vol. 205, p. 05020). EDP Sciences. https://doi.org/10.1051/e3sconf/202020505020

Stewart, M. A., and McCartney, J. S. 2013. Centrifuge modeling of soil-structure interaction in energy foundations. J. Geotech. Geoenviron. Eng., 10.1061/(ASCE)GT.1943-5606.0001061, 04013044.

SOLVING VARIABLE BEDROCK CHALLENGES WITH DUCTILE IRON PILES

Bryan Plaskett, E.I., Earth Tech, LLC. 2620 Hunt Rd., Land O' Lakes, FL bplaskett@earthtech.com
Kevin R. Johnson, Ph.D., P.E., Earth Tech, LLC. 2620 Hunt Rd, Land O' Lakes, FL kjohnson@earthtech.com
Matt Caskey, P.E.' DuroTerra, LLC, 639 Granite Street, Braintree, MA mcaskey@duroterra.com

ABSTRACT

Working in urban environments presents unique logistical challenges beyond those typically associated with traditional geotechnical design and construction. These challenges can include constrained sites with limited access and laydown area, vibration concerns for adjacent structures, and the need to complete work within limited schedules. When challenging geotechnical conditions like variable bedrock elevations are present, the need to balance performance requirements with logistical constraints become the driving factor when selecting deep foundation systems. This paper focuses on the deep foundations used for a new elevated pedestrian bridge at a major health care facility in Tampa, Florida. In addition to the technical design requirements, the owner required a deep foundation system that could offer low vibrations, rapid production, and clean installation to limit impacts on their properties and the public roadway that separated them. The paper describes the project specific geotechnical and design conditions, describes Ductile Iron Pile design, and focuses on how installation of the system addressed the unique logistical challenges. In addition, the implementation of a full-scale load testing program to assess compression, uplift, and lateral design capacities is presented. This paper represents one of the few published references for Ductile Iron Piles on projects in the United States.

Keywords: ductile iron piles, variable bedrock, load testing

INTRODUCTION

Working in urban environments presents unique logistical challenges beyond those typically associated with traditional geotechnical design and construction. These challenges can include constrained sites with limited access and laydown area, vibration concerns for adjacent structures, and the need to complete work within limited schedules. When challenging geotechnical conditions like variable bedrock elevations are present, the need to balance performance requirements with logistical constraints become the driving factor when selecting deep foundation systems.

This paper focuses on the deep foundations used for a new elevated pedestrian bridge at a major health care facility in Tampa, Florida. The bridge connects a new patient tower to an existing clinic building. The pedestrian bridge is a steel-truss structure, supported by cast-in-place concrete columns and connects the second floor of the patient tower to the second floor of the clinic building. The 375-ft long bridge crosses a wide, busy arterial street and includes13 new foundations subject to heavy compression, tension, and lateral forces. Structural loads on the new foundations were reported to be as high as 1,100 kips (4,892 kN) in compression, 150 kips (667 kN) in tension and 150 kips (667 kN) laterally, driven by up to 3,000 ft-kips (4,069 kN-m) of overturning moment.

SUBSURFACE CONDITIONS

The subsurface conditions at the site as described by Terracon, Inc. (Terracon, 2021) consist of an upper stratum of loose to medium dense poorly graded sand (SP), poorly graded sand with clay (SP-SC), and poorly graded sand with silt (SP-SM) and was encountered to depths ranging from 10 to 19 feet (3 to 5.8 m) below the ground surface. The second stratum consists of soft to very stiff plastic clay (CH) and was encountered to depths ranging from 31 to 52 feet (9.4 to 15.8 m). The upper soil strata are underlain by limestone bedrock, which sampled as calcareous clayey sand with limestone fragments and was encountered until the boring terminal depths of 70 to 80 feet (21.3 to 24.4 m). The limestone bedrock has varying intervals of weathering, with weathered zone thicknesses ranging from 5 to 20 feet (1.5 to 6.1 m). A generalized profile of the subsurface conditions can be found in Figure 1.

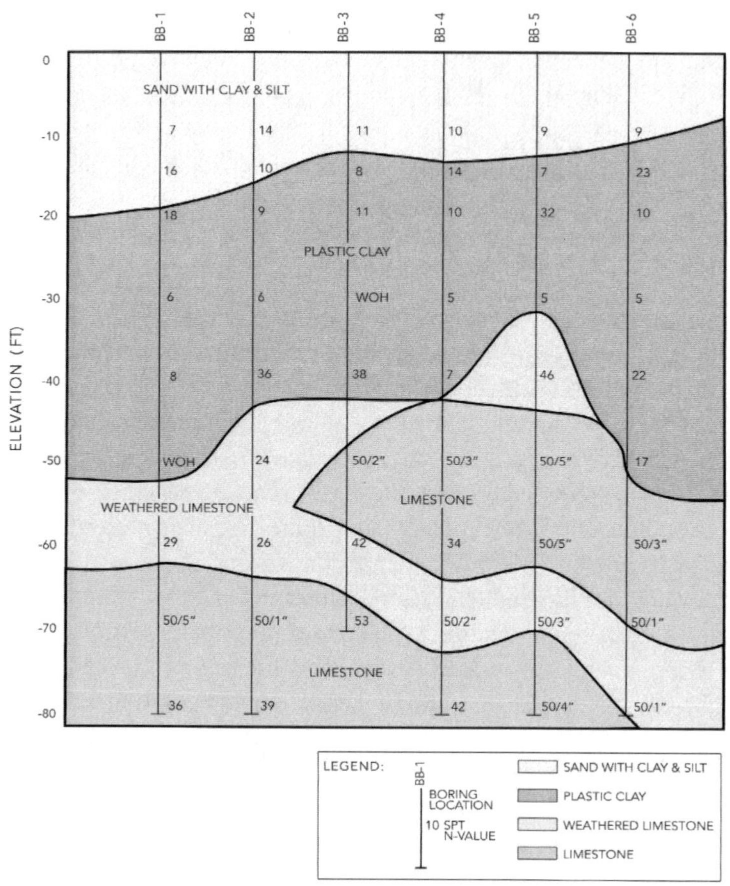

Figure 1. Generalized subsurface profile

FOUNDATION DESIGN OPTIONS

Due to the magnitude of loading associated with the new pedestrian bridge, a deep foundation system was required. In addition to the technical design requirements, it was important to the owner that the selected deep foundation system offer low vibrations, rapid production, and clean installation to limit impacts on both properties and the public roadway that separated them. Due to the variability of strength and depth of competent bedrock (45 to 65 feet) (13.7 m to 19.8 m), the depth to establish reliable pile capacities would vary widely. While cast-in-place deep foundations are normally ideal for sites with variable end-bearing layers, this option was ruled out due to concerns that high grout loss and overages could occur in the historically porous weathered limestone zones. Straight-shaft drilled piers were also dismissed due to variation in column loading conditions and the need for foundation redundancy at bridge column locations. It was believed that drilled piers would not offer an efficient design option to the project team. Pre-cast concrete piles were ruled-out due to the expected large amounts of cut-off waste and/or splicing needs that would result from the variable pile bearing elevations.

The design team decided to consider driven steel pipe piles and driven Ductile Iron Piles for deep foundation construction at the site. The geotechnical specialty contractor provided budget estimates for installing up to 95, 12.75 in (324 mm) diameter steel pipe piles and for installing up to 163, exterior-grouted 4.65 in (118 mm) diameter Ductile Iron Piles with an 8.7 in (220 mm) outer-diameter grouting shoe. The project team chose to use exterior-grouted Ductile Iron Piles due to a combination of cost and schedule benefits.

DUCTILE IRON PILE SYSTEM

For nearly a decade in North America and more than 30 years in Europe, engineers and contractors have used specially manufactured Ductile Iron Piles (DIPs) to provide a reliable and cost-effective alternative to more traditional deep foundation systems that address urban challenges. Ductile Iron Piles are a low vibration, driven pipe pile system. The modular, small-diameter piles are manufactured in standard 16.4 ft (5 m) or 18 ft (5.5 m) lengths to allow ease of installation on constrained sites. Ductile Iron Piles are manufactured using a centrifugal-casting process that transforms lamellar graphite cast iron into spheroidal graphite or ductile cast iron. The DIP material exhibits a design yield stress of 46.4 ksi (320 MPa) and modulus of elasticity of 24,600 ksi (170,000 MPa) (TRM 2014). Each pile is cast with a bell and spigot to form a Plug and Drive connection mechanism (Figure 2). The connection develops strength through a combination of the tapered compression fit as well as a friction (cold) weld that occurs during the high-frequency driving process, eliminating the need for field splicing. The piling material is delivered in bundles containing between 6 and 18 pile sections, depending on pile diameters.

Figure 2. Plug and Drive connection

The system is installed using a medium-sized excavator and a hydraulic breaker hammer (Figure 3). The breaker hammer uses a modified shank that seats into the pile bell for driving, using the hammer's percussive energy. The hammer size (energy) is matched with the size of the Ductile Iron Pile and the subgrade driving conditions to develop the most efficient installation while not overstressing the pile material. Hammer sizes typically range from 3,000 ft-lb to 8,000 ft-lb (415 kg-m to 1,110 kg-m) class hammers for the range of pile sizes. The hydraulic hammers operate at frequencies ranging from 250 to 500 cycles per minute, delivering a much higher frequency content than traditional high-impact, low frequency driven piling systems. Vibration measurements have been recorded on past project sites with different pile diameters, all at distances of less than 1.5 meters (4.9 feet) from installation. Results of the monitoring near the pile installation show that vibrations are generally less than 1 in/sec (25 mm/s) and that they rapidly decrease with distance from the pile location. (DuroTerra, 2016)

Figure 3. Typical Installation Equipment Set-up

Ductile Iron Piles have the flexibility to be installed to develop capacity in either end bearing or in friction. End-bearing DIPs generate the load-carrying capacity through bearing on very dense soil or rock after reaching the project "set" termination criteria. Installation begins with the placement of a flat driving shoe or rock point on the ground at the pile locations. A pile section is then positioned onto the shoe and driven into the ground using the vertical percussion energy. Using the same excavator, a second pile section is swung into place to make the Plug and Drive (bell and spigot) connection and then driven. This process is repeated until the pile termination criteria is reached. After driving, the pile is cut to the plan top of pile elevation and filled with neat cement grout. If higher structural capacity, or tensile resistance, is needed a center reinforcing bar can be placed into the wet grout.

Similar to micropiles in soil, friction DIPs generate compression or tension capacity by creating a grouted bond zone within a competent soil layer. (FHWA, 2005) Piles designed for frictional resistance are installed by first fitting an oversized conical grout shoe over the end of the hollow pile. Conical grout shoes range in diameter from 5.9 in (150 mm) up to 14.6 in (370 mm) depending on pile size and design requirements. The pile is then driven into the ground using the same excavator and hammer along with a grouting drive shank to continuously pump cement grout through the interior of the pile. The grout exits through the grout ports in the conical shoe (Figure 4), filling the annular space on the pile exterior during driving. The pile shaft becomes encapsulated in grout, developing a bond zone. Additional pile sections are driven until the pile extends to the design length within the bond zone. A center reinforcing bar can be inserted if required for tension or additional compression capacity.

Figure 4. Conical shoe with grouting ports

A unique feature of driven Ductile Iron Piles is the system's adaptability for sites with variable bedrock. When piles are driven to refusal or the established "set" criteria, some length of pile will remain above the pile cut-off elevation. This is particularly common at sites with variable bedrock elevations. The remaining free section of pile can be cut off with a chop saw and saved. An end-bearing cap or conical grouting shoe can then be placed on the bottom of the cut-section of pile which is then used as a starter section and driven to start the next pile location. Since the top of a cut pile section will always have a bell, a new pile section can easily be added to continue pile driving operations. For sites with variable bedrock, Ductile Iron Piles greatly reduce material waste and lead to the efficient use of pile materials.

Quality control/quality assurance monitoring is largely focused on pile alignment/verticality and recording penetration rates (amount of time to drive incremental lengths as marked on the pile section) during driving. The rates of penetration can be used as indicators of geotechnical resistance and pile capacity. For end-bearing piles, the pile is advanced until a particular rate of advancement or "set" is achieved. "Set" criteria, defined as the pile deflection measured over a time interval (typically 10 to 50 seconds), is a commonly used criterion that is either established or verified by full-scale load tests. (TRM, 2014) This criterion has developed through decades of experience and testing of the system and requires compatibility between hammer size (energy) and pile/cap dimensions for driving. For friction piles, the penetration rates can be used to identify the depth which corresponds to the start of the bond length. The pile is then advanced a particular distance into the competent layer based on the penetration rates.

DEEP FOUNDATION DESIGN FOR THE PEDESTRIAN BRIDGE

Design of the bridge involved supporting a 170 ft (51.8 m) curvilinear main span which, at the farthest reach, would be cantilevered 24 feet (7.3 m) beyond the plane of the supporting piers, and 20 feet (6.1 m) above McKinley Drive. This required foundation elements to resist significant overturning moments and permanent net tension loads at several locations.

Coordination between the structural, geotechnical, and specialty pile design teams led to a foundation plan which incorporated two 35 x 22 x 5 ft (10.7 x 6.7 x 1.5 m) concrete pile caps, with 35 piles each, for the main span piers, and 11 other pile caps with between 5 and 20 piles each, for a total of 184 piles. All piles were designed for a 75 kip (334 kN) allowable compression load and a 7.7 kip (34 kN) allowable lateral load. Additionally, 80 pile locations were identified as also requiring 30 kips (133 kN) of uplift capacity.

One of the initial challenges of pile design was estimating tip depths, both required and achievable, in the highly variable limestone across the site. While it was estimated that most piles would refuse between 45 – 65 ft (13.7 – 19.8 m), the wide of range of depths and degree of weathering meant that there was the possibility of either very early refusal, which would limit the amount of skin friction available for tension piles, or very deep refusal, which would lead to cost and schedule impacts to the project. Since the tension piles, which accounted for almost half of production, would require external grout to achieve skin friction, it was determined that there would be economy to gain by designing all piles to be externally grouted. This mitigated the risk posed by deep refusals, as the piles could be terminated at a prescribed maximum depth which would provide the full compression capacity in friction alone. This also simplified construction and QC by standardizing the driving process for all piles.

Assuming a final set of 1-in/10s and a grout-to-ground bond strength of 5 psi (35 kPa), or approximately 1.6 klf (23 kN/m), for the entire length of pile below 15-ft (4.6 m), it was estimated that an average tip depth of 55-ft (16.8 m) would satisfy all performance requirements. Using an Atlas Copco MB1700 hydraulic breaker, the set criteria of 1-in/10s was somewhat aggressive compared to the general recommendation of 1-in/50s for the size pile being used, however it was justified by the design team's experience driving piles in the local geology, as well as by the use of external grouting. Load testing would be imperative in verifying this assumption, as well as assessing actual grout-to-ground bond. The results obtained from installation and testing of sacrificial test piles prior to production would be used to determine final geotechnical design and installation criteria.

Structural design of piles was performed using International Building Code (IBC) allowable stresses and manufacturer supplied material properties. Utilizing a 4,000 psi (28 MPa) sanded grout, with an average diameter of 8.7 in (221 mm), provides an allowable compression strength 168 kips (747 kN) for an unreinforced pile. However, in uplift piles, a full-length 1-in (25.4 mm) diameter high-strength threaded bar was required to transfer tensile loads across the bell and spigot spliced joints. The threaded bar would also extend into the pile cap and terminate with a 4 in x 4 in (102 mm x 102 mm) plate at the cap upper reinforcement layer to serve as anchorage. Reinforcement for compression piles consisted of only a 10-ft (3 m) hooked #4 bar which served to hold a compression bearing plate in place at the pile head. Details of the pile design are provided in Figure 5.

Both single pile and group lateral analyses were performed using LPILE and DeepFND, respectively. The piles were designed to be embedded just 6-in (152 mm) into the pile cap, below the bottom reinforcement layer, which would generally classify the degree of fixity at the pile cap as a free-head condition. However, as is best practice, the piles were analyzed assuming both free and fixed-head cases and were determined to be acceptable under conditions of up to 75% fixity with a combined axial-bending interaction less than one.

Lateral capacity was governed by the internal bending resistance of the pile and not by displacement. Usually, this means that care must be taken to ensure a splice does not occur within the high moment region below the pile cap. This would be a challenge at the subject site though, as the variable subsurface conditions meant that exact splice locations may be nearly impossible to control. In Ductile Iron Piles however, previously published testing has shown that the bending resistance in the joint created by the bell-and-spigot connection is greater than that of the straight section (DuroTerra, 2022), thus splice location was not a concern for the design.

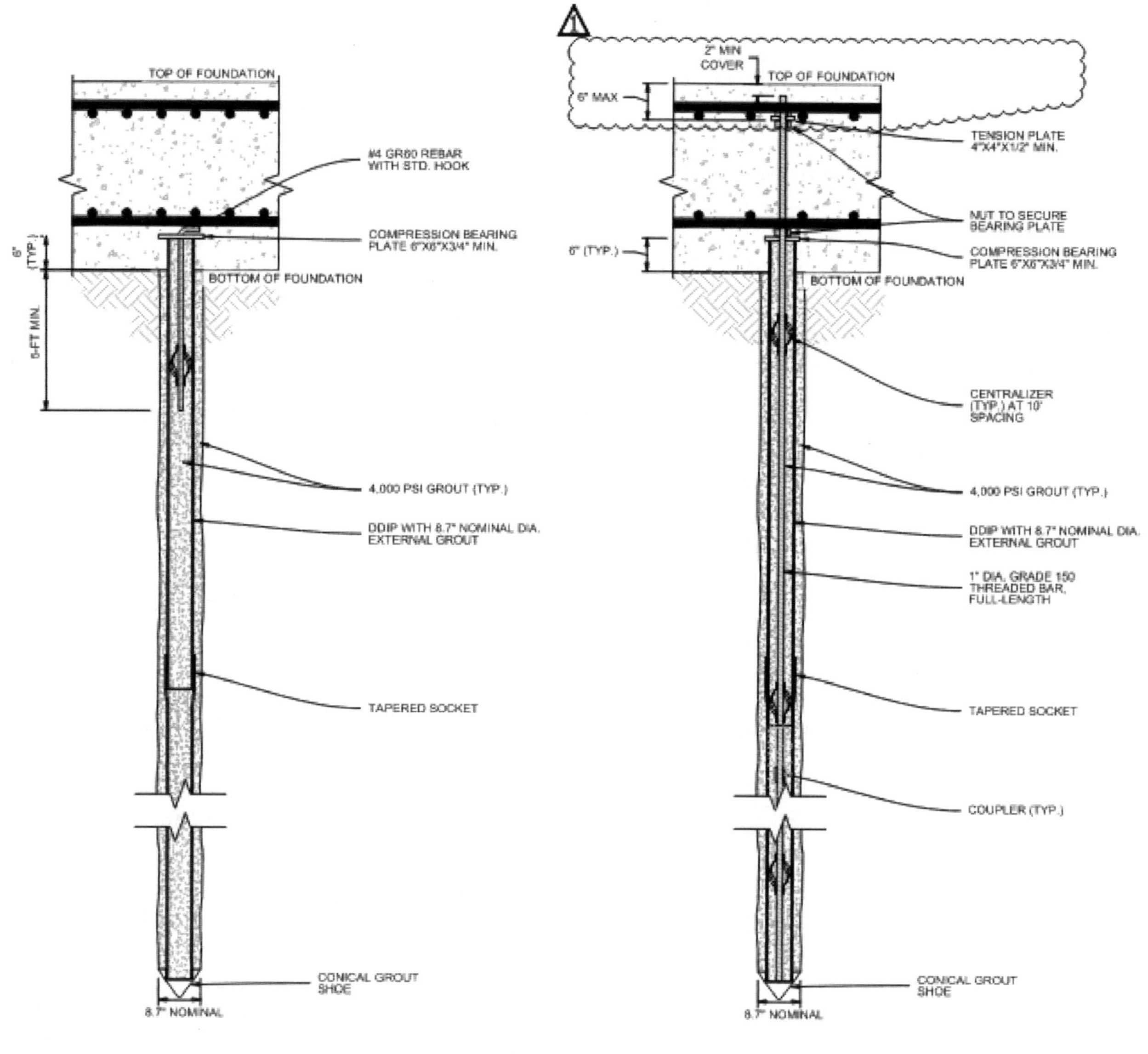

Figure 5. Compression pile (left) and tension pile (right) details

LOAD TESTING

To verify design assumptions and installation methods, a comprehensive test program consisting of compressive (ASTM D1143), tensile (ASTM D3689), and lateral (ASTM D3966) static load testing was implemented. The test program utilized five sacrificial test piles positioned as shown in Fig. 6, wherein the four corner piles would first serve as the subjects for tension and lateral testing, then serve as reaction anchors for compression testing of the center pile. Furthermore, following compression testing, the reaction piles could be used for additional tensile testing to higher loads or to failure.

The concerns associated with variable driving conditions were realized immediately upon installation of test and reaction piles. Several zones were encountered in which the pile would advance on the order of feet, rather than inches, in a single blow, or would advance with only a slight crowd force from the excavator. Piles TP-4 and TP-5 were driven to depths of over 70 ft (21.3 m) and 100 ft (30.5 m), respectively, before achieving the target set, and neither piles TP-1 nor TP-5 produced grout return to surface after exceeding a depth of about 40 ft (12.2 m). Conversely, Pile TP-2 just a few feet away, easily achieved set at less than 50 ft (15.2 m) depth and maintained grout to surface throughout installation.

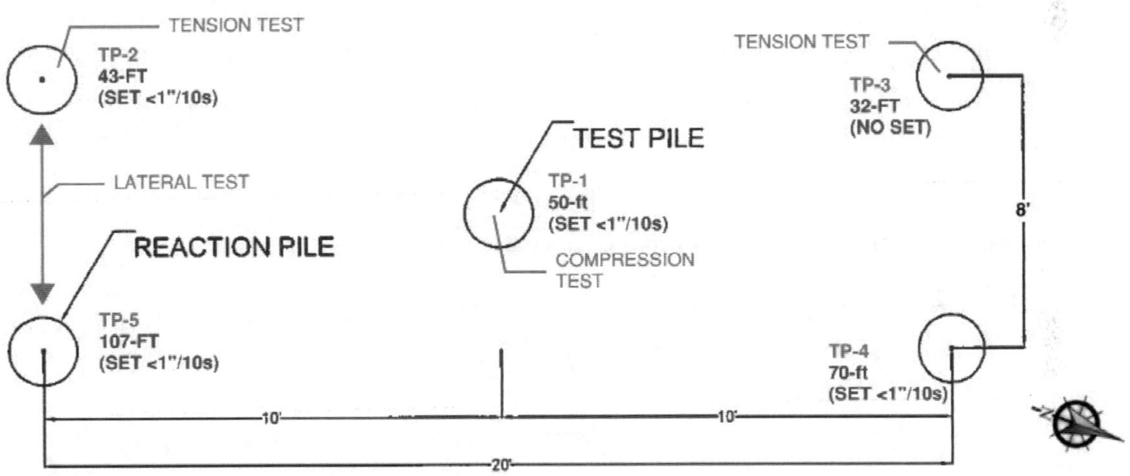

Figure 6. Load test program

A summary of test pile installation and load test results is presented in Table 1. The load-displacement data from all axial testing (both compression and tension) and lateral testing are provided in Figs. 7 and 8, respectively. Compression and tension tests were able to achieve over 4X the respective design loads and were only limited by the safe working capacities of the test equipment. Lateral testing showed that, at ground surface, the piles were able to achieve 2X design load in a free-head condition, with just under 1-in (25.4 mm) displacement as required per IBC standards.

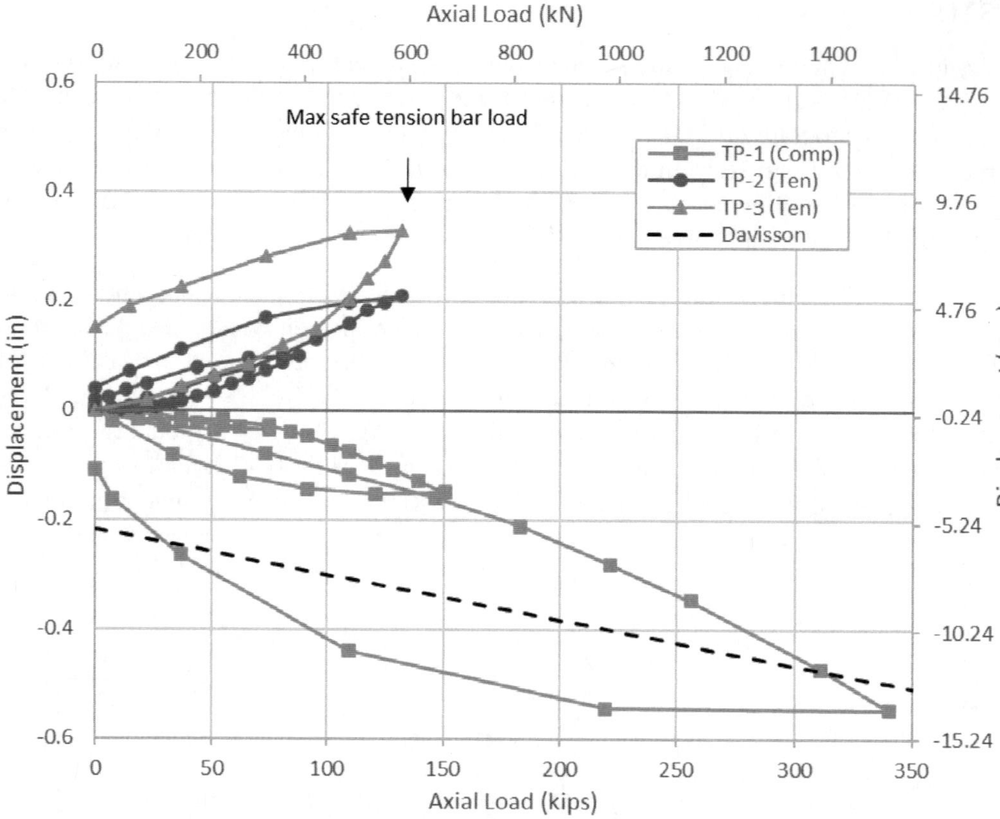

Figure 7. Compression (negative) and tension (positive) displacement vs. load

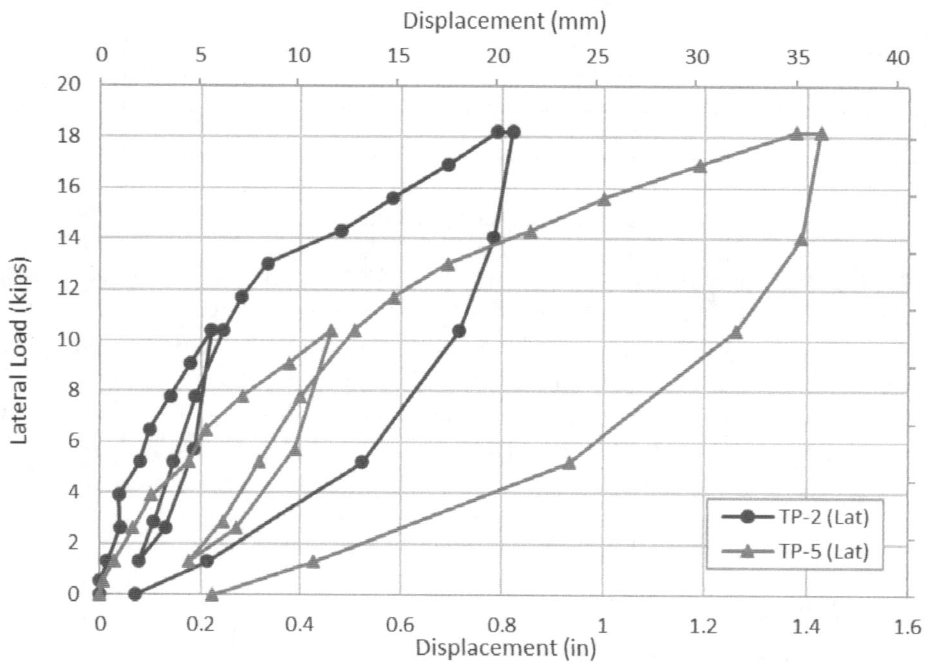

Figure 8. Lateral displacement vs. load

Table 1. Summary of load test results.

Test Pile	TP-1	TP-2	TP-3	TP-4	TP-5
Length, ft (m)	50 (15.2)	44 (13.4)	34 (10.4)	72 (22.0)	107 (32.6)
Target Set Achieved	Y	Y	N	Y	Y
Grout Volume, CY (m^3)	2.1 (1.61)	0.64 (0.49)	0.45 (0.34)	0.92 (0.70)	1.7 (1.30)
Grout Return	N	Y	Y	Y	N
Load Test	Comp	Ten, Lat	Ten	-	Lat
Compression					
Realized Ult. Compression Load, k (kN)	340 (1512)	-	-	-	-
Displacement at Ultimate, in (mm)	0.15 (3.8)	-	-	-	-
Failure	N	-	-	-	-
Realized Factor of Safety	4.5	-	-	-	-
Tension					
Realized Ult. Tension Load, k (kN)	-	130 (578)	130 (578)	-	-
Displacement at Ultimate, in (mm)	-	0.05 (1.3)	0.08 (2.0)	-	-
Failure	-	N	N	-	-
Realized Factor of Safety	-	4.3	4.3	-	-
Realized Unit Friction, klf (kN/m)	-	3.0 (43.1)	3.8 (55.8)	-	-
Lateral					
Realized Ult. Lateral Load, k (kN)	-	18.2 (81)	-	-	15.6 (69)
Displacement at Ultimate, in (mm)	-	0.58 (14.7)	-	-	0.99 (25)
Failure	-	N	-	-	N
Realized Factor of Safety	-	2.4	-	-	2.0

CONSTRUCTION

As with many construction projects, especially those in urban areas, engineering had to be carefully coordinated with site and schedule constraints for successful execution of the design. Vibration concerns, restricted headroom areas, and a firm completion deadline were all challenges that had to be overcome.

Piles driven on the west side of McKinley Drive were to be installed immediately adjacent to a fully operational hospital, meaning that vibration and noise levels typically associated with other driven pile types were not acceptable. Driven Ductile Iron Piles, however, produce comparatively minimal vibrations and noise, therefore all piles were able to be successfully installed with no impacts to the existing hospital throughout the entire project. Vibration and sound monitoring instrumentation setup along the edge of the building recorded maximum Peak Particle Velocities reaching 0.295 in/sec (7.49 mm/s) at 25 Hz, or just 24% of acceptable threshold values, at distances as close as 25-ft from pile driving. (Fig. 9)

The newly constructed treatment facility on the east side of McKinley Drive also posed a challenge. The upper floors created an overhang which limited the usable headroom to just 30 ft (9.1 m) at six pile locations. A common obstacle for piles installed with a crane or fixed-mast rig, often requiring design revisions, the headroom constraint was not an issue for the medium-sized excavator and the piles were easily able to be installed as designed.

The ability to drive piles using only a medium-sized excavator also had direct benefits to schedule constraints. When it came time to re-mobilize from the east side of McKinley Drive to the west side, it provided for a very quick transition, as the excavator could simply be loaded onto a lowboy trailer and driven to the other side of the highway. Not requiring complete breakdown and setup of a crane or other large rig allowed for minimum downtime and practically immediate continuation of production work.

Furthermore, while no production piles required depths of 100+ feet (30+ m) to achieve set, as in the case of Test Pile TP-5, the variable bedrock did result in tip depths ranging anywhere from 32 ft (9.8 m) to 70 ft (21.3 m), meaning that for any given pile the possible number and locations of required splices is unknown. With Ductile Iron Piles however, unplanned splices can be performed quickly and easily without sacrificing capacity, drivability, or excessive time. Additionally, the cut-off lengths at the end of driving each pile were used as the lead section to start the next pile, making efficient use of materials.

With production rates as high as 20 piles per day, the deep foundation scope was able to be completed within budget and schedule. Overall, the efficiencies realized by using DIP allowed for an early completion, reducing the originally anticipated duration by more than 25%.

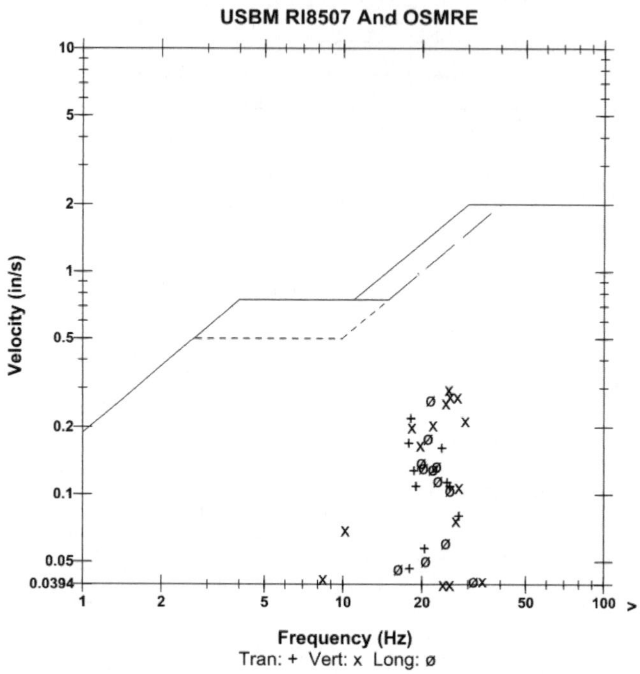

Figure 9. Recorded vibrations during pile driving.

CONCLUSION

Driven Ductile Iron Piles provided a deep foundation option that met the owner's needs for a solution that would support the heavy structural loads of the new bridge with a low-vibration piling system that would have minimal impact on the adjacent structures and roadways. The Ductile Iron Pile system was designed to accommodate complex loading conditions, and the construction methods were adaptable to variable bedrock strengths and elevations encountered at the site. Construction sequencing met the owner's need for flexibility to construct deep foundations on both sides of the busy roadway and installations were performed quickly and safely. Full scale compression, tension, and lateral pile load testing verified capacities up to and beyond design requirements, as well as helped add to the growing database of achievable pile performance with DIP in variable soil conditions.

REFERENCES

DuroTerra, LLC. 2022. "Technical Brief: Bending Resistance of Ductile Iron Piles." Dated February 17, 2022.

DuroTerra, LLC, 2016 "Technical Brief: Vibrations and Sound." Dated December 28, 2016.

Federal Highway Administration (FHWA) (2005). "NHI Course No. 132078: Micropile Design and Construction- Reference Manual. Publication No. FHWA-NHI-05-039. FHWA. Washington, D.C. P. 5-68, 5-69.

International Code Council, Inc. (ICC). 2018. International Building Code 2018. Country Club Hills, IL.

Terracon Consultants, Inc. 2021. "Revised Geotechnical Engineering Report." Dated January 6, 2021.

Tiroler Rohre Gmbh (TRM). 2014. "Piling Systems for Deep Foundations." October 2014

SOIL IMPROVEMENT THROUGH RAMMED AGGREGATE PIERS TO MITIGATE LIQUEFACTION RISK IN STORAGE TANKS

Francisco A. Flores, IGM Ingenieros Geotecnistas Mexicanos, Mexico city, Mexico, +52 55 3576 0734, alonso.fafl@igmmexico.com
Jorge Arroyo-Esqueda, Geopier Foundation Company, Mexico, +52 33 1438 7926, JArroyo@geopier.com
Héctor A. de la Fuente, Soilsolution, Tabasco, Mexico, +52 99 3242 2797, hdelafuente@soilsolution.com.mx
Erika Bernal, Soilsolution, Tabasco, Mexico, +52 99 3347 6605, ebernal@soilsolution.com.mx
José L. Aguirre, Soilsolution, Jalisco, Mexico, +52 66 9246 7075, j.aguirre@soilsolution.com.mx

ABSTRACT

This paper presents the analysis, design, and construction of soil improvement for storage tanks foundation support at a site in Tabasco, Mexico. A total of 980 Rammed Aggregate Piers of 12.5 m in length and 0.5 m in diameter were designed and constructed to support four storage tanks. The soil conditions at the site generally consist of loose to medium dense silty/clayey sand to about 8m depth, underlain by very loose to medium dense silty/clayey sand with seams of very soft clay. The use of Rammed Aggregate Piers induced densification of the soil to increase its stiffness and reduce the liquefaction potential to control both static and dynamic induced settlements. Standard Penetration Test (SPT) and Cone Penetration Test (CPTu) explorations were performed prior to and post soil improvement to compare the liquefaction potential and the Factors of Safety associated with the two conditions. Finally, the soil improvement installation rate is presented. Soil improvement using Rammed Aggregate Piers represented a suitable alternative to support storage tanks foundation in potentially liquefiable areas.

Keywords: Soil improvement, Rammed Aggregate Piers, Liquefaction

INTRODUCTION

In the hydrocarbon sector, storage tanks represent an important geotechnical challenge due to the high bearing pressures resulting from the stored material. In addition, in areas with potentially liquefiable soils, this challenge is even greater. In these cases, deep foundations are typically used to support the service loads from the tank structure. An alternative to deep foundations is the use of Rammed Aggregate Piers to mitigate the liquefaction potential in sand-like behavior soils, thus increasing the resiliency of the structure. In cases where the soil or project conditions do not allow the use of shallow foundations, it is common to design deep foundations by evaluating the axial load capacity by analytical methods and in some cases, depending on the importance of the project, in-situ load tests are implemented. Also, the tolerance of the total and differential settlements is restrictive. Therefore, pile groups are frequently necessary. Due to the geometry of storage tanks, pile groups are placed following a circular array in order to distribute both static and accidental loads as uniformly as possible (Vega *et al.*, 2021).

This paper presents the analysis, design, and construction of soil improvement for storage tanks foundation support at a site in Tabasco, Mexico. Standard Penetration Test (SPT) and Cone Penetration Test (CPTu) explorations were performed prior to and post soil improvement to compare the liquefaction potential and the Factors of Safety associated with the two conditions.

CASE STUDY DESCRIPTION

The project consists in the construction of hydrocarbon storage tanks in Paraíso, Tabasco, Mexico. The tank diameters are 30.5 m for one tank, 18.3 m for two tanks, and 4.6 m for one tank with heights ranging from 7 to 12 m. Figure 1 shows a schematic of the tanks and the geotechnical conditions encountered at the site. A total of 980 Rammed Aggregate Piers (RAPs) of 12.5 m in length and 0.5 m in diameter were designed and constructed to support the four storage tanks.

The soil conditions found in the tank's footprint consists of 0.0 to 2.8 m, sandy material fill with silt (dredged); 2.8 to 6.4 m, very soft fat clay; 6.4 to 12.6 m (liquefiable layer), loose to medium dense sand/silty sand with seams of soft clay from 14. 7 to 17.5 m, very soft clay with organics; 17.5 to 24.4, very loose to medium dense clayey sand with seams of clay; 24.4 to 31.5 m, medium dense to dense clayey sand; 31.5 to 42.0 m, medium to very stiff clay (See Figure 1). The groundwater table was found at approximately 3 m depth. Figure 2 shows the geometrical configuration of the RAPs proposed for each storage tank.

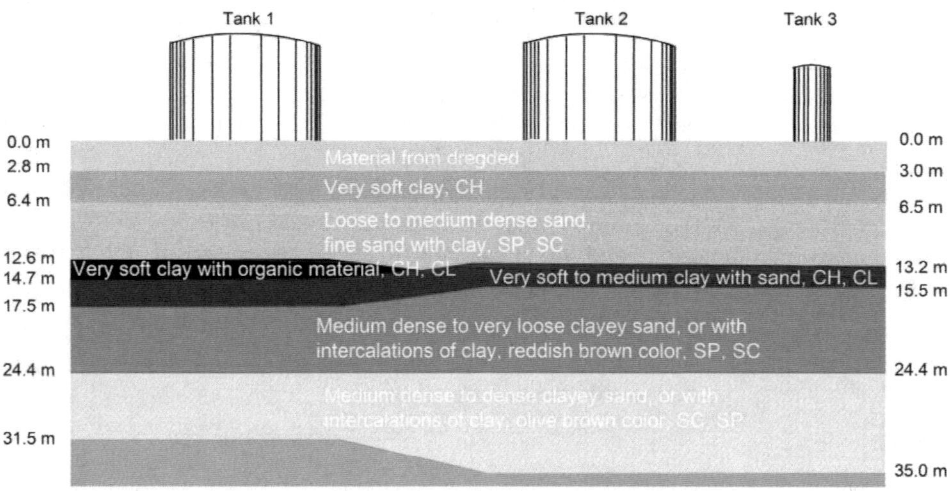

Fig. 1. Diagram of the storage tanks and stratigraphic cross section

GEOTECHNICAL SOLUTION

According to the geotechnical analysis and design, the post liquefaction induced settlement was reduced from 40 cm in free-field conditions, to 5 cm in improved conditions with the RAPs solution. The RAPs array was reviewed to meet the failure and serviceability criteria, such as, edge instability failure (static and dynamic condition) and static settlement. RAPs have been subjected to rigorous studies on their effectiveness in mitigating the risk of liquefaction of silty sands. For details of the study see Rollins *et al.* (2021).

Rammed Aggregate Pier (RAP) displacement installation process

The RAPs displacement construction process consists in:

The mandrel is driven into the ground using a strong static crowd force provided by the mast and augmented by dynamic vertical impact energy imposed by the vibro-hammer. In each installation, clean aggregate of nominal size from 12.7 to 38.1 mm (½ to 11/2 inch) is deposited though the hopper and mandrel. The equipment will rise 1.2 meters at a constant rate. The equipment is then driven back down providing static vertical thrust and high vertical impact energy provided by a vibratory -driving hammer, producing a rammed aggregate layer approximately 0.30m (1 ft) thick. Hydraulic thrust pressure will be recorded to obtain a measure of vertical densification force. Additional vertical impact energy can be used depending on the specific site and according to installation procedures. This process is repeated until it reaches the surface, resulting in a high-density Rammed Aggregate Pier, and in the process densifying the surrounding soil by expanding the cavity by the displacement produced during the process (Figure 3).

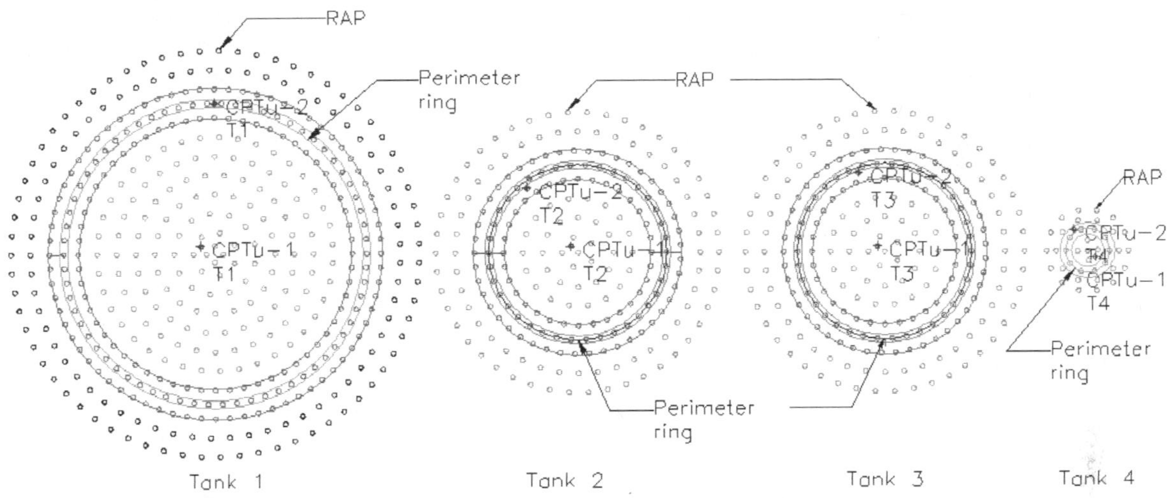

Fig. 2. Distribution of Rammed Aggregate Piers (RAPs) in each storage tank

Settlement control and design methodology

RAP soil reinforcing elements are designed to control foundations settlements to meet the project design criteria. Foundation settlements are estimated by summing the estimated settlement in the RAP-reinforced zone (the "upper zone") and the estimated settlement in the zone of soil below the bottoms of the RAP elements (the "lower zone") in accordance with the methodology described by Lawton *et al.*, 1994. These zones are depicted in Figure 4.

Upper zone settlement – spring analogy

Upper zone settlement calculations implement a spring analogy as shown in Figure 5. The RAP elements act as stiff spring; the matrix soil between the pier acts as softer springs. The stiff RAP elements attract a larger percentage of foundations-bottom stress than the soft springs. By assuming that the foundation is perfectly rigid, the top-of-RAP stress (q_g) may be computed using Equation 1:

$$q_g = q\frac{R_s}{R_a R_s - R_a + 1} \qquad [1]$$

where q is the average foundation-bottom stress, R_s is the ratio of the stiffness of the RAP elements and the matrix soil, and R_a is the ratio of the area coverage of the RAP elements to the gross footprint area.

The stiffness ratio, R_s, may be defined as the ratio of the spring constants of the RAP (k_g) and the matrix soil (k_m), equation 2. The RAP spring constant is defined as the relation between of the stress applied at top of pier element and the resulting vertical strain. The RAP spring constant is estimated using a database of spring constant established from past RAP modulus test and then verified with a modulus test performed at the project site.

$$R_s = \frac{k_g}{k_m} \qquad [2]$$

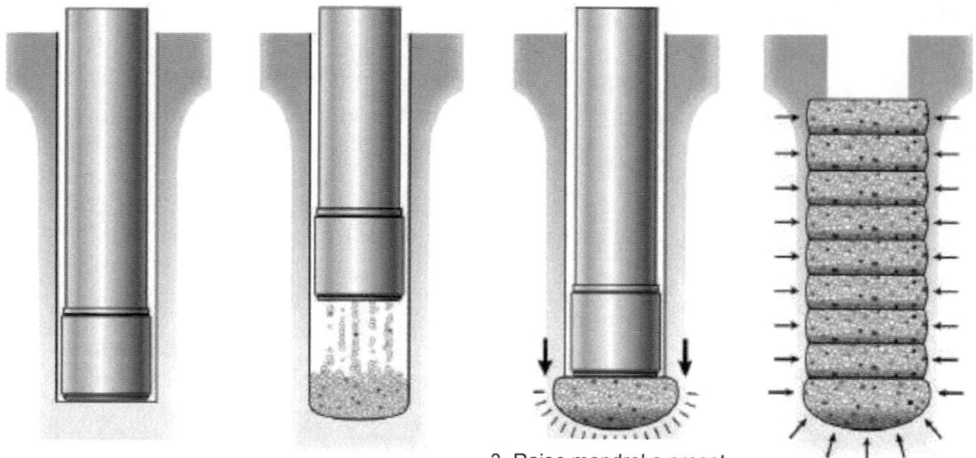

Fig. 3. a) Construction Method of the Rammed Aggregate Pier System and b) photograph of RAP installation at the site

The settlement in the Upper Zone (s_{uz}) is simply computed as the of q_g to the RAP spring constant (kg), Equation 3.

$$S_{uz} = \frac{q_g}{k_g} \qquad [3]$$

Upper zone settlement – composite modulus

Settlements for mat foundations and large footings, or large areas associated with embarkments, tanks, silos, or other structures may also be evaluated using a composite modulus approach where settlement within the upper zone of soil (zone of soil reinforced with the RAP system) is estimated using conventional settlement calculations as shown in the following equation (adapted from Terzaghi et al., 1996):

$$S_{uz} = \frac{q\ I_{\sigma} H}{E_{\ comp}} \qquad [4]$$

where q is the average bearing pressure, I_σ is the influence factor at depths within the compressible layer, H is the thickness of the compressible layer, and E_{comp} is the composite elastic modulus value within the aggregate pier-reinforced zone. The composite elastic modulus value is computed based on a weighted average of the elastic modulus values of the rammed aggregate pier elements and matrix soil using the following relationship:

$$E_{comp} = E_g R_a + E_m(1 - R_a) \quad [5]$$

where R_a is the area replacement ratio, E_g is the elastic modulus value of the aggregate pier, and E_m is the elastic modulus value of the matrix soil. The elastic modulus value of the aggregate pier is estimated in proportion to the stiffness modulus value of the pier verified by the modulus test. The elastic modulus value of the matrix soil is often estimated using correlations with undrained shear strength, Standard Penetration Test (SPT) blow counts, Cone Penetration Test (CPT) tip resistances or determined from consolidation test results.

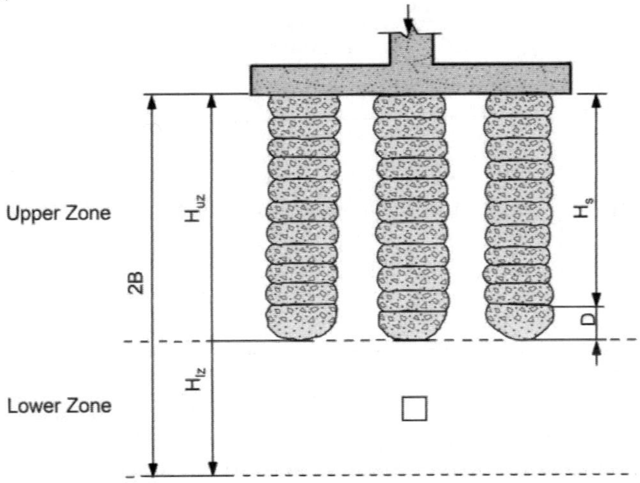

Fig. 4. RAP Two-Layer Settlement Approach

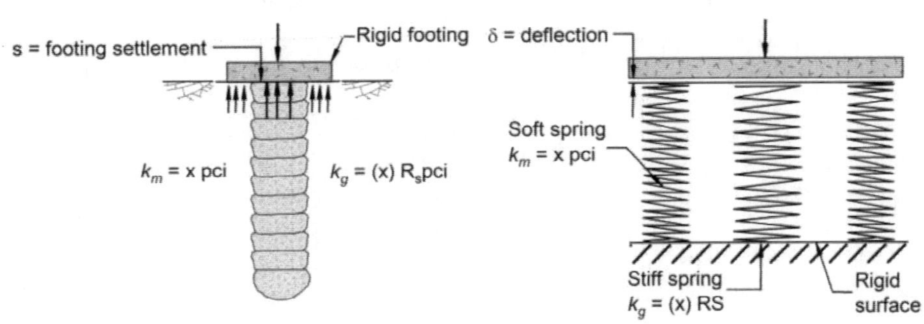

Fig. 5. RAP Design Spring Analogy

Lower zone settlement

Settlements within the "lower zone" (zone of soils beneath the upper zone which receives lower intensity foundation stress) are computed using conventional geotechnical settlement methods that involve: estimating the depth of stress influence below the foundation bottom (typically taken as twice the foundation width for square foundations), estimating the foundation-induced stress in the lower zone (established using conventional influence factor charts), and estimating the compressibility of the lower zone soils.

Lower zone settlements (s_{lz}) in granular soils are estimated using elastic settlement methodology with the equation:

$$S_{lz} = \frac{q\,H_{lz} I_\sigma}{E_s} \quad [6]$$

where q is the average foundation-bottom stress, I_σ is the influence factor in the lower zone, H_{lz} is the thickness of the lower zone, and E_s is the secant modulus of the soil in the lower zone.

Lower zone settlements (s_{lz}) in cohesive soils are often estimated using consolidation approaches as shown in the following equation:

$$S_{lz} = m_v * \sigma'_{v0} * H_0 \quad [7]$$

where σ'_{vo} is the vertical effective stress at the midpoint of the layer (H_0) and m_v is coefficient of volumetric variation. Alternatively, lower zone settlements in cohesive soils can be estimated using elastic settlement methodology.

Total settlements

The estimated settlement of RAP-supported foundations (s) is determined by summing the upper zone and lower zone settlement values:

$$S = S_{uz} + S_{lz} \quad [8]$$

Liquefaction susceptibility analysis

The liquefaction susceptibility analyses conducted for free-field conditions were performed with the semi-empirical methods proposed by Idriss and Boulandger in 2008 and Boulandger and Idriss, 2014. Therefore, a detailed explanation of the methods is not provided, this can be consulted in the cited references. For the case of liquefaction analysis of improved soils, the criteria adopted is detailed below.

Liquefaction susceptibility analysis with soil improvement based on Rammed Aggregate Piers

The installation of Rammed aggregate piers in liquefiable soils replaces loose soils with dense and rigid reinforcement elements, increasing the density of the surrounding soil significantly. The use Rammed aggregate Piers reduces the liquefaction potential by the following conditions:

The degree of densification generated by the installation of RAPs can be seen by an increase in the number of blow counts in the SPT test of the matrix soil in sands. This increase depends mainly on the percentage of fines content in sand.

$$N'_{SPT} = N_{SPT} + \Delta N_S \quad [9]$$

where N'_{spt} is the number of SPT blow counts improved when considering the densification of the matrix soil, ΔN_s is the increment of SPT blow counts due to densification and N_{spt} is the number of SPT blow counts on the soil during initial conditions (no improvement).

The vertical compaction-ramming of the RAP increases the lateral stress of the adjacent soil. Hence, an increase in lateral stress results in an increase in the CRR cyclic resistance ratio (Rollins *et al.,* 2021).

POST-IMPROVEMENT REVIEW

As a result of the liquefaction susceptibility analysis for free field conditions, the amine water zone presents high liquefaction potential in the foundation soils. The potentially liquefiable sand layer is found from a depth of 6.5 m to 20 m, and a post-liquefaction induced settlement of 37 to 45 cm were estimated. With the results obtained in a punctual manner with the data from the borings performed in the area and according to their coordinates, isovalue curves were generated using the Kriging interpolation method, which assumes that the distance or direction between the sample points reflecting a spatial correlation can be used to explain the variation of the post-liquefaction induced settlements at ground elevation (Figure 6a). The soil improvement proposal with RAPs for the tank foundations consists of the following: installation of piers with a diameter of 0.50 m (installed) to a depth of 12.5 m from the existing ground elevation after site grading; on-center (o-c) spacing of 2 m (6.6 ft), in a concentric array for both, the tank foundation footprint and the confinement rows extending beyond the tanks footprint; o-c spacing of 1.5 m (5 ft) for the footing support or perimeter ring foundation. With the results obtained from the borehole data with the proposed improvement system and similar to the case without improvement, isovalue curves were generated at surface level (Figure 6b). This figure shows quantitatively the effect of the improvement on the decrease of post-liquefaction settlements.

The proposed solution using RAPs was constructed in a period of 50 days; under favorable working conditions, an efficiency of up to 575 ml/day was obtained. Subsequently, to validate its effectiveness, complementary geotechnical explorations were performed through CPTu and SPT tests. Comparison of the pre-improvement and post-improvement explorations showed the RAPs effectiveness in reducing the liquefaction potential of the site. To quantify the degree of soil improvement, pre- and post-improvement borings were performed at each tank location. The SPT results are shown in Figure 7, where a significant increase in the SPT blow numbers obtained can be observed in the four sites after applying the RAP improvement. In addition, CPTu tests were performed.

Figure 8 shows the increase in the cone tip resistance at the site where tank 1 is located. The same figure shows the Safety Factors calculated before and after soil improvement.

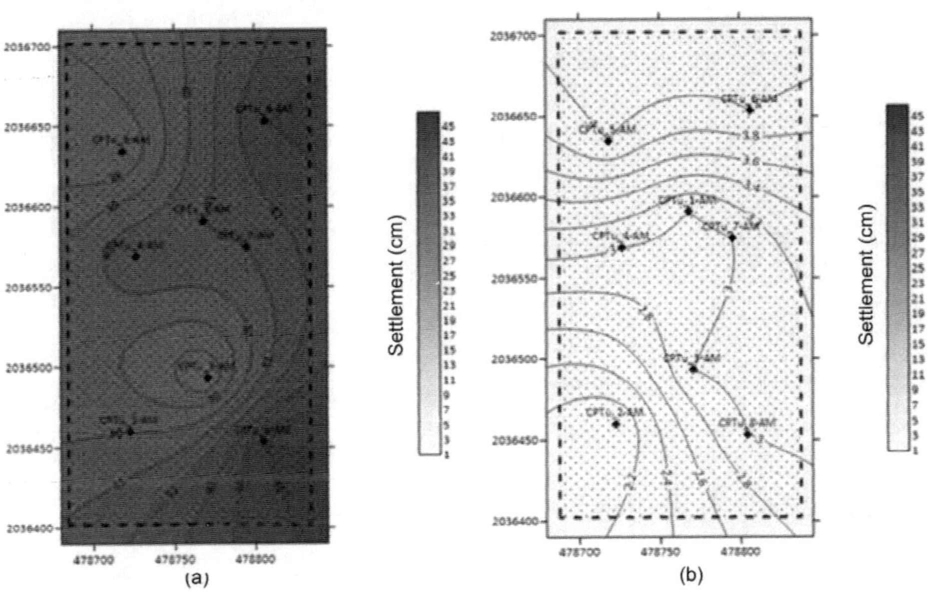

Fig. 6. Isovalue Curves of Post-Liquidation Induced Settlements at Surface Level for a) the Free Field Condition and b) for the Improved soil condition.

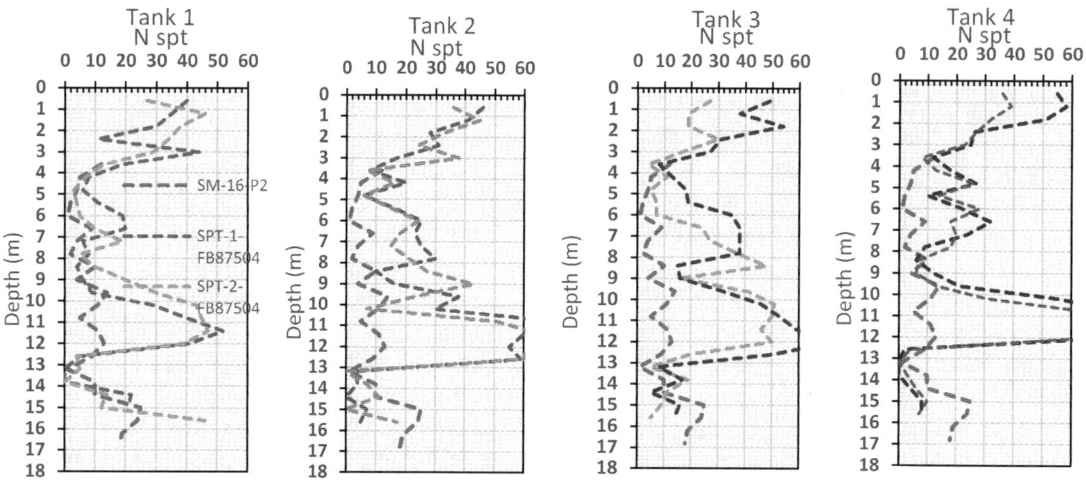

Fig. 7. SPT records in the four tanks before and after soil improvement with RAP

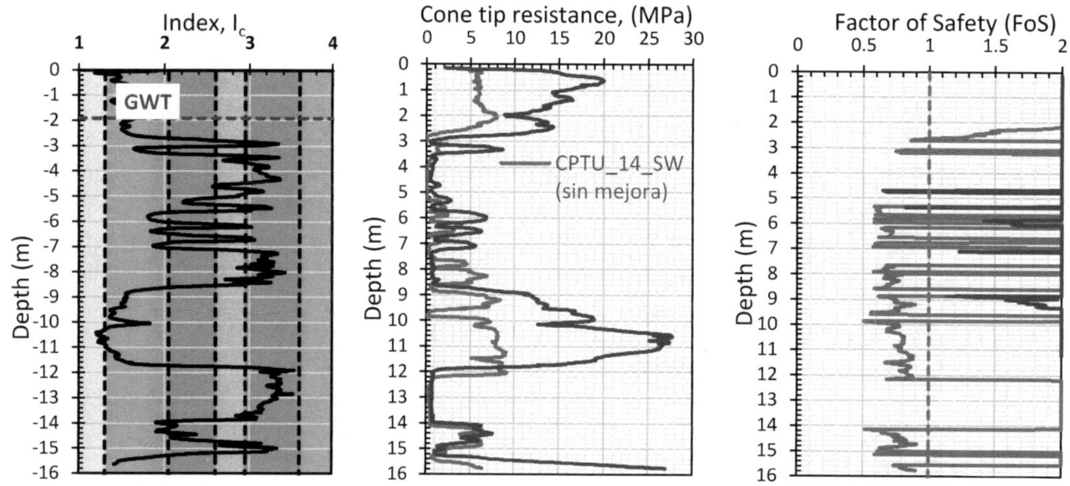

Tank 1

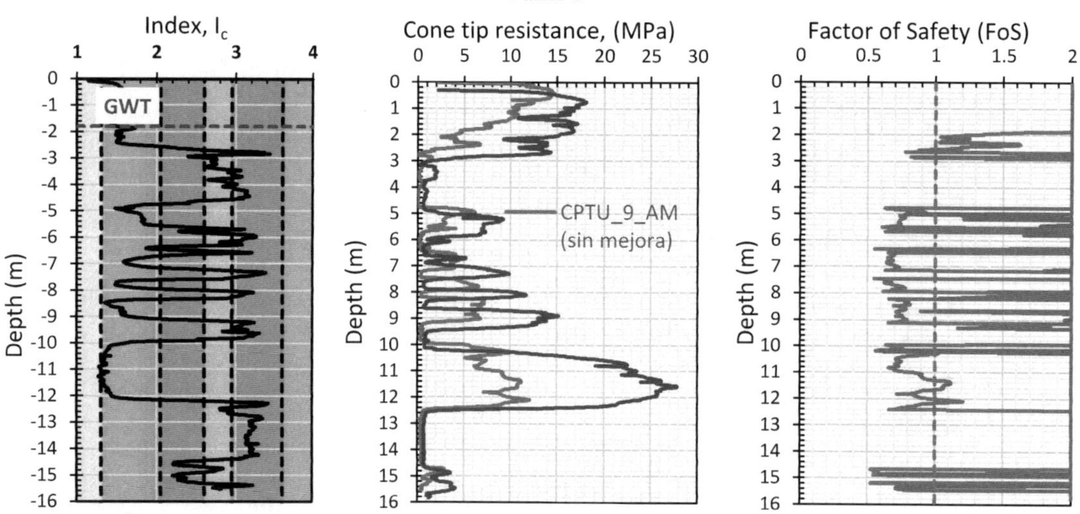

Tank 2

Fig. 8. CPT records in two of the four tanks before and after soil improvement with RAP

Figure 9 presents the daily installation rate obtained when constructing the 980 RAPs in less than 7 weeks of work. The average performance was 35 piles of 12.5 m in length per day.

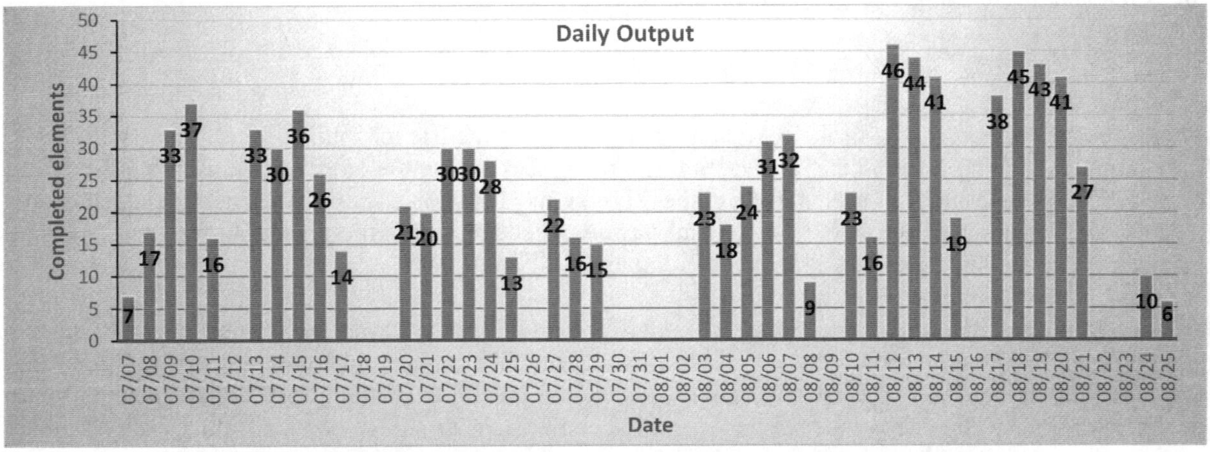

Fig. 9. Plot of daily installation rate in the construction of RAP's

SUMMARY AND CONCLUSIONS

In the reinforced zone, the RAPs provided lateral compaction of the matrix soil, which induces an increase in lateral and confining stresses and in stiffness. In addition, they provide radial drainage of the matrix soil which allows for rapid excess pore pressure dissipation. Radial drainage provided by the RAPs also accelerates the consolidation process of clay layers. This causes settlements to occur relatively immediate in the reinforced zone. The results of the settlement analysis for tank loading conditions satisfy the allowable settlement criteria adopted in the design baselines. To obtain these settlements, a three-dimensional analysis was conducted using Rocscience Settle 3 software. The depth of RAPs reinforcement is 12.5 m from the finished floor elevation of the platform and with on-center element spacings of 2 m for the flexible bottom foundation of the tanks. For the ring footings, a 1.5m o-c spacing between elements and rings was used. The proposed RAP arrangement will be able to reduce the post-liquefaction induced settlements to 5 cm relatively uniformly for the improved area. The soil improvement effectiveness was verified by CPTu and SPT tests performed before and after soil improvement.

A Limit Equilibrium (LE) analysis was also performed to evaluate the edge stability under static loading condition considering the improved shear strength parameters due to the reinforcement provided by the RAPs. The composite shearing strength of RAP-reinforced soils is computed using the conventional method of calculating the weighted average of the shear strength components of the RAPs and matrix soil materials (FHWA 1999). The composite shear strength is expressed in the following equation (FitzPatrick and Wissmann, 2002):

$$\tau_{comp} = \sigma'_v \tan \phi'_{comp} + c'_{comp} \quad [10]$$

The composite cohesion intercept (c'_{comp}) is computed with the expression:

$$c'_{comp} = c'_g \, R_a + c'_m \, (1 - R_a) \quad [11]$$

Where c'_g is the cohesion intercept of the RAP aggregate, c'_m is the cohesion intercept of the matrix soils, and R_a is the ratio of the RAP area to the gross footprint area of the reinforced soil zone. Because the cohesion intercept of the RAP aggregate is zero, Equation (11) reduces to:

$$c'_{comp} = c'_m \, (1 - R_a) \qquad [12]$$

The composite friction angle (ϕ'_{comp}) is computed with the expression:

$$\phi'_{comp} = arctan\,[R_a \quad \tan\phi'_g + (1 - \; R_a)\tan\phi'_m] \qquad [13]$$

Where ϕ'_g is the friction angle of the RAP aggregate and ϕ'_m is the friction angle of the matrix soils. RAP friction angles ranging from 48 to 52 degrees have been measured in the laboratory and in the field, but for design purposes, slightly lower friction angles (typically 45 degrees) are often used. As a result of the LE analysis, the pressures produced by the tank at the edges are within the allowable range and do not generate edge instability, which is reflected in a Factor of Safety F.S>1.5.

For the dynamic condition, a pseudo-static limit equilibrium analysis was performed considering the maximum and minimum pressures corresponding to the hydrodynamic condition. In addition, the hydrodynamic model was subjected to a horizontal seismic coefficient taken as 50% of the Peak Ground Acceleration, PGA, taken from the seismic design spectrum (*PGA=0.25g*). As a result, it was obtained that the pressures produced by the tanks at the edges are within the permissible range and do not generate an edge instability as is reflected in a Factor of Safety of F.S.>1.1.

Soil improvement using Rammed Aggregate Piers represented a suitable alternative for foundations of storage tanks in potentially liquefiable areas.

REFERENCES

Boulanger, R.W. and Idriss, I.M. 2014. CPT and SPT Based Liquefaction Triggering Procedures. Earthquake Engineering Research Institute, Oakland, CA.

Federal Highway Administration, 1999. Ground Improvement Technical Summaries, Volume II. Demonstration Project 116. Publication No. FHWA-SA-98-086.

FitzPatrick, B.T. and Wissmann, K.J. 2002. "Technical Bulletin No. 5 – Geopier Shear Reinforcement for Global Stability and Slope Stability." Geopier Foundation Company, Inc., Blacksburg, Virginia.

Idriss. I.M., and Boulanger, R.W. 2008. Soil Liquefaction During Earthquakes. Earthquake Engineering Research Institute, Oakland, CA.

Lawton, E.C., Fox, N. S., and Handy, R. L. 1994. Control of settlement and uplift of structures using short aggregate piers. In-situ Deep Soil Improvement, Geotechnical Special Publication No. 45, ASCE, 121-132.

Rollins K.M., Amoroso S., Andersen P., Tonni L. and Wissmann K., 2021. Liquefaction Mitigation of Silty Sands Using Rammed Aggregate Piers Based on Blast-Induced Liquefaction Testing. J. Geotech. Geoenviron. Eng., 2021, 147(9): 04021085. https://doi.org/10.1061/(ASCE)GT.1943-5606.0002563

Terzaghi, K., Peck, R.B. and Mesri, G. 1996. Soil Mechanics in Engineering Practice. 3rd Edition, John Wiley and Sons, Inc., New York.

Vega E., Flores F.A. y Zarate I., 2021. Evaluation of pile groups effect under axial loading on foundations for tanks using numerical modelling, XXX National Geotechnical Engineering Meeting SMIG, Online, March 2021. [in Spanish]

EXPERIENCES USING A LEADER SYSTEM FOR LONG CFA(ACIP) PILES IN HAWAII

Franz-Werner Gerressen, BAUER Maschinen GmbH, Schrobenhausen, Germany, Franz-Werner.Gerressen@bauer.de
Damien Enright, Kelikai Inc., Honolulu, Hawaii, USA, damien@ssihawaii.com

ABSTRACT

CFA/ACIP pile is the most used single pass drilling method for deep foundation installations. It is constructed by rotating and driving a Continuous Flight Auger (CFA) into the ground at a controlled speed to the specified pile depth. Concrete is then pumped under continuous positive pressure through the hollow stem of the auger to fill the borehole as the auger is withdrawn. CFA/ACIP piles are preferred due to their high daily production capacity when used in suitable soil conditions. CFA/ACIP piles have been installed in the United States for decades and the industry has seen a huge development in equipment, installation process, monitoring and more. Nowadays, CFA/ACIP piles are installed using fully computer monitored drill rigs rather than a bulky crane and lead setup. Because of the torque and crowd offered by these rigs, piles can be installed to depths of more than 100 feet and diameters up to 48 inches. New age crane-based equipment with leader systems are than welcome to substitute old fashioned bulky cranes These new age crane-based systems combine the benefits of sufficient hydraulic power, fully equipped software, etc. with the additional benefits that they can do single pass drilling without extension systems, even when drill depths of more than 100 feet are required. The paper will explain the construction process and will discuss some jobsite references and experiences made using the new age crane-based equipment.

Keywords: CFA, ACIP, Single Pass Drilling, Leader System

INTRODUCTION

Nearly Everybody knows about piling using hydraulic drill rigs and using the so-called Kelly drill mode. Kelly drilling has a lot of advantages as it can work with a wide range of tools facing almost all types of soil and rock. It can also deal with diameters up to 10 ft (3.000 mm) and bigger or depth exceeding 328 ft (100 m). But there is a big concern about Kelly drilling as the installation process is relatively slow.

Therefore, a piling method like CFA/ACIP Pile provides a lot of potential in terms of installation performance and therefore cost, and even more important, time saving impacts.

Looking into the field of pile installation, significant progress has been undertaken in the last years with the development of increasingly powerful drilling rigs, new age crane-based equipment with leader systems and tools for more efficient quality control and safe pile installation. These developments provide piling companies an increased number of options for the execution of specialty foundation projects when using single pass applications like CFA/ACIP Piles.

SINGLE PASS METHOD CFA/ACIP

Single pass means in general, that the tool is pushed and rotated into the ground to the final depth in one stroke. After reaching the final depth, concreting is started, and concrete is pumped continuously while the tool is withdrawn. In the following body of this paper, the CFA/ACIP pile method will be explained in more detail.

CFA/ACIP piles are cast-in-situ piles, where a hollow-stem continuous flight auger is rotated and driven into the ground at a controlled speed to the specified pile depth (Figure 1, Steps 1-3). Concrete is then pumped under continuous positive pressure through the hollow stem of the auger to fill the borehole as the auger is withdrawn (Figure 1, Step 4). Once the auger is completely out of the borehole, a reinforcing cage can be inserted and driven into the fresh concrete to the designed depth (Figure 1, Step 5).

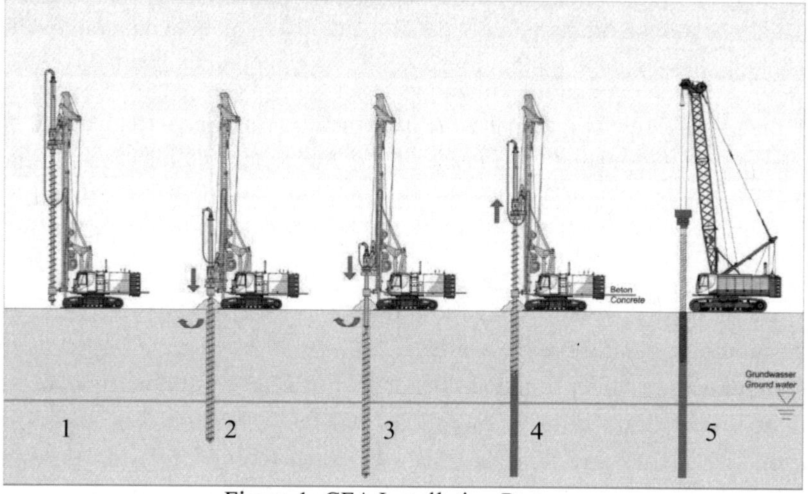

Figure 1: CFA Installation Process

As shown in Figure 1, Step 1-3, the use of an extension allows for a penetration depth which is deeper than the mast height of the drill rig. Even this can be seen as a benefit for reducing the size of the required equipment, there are also two other aspects to be considered. First, the penetration and concreting process is interrupted and therefore there is a time loss, second, the extension part is without flights and therefore no soil transport takes place in that area and stability of the borehole can become an issue.

To overcome these limitations, equipment like cranes with leader system are an option ensuring the process without interruptions with no risk of borehole instability. The installation process using such system is shown in figure 2.

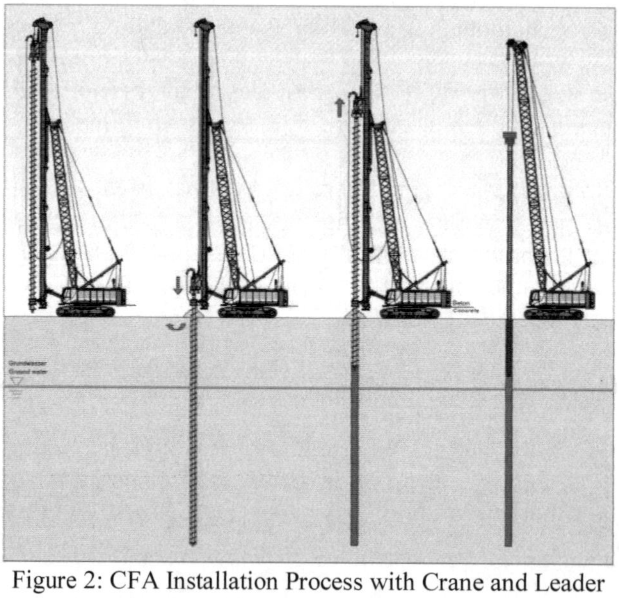
Figure 2: CFA Installation Process with Crane and Leader

This method is suitable for cohesive soils and granular soils with sufficient cohesive content to allow the soil transport on the auger, where the undrained shear strength "cu" should be above 15 kN/m² (313 Pounds/ft²) and in soils where no boulders (or obstructions) are present.

The main application for CFA piles is the use for foundations; the use of these types of piles for walls are limited.

ADVANTAGES, LIMITATIONS, AND CONSIDERATIONS

Single pass methodologies provide a lot of chances in terms of installation performance and therefore cost, and even more important, time saving potentials. Reason for that is that the penetration is taking place in one step, therefore, much faster production is possible, then using the well-known Kelly drilling method.

Also, the concreting process is much faster, as the drilling tool acts like a tremie pipe during the extraction process, where concreting and extraction of the tool runs simultaneously.

Besides the fast installation process, there are some other important topics to be

Assistance Systems allow the control of the installation process. As a specific feature, drilling equipment can be equipped with electronic software for use with the so called "Drilling Assistant", which modulates the optimum rate of penetration and crowd force during the drilling process for an optimal speed of rotation of the drilling tool, especially working in single pass methods like CFA/ACIP. The desired initial parameters can be input by the rig operator with the help of simple onscreen menus.

For instance, the system avoids for a system like CFA/ACIP-Drilling unwished situations like the well-known over-rotated and excessive soil removal, known as "over-flighting" (Figure 3). That means penetration speed is to low compare with rotation speed. This leads to a situation, where the surrounding soil get loosened, and the load capacity of the pile might be reduced. In addition, beside the increased amount of spoil, also significant overconsumption of concrete can be observed. Both leads to increased cost for pile installation. Moreover, as a worst-case scenario, this can cause ground around the tool to become unstable and ground settlement to be induced.

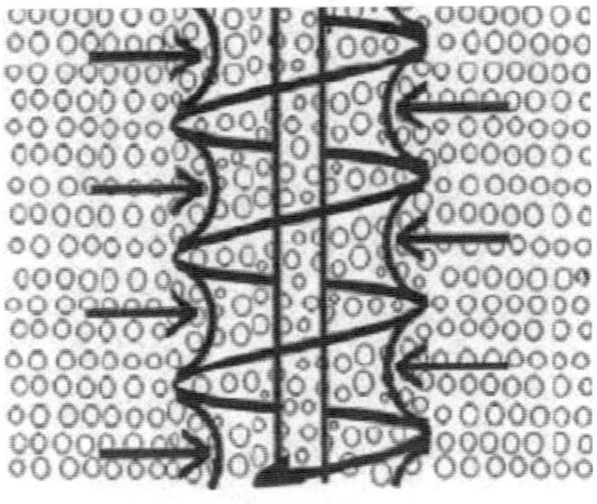

Figure 3: principle of "over-flighting"

By using the "Concreting Assistant", the speed of extraction and volume of concrete can be adjusted. This involves measuring the volume of concrete continuously throughout the concreting process and automatically regulating the speed of extraction based on the concrete flow rate. Of course, also the concrete pressure is monitored and controlled during the concreting process. As control of all relevant parameters takes place, concrete overconsumption can be reduced, and blockages can be avoided.

The use of these assistant systems can support a safe and repeatable installation at a reduced concrete overconsumption.

Concrete/Grout and Reinforcement

Concrete/Grout and reinforcement need to be well designed and prepared when using single pass applications. Concrete, or grout, need to have a certain workability, especially to allow the installation of the reinforcement cage immediately after finishing the concreting process. Typically, the max. aggregate size is limited. In terms of the reinforcement, the cage needs to have a specific design and stability allowing the installation into the fresh concrete without damages, even when using a small vibrator in case self-weight is not sufficient.

Range of application

Besides the limitations defined by the soil conditions mentioned before, CFA/ACIP piles are also limited in depth and diameter. The limiting factor on the one hand is the mast height of the used equipment, which can be slightly increased by using extension systems on the other hand the required retraction force to get the tool out of the ground including the weight of the soil and the lateral friction.

As guidance it can be said that typical diameters for CFA are up to 3 ft (900 mm). In some cases, diameters of 4 ft (1,200 mm) could be seen as well. Typical depth is up to 100 ft (30 m), in some cases depth of 164 ft (50) m and more were carried out.

These numbers for sure are not fixed, as development of equipment and tools as well as project demand force further development.

EQUIPMENT

In the following we will see some of these developments going along with the described demands

Figure 4: new age crane base carrier with leader for CFA/ACIP piling

A combination of a new age crane base carrier and a 157 ft (48 m) leader enables the installation of CFA/ACIP piles in one stroke, without extension, to a depth of 130 ft (39 m) at a diameter of up to 2.5 ft (762 mm) as shown in figure 4.

JOBSITE EXAMPLE

Ko'ula Tower

The recently topped off Hawaii project called Ko'ula, is one of seven towers either under construction or fully completed within the Howard Hughes developed Ward Village Community (Figure 5) on the island of Oahu.

Figure 5: Overview Ward Village Community

Ko'ula (Figure 6) is a 40-story high rise built over the typically found Kaka'ako soils consisting of very soft, typically single digit blow count lagoonal deposit material, layered with loose sands and dense coral shelves, but often peppered with small to medium sized voids.

The first of its type, Bauer's MC-128 Duty-Cycle Crane was paired with a brand new Berminghammer lead and commissioned just in time to begin work at the Ko'ula project. Kelikai Inc., a Hawaii commercial drilling subcontractor is the lucky owner of this specialized, custom-made drill rig, and was tasked with the installation of over 300 CFA/ACIP piles to support the new tower and podium.

Figure 6: Ko’ula Tower

Piles were 24” (610 mm) in diameter and ranged in depth from 55’ (16,74 m) to 105’ (32 m) below grade (Figure 7). The MC-128 Crane and lead system allowed Kelikai to drill down to tip elevation, and through hard coral shelves with little to no resistance. A huge production benefit was realized due to their ability to single pass both the drilling and grouting cycles, without the need to engage a kelly extension system.

Figure 7: CFA/ACIP Installation

Stopping the drill progression in order to lock into the extension and then advance deeper, for each and every pile, dramatically effects the daily installed pile count, the project's critical path and the ability for the General Contractor to get out of the ground. Single pass drilling is one of the simplest ways for drillers to increase their production, increase their profit margins, but also get more competitive in their local markets.

CONCLUSION

Besides all the advantages that go along with single pass pile installations, there are other huge advantages realized with the combination of using a crane and lead system. For most crane and lead systems, the additional hydraulic requirements to power the rotary head/motor and drilling functions would also require a separate large hydraulic powerpack, with a huge footprint as well as a huge price tag. With this system, however that additional powerpack is not needed as the crane has enough capacity to power the crane functions as well as the lead system functions. Secondly, the stability of the crane, allows for a taller and stouter lead system out front, that can take the stresses of a stronger rotary head, with larger torque capacities, which will again provide a production increase and lessen the drill times per pile. Lastly, the user-friendly drill software for the CFA installation is embedded within the crane operating system itself, which is more commonly seen in typically hydraulic vertical drills.

For all the above reasons, the MC-128 Crane and Berminghammer Lead System allowed for a successful project with quality piles supporting the new 40-story building looking over Hawaii's skyline.

REFERENCES

Fleming, W. G. K. (1995), The understanding of continuous flight auger piling, its monitoring and control. Proceedings of the Institution of Civil Engineers - Geotechnical Engineering: 157 - 165

F.-W. Gerressen, 2018, Electronic support in installation of special foundation products - Ensuring safe production and increasing quality control, DFI-EFFC International Conference on Deep Foundations and Ground Improvement, Rome, Italy

F.-W. Gerressen, H. Kopfmüller, 2019, Alternative Piling Methods - Chances for Technical and Economical Solutions, XVI Panamerican Conference on soil mechanics and geotechnical engineering, Cancun, Mexico

HELICAL PILES PERFORMANCE IN GOLESTAN LOESS DEPOSITS VIA FULL-SCALE AND PHYSICAL MODELING TESTINGS

Mohammad Arabameri
Amirkabir University of Technology (AUT), Tehran, Tehran, Iran, m_ameri@aut.ac.ir
Abolfazl Eslami
Amirkabir University of Technology (AUT), Tehran, Tehran, Iran, afeslami@aut.ac.ir
Sara Heidarie Golafzani
Amirkabir University of Technology (AUT), Tehran, Tehran, Iran, s_heidari@aut.ac.ir
Fatemeh Rostami
Amirkabir University of Technology (AUT), Tehran, Tehran, Iran, fatemerostami20@aut.ac.ir

ABSTRACT

Collapsible loess is classified as problematic soil found in most continents. The loess immediately collapses under stresses induced by axial compressive loading or the soil weight when exposed to moisture. In Iran, vast areas of the Golestan province are covered with aeolian loess deposits. This paper investigates whether the simultaneous effect of inundation and pre-loading in helical pile installation are operative and effective in improvement of Golestan loess. The effect of installation conditions on loess collapsibility is studied via two main approaches, i.e., field investigations and physical modelling. The experimental studies include installation of eight single helix piles in the Amirkabir University of Technology Frustum Confining Vessel, i.e., FCV-AUT. Also, four single helix piles with embedment depth of 3 m and diameter of 250 mm were installed in Golestan province and full-scale pile static loading tests were performed. The results indicated that the simultaneous application of crowd loading and water pressure during installation of helical piles results in better soil shear strength parameters. Moreover, high crowd stresses or low crowd stresses accompanied by water pressure during pile installation brings about similar load-displacement behavior of model piles installed in FCV-AUT and full-scale helical piles.

Keywords: full-scale test, helical pile, Golestan loess deposits, physical modelling, FCV-AUT

1 INTRODUCTION

Unsaturated loess soils are widely distributed in arid and semi-arid areas includes china, Russia and Siberia, the USA, Europe, South America (i.e., Argentina and Uruguay), and South Africa. Collapsing soils have covered vast areas of the Golestan province in the north of Iran (Arabameri and Eslami, 2022a; 2022b). Loess includes primarily silt-sized particles with a low percentage of sand and clay, organized in the metastable and porous structures. A prominent property of loess soil is excessive volume reduction caused by wetting with or without stress (Houstan et al., 2001). This property of loess causes severe issues like land subsidence due to the rain and or irrigation, cracking roads or sites, building distortion, and building cracks (Mihalache and Buscarnera, 2015). Although these soils are somewhat resistant in natural and semi-saturated conditions, they suddenly disintegrate when exposed to loading and wetting, unlike consolidation settlement. To deal with this unwanted feature of loess soils in construction projects, engineers have employed various techniques for improvement; like densification, presoaking, reinforcement, chemical improvements, etc. Wang et al. (2014) used flooding to improve a 36.5-meter thick loess soil layer in Heping town of Lanzhou and demonstrated that it is a good technique (based on convenience and expenditure) to improve thick loess soils. Feng et al. (2014), applied the dynamic compaction (CD) method about loess soils in Gansu province, China. They evaluated the executive parameters of this improvement method. They found allowable bearing capacity and depth of improvement about 250 kPa and 10 meters, respectively. Field testing after modification depicted that soil collapsibility has decreased and vanished in some areas.

Nearly fifty methods have been proposed for soil improvement, classified as earthworks, compaction, physical-chemical stabilization, hydraulic improvement, and reinforcement. Application of helical piles is one of the improvement techniques for Golestan loess soils and it has gained popularity among geotechnical engineers for their fast installation and optimal construction material (Sakr, 2015; Motamedinia et al. 2019). The spiral motion of the pile helixes disturbs the surrounding soil, resulting in vertical and radial soil displacement during installation. The disturbance alters the geotechnical parameters of soils and changes the bearing capacity of the soil under the helix. During the installation of the helical pile, the soil surrounding the shaft is pushed down and compressed under the helix plates (Perko, 2009; Tsuha et al., 2012).

Therefore, in this study, the improvement of loess soils is evaluated by performing helical piles in two methods, i.e., dry and wet installation approaches, and by applying crowd and torsion when driving helical piles. First, the geotechnical parameters including void ratio, density, cohesion, collapsibility index, and internal friction angle for the samples are measured, before and after compaction in different amounts of crowd loading. Samples were divided into series of undisturbed and remolded for both natural and saturated conditions. In the next step, the load-displacement behavior of helical piles is investigated. Regardingly, eight single helix model pile are installed in the physical model apparatus i.e., FCV-AUT and four single helix piles are installed in the Golestan province via wet and dry methods, under different crowd loadings.

2 SITE LOCATION AND AREA OF STUDY

Golestan loess is spread out over a vast area in the north of Golestan Province, Iran, on the Turkmenistan border. The 2019 flood of Golestan Province severely damaged local facilities and the railway built on the Golestan loess. Figure 1 is an example of the Golestan loess failure and the consequence of its failure on a road and culvert located in this area. The present study focuses on the west side of Incheh-Borun, a village on the Iran–Turkmenistan border, 150 km from Gorgan, given the important facilities in the area (Figure 1). Observations and experiments performed on the region's soil illustrated that the soil of this area is composed of silt and clay. During the floods of 2019, the joint railway line between Iran and Turkmenistan, a large number of medium-span bridges, and material storage silos built on collapsed loess soil were damaged. The block sample was removed from the excavated trench surface to collect and prepare remolded and undisturbed samples for soil characterization. Loess soil samples were packed with plastic to preserve their natural moisture and placed in wooden boxes to prevent them from disturbing and crushing. Figure 2 illustrates the grain size distribution of the investigated loess soil. It shows that the loess soil is ML type. The properties of the loess soil of the region are shown in Table 1.

Table 1. Incheboron loess soil properties

Properties	value	Properties	value
Soil type	ML	Specific gravity	2.68
Adhesion (kPa) (C)	15	Friction angle (φ)	19°
Adhesion (kPa)(C′)	4	Friction angle (φ')	8°
Natural water content (%)	5	Plasticity Index (%)	12.8
Liquid limit (%)	28.5	Plastic limit (%)	15.7
Initial void ratio	1.097	Density (gr/cm^3)	1.28

3 TESTING PROGRAM

This study aims to investigate the effect of crowd loading in dry and saturated loess soil conditions to study the variation of geotechnical parameters of this soil under different situations and their effect on the behavior of structural elements in these soils. This study performed three experimental tests on loess soil samples (i.e., remolded and undisturbed) prepared from the site. The first experiments were carried out for loess soil characterization, such as mechanical tests, determining collapsibility potential, density, and shear strength parameters from the direct shear test. Moreover, dry and saturated samples were compacted under different crowd loadings and their mechanical properties were measured to study changes in pores and particle size during disturbance due to loading and wetting. The second experiments include the installation of model helical piles in the FCV-AUT apparatus via dry (i.e., by applying torque and crowd) and wet (implementing torque, crowd, and water injection simultaneously) conditions. The bearing capacity of the installed piles was calculated from the load-displacement curves to investigate the effect of different crowd loadings and water injection during the installation of helical piles in FCV. The third experiments are static pile load tests performed on the wet and dry installed helical piles in the field.

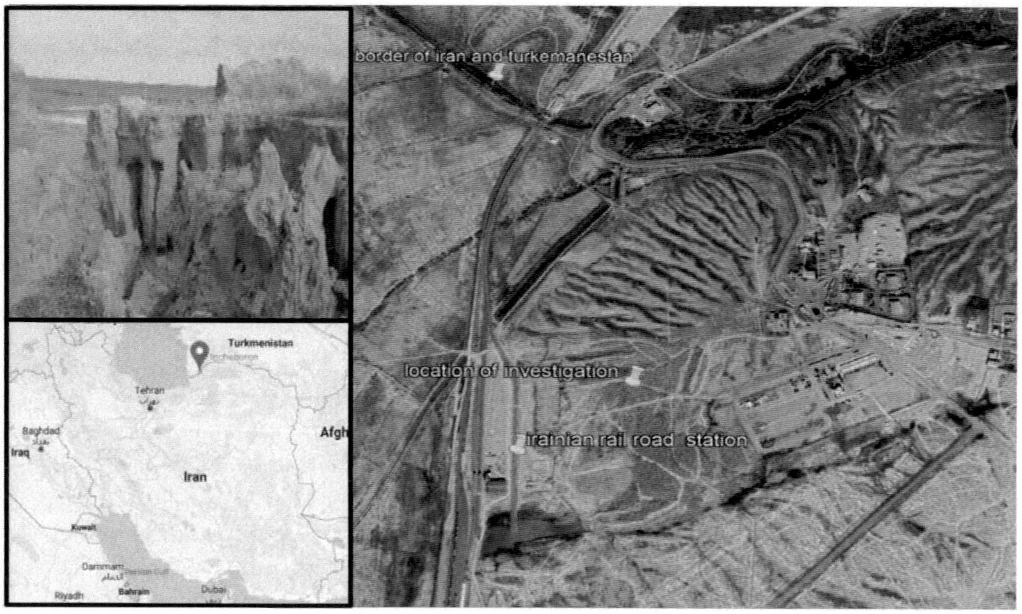

Fig. 1. Study area

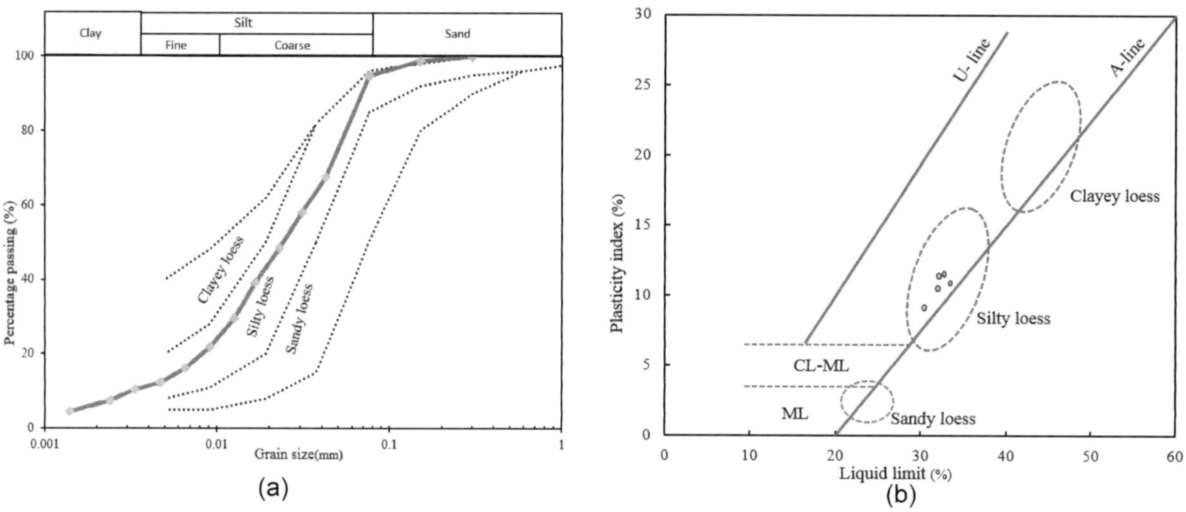

Fig. 2. Golestan loess physical properties; (a) Particle size distribution (b) Plasticity

Sampling and Laboratory Tests

The remolded and undisturbed loess soil samples were prepared for the experiments in dry and saturated conditions. The cylindrical mold of the oedometer has a diameter of 5 cm and a height of 9.1 cm, and the straight-cut cube mold size has a width and height of 6 cm and 1.8 cm, respectively. The cubic samples with approximately 40 * 40 * 40 cm dimensions were cut with a saw and gently split into smaller pieces to prepare an undisturbed sample. Sampling molds are placed on loess soil blocks, and by slow cutting with a cutter and brush, loess soil located under the sampling wall of the undisturbed sample is placed inside the mold according to Figure 2a. Due to the brittleness of loess soil, the samples might be destroyed when preparing an undisturbed sample, according to Figure 2b. To make the remolded sample, first, according to the density of loess soil on-site, the weight of loess soil in proportion to the volume of the desired test mold was obtained. Then, the loess soil was poured in three steps into the sampling mold and compressed to the specified height as depicted in Figure 2c. Saturated loess samples were prepared by immersion in a mold for 48 hours before each test.

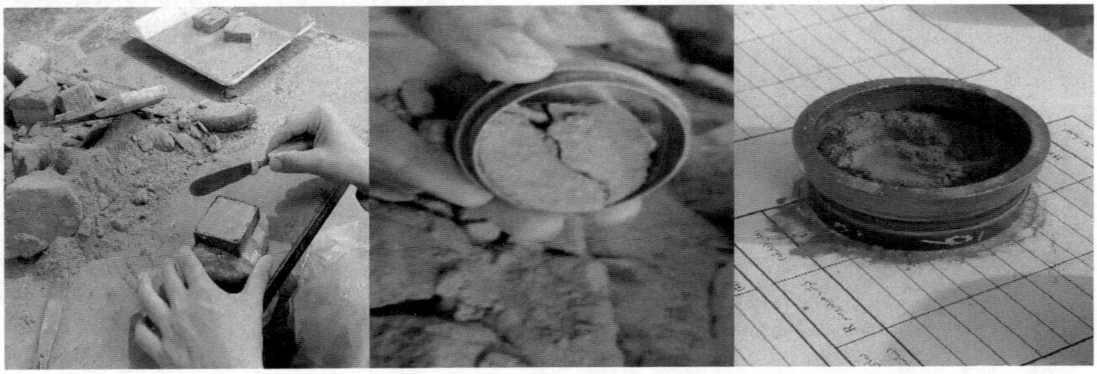

(a) (b) (c)

Fig. 3. Sample preparation (a) Undisturbed sample (b) Destroyed sample (c) Remolded sample

For naming samples, D, S, U and R stand for dry, saturated, remolded and undisturbed samples, respectively. For instance, the name MDU is used for the undisturbed dry sample, and MSR is used for the remold saturated sample. After preparing the samples, the dry and saturated samples were compacted under crowd of 50, 100, 150, 200, 250, 300, 350, and 400 kPa. According to the test schedule, dry samples were subjected to vertical loads initially, but saturated loess soil samples were subjected to vertical loads after 72-hour-saturation. Determination of collapsibility index is a significant test in recognizing the loess subsidence potential (ASTM D5333). Also, other tests such as compaction (ASTM D792) and direct shear (ASTM D3080) were performed to determine internal friction angle and cohesion parameters.

Physical Modeling

Physical modeling is one of the important and complex issues in geotechnical engineering. The FCV is a unique laboratory instrument that creates stress distributions similar to local conditions and is almost linear. The most important advantage of FCV is the simulation of local conditions, i.e., the stress is zero at the top of the device, and at the bottom, it is maximum. Accordingly, the model behavior can be generalized to the actual full-scale behavior with a minor error. As shown in Figure 4, the FCV is a conical vessel capable of modeling deep foundations. So far, three FCVs have been built around the world. The first was built in 1996 at McMaster University (Horvath et al., 1996; Sedran, 1999), the second device in 1999 at South Florida University (Mullins et al., 2001), and the third FCV device in 2012 at the Amirkabir University of Technology (Zare and Eslami, 2014). Amirkabir University FCV device has a height of 130 cm, the floor diameter is 135 cm, and its upper cross-section diameter is 30 cm. The main body of the device is made of steel sheet with a thickness of 10 mm and in the form of two detachable pieces at a distance of 70 cm from the floor for ease of prototyping. At the bottom, a 15 mm thick steel plate is placed so that a rubber membrane is placed between it and the device's floor, through which the floor pressure can be applied. Twenty-four screws with a diameter of 22 mm have been used to join the mentioned plate to the device's floor. The large size of the device makes it possible to model large-sized helical piles. As a result, it reduces errors related to the effects of scale and boundaries. Soil modeling in FCV was done by pouring soil in 5 cm layers with a natural unit weight of 12.5 kN/m^3 and the least possible compaction energy. The dimensions of the helical pile of the laboratory model for this research are proportional to the helical pile implemented in the field, installed at a depth of three meters and six meters in the soil. Stress condition in the field at depths of 3 meters was simulated via the hydraulic pressure of 0.5 bar applied from the floor to the modeled soil inside the FCV-AUT device. The details of the helical pile modeled in FCV are illustrated in Figures 5a and 5b.

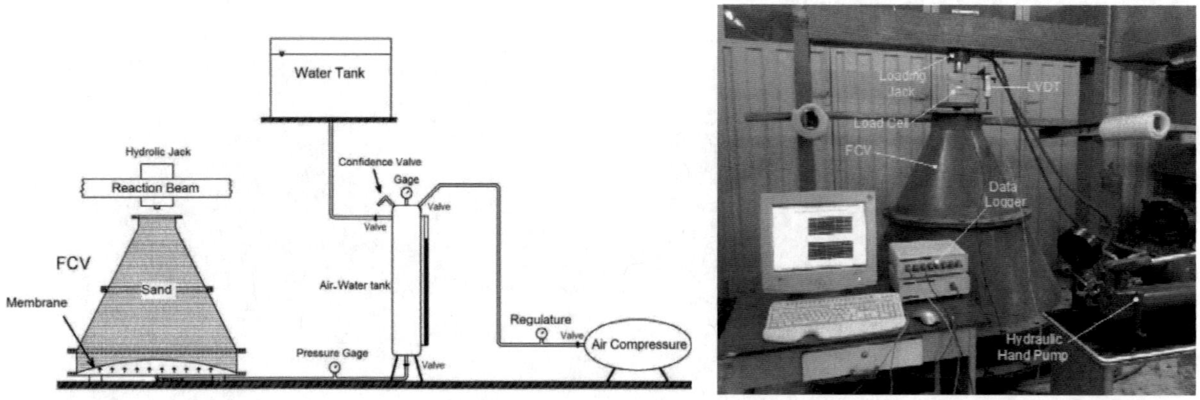

(a) (b)

Fig. 4. FCV-AUT; (a) schematic scheme (b) device equipment (Zare and Eslami, 2014)

Full-scale testing

Four single helix piles with embedment depth of 3 m and helix diameter of 250 mm were installed in the area of study via wet and dry installation methods. Figure 5c depicts the detail of these single helix piles. In the dry method, the piles were driven into the ground by applying crowd. In the wet method, in addition to crowd, water was injected into the soil through the pile shaft and the nozzles placed on the top and bottom of the helix. The helical pile was installed into the ground by applying torque and crowd provided by a hydraulic machine (drilling machine).

4 RESULTS AND DISCUSSION

Mechanical tests

Figure 6a shows the void ratio changes of dry and saturated loess soil samples compacted under different crowd loadings for undisturbed and remolded samples. As illustrated in Figure 6a, the void ratio of the dry loess soil samples undergo minimal changes with increasing stress; for instance, the soil experiences up to 10% change in both remolded and undisturbed samples. The changes in the void ratio of undisturbed loess decreased before 150 kPa stress and remained nearly constant by increasing the stress. While in dry remolded loess, there is a relatively linear decreasing trend. In saturated loess soil, the void ratio changes in both undisturbed and remolded states with the application of 100, 50, and 150 kPa crowd loadings were significant, and their amount reduced to 34% and 31%, respectively. The small changes in the void ratio in dry loess samples compared to the saturated state are due to friction between the particles and the high structural strength of the soil in the dry form. In the saturated form, a significant volume change occurs due to the reduction of friction between the grains, waterworn of clay and carbonate bonds, soluble bonds, and the reduction of matrix suction in the loess soil structure. The void ratio varies at lower rates for undisturbed saturated samples versus remolded at low stresses, which can be due to the more structural strength of undisturbed soils.

In saturated loess, no significant void ratio change occurs at high crowd stresses because the initial structure is disturbed due to moisture and crowd loading leading to the formation of a denser structure. Further evaluation of the diagram shows that the void ratio of loess soil for the two undisturbed and remolded samples has an almost similar trend for the dry and saturated states. Figure 6b shows the density of dry and saturated loess soil samples under different stresses for undisturbed and remolded samples. Given the small changes in the void ratio against increasing load in dry loess soils, it is clear that the sample density changes are also inconsiderable.

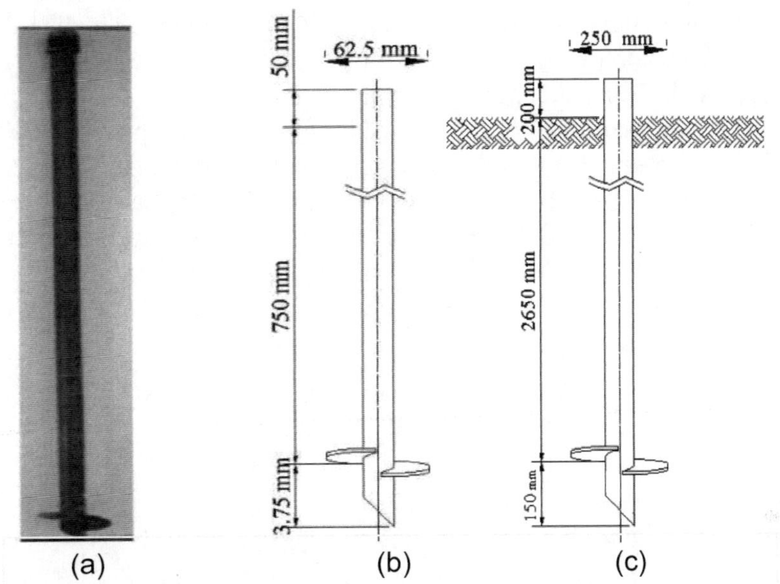

Fig. 5. Investigated helical piles; (a) installed single helix pile in FCV-AUT, (b) details of helical piles modeled in FCV, and (c) details of installed helical pile in field

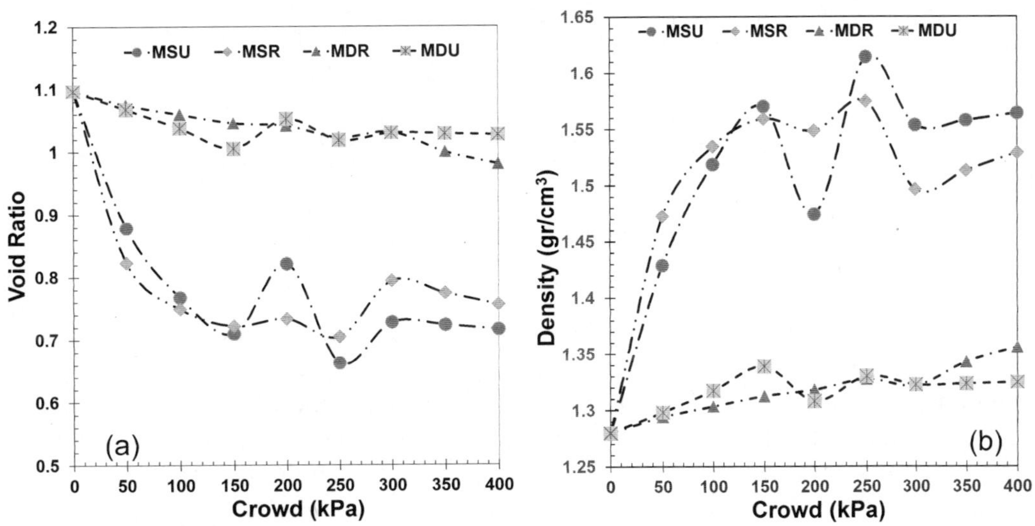

Fig 6. The effect of crowd on variation of soil properties; (a) void ratio and (b) density

In the remolded dry samples, the density change has a partly linear slope, and as stress increases, density gradually increases. Dry loess soil density increases with the load growth, leading to the pore size reduction via filling them with Aggregates due to the collapse and compaction. At low stresses, the density changes at a lower rate in undisturbed saturated samples compared to remolded samples. The density increase for remolded and undisturbed samples in saturated conditions are 20% and 22%, respectively. The variation of void ratio and density with crowd loading for undisturbed and remolded samples, and saturated and dry samples, is the same.
One of the most important properties of loess soil is the collapse index. As shown in Figure 7, the collapse index is presented for undisturbed and remolded loess soil under various stresses in dry and saturated conditions. According to Figure 7, in dry loess soil, with increasing stress, the changes in the collapse index are small and about 6%, while saturated loess soil is about 77%. These numbers indicate the effect of water on collapse loess is severe. The change in the collapse index has an almost linear slope in undisturbed and remolded dry loess and gradually decreases with the increasing crowd. The collapse index falls drastically, i.e., about 75%, in undisturbed and remolded saturated loess samples in low crowd loadings. After applying 100kPa stress in the saturated loess soil, the collapse potential changes are negligible, about 2%. The significant changes in the collapse index in saturated samples compared to dry samples are the considerable reduction of friction between the grains due to wetting their surface and the break of soil particle bonding factors. Collapse potential is 14 and 4 for dry and saturated loess, at 400 kPa, respectively. The small collapse index in saturated loess is due to the simultaneous effect of wetting and crowd on soil structure disturbance.

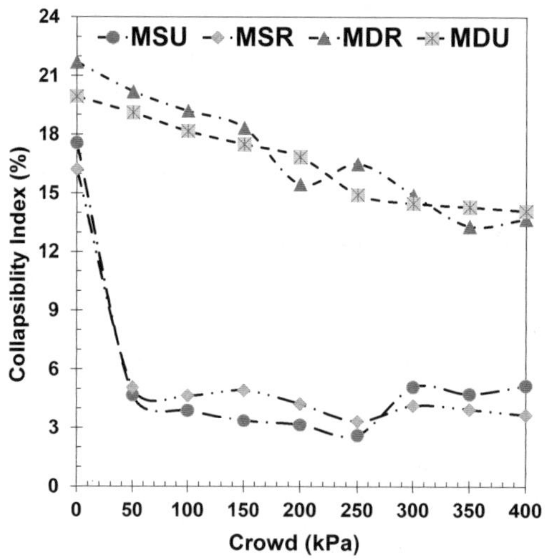

Fig 7. Changes in collapsibility index versus load increase

Figure 8a shows the friction angle of undisturbed and remolded samples of loess soil under different stresses for both saturated and dry states. The stress increase brings about the collapse of the structure and eventually soil compaction, leading to the increase of face contact among soil particles. Therefore, the friction angle of the loess soil increases. According to the diagram, the friction angle increases with stress by about 45% and 100% in the undisturbed and remolded dry samples, respectively. Increasing the friction angle with increasing stress in saturated loess soil is 100% for an undisturbed sample and 120% for a remolded sample. The friction angle remains almost constant after the stress of 150kPa and 200kPa in dry and saturated loess, respectively, for remolded and undisturbed samples. In compacted dry loess soil under stresses less than 150kPa, the friction angle of undisturbed samples is higher than that of remolded samples due to the bonds between the particles and the high structural resistance of the undisturbed samples compared to the remolded samples. However, as the stress and density of the samples increase, the structure of the remolded and undisturbed samples becomes similar and tends to the same value, i.e., the difference between internal friction angles reduces. In the lower stresses, the friction angle of the saturated undisturbed sample is greater than the saturated remolded sample because of the bonds between the particles in the undisturbed samples. The lower internal friction angle in saturated loess soil than in dry loess soil is due to water in the pores and among the particles. Figure 8b shows the cohesion of dry and saturated loess soil samples under different stresses for undisturbed and remolded samples. Cohesion in the loess soil resistance is not determinant because of the low percentage of clay. Compared to the dry form, the decreased cohesion in the saturated condition is due to the disappearance and dissolution of clay and carbonate bridged in undisturbed loess soil. As crowd stress increases, the structure is compacted, and the behavior of undisturbed and remolded samples becomes similar, which occurs in the presence of water at lower stress levels.

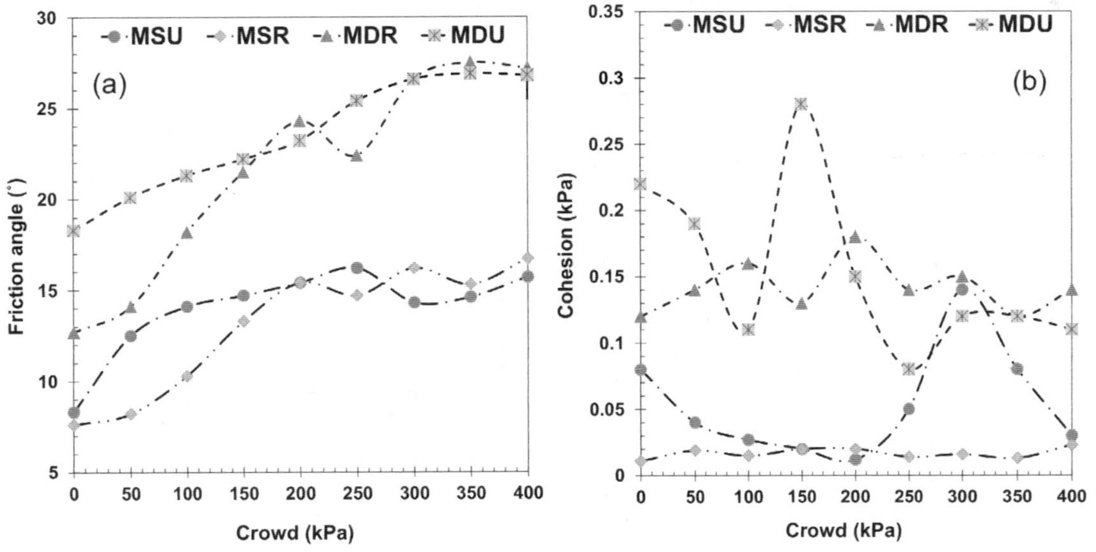

Fig 8. Variation of shear strength parameters with crowd loading; (a) friction angle (b) cohesion

FCV Results

Load-displacement curves obtained from the static test on helical piles were implemented by both dry and wet methods in the FCV-AUT device, shown in Figures 9 and 10. Also, to better interpret the load-displacement diagrams, the bearing capacity of the helical pile was calculated. Terzaghi (1942) defined the axial compressive capacity of piles, as the load corresponding to a displacement equal to 10% of the pile diameter. This definition ensures that there is enough displacement to mobilize the end-bearing component of the pile capacity. Similarly, the axial compressive capacity of helical piles can be defined as the load corresponding to a displacement level of 10% of the average helix diameter. As provided in Fig 9b and 10b, the bearing capacity of helical piles was interpreted at 10% of the helix diameter. According to the calculated bearing capacity for both dry and wet methods, it can be seen that as the crowd increases, the bearing capacity of the pile increases. The rate of increase for 25 kPa, 100 kPa 300, and 400 kPa crowds compared to no crowd condition are 25%, 43.6%, and 65.45%, respectively. Also, for wet conditions, load increase for crowds of 100kpa, 300 kPa, and 400kPa compared to no crowd state are 31%, 37%, 60%. Pushing helical piles into the ground in a circular (rotational) motion, breaks down the cementation, and soil collapses, which in turn causes the rearrangement of soil particles and eventually the formation of a denser structure (Arabameri and Eslami 2022a; 2022b). Therefore, the soil around the pile becomes denser after installation due to collapse, and its geotechnical properties are improved.

Figure 11 compares the performance of helical piles installed in FCV-AUT and field via wet and dry methods under the crowd loadings of 50 kPa and 400 kPa. UD 3.1 and UW 3.1 stands for full-scale single helix piles with embedment depth of 3 m installed via dry and wet methods, respectively. FUD 3.1 and FUW 3.1 indicate the model helical piles installed under hydraulic pressure of 0.5 bar via dry and wet methods in FCV-AUT. The bearing capacity of the piles at 10% of the helix diameter is considered. The bearing capacity of helical piles under 50 kPa crowd is 25 kN and 19 kN for the dry condition in field and laboratory, respectively, and 11 kN and 9 kN for wet samples. For helical piles under crowd of 400 kPa for the dry state in the field and laboratory is 38 kN and for the wet form is 13 kN. The results indicated that the bearing capacity of full scale and physical scale piles under 50 kPa crowd for dry and wet conditions are different from each other. However, this difference is less in wet conditions. This difference between field and laboratory load-displacement behavior at 50 kPa crowd is due to the structural difference between the remolded loess in the FCV and the undisturbed loess in the field, as well as the effect of scale. In the wet condition, more disturbance of the loess structure occurred due to water pressure. Under 400 kPa crowd loading, the differences were reduced between full-scale and model piles. Higher crowd loading brings about more disturbance of loess structure and crushing of loess particles. Accordingly, the similar load-displacement behavior of model and full-scale piles indicated that under high crowd loadings, the remolded soil in the laboratory behaves in a similar way to loess deposits in natural environment during penetration of helical piles. Therefore, despite their sensitive structure, modeling of loess soils in the laboratory are possible under high stresses or low stresses with water.

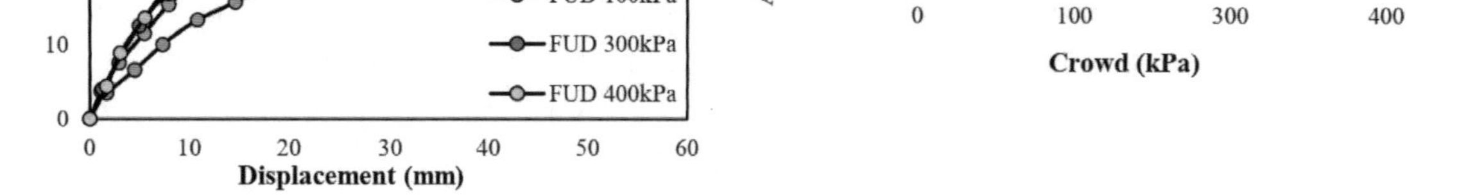

Fig 9. Dry installed helical pile behavior in FCV-AUT; (a) Load vs. Displacement (b) Axial bearing capacity vs. crowd

5 CONCLUSIONS

An investigation was performed to study the effect of wet and dry installation method on the improvement of the geotechnical properties of Golestan loess soil deposits and load-displacement behavior of single helix piles. Regardingly, eight model piles and four full-scale helical piles were installed in FCV-AUT and field, i.e., the Golestan province. Laboratory tests on remolded and undisturbed samples were performed in both dry and saturated conditions. The results indicated that with wetting and increasing the stress level, the bonding factors were destroyed, and due to particle rearrangement and crushing of aggregates, the void ratio was reduced, and the soil was compacted. For saturated samples, severe soil structure destruction under low stresses occurred due to simultaneous loading and wetting effect (similar to high-stress levels). As stress increased in two series of undisturbed and remolded samples, the void ratio and the collapsibility potential decreased. Also, the density and the internal friction angle of soil increased for the remolded and saturated samples. The cohesion parameter was less considered in this research because of the low percentage of clay in the investigated samples. Moreover, the installation of helical piles led to soil collapse due to the simultaneous application of torsion and loading. As a result, a denser structure is formed, and soil shear strength parameters improved. Therefore, with the increase of crowd loading, the bearing capacity for the dry and wet condition increased, compared to the no-crowd state for model piles. Comparing the helical pile performance in FCV and field, it was illustrated that the load-displacement behavior under high crowd stresses or in lower crowd stresses with application of the water pressure, is similar for model piles and full-scale helical piles.

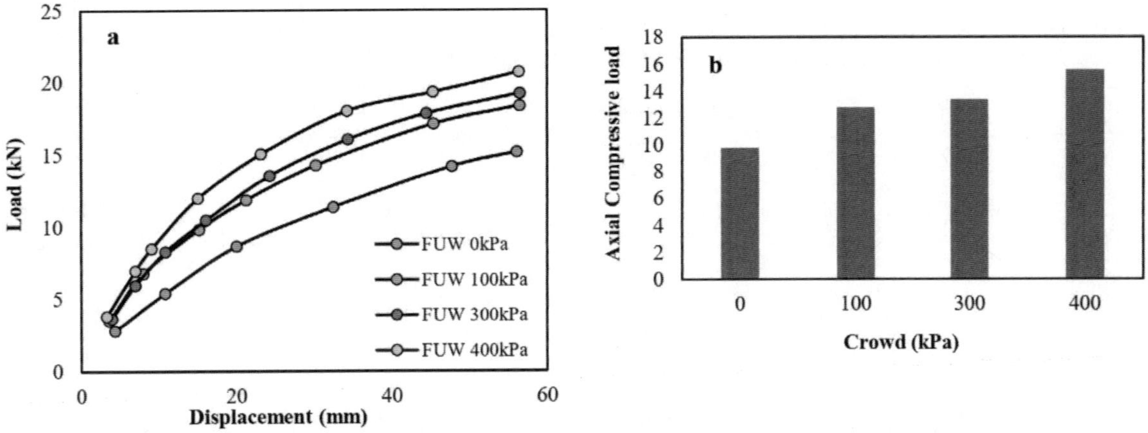

Fig 10. Wet installed helical pile performance in FCV-AUT; (a) Load vs. Displacement (b) Axial bearing capacity vs. crowd

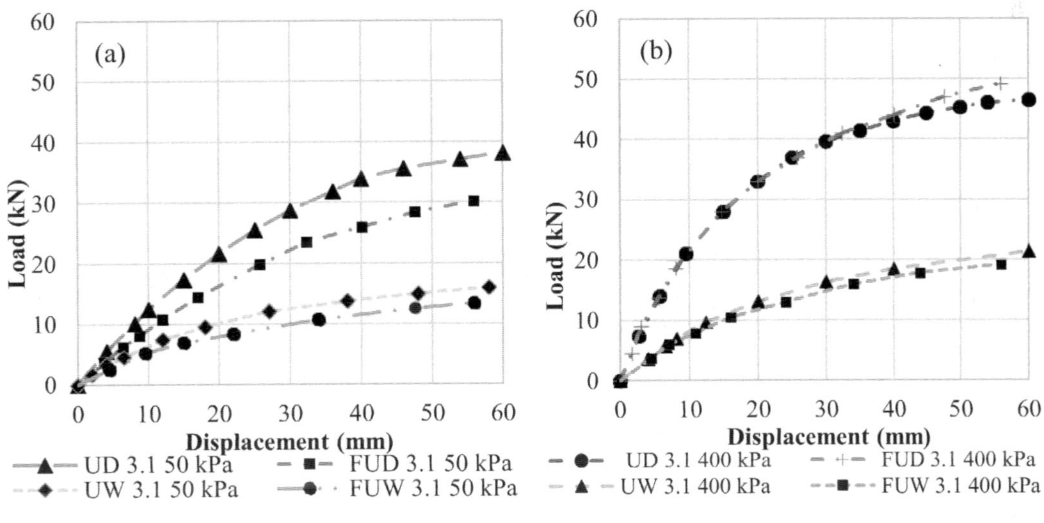

Fig 11. Helical pile performance in field; Load vs. Displacement curve (a) under stress of 50 kPa (b) under stress of 400 kPa

REFRENCES

Arabameri, M., & Eslami, A., 2022a. Wet and dry installation effects on the capacity of helical piles in loess deposits. International Journal of Geotechnical Engineering, 16(4), 486-498.

Horvath, Robert G., and Dieter Stolle.,1996. Frustum confining vessel for testing model piles. Canadian geotechnical journal 33.3, 499-504.

Arabameri, M., & Eslami, A., 2022b. Wet and dry installation effects on the capacity of helical piles in loess deposits. International Journal of Geotechnical Engineering, 16(4), 486-498.

ASTM D3080 / D3080M-11, 2011. Standard Test Method for Direct Shear Test of Soils Under Consolidated Drained Conditions (Withdrawn 2020), ASTM International, West Conshohocken, PA. DOI: 10.1520/D3080_D3080M-11.

ASTM D792-20, 2020. Standard Test Methods for Density and Specific Gravity (Relative Density) of Plastics by Displacement, ASTM International, West Conshohocken, PA. DOI: 10.1520/D0792-20.

ASTM D5333-03, 2003. Standard Test Method for Measurement of Collapse Potential of Soils (Withdrawn 2012), ASTM International, West Conshohocken, PA. DOI: 10.1520/D5333-03.

Eslami, A., Akbarimehr, D., Aflaki, E., & Hajitaheriha, M. M., 2019. Geotechnical site characterization of the Lake Urmia super-soft sediments using laboratory and CPTu records. Marine Georesources & Geotechnology, 1-12. doi.org/10.1080/1064119X.2019.1672121.

Fateh, A. M. A., Eslami, A., & Fahimifar, A., 2017. Direct CPT and CPTu methods for determining bearing capacity of helical piles. Marine Georesources & Geotechnology, 35(2), 193-207. doi.org/10.1080/1064119X.2015.1133741.

Houston, S. L., Houston, W. N., & Lawrence, C. A., 2002. Collapsible soil engineering in highway infrastructure development. Journal of Transportation Engineering, 128(3), 295-300.

Feng, S. J., Du, F. L., Shi, Z. M., Shui, W. H., & Tan, K., 2015. Field study on the reinforcement of collapsible loess using dynamic compaction. Engineering geology, 185, 105-115.

Liu, Z., Liu, F., Ma, F., Wang, M., Bai, X., Zheng, Y., ... & Zhang, G., 2016. Collapsibility, composition, and microstructure of loess in China. Canadian Geotechnical Journal, 53(4), 673-686. doi.org/10.1139/cgj-2015-0285.

Mihalache, C., & Buscarnera, G., 2015. Is wetting collapse an unstable compaction process? Journal of Geotechnical and Geoenvironmental Engineering, 141(2), 04014098.

Motamedinia, H., Hataf, N., & Habibagahi, G., 2019. A study on failure surface of helical anchors in sand by PIV/DIC technique. International Journal of Civil Engineering, 17(12), 1813-1827.

Perko, H.A. 2009. Helical Piles: A Practical Guide to Design and Installation. In Helical Piles: A Practical Guide to Design and Installation. John Wiley & Sons.

Sakr, M., 2015. Retracted: Relationship between Installation Torque and Axial Capacities of Helical Piles in Cohesionless Soils. Journal of Performance of Constructed Facilities, 29(6), 04014173. doi.org/10.1061/(ASCE)CF.1943-5509.0000621.

Sedran, Gabriel., 1999. Experimental and analytical study of a frustum confining vessel. Diss.

Shao, X., Zhang, H., & Tan, Y., 2018. Collapse behavior and microstructural alteration of remolded loess under graded wetting tests. Engineering Geology, 233, 11-22.

Tsuha, C.H.C., Aoki, N., Rault, G., Thorel, L., and Garnier, J., 2012. Evaluation of the efficiencies of helical anchor plates in sand by centrifuge model tests. Canadian Geotechnical Journal, doi:10.1139/T2012-064.

Wang, Changdan, et al., 2014, Experimental analysis on settlement controlling of geogrid-reinforced pile-supported embankments on collapsible loess in highspeed railway. International Journal of Pavement Engineering 15.9: 867- 878.

Zare, M., and A. Eslami., 2014, Study of deep foundation performances by frustum confining vessel (FCV). International Journal of Civil Engineering 12.4: 271-280

THE IMPACT OF END BEARING PLATE ON MICROPILE CAPACITY CASTED IN COHESIVE SOIL

Soorkty[1], Elarabi[2]
[1]B.Sc., MSc., PhD Candidate at Building and Road Research Institute, U of K, Fellowship of Sudanese Engineering Society
Email: aminabbas65@yahoo.com, husseinelarabi@yahoo.com

[2] Full Professor at Building and Road Research Institute, University of Khartoum

ABSTRACT

The aim of this paper is to discuss the impact of end bearing plate on the capacity of micropile. Preliminary studies have been made in Sudan in specific area of cohesive soil. Experimental works were carried on a number of micropiles with and without end bearing plate. Same diameters, lengths and techniques of placing cement grout are considered, there after the axial load test applied to find out the capacities of micropiles in both cases. Deduced results in each case show that the end bearing plate has no positive impact on micropile capacity.

Keywords: *micropiles, end bearing plate, cohesive soil, Sudan, capacity.*

1.0 Introduction

Micropile was first introduced to the geotechnical engineering practice in early 1950's in Italy and mainly used for underpinning the historic buildings or structures with settling foundations. micropiles are increasingly used in wide range as a preferred foundation option in different construction projects to carry a sizable axial load and lateral load. Federal Highway Administration (FHWA 2005) has classified micropiles into four main categories accordingly placing grout technique as follows. Type A, the grout has placed under gravity, Type B, the grout installs under pressure ranging from 0.5 to 1 MPa, Type C, initial grout pours under gravity and then similar grout pressurizes before the hardening of the initial grout, and Type D, where initial grout has placed under gravity then before hardening of the initial grout, additional grout pressurizes at 2 to 9 MPa through sleeved pipe by using packers. Recently, hollow bar micropile was introduced as an efficient micropile construction technique where drilling and grouting are performed in one step, so there is no need for casing. The installation of hollow bar micropile starts with advancing a threaded hollow bar utilizing a drilling bit and drilling grout to the predetermine level and then the grout is going to pressurized through the nozzles of the drill bit.

Micropiles, also known as mini piles, needle piles and root piles, are a type of deep foundations composed of placed or injected grout and steel reinforcement or steel pipes with diameters of typically less than 300 mm. Micropile has recently gained higher acceptance in geotechnical engineering practice and has been utilized as a foundation solution in many applications such as resisting uplift loads, seismic retrofitting, underpinning of existing foundations, control of settlements.

Growth of the micropile industry is made evident by the proliferation of technical publications, trade organizations, and design guideline publications. More importantly, growth is evident in the variety of projects and applications where micropiles have solved some difficult problems. There has been a steady increase in the number and scale of micropile projects, as well as in the design capacities and applications of micropiles in world wide.

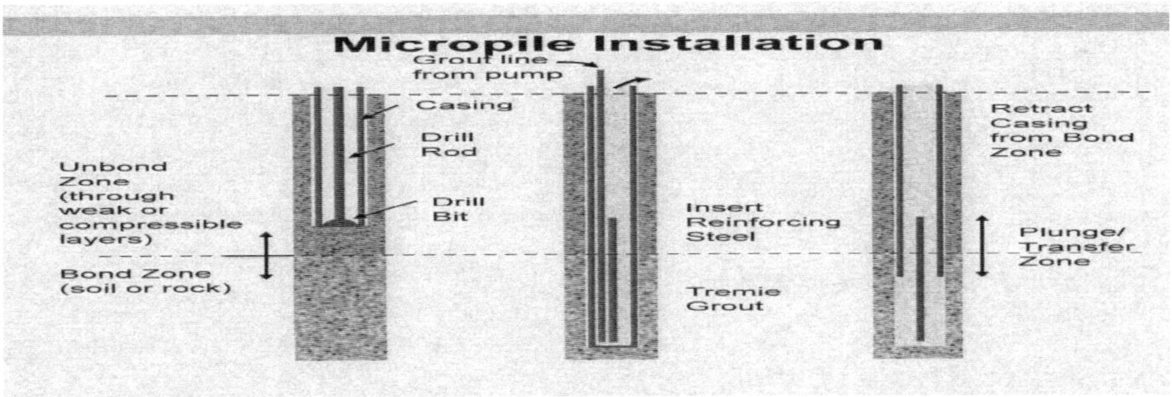

Plate 1. Micropile Installation

3.0 Experimental work for Micropiles constructed in Sudan.

3.11 Soil Investigation:

The area of (20×20 m2) in the land of Building and Road Research Institute, University of Khartoum (B.R.R.I, U OF K) located in Soba, south of Khartoum was used for carry the experiments. The soil is identified and classified according to the British Standard BS 1377 with the Unified System for Classifying Soils (USCS).

The boreholes drilled at the investigated area, revealed that the soil profile consists of three layers. A layer of Silty CLAY of low plasticity extended from surface to 5.0 m. This layer underlained by a layer of Clayey SAND which extends to the average of 14.0 m. A layer of Silty CLAY of high plasticity was encountered at 14.0 m and extended to the bottom of the profile.

Summary of the soil profiles are given by Table 1.

Table 1.Soil Profile B.H # 1 Location: Khartoum- Soba Coordinates: E: 645802 N: 1303779

Depth (m)	Description	Group Symbol	Atterberg Limits			N.M.C %	%pass sieve #200	SPT-Results	
			L.L %	P.L %	P.I %			Blows Each 6,3,3,3,3in	N-Value
	Hard, Light brown, dry, Silty CLAY of low plasticity, Calcareous materials.	CL				20.98			
1.5			49	16	33	10.93	51	10,7,9,11,11	38
						10.42			
3.0			33	22	11		59	16,7,10,11,1	40
4.5	Very dense, Light brown, dry, Clayey SAND.	SC	34	20	14		43	13,12,22,17	>50
6.0			61	21	40		48	38,51	>50
7.5			54	19	35		44	24,51	>50
9.0			55	18	37		44	39,51	>50
10.5			51	18	33		43	41,51	>50
14.0									

Water table was not encountered

3.12 Description:

Micropiles for field work were divided into two groups (G1 and G2). Each group containing three piles with diameter D10 and same length equal 4m, micropiles of G2 have an end bearing plates, see Plate 2. The total number of micropiles is 6. A general plan procedure for the field works was planned. Each of the above groups was casting as micropile type 1B, the both groups were casted by pressure with amount of 1 MPa (10 Bar) see Plate 3. This was done after executed studies work such as excavations, laboratory tests of soil and the initial design. All of these works give room for comparing the capacities of the two groups of micropiles after conduct a compression load test for G1 and G2 respectively.
fcu = 55 N/mm^2 (was used grouting cement of 10 N/mm^2 bending resistance).

Grout was poured for micropile (G1, G2) under pressures of (10 Bar were considered (1 bar=0.1 MPa)) (accordingly the classifications Type 1B).
A pressure machine (comperson) of 16 Bars was used to produce the pressure.

Plate 2. Steel reinforcement with and without end bearing plate.

3.2 Testing Programs of Micropiles:
3.2.1 Preliminary Design:

Micropiles drill hole diameter is 10.16 cm (4 inch) with temporary casing diameter 9.8 cm (thick walls = 3.6 mm).

Preliminary design of micropile was made for testing program. The data used for the design:
fcu = 55 N/mm2, fy = 420 N/mm2.

The description of the soil through which the bond length will be calculated is→ Hard, Light brown, dry, silty CLAY of low plasticity with calcareous material.
From the summary of typical ∂bond (grout to ground bond) for preliminary micropile design and feasibility evaluation (Tables (3-2)).
For Micropile type 1A (50----120kPa)
For Micropile type 1B (70----190kPa)
Accordingly the soil test result we will take ∂bond = 120 kPa
Design of Micropile type (1A):

The allowable compression load for the uncased length of a micropile is given as:

*Pc-allowable = [0.4f'c-grout*Agrout + 0.47*Fy-bar*Abar]...*

For strain compatibility between the concrete and steel reinforcement accordingly PTI (Post Tensioning Institute, Farmington Hills, USA)
Load taken on grout = area of concrete × strength of concrete/4
Load taken on steel = area of steel × yield of steel/2 (as assumption of preliminary design).

Although rebar is not a typical for micropiles reinforcement; it was used in this study, this is because it is widely used in other types of foundations in our area; it makes us look for a practical comparison.
Cross section area of steel pile is:
Ag10 = 7857.14mm^2 use 3Φ10 Aast = 235.71mm^2.
*Pc10-allowable =[0.4*50*(7857.14-235.71)*0.25 + 0.47*420*235.17*0.5] = 61.32 kN*

PG-allowable = (∂bond/FS) × π × *Db* × *Lb*
Where:
∂ Bond = grout to ground ultimate bond strength
FS = factor of safety applied to the ultimate bond strength
Db = diameter of drill hole
Lb = bond length
PG-allowable = micropile Geotechnical Capacity
Lb10 = (61.32 × 2) / (3.14 ×0.1×120) = 3.3 m
Load structural capacity:
From FHWA Design Guide Line: Pult-compresion = [(0.85f'c-grout × Agrout) + (fy-bar × Abar)] where:
Pult= the ultimate load compression on micropile.
Pult10-compresion = 94 KN.

Accordingly the length of drill hole was taken 4m; bond length of micropile 3.5 m.

3.2.2 Installation method:

For the construction of micropiles type B, the ground was drilled using a temporary casing, after reaching the 4 m depth, the drill bit was removed, and steel reinforcement was placed inside the hole, then during the temporary casing removal the cement grout was poured under pressure of 10 bars.

Plate 3. Grouting of Micropile.

3.2.2 Field Compression Load Test procedure:

For this compression load test was using a reaction frame or kentledge system is undertaken on a test concurrent with construction of the micropiling works.

Kentledge system has assembled with sufficient care (By using a truck which was loading with additional weight, the total weight was measured over 30 tons) Fix three (3) nos. dial gauges on an independent frame to measure the micropile head displacement. Two scale rules have been fixed on a top of the micropile head on diagonally opposite sides. Another scale rules have been fixed as independent bench marks to be rested on a permanent structure inside firm ground. Hydraulic jack machine was used to measure the load act on micropile by reaction frame, with data logger device to measure the output. All testing equipment has been protected from unnecessary disturbance prior to and throughout the load test.

The test was used to validate the micropile design. In this case the micropile was loaded to 3 times the design working load see Plate 4. Controlled-stress method has been used with quick maintained load test (QML).

In this procedure, 15% of design load was added at a time and held for 5 minutes. The maximum load is equal to 3 times the design load.

Loading was conducted by incrementally loading and unloading the micropile and measuring the movement of the micropile head at each load increment. The micropile-head movement reading is recorded just after the load has been applied and a second reading after the load has been maintained for a sufficient amount of time to ensure that pile-head movement has stabilized.

When the load is reached, unloading is then done with four equal decrements allowing 5 minutes between each two decrements. This method has advantage of being fast but it cannot be used where the pile is installed in strata with significant creep properties because load may be shed from the sides to the base without changes in the applied load (O'Neill and Hawkins, 1982).

Plate 4. Micropiles Compression load test.

3.3 The result of the experimental work:

Table 2. Failure loads of micropiles of G1 & G2.

G1		G2	
Without end bearing plate		With end bearing plate	
Pn	**Failure load KN**	**Pn**	**Failure load KN**
P3	194.3	P2	177.55
P5	159.89	P4	154.02
P7	202.1	P8	195.4
P11	201.41	P10	185.59

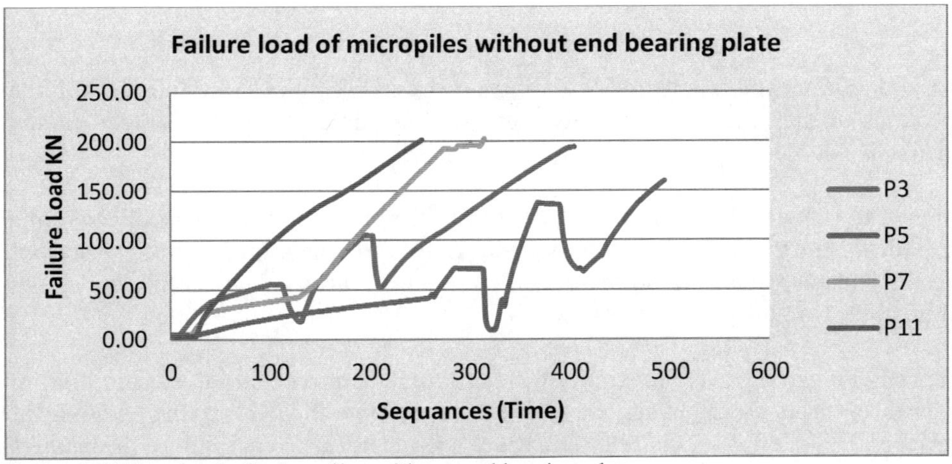

Figure 1. Failure load of micropiles without end bearing plate.

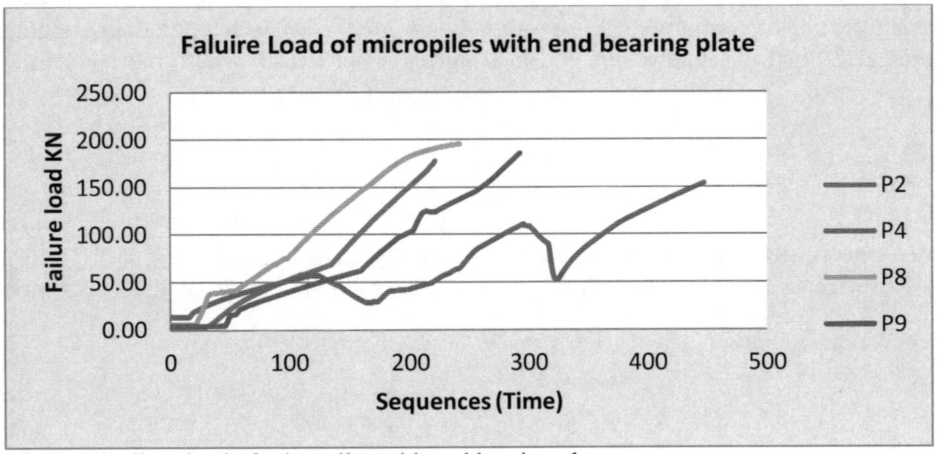

Figure 2. Failure load of micropiles with end bearing plate.

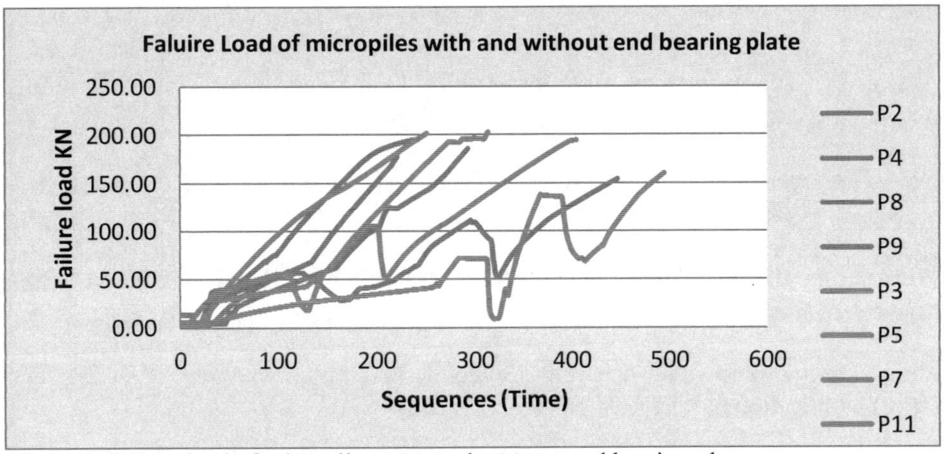

Figure 3. Failure load of micropiles with and without end bearing plate.

4.0 Discussion

By going through the results of the experiments for two groups (G1 and G2), it becomes clear that there is a discrepancy in the values of the failure loads for two groups see Table 2, Figure 1, Figure 2 and Figure 3 but this discrepancy is not much.

All micropiles are designed to transfer the load through the shaft friction along the length of micropile shaft to the surrounding medium, the end bearing at Micropile tip is generally negligible for the reasons of small base bearing area, this enhances that the possibility of using end bearing plate on micropiles has very few impact, almost negligible on micropiles capacities.

Tow failure mechanisms can be expected as the strain of the reinforcement reaching the strain limit of grout. First is the crushing of grout body under excessive compression. Second is the yielding at reinforcement/grout interface. (S.S Liew & C.C Fong, 2003) are seen that the second failure mechanism will likely to happen, because the adhesion of most normal material is always lower than the cohesion. The authors opinion in case of using end bearing plat at micropile tip, that the first failure mechanism will likely happen, because in the case of (G2), there is a restriction in the movement of pile tip downward (having a plate) after passing the elastic term crushed in grout is likely possibility to happen than yielding at reinforcement/grout interface. For G1 a micropiles end bearing plate not exists, where the end of the pile tip is in the form of a cone shape, which does not restrict movement downward.

5.0 Conclusions

The end bearing plate has no positive impact on micropile capacity. Crushing of grout body under excessive compression is expected as a failure mechanism in case of using end bearing plate at micropile tip.

6.0 References

[1] A. SOORKTY U OF K, AND H. EARABI U OF K., "IMPROVMENT OF MICROPILE CAPACITIES CASTING WITH PRESSURE AND CONSIDERING A BIG INTERVAL OF TIME" 4TH GEO-CHINA INTERNATIONAL CONFERENCE, AND ASCE 2016.

[2] Bruce, D., "The Basics of Drilling for Specialty Geotechnical Construction Processes," Grouting and Ground Treatment, ASCE, LA, 2003.

[3] Dipl.-Ing, Andreas Brandner, IB-Brandner, Karl-Schoenherr-Strasse 8, A-6020 Innsbruck – Austria," MICROPILE FOUNDATION – THE EASY WAY".

[4] Elarabi, Abbas, "Capacities of Micropiles Cast In with Pressure" Vol. 19, 2014 Bund. X, pp 3871-3878.

[5] FHWA, "Micropile Design and Construction Guidelines" Implementation Manual, US Department of Transportation, Publication No. FHWA-SA-97-070, 2000.

[6] Ir. Liew Shaw Shong & Fong Chew Chung Gue & Partners Sdn Bhd, Kuala Lumpur, Malaysia "Design & Construction of Micropiles", 2003

[7] John R. Wolosick, P.E., D.GE and Robert F. Scott, Jr., P.E. "Loading Effects on Battered Micropiles Supporting Lateral and Compression Forces," 2017 International Workshop on Micropiles 9th Lizzi lectures.

[8] Maged Abdlrahem1, M. Hesham El Naggar, F. ASCE2 and Wenbing Wu3
"LATERAL PERFORMANCE OF HOLLOW BAR MICROPILE IN COHESIONLESS SOIL" 2017 ISM.

[9] M. Deriszadeh, M.Sc., Ph.D.1 and M. J. Kokan, M.A.Sc., P.Eng.2
"A Review of Micropiles Beneficial Aspects through a Case History near False Creek, Canada" 2016 ISM.

[10] Paul Liles, P.E, John Wolosick, P.E and Tony sak, P.E "Design and construction of High-Cpacity Micropiles for ABC Projects" ABC-UTC Monthly webinar 2020.

WHAT IT TAKES (GEOSTRUCTURALLY) TO EXTEND THE PLANE TRAIN TUNNEL UNDER THE WORLD'S BUSIEST AIRPORT

Alexander J. Darnell, PE, Keller North America, Inc., Alpharetta, Georgia, USA, 770-442-1801, alex.darnell@keller-na.com
Ryan T. Smith, PE, Keller North America, Inc., Alpharetta, Georgia, USA, 770-442-1801, rtsmith@keller-na.com
Anthony Sak, PE, Keller North America, Inc., Alpharetta, Georgia, USA, 770-442-1801, asak@keller-na.com

ABSTRACT

Every second can count when trying to board a departing flight or catch a ride-share to your final destination. Hartsfield-Jackson Atlanta International Airport is in the midst of a project that aims to cut about 13 seconds between automated people mover trains (Plane Train) that move passengers between the terminals and concourses. The project involves extending the Plane Train tunnel west from its existing terminus under the domestic terminal. Working from inside non-public areas of the terminal, Keller North America, Inc. (Keller) incorporated systems of micropile-supported steel beams and grillage to both underpin existing columns prior to tunneling and support open-cut excavation at the tie-in of the new to the existing tunnel. Dewatering aided in mining a challenging section at the start of the drive of the mixed-face, sequentially excavated tunnel and an automated instrumentation system monitored the performance of the construction and its impact to the terminal. Cut and cover egress and ventilation tunnels required braced support of excavation.

Keywords: micropiles, underpinning, dewatering, instrumentation, soldier piles, bracing

INTRODUCTION

In 2021, the Hartsfield-Jackson Atlanta International Airport (HJAIA) in Atlanta, Georgia regained the title of the world's busiest passenger airport, measured by both total number of passengers and aircraft takeoffs/landings (ACI, 2022). Composed of both a domestic and international terminal bookending six concourses, passengers and staff move between these structures in an underground, three-passage tunnel complex of a center pedestrian walkway flanked on either side by single-track, automated people mover (APM) tunnels. The APM, known as the Plane Train, carries approximately 200,000 people per day on four-car trains that cycle between stations about every 2 minutes (HJAIA 2022).

HJAIA embarked on a project with Clark / Atkinson / Technique JV (CAT JV) in a progressive design-build contract, with McMillen Jacobs Associates (MJA) as the lead designer, to extend the Plane Train tunnel approximately six hundred feet to the west. The completed project aimed to cut approximately 13 seconds from the two-minute Plane Train wait time.

PROJECT OVERVIEW

The project consists of extending the existing parallel Plane Train cut-and-cover tunnels to a bifurcation where the then single tunnel continues to a new shaft that connects the tunnel to new at-grade emergency egress and ventilation structures up to the ground surface, see Fig. 1. The terminal in the project area is a concrete slab on grade with individual, spread footing-supported columns. A close up of the parallel tunnel alignment, Fig. 2, shows that nineteen column footings were within the plan limits of the tunnels. The distance between the bottom of column foundations and the tunnel excavation line (cover) is as little as seven feet, see Fig. 3. Thus, these columns required underpinning.

For the breakthrough into the existing Plane Train tunnel, temporary, braced excavation support utilized micropile soldier piles.

Drilled dewatering wells with submersible pumps lowered the groundwater, facilitating tunnel mining from breakout at the construction shaft eastward to the existing Plane Train tunnel.

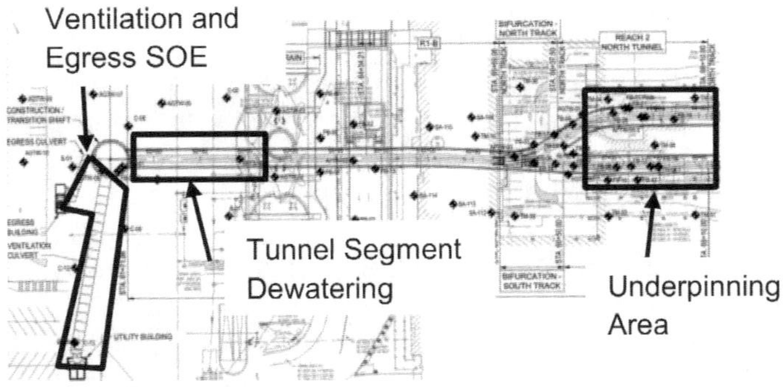

Fig.1. Project layout

Construction of the egress and ventilation access from the construction shaft required support of excavation. With no overhead restrictions, the soldier pile installation proceeded using typical drilled shaft construction techniques with wide-flange section piles.

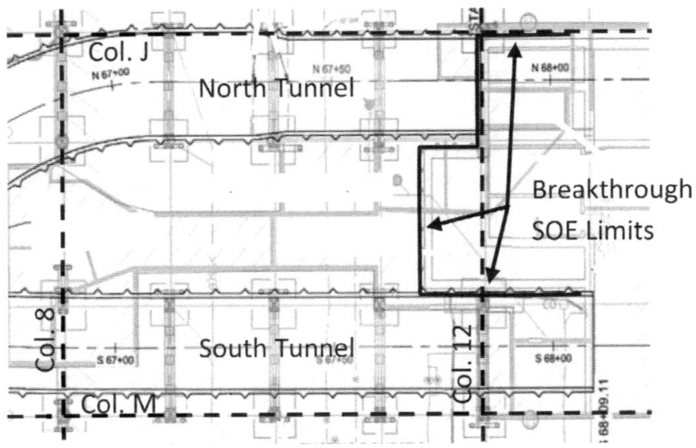

Fig. 2. Closeup of underpinning area.

MJA designed the underpinning and excavation support. CAT JV subcontracted Keller North America, Inc. (Keller) to design and construct the underpinning system, install the breakthrough support of excavation soldier piles, design and install the dewatering system, and construct the egress/ventilation excavation support. GEO-Instruments, Keller's sister company, provided real-time geo-structural monitoring to monitor the performance of the various systems, under a separate contract to CAT JV.

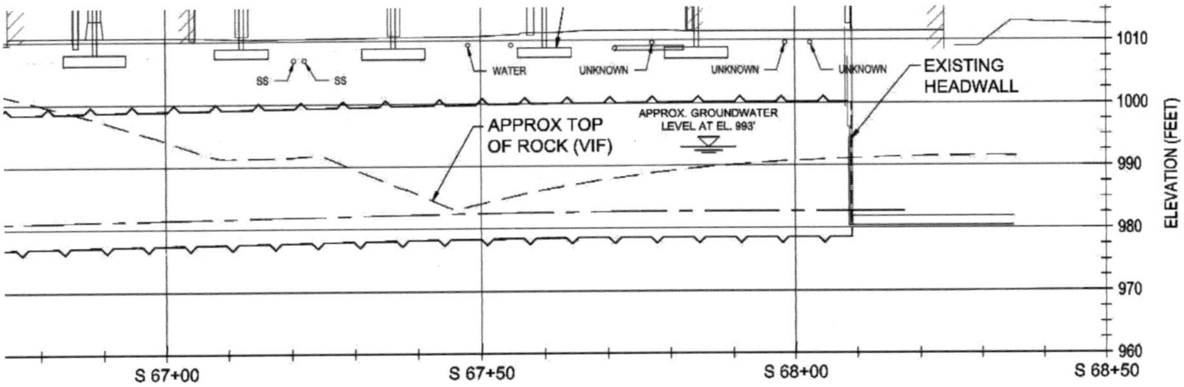

Fig.3. Profile through South Tunnel under Terminal

SUBSURFACE CONDITIONS

The subsurface profile throughout the project area consists of fill underlain by residual soil, partially weathered rock (PWR - saprolitic soil with standard penetration test resistance greater than 100 blows per foot), underlain by competent biotite gneiss bedrock. The depth to and thickness of the overburden materials varied across the project. Within the terminal underpinning area, the approximate top of bedrock varied between 12 and 21 feet below the slab subgrade and, in all locations, above the excavated tunnel invert, see the profile in Fig. 3. Excellent quality bedrock, as measured with the rock quality designation (RQD), was encountered within five feet of the top of bedrock. Over the length of the mixed-face tunnel section, the top of bedrock varied from below tunnel invert to above tunnel crown.

TERMINAL UNDERPINNING DESIGN

Nineteen, concrete-encased, steel columns of the terminal required underpinning, due to their proximity to their respective tunnel. The columns carried combined loads between 166 and 215 kips to square footings measuring between 7.5 feet to 10 feet on a side. Steel beams attached to the columns spanned over and extended outside the future tunnel excavation line transferring the load of seven column pairs to steel grillage supported on micropiles. At three locations steel grillage and micropiles carried their respective column loads. For the remaining two columns, separate load-balanced, cantilevered steel beams carried the loads. A load-balance cantilever beam consists of center compression micropiles with the column load and tension micropiles on opposite ends.

The completed underpinning assemblies were temporary. After the tunnel excavation and installation of the initial tunnel support had passed a column location, transfer of the loads back to their respective footing could occur and the steel grillage/beams removed. One column received a new spread footing and one column received a new pile cap supported on permanent micropiles. Overall, the underpinning operation required forty-eight micropiles. Support of excavation for the breakthrough incorporated two underpinning micropiles.

MJA designed three micropile types using American Petroleum Institute (API) N80 casing: 9-5/8-inch diameter at both 0.75-inch and 0.545-inch wall thickness, and 7-5/8-inch diameter at 0.5-inch wall thickness with a #14 Grade 75 full-length center bar. Low headroom conditions, less than 10 feet, necessitated three or five foot-long, threaded casing. MJA established the top of the bond zone at an elevation below the tunnel invert, well into the excellent quality bedrock encountered on the site. The MJA design also required unbraced pile lengths of 31 feet. Since threaded micropile casing joints offer approximately 50 percent bending capacity compared to the full section, all compression micropiles required full threaded-joint reinforcement. Micropiles immediately adjacent to the tunnel excavation required a 0.5% out-of-verticality tolerance and plan location tolerance of 0.5 inches or 1-inch of the plan location, depending on the column.

Keller value-engineered the micropiles to two types, both using 9-5/8-inch diameter, API N80, structural grade casing with a 0.545-inch wall thickness. For compression piles with a maximum design load of 185 kips, casing extended the full pile length to the bottom of the nominal 10-inch diameter drill hole. Staggering the threaded joints of a 7-inch diameter, 0.453-inch wall thickness, API N80 inner casing with the threaded joints of the 9-5/8-inch diameter outer casing provided the joint reinforcement. While this double-cased compression micropile design was heavier than the MJA design, the construction efficiency of using this micropile detail outweighed constructing other threaded joint reinforcement, particularly due to the short casing section demanded by the low headroom conditions.

As there were only four tension piles, with a maximum design load of 80 kips, Keller elected to use the same nominal 10-inch diameter tooling for the tension pile as used for the compression piles, though the casing terminated at the top of the bond zone and tension reinforcement consisted of a full-length, 1-1/4-inch, Grade 150, all-thread bar.

Keller conservatively designed an 8-foot-long bond zone for both micropile types, using tremie-installed, 5,000 pound per square inch (psi) mix design and an ultimate grout-to-bedrock bond stress of 200 psi.

To simplify the steel beam and grillage construction, where possible, Keller used the same steel sections throughout. For instance, the steel beams used for the seven column pairs were 2-W18x86 sections with spans between 27.5 feet and 30.75 feet in lieu of MJA's different beam sections specific to individual column pairs.

UNDERPINNING INSTALLATION

The underpinning installation sequence, in coordination with CAT JV, included:

- Remove column concrete cover
- Core through the existing slab and footing (Keller).
- Install micropiles, collecting spoils using a diverter swivel into a hydrovac truck, to the final depth with the specified bond zone length (Keller).
- Cut through the slab and expose the steel column base plate. Back off the base plate anchor bolts. Backfill the excavation.
- Assemble the underpinning frame assembly (Keller).
- Lift the columns using hydraulic jacks to either one-eighth-inch movement or design gravity load, then transfer to the assembly (Keller).
- Remove load from the assembly, at the direction of CAT JV, transfer back to the column footings. Remove the assembly (Keller).
- Excavate backfill over footings, reattach column base plate, cutoff micropiles, backfill excavation, and repair slab.

Keller mobilized Comacchio MC-4D and MC-8D hydraulic-powered drill rigs to install micropiles. The hydraulic unit power pack is detachable from the rig, which allows the rig to enter and maneuver in tight spaces and low overhead clearance (eight feet for the MC-4D and about 10 feet for the MC-8D). Keller advanced casing through the footing core holes to the top of bedrock using rotary duplex tooling. Rotary percussive tooling drilled the bedrock to completion of the drill holes.

The first micropile installed was a load test pile. Sacrificial piles are typically load tested. However, given the access within the footprint of the existing terminal, Keller installed a production tension micropile as the test pile, except a 2-1/2-inch, Grade 150 bar replaced the design 1-1/4-inch bar to allow testing the bond zone with the compression-only micropile design load of 185 kips. The quick load test, in accordance with the ASTM D3689, resulted in a measured 0.165-inch deflection at the maximum test load of 370 kips on a 56-foot-long pile with a 48-foot free length and 8-foot bond zone, well within the test acceptance criteria of 1-inch deflection at the design load.

The tight working conditions, working adjacent to critical airport mechanical, electrical, and communication systems meant Keller had to plan in detail the access and execution of the micropile installation and carefully execute the work to avoid shutting down the busiest airport in the world. Views of the some of the restrictive access available for Keller's crews and their drill rigs are shown in Figs.

Fig. 4. Low headroom access through a double-entry door.

4-5. Micropile installation was mostly uneventful, except for:

- Finding the occasional unknown utility lines between the slab and the top of footing.
- Off-perpendicular cuts in the micropile casing manufacture resulted in threading defects.

Fig. 5. Drilling adjacent to an existing wall, with casing and drill rods standing by.

Where possible, Keller adjusted micropile locations to manage the occasional utility, if CAT JV could not relocate them. Dealing with improperly cut micropile casing fell to Keller. Since all casing material had already been delivered to the site, working under a tight schedule, and with the supply-chain shocks caused by the Covid-19 pandemic, Keller elected to fight through the casing thread defects, that is, using the casing that was not straight when threaded together, even under the strict verticality tolerance required of the project.

Two unusual micropile installation results, unusual for the Atlanta area, were both driven by the need for the bond zone to begin below the tunnel invert:

- Bond zones for all 48 micropiles were at basically the same elevation.
- The cased, unbond zone penetrated competent rock, in cases up to 25 feet thick, before starting the bond zone.

Fig. 6. Verticality measuring in process.

Twenty-three of the micropiles, specifically those closest to the tunnel excavation line, required as-built verticality measurements. A DeviGyro, see setup in Fig. 6, profiled the micropile casing location along its length, between the top of casing and to the invert of the excavated tunnel. With practice, the measurement process, not including set up and take down, took less than 30 minutes to complete. Even with the miscut micropile casing, all micropiles were within the 0.5% specified tolerance.

Fig. 7. Welder at work on grillage.

Keller's welding crews began work on the steel grillage as micropile installation finished at each column, see Fig. 7 for grillage assembly. Up to eight certified welders worked for 10 weeks on the underpinning frame assemblies. A pair of 100-ton rams lifted the columns from their respective footings, see Fig. 8, and transferred to the grillage using Grade 150 all-thread bars and nuts. CAT JV installed fire protection coating on the grillage and beams upon completion of the underpinning work.

Regarding the removal of the load and assembly, the MJA design specified this can occur after the excavation of the full tunnel cross section and complete installation of the initial support, as opposed to waiting for the completion of the final tunnel liner. However, this was at the discretion of CAT JV.

Earlier mention of micropile unbraced length and construction of a new spread footing came together for one column in particular. Due to site constraints, a pair of micropiles, by design, breached the excavated tunnel envelop. As tunnel mining advanced through this column location the two micropiles stood inside the tunnel, see Fig. 9. After construction of the new footing and once the initial tunnel support achieved design strength, the new footing received its column and CAT JV cut the proud pair of micropiles flush to the initial support shotcrete.

Fig. 8. Load transfer to grillage.

Fig. 9. Tunnel mining exposes micropile (highlighted).

Initially estimated to take approximately 10 months, the complete underpinning installation was scheduled for six-hour night shifts, assuming actual production of about four hours per shift. Due to the drastic reduction in air travel brought on by the Covid-19 pandemic HJAIA allowed work during normal business (daylight) hours and shifts up to 10 hours per day, resulting in completion of the underpinning in approximately four months.

INSTRUMENTATION MONITORING

Keller installed strain gages in four underpinning micropiles. Instrumentation consisted of a set of four gages tied to four, full-length, #4 rebar at four levels corresponding to the following locations, five feet below bottom of footing, at the top of bedrock and plus/minus five feet from the top of bedrock. The wiring extended from the gages to the top of the micropile, see Fig. 10. Geo-Instruments then installed data loggers to collect measurements for wireless transmitting real-time data through a web-based management system.

Fig. 10. Micropile strain gages wired to data logger.

DEWATERING

Approximately 150 feet of the sequentially excavated and shotcrete lined tunnel alignment would pass through a depression in the PWR/bedrock surface and necessitated dewatering to provide adequate stand-up time for the residual soils expected in the tunnel face. The deepest rock surface was within several feet of the tunnel invert, a condition which makes a completely dry tunnel face unlikely. The intent was to provide stable and workably dry ground conditions for tunneling through overburden soils. The design groundwater elevation was approximately 10 feet over the tunnel crown.

Based on the experience of previous Atlanta area open face tunneling projects, Keller anticipated that the upper bedrock was the primary water bearing unit and the stratum from which the overlying residual soil must be dewatered. Additionally, because the wells would draw from intersecting fractures in the bedrock, Keller expected erratic well yields. The design consisted of parallel rows of wells spaced on 10-foot centers to accommodate the variability of flow. The wells were advanced to a depth of 10 feet below tunnel invert or 15 feet into competent rock, whichever occurred first. Rotary duplex drilling methods advanced a casing to the top of rock and a down-the-hole hammer advanced the well boreholes to the termination depth. Although the anticipated low well yields would typically suggest the use of ejectors, the design utilized deep wells (with submersible electric pumps) as this system was more amenable to direct burial.

As anticipated, 20 percent of the wells resulted in 80 percent of the total system groundwater extraction. The system proved adequate to permit the mining of the sequentially excavated tunnel without any ground instability incidents.

SUPPORT OF EXCAVATION

MJA designed braced support of excavation (SOE) systems, one located at the breakthrough into the existing tunnels and the other required at the separate ventilation and egress cut and cover tunnels.

At the breakthrough, Keller installed micropiles of the same section as the compression micropiles for the underpinning. CAT JV excavated the material between the SOE and the existing tunnel, welded lagging brackets to the micropiles, installed the lagging boards, and braced the SOE against the existing tunnel structure, see Fig. 11.

Fig. 11. Braced excavation support at the breakthrough. Note the micropile-supported grillage encased in fire protection in upper righthand corner.

For the ventilation/egress tunnel SOE, Keller's scope consisted of installing the soldier piles and bracing system, CAT JV excavated and lagged the exposed cut. Soldier piles consisted of W21x55 beams set on 7-foot centers in 30-inch diameter drill holes, with the bottom 14 feet of the hole backfilled using 4,000 psi concrete and the balance of the hole with 500 psi flowable fill. Soldier pile lengths varied between 35 and 40 feet. Casting the tunnel concrete against the SOE system necessitated SOE pile tolerance of 1-1/2-inch in plan and maximum 1% from vertical. Keller returned to the site for installation of the single-level internal bracing once the excavation subgrade reached two feet below the bracing level, approximately nine feet below top of wall grade. A bird's eye view of the ventilation/egress tunnel SOE installation is shown in Fig. 12.

CONCLUSIONS

Construction at the world's busiest airport has its challenges. A project to extend the existing Plane Train tunnel by mining under the existing terminal with less than 10 feet of cover between the tunnel crown and terminal slab on grade perfectly fits into that category. Working with the progressive design build team of Clark/Atkinson/Technique JV and McMillen Jacobs Associates, Keller installed a temporary underpinning system to relieve the stress of individual column spread footings that overlie the tunnel extension alignment. Keller also installed a temporary dewatering system and elements of support of excavation systems.

Fig 12. Bracing system in process for the ventilation (left) and egress tunnel SOE.

Keller overcame potential obstacles to success, such as low headroom conditions, tight working conditions, working adjacent to critical airport mechanical, electrical, and communication systems, skewed micropile casing threads, and tight verticality tolerances, all during the Covid-19 pandemic. Of course, the travelers through Hartsfield-Jackson Atlanta International Airport had no idea this work was going on beneath their feet, as they walked over area. But a complicated system of geostructural construction facilitated future passengers making their flight, with a few seconds to spare.

REFERENCES

Airports Council International (ACI), Press Release, April 11, 2022, "The Top 10 Busiest Airport in the World Revealed", website address aci.aero/2022/04/11/the-top-10-busiest-airports-in-the-world-revealed/, accessed April 18, 2022.

Hartsfield-Jackson Atlanta International Airport (HJAIA) webpage, "ATL Fact Sheet", www.atl.com/about-atl/atl-factsheet/, accessed April 18, 2022.

THE MINIATURIZATION OF DIAPHRAGM WALL TECHNIQUE – THE CUBE SYSTEM

Franz-Werner Gerressen, BAUER Maschinen GmbH, Germany
Alexander Blatt, BAUER Maschinen GmbH, Germany
Leonhard Weixler, BAUER Maschinen GmbH, Germany

ABSTRACT

Diaphragm wall technique is usually used to construct structural elements in the ground, commonly used for retention systems, permanent foundation elements or deep ground-water barriers, named Cut-Off-Walls (COW).
It can be anticipated that global urbanization and increasing demands on environmental considerations will need to be accommodated in underground space in the future.
These trends show an increasing requirement for diaphragm walling in even more complex conditions, especially in inner-city applications with limited space. Therefore, there is the demand to minimize the size of equipment to accommodate such projects.
Following this demand, a new trench cutter system, the so-called CUBE System, was recently developed. The complete trench cutter system has been fit into container size dimensions so that it can work in very limited space conditions.

In addition, complex conditions in terms of space limitations require specifically adapted solutions for slurry, spoil, reinforcement and concrete handling and the related logistics to ensure smooth production.

Due to the limited space requirements, the impact on the surrounding environment is limited. As the system is electric driven, an additional reduction of carbon footprint can be expected.
First tests of the system have been carried out and the results will be shared.

The paper will explain the construction process and the specific challenges and experiences made for these extreme conditions.

Keywords: Diaphragm wall, Trench Cutter

Urbanization and complex conditions

Trends showed an increasing population in recent years. Hand in hand the number of peoples living in cities is growing subsequently. To date, over 50 % (around 4 billion people) of the world's population already lives in cities and urban areas. The expand of the cities calls for new infrastructure like new underground metro lines, water systems, logistic centres, underground parking and much more. This and rising prices in cities ask for solutions for future projects. Further, the demand for more resources and existing structures in the urban areas request technology, which can deal with great depths and challenging soil conditions.

Machine suppliers are confronted with competing trends of developing small equipment for inner city jobsites but also provide solutions for increasing depths. The challenge to develop machines with low emissions and solutions to reduce impact on daily lives sets restriction to fundamentally rethink methods and processes. In addition, the developments must be more and more powerful, to meet tight schedules and therefore are able to carry out cost effective projects.

One chance to confront these trends is the diaphragm wall method. It combines many features like the ability to be used in various soil and rock conditions with high performance or the possibility to use it in inner-city regions.
As the following examples show, diaphragm wall equipment already can reach depth up to 250 m on the one hand. On the other hand, recent developments show a total redesign of the technology for confined space situations.

This total redesign was presented under the name CUBE-system, which fits all components of the diaphragm wall cutter technology into container sized dimensions.

THE WORKING PRINCIPLE

Construction sequence

The working steps follow a pre-defined sequence. The main steps are shown in Fig. *1*:

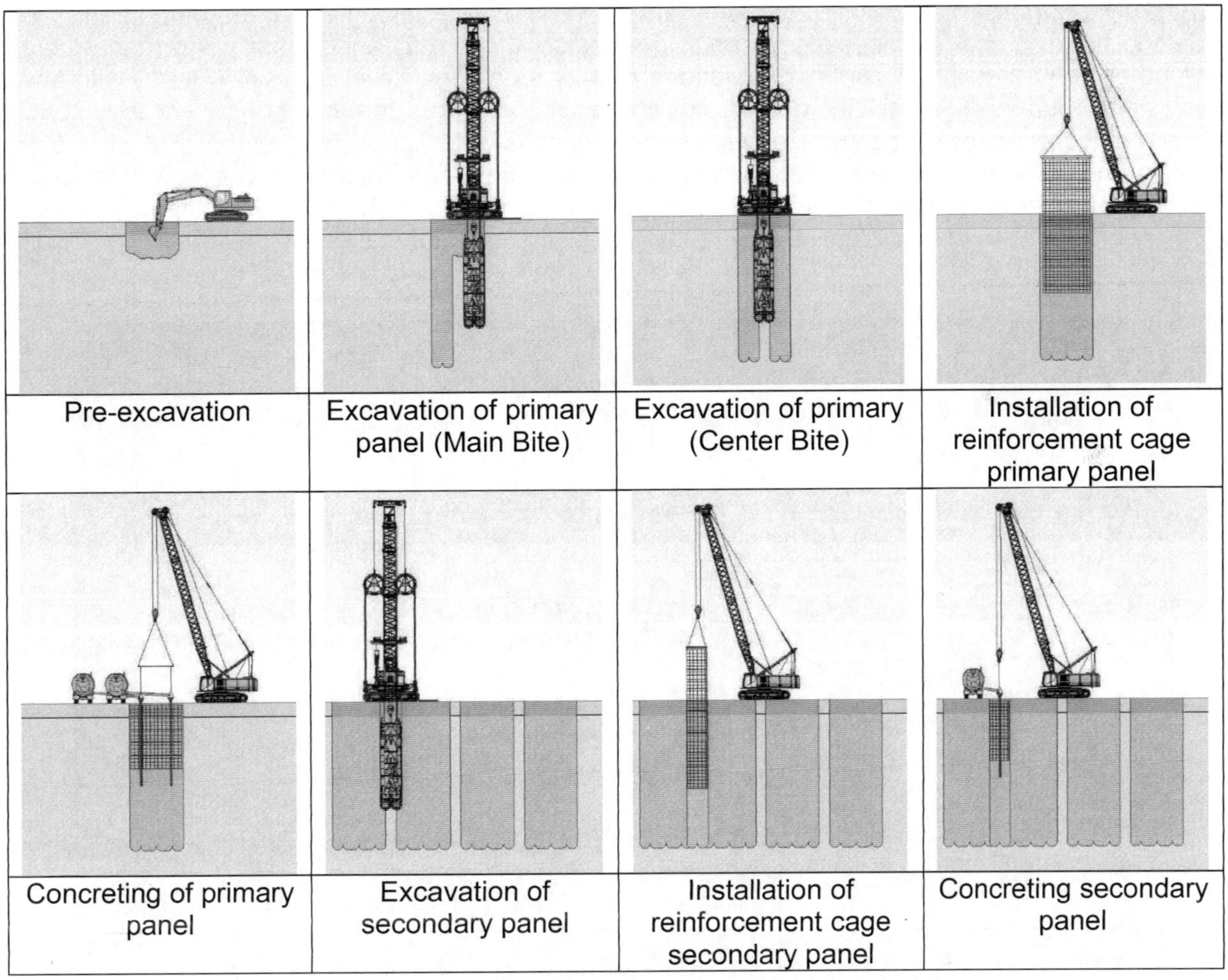

Fig. 1: Construction sequence

In confined conditions the use of special auxiliary equipment (e.g., for installation of reinforcement cages) might become necessary. During excavation the open trench is stabilized by supporting fluid.

Site set-up and logistics

For diaphragm walling, the handling of the slurry is crucial and compared to other methods, a more complex site set-up becomes necessary.
Beside the cutter and its base carrier, other equipment such as desanding plant, mixing plant, different pumps and storage of the slurry requires space. Further access areas, for the supply of reinforcement cages, concrete and the transport of the excavated material are needed as well as storage of spare parts etc.

As the technique possesses many advantages, the demand to use it also in limited site conditions already led to a special carrier of reduced hight, for so called low-headroom applications.

Development of equipment

Currently it is possible to monitor two different trends in the development of the equipment. On the one hand, it shall be possible to reach greater depth (Fig. 2) and higher performance. On the other hand, small and confined situations require small flexible equipment (Fig. 3) with low emissions. Especially inner-city jobsites set challenges with the demand to new infrastructure. Over the years different contrary developments to satisfy demands took place. But all these developments had in common, that it was possible to identify the diaphragm walling equipment. A completely new approach to widen the range of application is the so-called CUBE system, shown in figure 4. The system is the next step in the miniaturization of the diaphragm wall technique.

Fig. 2: Cutter system for great depth up to 250 m

Fig. 3: Low-headroom cutter system

Fig. 4: CUBE-System

Aim of the system

The aim of the system is to provide additional possibilities. For example, it is possible to create walls out of underground structure, such as metro stations or tunnels. This new flexibility widens the range of possibilities during the planning stage of new infrastructure (for example underground car parks, water reservoirs, storage areas, etc.). Currently new planned projects often cannot be planned and executed where it would be needed, as existing structures set restrictions. So, in some cases projects are not even started because of lack of technical solutions or enormous efforts in case of execution. In those cases, compromises must be made, e.g., find new not ideally locations, postpone or even cancel projects. With compromises always come additional costs over the lifetime of such a project.

Now, the system allows planning of underground structures exactly where it is needed in conjunction with minimum impact on the environment above ground level. This creates new options and solutions for planers and urban development. Execution of D-Wall out of a tunnel as shown in figure 5 is one of the options. The depth of the tunnel is flexible and can be adjusted to the existing structures. To avoid additional works like e.g., groundwater lowering, it is recommended to have the tunnel above groundwater level.

Fig. 5: Cube system working in tunnel

The CUBE system

As the name indicates, the components are aligned in high cube container. The dimensions of the system can be seen in figure 6.

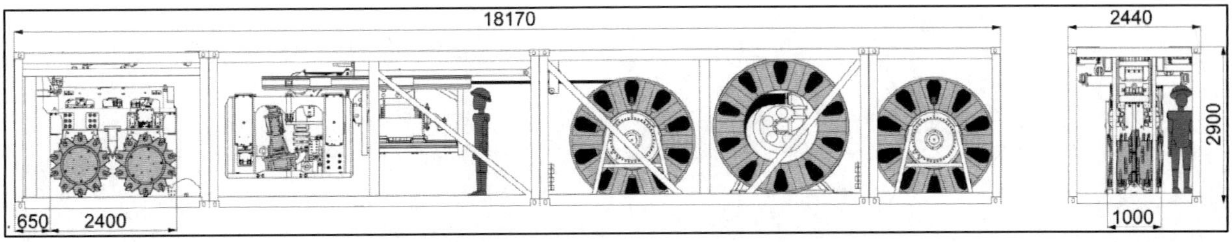

Fig. 6: CUBE system - dimensions in mm

Within 4 high cube container it contains all known components of a diaphragm wall cutter. As displayed in figure 6, panels with a length of 2400 mm and widths of up to 1000 mm are possible. To fit the cutter into the containers, it needs to be separated into two pieces. The first container accommodates the cutter head, where the cutter wheels and gearboxes are mounted. Within the second container the pumping unit is placed. To start the cutting process, this unit needs to be placed over the cutter head. For that a pre-excavation of 1,5 m is necessary to lower the first part and therefore enable the second unit to move over. After that both parts can be connected, and the cutter unit is able to work. The working principle is identical to standard trench cutting systems.

Because of the containerized dimensions, the need of heavy transports become irrelevant, which makes the system flexible and easy to transport. By connecting the containers with so-called quick ties, a fast and uncomplicated assembling can be ensured.

On both sides of the containers, passages make it possible to walk through. This is an important factor to meet the safety standards for first aid access and material transport within a tunnel. Further it guarantees a safe environment for service tasks, such as teeth change on the wheels.

Functionality test

To understand the processes and restrictions as well as to get a feeling for very confined conditions, a test was carried out. Scope of the test was the construction of 7 single bite panels. Therefore, a complete jobsite set-up was assembled (Fig. 7). In addition, a working platform, which included the rail system was constructed. Further a tunnel was simulated to work within height and width restrictions.

In total five primary and two secondary panels were executed, whereby one primary was constructed to the maximum depth of 40 m. All other panels had a depth of 15 m. All panels have been concreted using a concrete pump. Different reinforcement cages have been tested. Further different equipment tests have been carried out as well.

Fig. 7: Jobsite layout

As the cutter system is not mounted on a carrier with tracks, a new idea for the movement of the system had to be developed. The movement now is generated by a clamping device and by hydraulic cylinders as shown in figure 8. The device clamps hereby on a rail system and a cylinder pushes or pulls the system in the desired direction. In the meantime, the opposite clamping device moves into position and ensures a continuous movement.

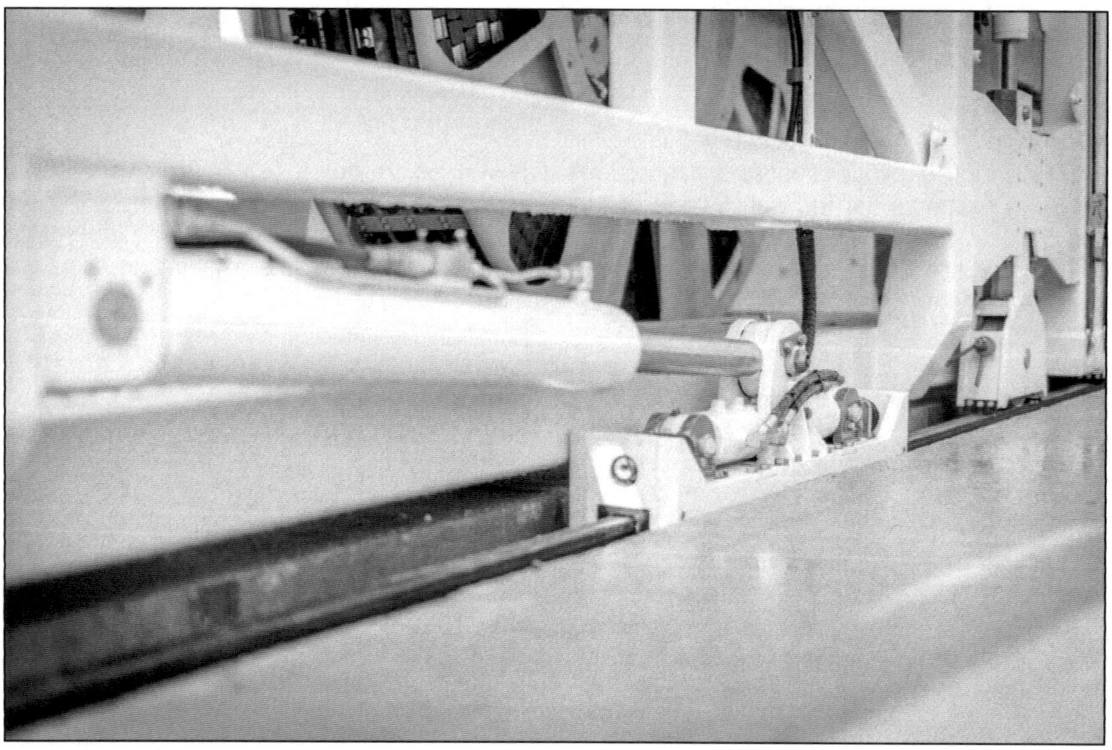

Fig. 8: Clamping device

Like any other cutter system, a pre-excavation is required. The system can work with a reduced pre-excavation depth of 1.5 m, which can be carried out by a small excavator also within confined conditions.

The cutting process itself is not different compared to standard equipment. The loosened material is transported via the stabilizing fluid to a desanding plant. Here the solids are separated from the liquid medium, which is afterwards ready for reuse.

Even the cutting process is quite like the standard procedure, reinforcing and concreting is a bit more challenging due to the very limited height restrictions. Therefore, two different types of reinforcement cages were tested to assess the handling and installation during the construction within a tunnel. Each segment of the cages had a height of 1.85 m and were transported to an independent container. This container (Fig. 9) contains all handling devices for the installation of the reinforcement cages and tremie pipes. Thus, it is also responsible for the concreting process.

During the test, it was found that the use of prefabricated cages in conjunction with the use of Lenton couplers as shown in figure 10 was the most favourable solution for practical application.

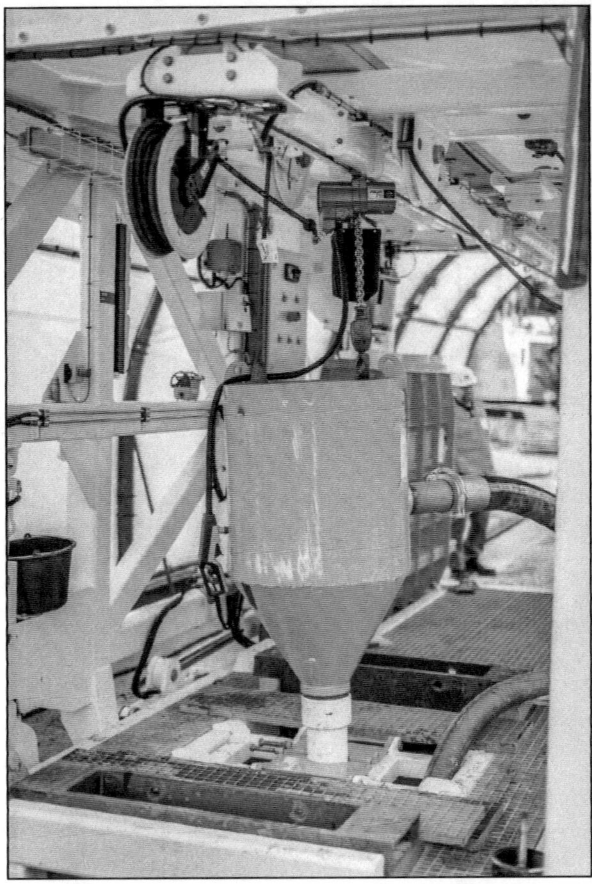

Fig. 9: Reinforcement and concreting container

Fig. 10 Cage coupling

A concrete pump (Fig. 11) supplied the concrete through the simulated tunnel to the trench.

Fig. 11: Concrete supply via concrete pump

CONCLUSION

With an increasing number of people, moving into urban and inner-city areas, the demand for new infrastructure will also soar in the future. Existing buildings and structures as well as high prices for areal spread, set restrictions and for these new infrastructure facilities. In addition, environmental aspects will become more important and will raise the question for new strategies. The impact of these demands will further influence for the construction process and the design of equipment.

The importance to provide solutions for confined space jobsites with challenging soil conditions already rises difficulties for logistics, storage, and methods as well as equipment to solve the restrictions. Today, suppliers already provide compact machines of high performance.

Further, efforts to reduce noise and fuel emissions will take place to prepare the equipment for wider spread of inner-city applications.

The CUBE System with its new design, new thinking of system and its reduced dimensions will make numerous projects possible, which lack of solutions until today.

The test proofed the functionality of the system and therefore provides opportunities for projects which never thought before.

REFERENCES

Gerressen, F.-W.; Blatt, A.; 10th Anniversary Conference on Deep Foundation Technologies for Infrastructure Development in India, November 2020; Confined space and inner-city projects – future challenge and opportunity for diaphragm walling

Gerressen, F.-W.; Schwank, S.; Blatt, A.; DFI 45 Annual Conference on Deep Foundations, October 2020; Diaphragm wall technique used in a different way – cutting trenches up to 823 ft. (251 m) for an exploration and bulk sampling program

PERFORMANCE OF OPEN AND CLOSE-END SCREW PILES IN BEARING LAYER

Faisal Shahzad (corresponding author)
Master Student, Department of Civil and Environmental Engineering, Saitama University, Japan
faisalshahzad352@gmail.com

Adnan Anwar Malik
College of Engineering, Science and Environment, Newcastle Australia Institute of Higher Education, Singapore
adnananwar.malik@newcastle.edu.au

Jiro Kuwano
Department of Civil and Environmental Engineering, Saitama University, Japan
jkuwano@mail.saitama-u.ac.jp

Tadashi Maejima
ASAHI KASEI CONSTRUCTION MATERIALS CORPORATION
maejima.tc@om.asahi-kasei.co.jp

ABSTRACT

Single helix screw piles carry most of the structural load through end-bearing if the helix is embedded in the dense bearing layer. The tip condition of the central shaft of the screw pile can be closed, closed with cutting blades, opened with flat cut, and opened with wedge shape cut and its selection is based on the in-situ ground conditions. These different end conditions of the central shaft disturb the ground differently, leading to the change in bearing capacity. In this study, an attempt is made to investigate the effect of open and close-end screw piles on installation effort (installation load and torque) and bearing performance in the dense bearing layer. The model steel screw pile with a pile shaft diameter of 21.7 mm and helix diameter of 43 mm is used in this study. Test results show that the open-end screw pile required less installation effort than the close-end pile that penetrated the bearing layer by $1D_h$ with an unplugged condition. Further, the open-end screw pile shows a higher bearing response than the close-end screw pile if it enters the bearing layer with an unplugged condition.

Keywords: Open-end screw pile, Close-end screw pile, Model pile load test, Bearing capacity

INTRODUCTION

When the soil layer at shallow depth cannot provide sufficient resistance to support the superstructure, a pile foundation is usually used to transfer the load to a greater depth (Meijuan et al. 2018). The pile material can be concrete, steel, timber, plastic, or composite. Its selection depends on material availability, the magnitude of loading, the type of soil, and the site conditions in which the pile is installed (Budhu 2010). The piles can be either straight pipe piles or screw piles. Screw piles consist of a central steel shaft with one or more helices attached at the various length of the shaft (Malik et al. 2017). The construction advantages of screw piles over other piles are its ease of installation, access to narrow places, reusable, and noise control (Sakr M, 2011). The tip condition of the central shaft of the screw pile can be closed, closed with cutting blades, opened with flat cut, and opened with wedge shape cut and its selection based on the in-situ ground conditions. Previous studies show that the behavior of open-end piles differs from that of close-end piles (Szechy 1961; Carter et al. 1979; Randolph et al. 1979; Klos and Tejchman 1981; Lu 1985; Smith et al. 1986; Paikowsky and Whitman 1990; Lee et al. 2003). Different end conditions of the central shaft disturb the ground differently, leading to the change in bearing capacity. Further, when an open-end pile is driven into the ground, a soil plug may develop within the pile during driving, preventing or partially restricting additional soil from entering the pile (Paik et al. 2003). Previous studies suggested the three possible pile driving modes during the pile installation and pile load test. According to Kumar and Kumawat (2016), "If a pile is driven in a fully coring (or fully unplugged) mode, soil enters the pile at the same rate as it advances. On the other hand, if a pile is driven under plugged or partially plugged conditions, a soil plug finally attaches itself to the inner surface of the pile, preventing additional soil from entering the pile". A pile driven in the plugged mode behaves similarly to a closed-ended pile (Paik et al. 2003). The formation of a plug inside an open-end pile during its installation greatly depends on the installation mechanisms (Jacking, Vibratory, and Impact), ground density (Dense or Loose), pile inner diameter (D_i) to pile wall thickness (t) ratio (D_i/t) and pile penetration depth as suggested by Li et al. (2021). During installing an open-end pile, plug length increases with the increase in pile penetration depth. When the pile reaches the bearing layer, it is usually either partially plugged or fully plugged. Due to a greater degree of soil plugging at larger depths, the behavior of an open-ended pile is quite similar to a closed-ended pile.

In this study, the main focus is to investigate the effect of the fully open and fully closed end conditions of a screw pile entering the dense bearing layer. Previous studies stated that penetration equal to the diameter of a helix ($1D_h$) into the bearing layer is optimum to achieve high ultimate resistance (Ikeda et al. 2010). So, the open and close-end screw piles are penetrated the dense bearing layer by $1D_h$, and installation effort (installation load and torque) is monitored during the pile installation. The effect of open and close-end conditions of the screw pile on the bearing response is also observed in this study.

TESTING EQUIPMENT

This study used a steel screw pile having a pile shaft diameter (D_s) of 21.7 mm, helix diameter (D_h) of 43 mm, and length of 660 mm, as shown in Fig. 1. The end condition of the central shaft of the screw pile was open and close-end (flat). The thickness of the helical plate was 1.75 mm, and the helix pitch was 19.7 mm. A cylindrical steel container having a diameter of 1000 mm, and a depth of 1100 mm was used to prepare the model ground for pile installation and pile load tests (Saleem et al. 2020). In model scale testing, the size of the container with respect to pile diameter is critical. Test results can be significantly affected if the soil influence zone due to the pile installation intervened with the container walls.

As per previous studies, the zone where the soil is affected by loading is usually 3-8 times the pile tip diameter (Kishida, 1963; Robinsky et al. 1964). The zone of influence around the pile suggested by Randolph and Wroth (1978) is 0.9-1.4 times the length of the pile. Further, for the piles in clean sand, the influence zone above the pile tip is between 1.5 and 2.5D, and the zone below the pile tip ranges from 3.5 to 5.5D (Yang, 2006). Previous studies (Malik et al. 2019, 2020; Ho et al. 2021, 2022) showed that the container diameter of 592 mm is sufficient for a 43 mm helix diameter to avoid boundary effects on load settlement response. In the present study, the diameter of the cylindrical container was 1000 mm, and the depth was 1100 mm, as shown in Fig. 2. The clearance around the pile was about 478-489 mm (11-22 times the pile diameter, 43 mm), and clearance below the pile tip was 700 mm (16-32 times the pile diameter, 43 mm). Dry Toyoura sand was used to prepare the model ground. The relative density of the dense bearing layer (layer 1 to 7, i.e., bottom 700 mm) was kept at about 70 %. In contrast, the top three layers (layer 8-10, i.e., top 300 mm) were prepared as loose-to-medium dense ground with a relative density of about 45 %. The properties of Toyoura sand are as follows; specific gravity = 2.65, D_{50} (50% pass particle size) = 0.20 mm, e_{max} (maximum void ratio) = 0.98 and e_{min} (minimum void ratio) = 0.60. The displacement-controlled pressing system was used for pile penetration, and the rotation system was used to rotate the pile during its installation. A load/torque cell was used to monitor the installation effort (load and torque) required for pile driving during pile installation and pile load test. A displacement transducer was used to monitor the penetration of the pile during pile installation and pile load test.

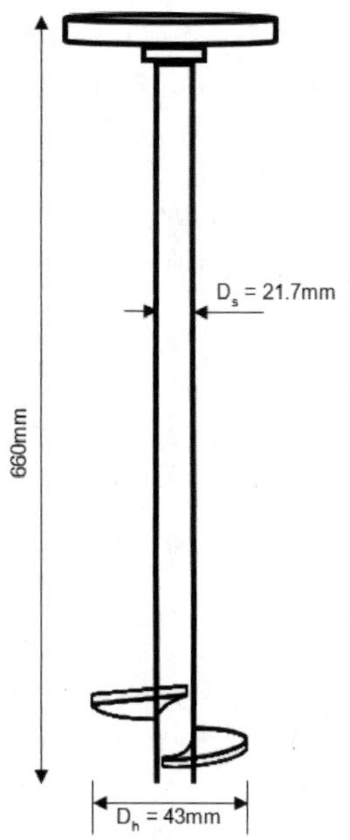

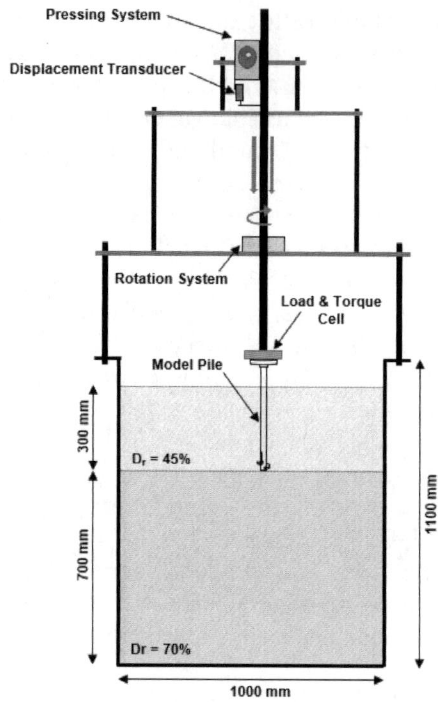

Fig. 1. Schematic diagram of model screw pile

Fig. 2. Schematic illustration of testing equipment

TESTING METHODOLOGY

Model Ground Preparation

At first, the ground was prepared in the model container in layers, each having a thickness of 100 mm after compaction by a handheld rammer. In test series 1, the model ground was prepared first up to 10 layers (bottom 7 layers at 70 % relative density, top 3 layers at 45 % relative density). Then pile (open/close-end) was installed from the surface of the model ground, as shown in Fig. 3. In test series 2, the model screw pile (open/close-end) was placed on the model ground after the compaction of the 7th layer (700 mm). Loose-to-medium dense sand (45 % relative density) was placed around the pile in the remaining three layers (Layers 8, 9, 10), as shown in Fig. 4.

Model Pile Installation

In test series 1, the screw piles with open and close-end shaft conditions were installed from the top surface of the model ground by simultaneous pressing and rotation, keeping the penetration (S) to helix pitch (P_h) ratio equal to 1 (refer to Fig. 3). In test series 2, the screw piles with open and close-end shaft conditions were installed in a dense bearing layer (penetration of one time the helix diameter, $1D_h$) by pressing and rotation (refer to Fig. 4). The penetration (S) to helix pitch (P_h) ratio (S/P_h) was equal to 1.

Pile Load Test

After installing the piles in both test series, pile load tests were conducted. The pile load test used a 2 mm/min load rate. Load on the pile and pile settlement was recorded until a plunging resistance state (pile settlement increases without an increase in load) was observed. Figure 5 shows the schematic illustration of the pile load test setup for a single helix screw pile. In the case of an open-end screw pile, the plugging length was checked after the pile load test.

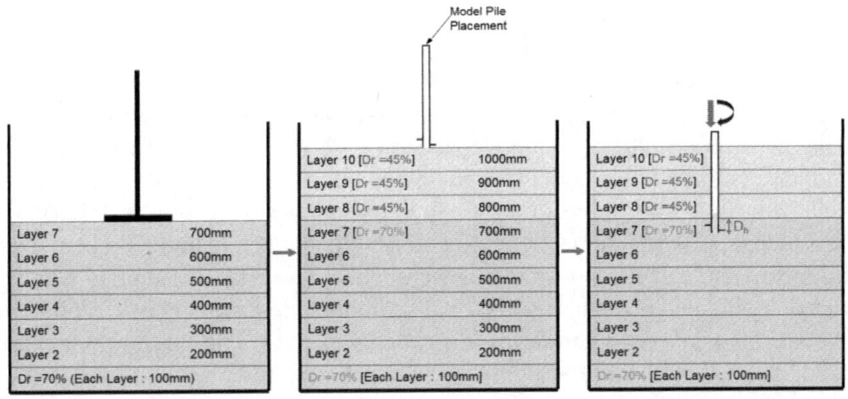

Fig. 3. Test Series 1, schematic illustration of model ground preparation and pile installation

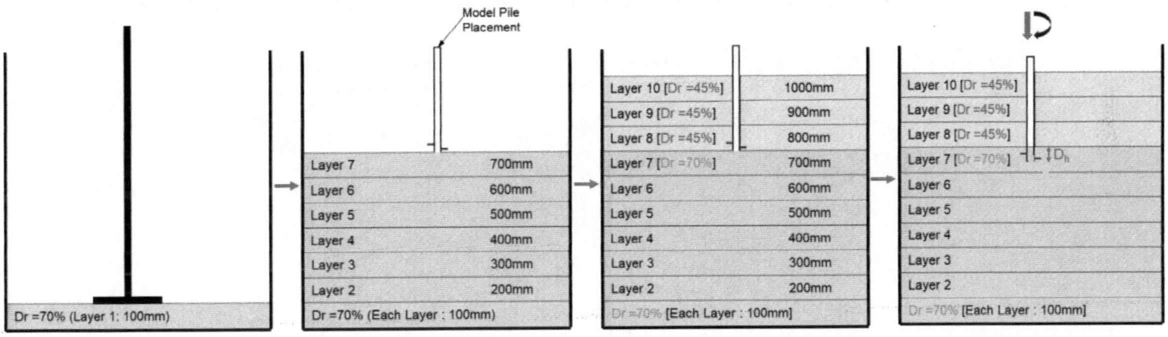

Fig. 4. Test Series 2, schematic illustration of model ground preparation and pile installation

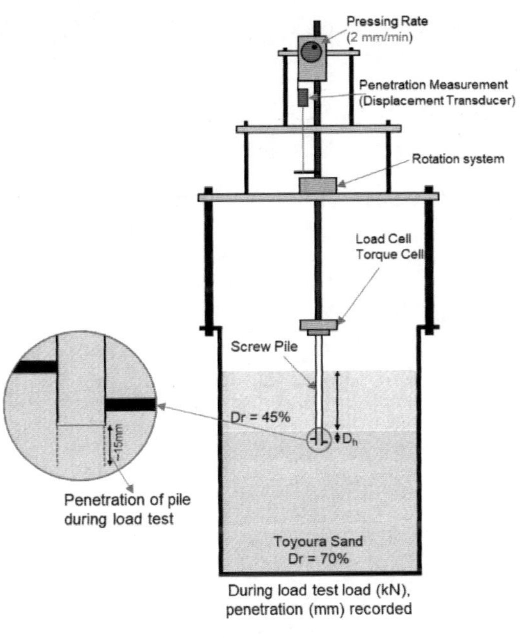

Fig. 5. Schematic illustration of pile load test setup

TEST RESULTS

Test Series 1: Pile Installation from Top of Model Ground

In these tests, open and close-end screw piles were installed from top to 1D_h depth into the bearing layer. Most previous studies with open and close-end conditions are on steel pipe piles, and limited data is available for screw piles. As stated by previous studies, open-end steel pipe pile behavior is quite similar to close-end pipe pile at greater depth due to continuous plugging during pile installation; this series of tests were conducted to check the statement mentioned above for screw pile. Figure. 6 shows that the installation load and torque are quite similar for open-end and close-end screw piles when installed from the surface of the model ground. In addition, the load settlement response is also quite similar for both types of piles, as shown in Fig. 7. This similarity in installation effort and bearing response is due to the plugging effect that originated at the initial penetration depth, resulting in a similar end condition, i.e., close-end when entering the bearing layer. The plug length ratio ($PLR = \frac{h}{H}$), which is the ratio between soil penetrating the central shaft (h) over total penetration depth (H) is 0.15, indicating a close to fully plugged condition. The soil height entering the central shaft was checked after the pile load test.

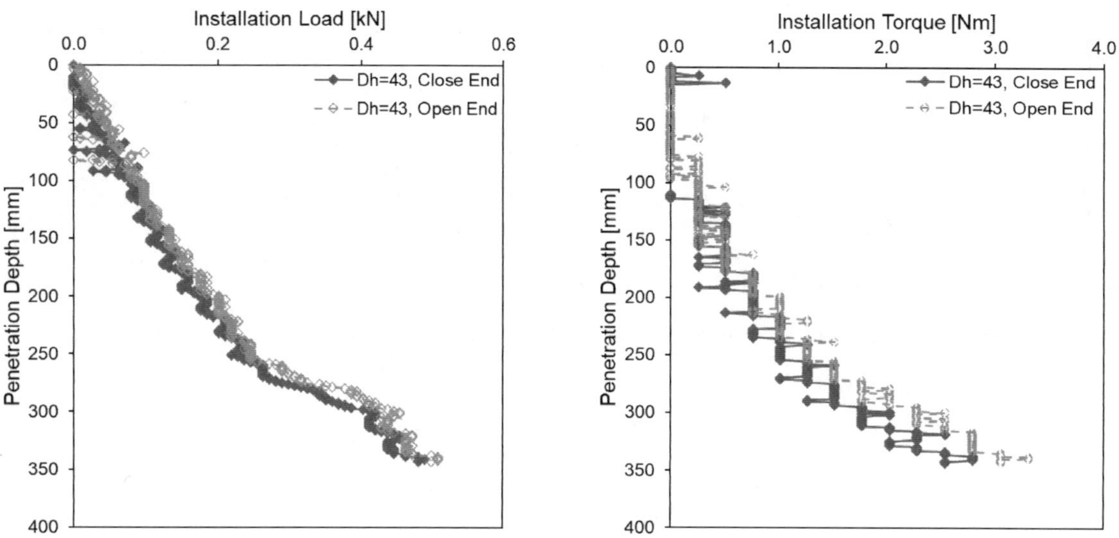

Fig. 6. Installation effort comparison for open and close-end screw piles installed from the top

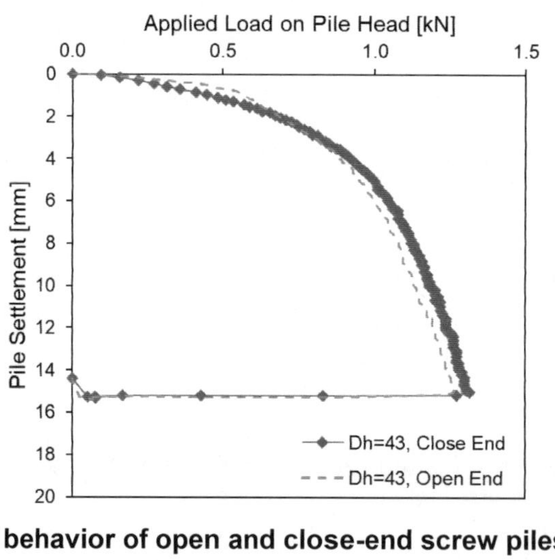

Fig. 7. Load-settlement behavior of open and close-end screw piles installed from the top

Test Series 2: Pile Installation into Bearing Layer Only

In these tests, an open or close-end screw pile was placed on the top of the 7th layer, and then the top three layers were placed around the model pile in a loose-to-medium dense state. The pile was installed into the bearing layer only by $1D_h$ (43mm) depth, as suggested by Ikeda et al. (2010), to achieve a high ultimate resistance. It is observed that the installation torque is quite similar for both types of piles; however, the open-end screw pile showed less installation load than the close-end screw pile, as shown in Fig. 8. This reduction in installation load is due to less resistance from the ground underneath the central shaft, and this is because soil entered the central shaft. Moreover, the load settlement response of the open-end screw pile is better than the close-end screw pile, as shown in Fig. 9. The plug length ratio ($PLR = \frac{h}{H}$) is 0.79 indicates a close to unplugged condition.

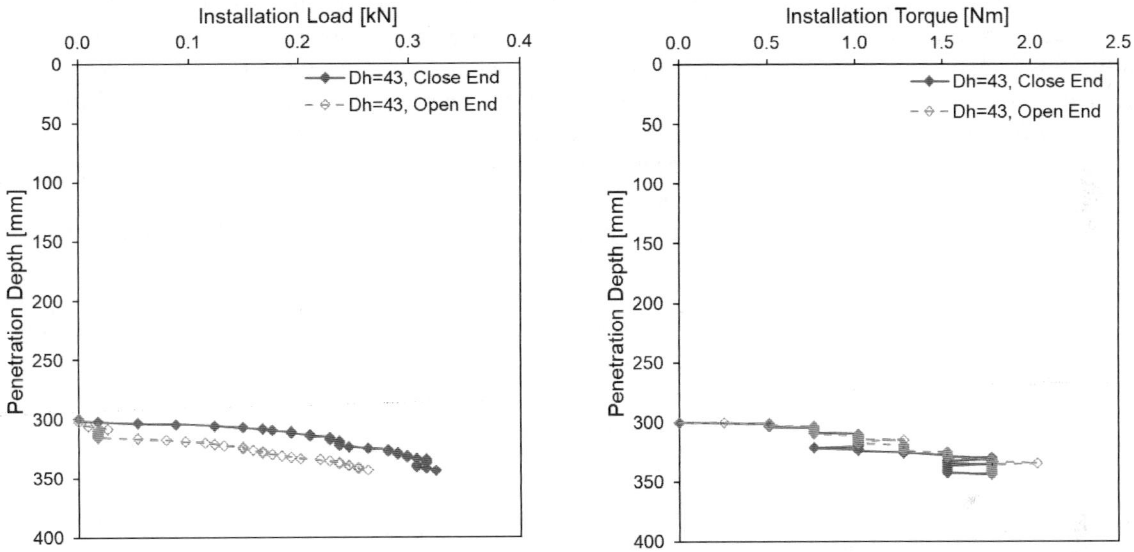

Fig. 8. Installation effort comparison for open and close-end screw piles installed in bearing layer only

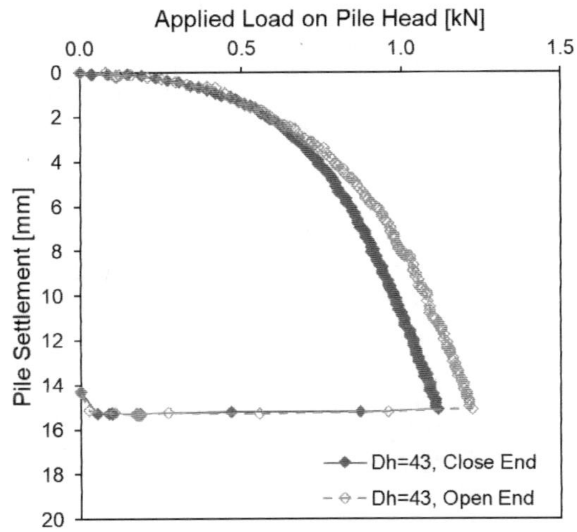

Fig. 9. Load-settlement behavior of open and close-end screw pile installed in bearing layer only

SUMMARY AND CONCLUSIONS

In this study, the effect of the open and close-end condition of a screw pile on the installation effort (installation load and installation torque) and bearing response was investigated for two different test series. A model screw pile with a pile shaft diameter (D_s) of 21.7 mm and helix diameter (D_h) of 43 mm was used. Dry Toyoura sand was used to prepare the model ground.

In test series 1, a screw pile was installed in a dense bearing layer from top to $1D_h$ into the bearing layer. Test results confirmed that the behavior of an open-end screw pile is quite similar to a close-end screw pile at a larger depth because of soil plug formation inside the open-end screw pile shaft during the initial stage of installation, resulting in similar end conditions, i.e., close-end when entering the bearing layer.

In test series 2, the screw pile was installed by $1D_h$ in a dense bearing layer only. Test results showed that the performance of the open-end screw pile is better compared with the close-end screw pile in terms of its installation effort and bearing response. During the installation, the open-end pile behaves as a replacement pile because of soil moving inside the pile shaft. Thus, lesser resistance is offered from the ground beneath for an open-end pile, resulted in a lesser installation load to install the pile. During a pile load test, the open-end pile shows a higher bearing response than the close-end screw pile. As the pile penetration depth is just $1D_h$ (43mm), the open-end pile less disturbs the ground underneath the helix because soil enters the central shaft during penetration. As helix contribution in bearing response is quite significant, and in the open-end condition, the soil is less disturbed under the helix, the higher bearing response is observed. Whereas, in a close-end screw pile, the central shaft pushes the soil in a radial upward direction that causes more ground disturbance underneath the helix, which results in a lower bearing response for a close-end screw pile. To reinforce this outcome, more experimental testing is recommended.

REFERENCES

Carter, J.P., Randolph, M.F., and Wroth, C.P. 1979. Some aspects of the performance of open and closed-ended piles. Numerical method in offshore piling, London, 165–170.

Ho, H.M., Malik, A.A., Kuwano, J., & Rashid, H.M.A. 2021. Influence of helix bending deflection on the load transfer mechanism of screw piles in sand: Experimental and numerical investigations, Soils and Foundations, 61, 8740-885.

Ho, H.M., Malik, A.A., Kuwano, J., Brasile, S., Tran, T.V., & Mazhar, M.A. 2022. Experimental and numerical study on pressure distribution under screw and straight pile in dense sand, International Journal of Geomechanics, 22(9), 0422139-1-10.

Ikeda, A., Tsuchiya, T., Shimada, M., and Nakazawa, F. 2010. The effects of partial embedment of a helical screw pile into a bearing layer on its penetration and load-settlement behavior, J. Struct. Constr. Eng., AIJ, 75(651), 951-956.

Kishida, H. 1963. Stress distribution by model piles in sand. Soils and Foundations, 4(1), 1-23.

Klos, J., and Tejchman, A. 1981. Bearing capacity calculation for pipe piles. Proc., 10th Int. Conf. on Soil Mechanics and Foundation Engineering, Stockholm, Sweden, 2, 751–754.

Kumar, R., & Kumawat, V.K. 2016. Study on open and closed end pile foundation (comparative study in open and closed end pile foundation). International Journal of Engineering Research & Technology, 4(23), 1-4.

Lu, S.S. 1985. Pile driving practice in China. Proc., Int. Symposium on Penetrability and drivability of piles, San Francisco, 1, 13−20.

Li, L., Wu, W., Liu, H., and Lehane, B. 2021. DEM analysis of the plugging effect on open-ended pile during the installation process. Ocean Engineering, 220.

Malik, A.A., Kuwano, J., Tachibana, S., & Maejima, T. 2017. End bearing capacity comparison of screw pile with straight pipe pile under similar ground conditions. Acta Geotechnica, 12(2), 415-428.

Malik, A.A., Kuwano, J., Tachibana, S., & Maejima, T. 2019. Effect of helix bending deflection on load settlement behaviour of screw pile. Acta Geotechinca, 14, 1527-1543.

Malik, A.A., & Kuwano J. 2020. Single helix screw pile behavior under compressive loading/unloading cycles in dense sand. Geotechnical and Geological Engineering, 38, 5565-5575.

Paikowsky, S.G., and Whitman, R.V. 1990. The effect of plugging on pile performance and design. Can. Geotech. J., 27, 429–440.

Paik, K., Salgado, R., Lee, J., and Kim, B. 2003. Behavior of open- and closed-ended piles driven into sands. Journal of Geotechnical and Geoenvironmental Engineering, 129(4), 296-306.

Paik, Kyuho, and Rodrigo Salgado. 2003. Determination of bearing capacity of open-ended piles in sand. Journal of Geotechnical and Geoenvironmental Engineering 129.1 (2003): 46-57.

Robinsky, E.I., & Morrison, C.F. 1964. Sand displacement and compaction around model friction piles. Canadian Geotechnical Journal, 1(2), 81-93.

Randolph, M.F., & Wroth, C.P. 1978. Analysis of deformation of vertically loaded piles. Journal of the geotechnical engineering division, 104(12), 1465-1488.

Randolph, M.F., Steinfelt, J.S., and Wroth, C.P. 1979. The effect of pile type on design parameters for driven piles. Proc., 7th European Conf. on Soil Mechanics, British Geotechnical Society, London, 2, 107−114.

Saleem, M.A., Malik, A.A., & Kuwano, J. 2020. End shape and rotation effect on steel pipe pile installation effort and bearing resistance. Geomechanics and Engineering, 23(6), 523-533.

Szechy, C.H. 1961. The effect of vibration and driving upon the voids in granular soil surrounding a pile. Proc., 5th Int. Conf. on Soil Mechanics and Foundation Engineering, Paris, 2, 161−164.

Smith, I.M., To, P., and Wilson, S.M. 1986. Plugging of pipe piles. Proc., 3rd Int. Conf. on Numerical Method in Offshore Piling, Nantes, France, 53−73.

Xu, M., Ni, P., Mei, G., & Zhao, Y. 2018. Load-settlement behavior of bored piles with loose sediments at the pile tip: Experimental, numerical and analytical study. Computers and Geotechnics, 102, 92-101.

Yang, J. 2006. Influence zone for end bearing of piles in sand. Journal of geotechnical and geoenvironmental engineering, 132(9), 1229-1237.

DESIGN, INSTALLATION AND PERFORMANCE OF EARTH RETENTION SYSTEM TO LIMIT GROUND MOVEMENTS AROUND SENSITIVE STRUCTURES IN WASHINGTON, DC

Joanna Mason, Schnabel Geostructural Design & Construction, Bethesda, MD, USA, 301-657-3060, joanna.mason@schnabel.com

Stanley L. Worst, Schnabel Geostructural Design & Construction, Cary, IL, USA, 847-639-8900, stanley.worst@schnabel.com

Douglas R. Hardin, Schnabel Geostructural Design & Construction, Cary, IL, USA, 847-639-8900, doug.hardin@schnabel.com

ABSTRACT

A challenging earth retention project presented itself in Washington, DC in the backyard of the historic House of the Temple (The Scottish Rite of Freemasonry, Supreme Council 33°). Located at 15th St NW and S St NW, a forty-foot-deep underground basement was to be installed adjacent to the existing Temple, a historic Carriage House structure dating back to the 1800's and across a narrow alley from a four-story brick apartment building. The unique geology of the site included typical urban fill overlying a mixture of loose alluvial soils underlaid by dense marine deposits of the Potomac Formation. The clay, sand and gravel of the Potomac included nested boulders up to three feet in diameter. Design of the earth retention system accounted for ground conditions, installation methods and estimated movement of the system and adjacent structures. Anticipated ground water and unanticipated water from leaky sewers also proved obstacles to overcome during construction. Soldier beams were installed using a variable moment-vibratory hammer to minimize disturbance to the historic and sensitive structures. Monitoring of the existing structures included both seismograph and optical targets during installation of soldier beams and as the excavation progressed. Micropiles drilled into the Potomac Formation were a part of the permanent foundation for the new structure. This local case history will delve into the details of the successful design, installation, and performance of an earth retention system and micropiles in a historic and highly congested urban environment.

KEYWORDS

Potomac Formation, Earth Retention, Washington DC, Micropiles

PROJECT BACKGROUND

Schnabel Geostructural Design & Construction (Schnabel) was hired to design and install the Temporary Earth Retention System (ERS) and Permanent Foundation Piles for the Scottish Rites Apartments (SRA) Project. Located directly behind the existing House of the Temple (HOT) of the Scottish Rite, the new apartment complex will have a tiered two-story below ground parking garage. The historic one-story brick Carriage House was also on-site and within proximity of the excavation.

GEOTECHNICAL INFORMATION

Located approximately half a mile east of DuPont Circle and one mile north of the White House, the site soil profile was typical for northern downtown Washington, DC. Figure 1 shows a geologic map of the area. Upper soils consist of man-made fill in addition to alluvium deposits. The alluvial terrace deposits are made up of loose to medium dense sands, silts, and gravels and are approximately twenty-five feet in thickness. The soil borings show blow counts in the single digits in the fill and river deposits. Below the alluvial layer are marine deposits from the Potomac Formation, Cretaceous in age. These marine deposits are made up of stiff to hard clay and medium dense to very dense sand, and gravel. Soil borings show that at a depth of approximately forty feet the blow counts exceed 50-blows per six inches. This high blow count zone includes gravel, clay, and sand in addition to occasional boulders. The groundwater table is roughly at the Alluvium-Potomac interface and varies throughout the year.

The soil profile described above makes for a challenging ERS installation. The vertical component of the ERS must penetrate the dense, abrasive Potomac Formation. The abrasiveness of the sands and gravels results in excessive wear to drill tooling. Cobbles and boulders obstruct drilling and installation of ERS components. As the excavation progresses, loose sands and silts are at risk of running into the excavation during placement of lagging. Groundwater must be dealt with through dewatering, yet fine grained soils limit the effectiveness of a dewatering program.

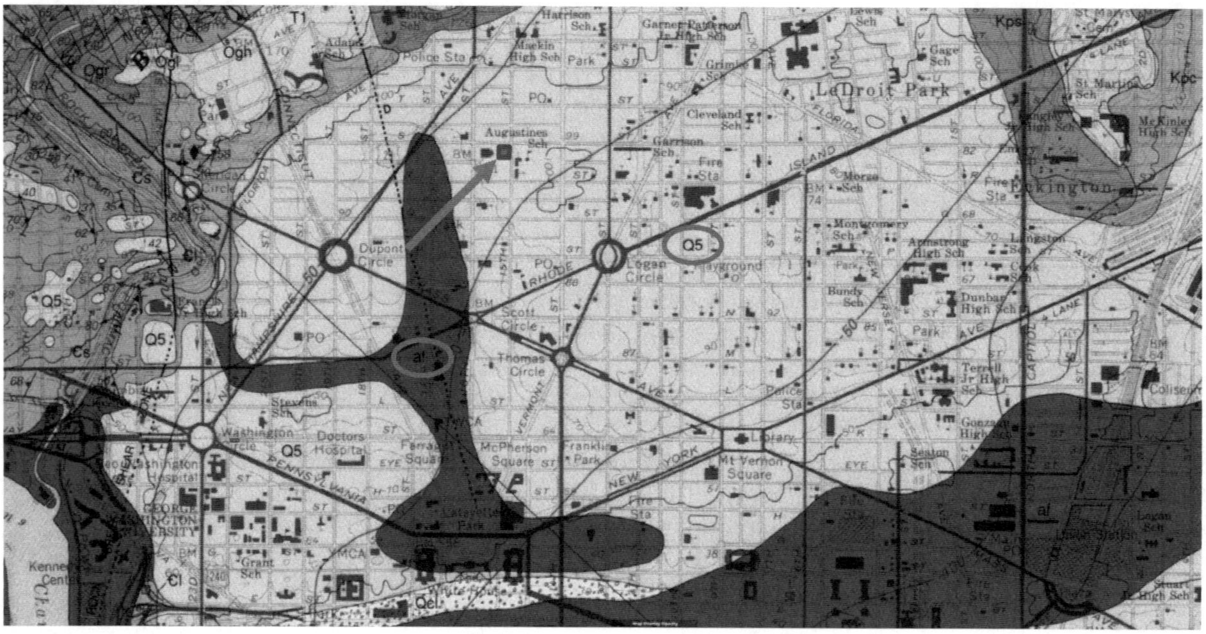

Fig. 1 - Geologic Map of Downtown Washington D.C.

DESCRIPTION OF MAP UNITS

Q5 **Gravel, sand, silt, and clay (middle Pleistocene)**—Gray to gray-brown; crudely to well bedded. Found mainly beneath an irregular surface between 40 and 105 ft in elevation that extends northward from the White House for about 2 miles. Swamp deposits with "dead" ^{14}C occur in this unit in Lafayette Park (Ives and others, 1967; Knox, 1969). Q5 mineralogy is similar to that of units Q3 and Q4 except that unit Q5 is more deeply weathered. Weathered sediment has been stripped from the surface by natural or artificial means in many places, so relatively fresh minerals are exposed. Unit Q5 typically 10–20 m thick. Unit Q5 correlates with the Chicamuxen Church Formation of McCartan (1989b, c) in Charles and St. Marys Counties, Maryland, and with Charles City and Chuckatuck Formations in Virginia (Mixon and others, 1989). Age estimated to be 450 ka on basis of stratigraphic position and possible correlation with marine units in North Carolina.

af **Artificial fill**—Heterogeneous composition and texture. Largest areas of low-lying land filled for building purposes, determined from examination of historical topographic maps and a report and map by Froelich and Hack (1975). Most land surface in the urban area has been cut or filled artificially.

EARTH RETENTION SYSTEM

Scope of Work

The project required an Earth Retention System to safely excavate to the basement subgrade of a new apartment building. The ERS was installed in tandem with a dewatering system designed and installed by another contractor. The Scottish Rites Apartments has two different subgrade elevations, separated by an intermediate earth retention system. The upper subgrade is approximately fifteen feet below ground while the lower subgrade is approximately forty feet below ground.

Adjacent Structures

The House of the Temple (HOT) of The Scottish Rites, a one-story brick Carriage House, and a privately owned Four-Story Brick Townhouse (Campbell Townhouse) were all considered adjacent structures to the excavation as shown in Fig. 2 and Fig. 3.

Fig. 2 - House of Temple and Carriage House

Fig. 3 - Adjacent Private Townhouse

The House of the Temple is an historic structure built in 1915 and founded on massive cast in place concrete footings. Surrounding an 80-foot tall, bifurcated dome, this structure was deemed very sensitive to any movement from construction activities. The Northwest corner of the HOT was damaged in the 2011 earthquake and thus engineers would not permit the structure to experience any construction-related vibrations. The HOT is located outside of the excavation zone of influence of the ERS but within twenty-five feet of the soldier beam installation. In addition to vibration impacts, there were concerns that differential settlement due to dewatering drawdown at the project site would negatively impact the HOT.

The Carriage House is a one-story brick structure located approximate eight feet away from the 40-foot deep ERS. This historic structure was built in the 1800's and investigations showed it did not have a typical continuous spread footing, but rather a continuation of the exterior brick wall below grade served as the building footing. Project plans indicated that a new strip footing would be constructed in sections to carry additional loading of new construction. Those plans were not finalized at the date of this submission. The Carriage House is to be incorporated into the new Scottish Rites Apartment Building with interior steel façade reinforcement and new footings constructed to support three additional levels. During the construction of the ERS, the Carriage House was gutted to accept the new interior structure. The exterior brick was also pressure washed, exposing mortar joints and existing cracks. This exterior cleaning allowed for easier pre-construction documentation of existing conditions. Figure 4 shows cracking around windows prior to the new construction. This cracking pattern can be misconstrued as resulting from excavation induced movements and differential settlement, making it important to document before significant construction work occurs.

Fig. 4 – Pre-Existing Cracks at Carriage House

The Four-Story Brick Townhouse (Campbell Townhouse) to the south of the project site was built in 1925 with a 1-story below grade basement in addition to three stories above grade. The Campbell Townhouse is a privately owned structure across a 15-foot-wide public alley. It was assumed that the Campbell Townhouse is founded on a typical brick structure on an unreinforced concrete footing. The interior preconstruction surveys indicated this structure was in good condition without major cracks or deterioration prior to start of construction. The exterior of the townhouse was brick with minor cracks in the mortar joints. The HOT, Carriage House and Campbell Townhouse vibration and movement thresholds were as follows:

Table 1 – Existing Structure Thresholds

Location	Vibration	Movement / Settlement
House of Temple	0.5 in/sec	1/8 inch
Carriage House	0.5 in/sec	1/4 inch
Townhouse	0.5 in/sec	3/8 inch
Elsewhere	2.0 in/sec	0.4% H

All limits of structures within the zone of the influence of the excavation were determined by limiting acceptable strain that would not cause structural damage to the adjacent structures. The limits determined for the ERS are determined based anticipated movement for the height of excavation (H).

Design Basis and Concepts

The ERS chosen for this project consisted of driven soldier beams, wood lagging, and tiebacks. This type of ERS is considered a relatively flexible shoring system, as compared to a stiffer concrete secant pile or slurry wall. Flexible systems allow for the small amount of movement required for active earth pressures to develop. Consistent with the derivation of apparent earth pressure diagrams from active pressures,

Schnabel used a trapezoidal loading as described in the book, "Tiebacks in Foundation Engineering and Construction," to design the ERS.

The HP12x53 and HP14x73 soldier beams were driven at a typical eight-foot center-to-center spacing with lagging placed behind the front flange. The wood lagging acts to brace the soldier beam laterally in the plane of the wall. The maximum midspan moment was designed assuming a hinge at subgrade for the soldier beams.

Tiebacks with design loads ranging from 69 Kips to 176 Kips were designed based on our 60 plus years of experience in the area. Regroutable anchors were drilled with a 6" diameter cased drill hole in the silty sands. Although granular material can sometimes be grouted under pressure and achieve a higher bond stress than gravity grouting, the silt content on this project lent itself to using regrouts to ensure capacity was achieved. Grout takes and pressures achieved during regrouting of the tieback bond zones were consistent and indicative of achieving the design bond stress. Regrouting a tieback bond zone is never a guarantee of a higher capacity, but typically results in a larger grout body, higher normal stresses acting at the interface of the ground and the grout body or larger asperities or bulbs of grout, increasing the overall capacity. A bond stress of 25 psi was used to design the bond lengths on the project. As is typical for tiebacks, every tieback on this project was tested to at least 120% of the design load. Based on test results, the 25psi used for bond length design was conservative, and a higher bond stress was achieved. The graph of a performance test in Fig. 5 shows that only a percentage of the bond length was used.

Figure 5 is a plot of the load versus movement during the tieback load test. The upper line indicates the theoretical elongation of the structural steel member of the tieback if all the load was transferred at the start of the bond length. The lower line indicates the same elongation if all the load was transferred at the bottom of the bond length. It is typical for the data to plot in between the lines, which shows visually that only part of the bond length is being used. A further analysis per the Post Tensioning Institute's Manual for Prestressed Rock and Soil Anchors leads to the conclusion that a bond stress of over 40% higher than used for design was achieved.

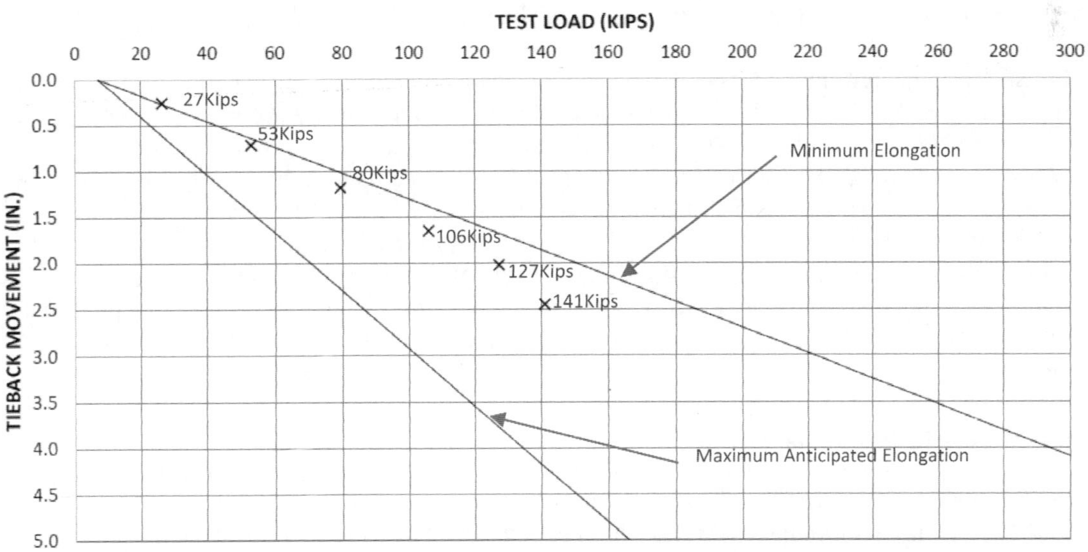

Fig. 5 - Tieback Test Graph of Load vs. Movement for 115 Kip Design Load

Away from adjacent structures, the design vertical and lateral movement thresholds were typical for temporary ERS in the area, with 0.004H as the anticipated vertical and lateral movement. During excavation, only 0.5 inches of lateral soldier beam movement was recorded in the monitoring data. This corresponds to approximately 0.001H to 0.0015H based on the depth of excavation.

Based on the settlement thresholds for the Carriage House and the Campbell Townhouse, a movement-based design was checked against the typical apparent earth pressure diagram design. A semi-empirical method after Cording (1981) was used to evaluate the excavation induced movement of the ERS. Figure 6 shows the relationship between normalized volume of lateral wall displacement (y-axis) and the logarithm of soil/wall stiffness (x-axis). The variables in Figure 6 are as following:

V_L = Lateral volume of displaced soil (in^3/in)
E_S = Soil modulus (psi)
ΔH = Average Vertical Strut Spacing (ft)
H = Height of Excavation (ft)
γ = Unit Weight of Soil (pcf)
L = Average Vertical Excavation prior to installing next Strut (ft)
I = Moment of Inertia of the Wall (in^4)

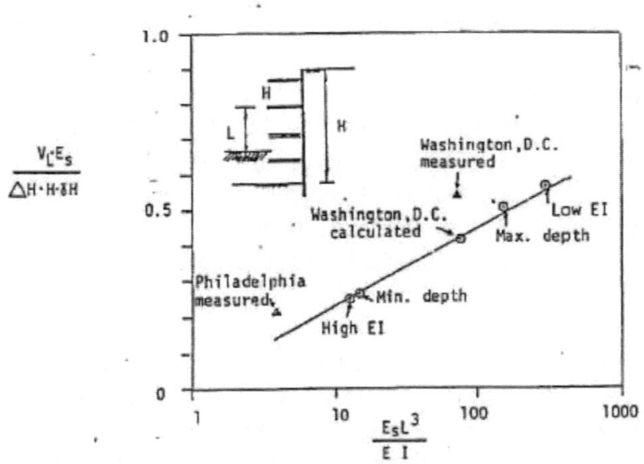

Fig. 6 - Relationship Between Normalized Volume of Lateral Wall Displacement and the Logarithm of Soil/Wall Stiffness

As shown in Figure 6, this method emphasizes the soil modulus and the vertical distance between lateral supports. This leads to the conclusion that stiff or dense ground will limit movement of the ERS regardless of the systems flexibility, and the most influencing factor for reducing movement of the ERS is the number of tiers of lateral support, or the distance between them. Adding an additional tier of tiebacks or braces has a greater influence on limiting movements of the ERS than reducing the soldier beam spacing or using a larger section for soldier beams.

The movement-based design method assumes good workmanship and limited loss of ground during the installation of the ERS. For the movement-based design method to provide valid results, sound construction procedures which limit loss of ground and excessive ground movements need to be used during construction of the ERS. The method of soldier beam installation can greatly influence overall movements if ground is lost during drilling, or if vibrations consolidate soils during driving. Lagging installation always has the potential to allow for ground loss if not performed under control and with the consideration of the soil conditions. The granular soils at this project site had enough silt and clay content that a five-foot vertical lift could stand up without unravelling. Drilling tiebacks has the potential to allow for ground loss if positive pressure is not kept inside the hole or if full depth casing is not used and the drillhole collapses.

Installation Methods – Soldier Beams

Initially, soldier beams were to be installed with a drill-and-set method due to the concern over vibrations. After discussion of the advantages with using a variable moment hammer, it was agreed that all soldier beams could be driven if continuous vibration monitoring was performed. A variable moment hammer minimizes the change in frequency of the vibrations at start-up and shutdown of the equipment, limiting the spikes in vibration levels seen at these times with typical vibratory hammers. The adjacent buildings were all monitored using Automated Vibration Monitoring. At the HOT building, vibrations were minimal, with readings of 0.01 to 0.05 in/sec recorded at distances of 50 to 100 feet from the soldier beam operation. At the Carriage House, which was only 8 feet away, vibration readings stayed below the 0.5

in/sec limit during installation with one exception. While installing a few different soldier beams, the vibration limit spiked to a value of 1.2 in/sec for a short duration when the beams encountered a very dense layer approximately two feet thick at a depth of approximately 35 feet. When this occurred, the existing cracks were monitored real-time, and installation proceeded while closely observing the penetration rate of the beam and the existing conditions of the structure. The Townhouse, approximately 19 feet away, also experienced similar vibrations and monitoring, with a peak value of 1.0 in/sec recorded. There were four instances at the Townhouse where values exceeded 0.5 in/sec and it was collectively decided to proceed under close observation of existing cracks and beam penetration rate. Overall, there were no new cracks that developed on the existing structures, and existing cracks did not propagate or widen.

Difficulty was encountered while driving the soldier beams into the very dense layer approximately 35' to 40' below existing grade. Based on local experience in the area, this layer was likely a cemented sand at the bottom of the river deposit. The soldier beams had driving shoes installed on them and were driven until the vibration thresholds were met. After being allowed to sit overnight, the beams were able to be driven to their final depth with less difficulty and lower vibrations the next day. One possible explanation for this phenomenon is that ground water was able to percolate along the sides of the soldier beam and into this layer from above and loosen the cemented sands. Cemented sands, cobbles and boulders up to three feet in diameter were confirmed during the excavation to grade.

Installation Methods – Tiebacks

The tiebacks were installed using double head rotary drilling and a down the hole hammer where necessary to penetrate the gravel, cobbles, and boulders at depth. The sands and gravels were extremely abrasive, and the casing shoes were replaced every few days. Compressed air was used as the flushing medium. Water was added to the compressed air stream as needed to help remove softer cohesive deposits and to equalize water encountered in the granular layers.

Initial grouting was not under pressure and was pumped through a tremie tube inserted to the bottom of the hole. After the initial grout installation, the anchors were regrouted under pressure through valves located throughout the bond length the following day to increase capacity. Most of the tiebacks only accepted a few bags of grout.

Five percent of the tiebacks were performance tested to 133% of the design load, while the remaining tiebacks were proof tested to 120% of the design load. The load carrying capacity was very high, and minimal creep, on the order of 0.005 inches, was observed during testing, which is indicative of granular soils.

Earth Retention System Performance

The soldier beams and adjacent buildings were monitored using Automated Optical Prisms checking location in the X, Y, and Z directions. The soldier beams and adjacent buildings all stayed within the threshold with very little lateral and vertical movement. Values of up to 0.5 inches of lateral movement at the top of the soldier beams were typical on this project. This relatively small amount of movement can be attributed to the movement-based design, the good workmanship during construction, the equipment used during installation, and the relative density and stiffness of the ground itself.

Figure 7 shows three tiers of tiebacks for a 41-foot-deep excavation in front of the Carriage House. This tieback configuration limits the vertical distance between lateral support levels to approximately 12 feet. Had only two tiers been installed, the vertical distance between lateral support levels would have approached 17 feet. The concept of the movement-based design is that the cube of the vertical distance between support levels is related to the amount of lateral movement of the system. Thus, adding the third tieback tier to reduce the vertical distance between lateral support levels from 17 feet to 12 feet reduced the estimated movement of the ERS by almost a factor of three. The decision was made at the Carriage House and Townhouse to use three tiers of tiebacks to minimize potential movement of the ERS which in turn limits settlement of adjacent existing structures.

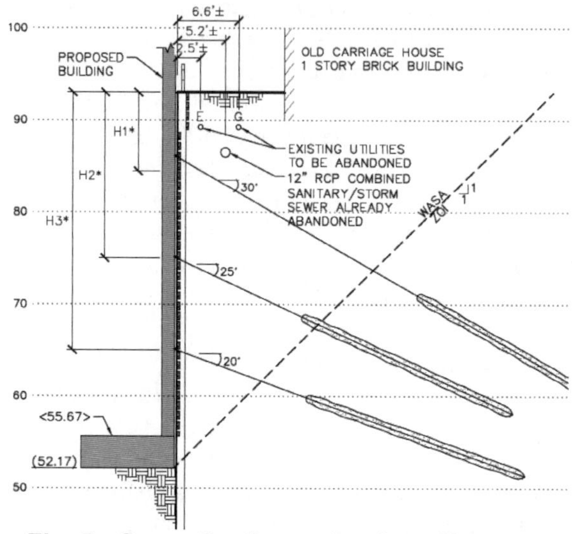

Fig. 7 - Cross Section at Carriage House

Soldier beam settlement was minimal, with most soldier beams experiencing less than 0.25 inches of settlement. This magnitude of soldier beam settlement shows the performance of the vertical toe resistance in addition to the wall friction acting above subgrade. Soldier beams which experienced more than 0.25 inches of settlement were in an area where a leaky existing sewer was encountered. Existing building movement was also negligible. The HOT and the Townhouse did not experience any settlement or tilt. The Carriage House experienced up to 1/32 inches of settlement. No creep movement was noted by the excavation monitoring data. Of most importance is that no visible signs of stress in the structures themselves were observed.

MICROPILES

Design Basis and Testing

The structural drawings specified 28 micropiles spread out over five different pile caps with a design working load capacity of up to 171 Kips in compression. This load was designed to be carried by a solid bar tendon in the center of a seven-inch diameter hole. Permanent casing was used in the upper 20 feet of the micropile. The permanent casing was used to locate the bond zone of the micropile below the lowest subgrade elevation for the project since the location of the micropiles was very near the intermediate ERS located at the change in subgrade elevations interior to the new building. The total length of the micropiles were 20’ permanent casing/free zone plus a 25’ bonded zone for 45’ total drilled length. The micropile bar extended up into the footings 4”.

After performing over 175 tieback tests, we were confident in the load carrying capacity of the soils, but only within a certain vertical stratum. The micropiles were to be bonded in soils at a greater depth than where the tiebacks were tested. A preliminary bond stress test was conducted to aid in the design of the micropile bond length. An eight strand tendon was installed in a vertical drill hole and tested in tension to its maximum structural capacity to determine a bond stress over an installed 25-foot bond length. The maximum test load for the 8-strand tendon was greater than twice the micropile design load. A bond stress of 66 psi was achieved.

Based on this preliminary test, an allowable bond stress of 30 psi was assumed for micropile bond design and a bond length of 25 feet was used for the micropiles. In accordance with ASTM standard D3689, a non-production micropile load test was also performed. This test was a tension test used to check that the bar tendon in the micropile bond zone had a capacity of at least twice the micropile design load.

Installation Method

Micropiles were installed using the same method used to install the tiebacks. As with the tiebacks, a down the hole hammer was used to penetrate through cobbles and boulders. The day after initial tremie grouting, one stage of regrouting was performed on each micropile, similar to the tiebacks.

SUMMARY AND CONCLUSIONS

The soils at the project site consisted of two main layers – the upper 30 feet of the excavation consisting of fill and alluvial river deposits of silt, sand, clay and gravels and the underlying dense marine deposits of the Potomac Formation consisting of sands, silts, and clays, cobbles, and boulders. Although challenging from both a design and installation perspective, the ERS was successfully installed without causing damage to the multiple adjacent sensitive existing structures. This was achieved utilizing a movement-based design, equipment that minimized vibrations and quality workmanship which did not allow for ground loss. Micropiles were successfully installed in the dense, cobbly soils to serve as permanent foundation elements. The relatively small drill diameter and ability to use a down the hole hammer made installation through the cobbles and boulders possible. The preliminary load test proved that the load transfer rate was higher than published data, allowing for efficiencies in both design and construction.

ACKNOWLEDGMENTS

We would like to acknowledge Harvey-Cleary Builders as the general contractor who managed the overall construction of this project. The long hours and hard work of the supervision and crew at Schnabel made this part of the project possible.

REFERENCES

Cording, Edward J., 1981, Use of Empirical Data for Braced Excavations and Tunnels in Soil, Lecture Series, Chicago ASCE, Chicago, IL.

ECS Mid-Atlantic, LLC, July 18, 2017, Revised June 4, 2020. Geotechnical Engineering Report. 15th and S Street, NW, Washington, DC.

Fleming, A.H., Drake, A.A., and McCartan, Lucy, 1994, Geologic map of the Washington West quadrangle, District of Columbia, Montgomery and Prince George's Counties, Maryland, and Arlington and Fairfax Counties, Virginia: U.S. Geological Survey, Geologic Quadrangle Map GQ-1748, scale 1:24,000.

Post Tensioning Institute Recommendations for Prestressed Rock and Soil Anchors, 2014.

Schnabel, Harry, Schnabel, Harry W., 1982, Tiebacks in Foundation Engineering and Construction. A.A. Balkema Publishers.

STATE OF THE PRACTICE IN FLORIDA ON DEEP FOUNDATION INSTALLATION-INDUCED GROUND DEFORMATIONS AND VIBRATIONS

Berk Turkel, Graduate Research Assistant, Dept. of Civil, Environmental, and Construction Engineering, University of Central Florida, Orlando, FL 32816, USA; e-mail: TurkelBerk@knights.ucf.edu
Jorge E. Orozco-Herrera, Graduate Research Assistant, Dept. of Civil, Environmental, and Construction Engineering, University of Central Florida, Orlando, FL 32816, USA; e-mail: Jeorozcoh@knights.ucf.edu
Luis G. Arboleda-Monsalve, Ph.D., Assistant Professor, Dept. of Civil, Environmental, and Construction Engineering, University of Central Florida, Orlando, FL 32816, USA; e-mail: Luis.Arboleda@ucf.edu
Boo Hyun Nam, Ph.D., Associate Professor, Dept. of Civil, Environmental, and Construction Engineering, University of Central Florida, Orlando, FL 32816, USA; e-mail: BooHyun.Nam@ucf.edu

ABSTRACT

Construction of buildings, roads, and traffic activities generate vibrations to urban infrastructure. The vibrations produced by construction activities have an impact on the soil through which they are transmitted. These effects become much more significant in granular soils. The soil conditions in Florida consist mainly of sandy soil deposits and this makes driven piles a commonly used method. In the design of deep foundation installations, it is significant to consider the potential construction issues related to the driving process. This process generates vibrations in the surrounding soils and thus, its effects on nearby soils and structures must be considered. This paper presents a web-based survey conducted among geotechnical engineers in Florida to better understand the common practice and experience regarding the effects of pile driving on surrounding soils and structures. The perception of different practitioners in the state of Florida on ground deformations and damages to adjacent structures induced by deep foundation installations are investigated. The engineers who have experienced ground vibrations and deformations were asked to define the susceptible soil conditions and main triggering factors of large ground deformations induced by pile driving. Sandy soils with loose to medium relative densities were found to be the most susceptible soil condition to generate ground deformations. Conclusions are drawn regarding the importance of considering ground vibration monitoring during the design phase of deep foundations.

Keywords: deep foundations, pile driving, ground deformations, ground vibrations

INTRODUCTION

Deep foundations have been commonly used as a compelling alternative in Florida because the geotechnical conditions allow the installation of these structures to carry large structural loads to deep competent strata. It has been reported in previous studies that deep foundation installations can cause ground deformations (i.e., settlement and heave). The soil behavior and the response of nearby structures to dynamic excitation induced by pile driving are influenced by the subsurface conditions and the interface properties of the soil-foundation system. In order to better understand the damage that occurred on the structures from the ground deformations induced construction vibrations, many research efforts have been made. The amount of ground deformation caused by pile driving depends on several factors including soil type and stratification, groundwater conditions and degree of saturation, pile type, and driving method (Massarsch 2004). Pile driving-induced ground deformations and vibrations were reported in the literature in numerous case histories by but not limited to Lambe and Horn (1965), Brenner and Viranuvut (1977), Mallard and Bastow (1980), Moore et al. (1995), Wong and Chua (1999), Kim and Lee (2000), Athanasopoulos and Pelekis (2000), Thandavamoorthy (2004), Massarsch and Fellenius (2008), Brandenberg et al. (2009), Seo et al. (2014), and Massarsch and Fellenius (2015).

Several studies have proposed methods to develop settlement estimation approaches. Some of these methods are summarized herein to guide researchers and practitioners. These methods have been derived from empirical methods, laboratory-based methods, or semi-empirical methods. Massarsch (2004) presented a simplified empirical method developed from several soil compaction projects. The author suggested that the maximum settlement (S_{max}) and average settlement (S_{avg}) within a distance of $3D + L/2$ from the center of the pile can be preliminarily estimated using the following equations:

$$S_{max} = \alpha(L + 6D) \tag{1}$$

$$S_{avg} = \frac{\alpha(L + 6D)}{3} \tag{2}$$

where, L is the effective length of the pile (i.e., length in the compressible layer), D is the diameter of the pile, and α is a compression factor that can be estimated for different soil density groups and driving effort as presented in the study. Note that the driving energy input value in this calculation depends on the pile installation method and the pile type.

Drabkin et al. (1996) proposed a method that consists of a polynomial model of vibration-induced settlements in sands that was validated with five case histories around the world. The model was developed based on several laboratory tests varying the following factors: vibration amplitude, deviatoric stress, confining pressure, soil gradation, number of cycles, relative density, and moisture content. The model of vibration-induced settlements for small to intermediate vibration levels is presented in the following equation:

$$\begin{aligned} \ln Y = 2.27 + 1.19x_1 - 0.71{x_1}^2 + 0.49x_2 - 0.68{x_2}^2 - 0.8x_3 + 1.09{x_3}^2 - 0.46x_4 \\ + 0.06{x_4}^2 + 0.45x_5 - 0.38{x_5}^2 - 0.19x_6 - 0.10x_7 \end{aligned} \tag{3}$$

where, Y is the settlement expressed in 0.0254 mm (1/1000 in) and the variables x_i are the major factors affecting the settlements. The testing ranges used for each variable are presented in the study. The tests conducted for the development of the mathematical model were performed by placing the soil sample under drained conditions in a triaxial cell that was placed on a shaking table with a vibratory frame. The polynomial model was developed by testing 150 mm (5.9 in) thick soil specimens, thus it needs to be extrapolated to be applicable to thicker soil layers. Drabkin et al. (1996) assumed that the settlement (Δ) of a layer with thickness H_t (in mm), is directly proportional to the settlement y (in mm) of the tested specimen as follows:

$$\Delta = \frac{y}{150} H_t \tag{4}$$

where input values are in millimeters. Additionally, at construction sites where the in-situ soil conditions are highly non-homogeneous, the authors proposed dividing the vulnerable layers into 10 equal thickness layers.

RESEARCH METHODOLOGY

A web-based survey was performed to investigate the effects of pile driving on adjacent structures. The scope of the survey was to understand the common practice and experiences of geotechnical engineers in Florida regarding the effects of pile driving on surrounding soils and structures. The survey link was sent out to geotechnical consultants in Florida. A total of 44 consultants participated in the survey. Some of the respondents did not answer all questions; thus, the total number of responses varies for each question.

A report was generated based on the responses. The survey results are presented and discussed in this paper.

PRACTITIONER EXPERIENCE IN CONSTRUCTION-INDUCED GROUND VIBRATIONS AND DEFORMATIONS

The intention of the authors was first to determine the perception of different practitioners in the state of Florida on ground deformations and damages to adjacent structures induced by pile driving. Figures 1 (a-b) present the answers of the practitioners when asked if they have experienced any ground deformations and structural damages induced by pile driving, respectively. It can be observed that the majority of the respondents have experienced ground deformations induced by pile driving as well as damages in nearby structures associated with these deformations. This result validates the need for a better understanding of this dynamic problem in order to be able to prevent it.

Fig. 1. Responses regarding experience on a) ground surface settlement induced by pile driving installations and b) damage to adjacent infrastructure during pile driving because of high vibration levels or large ground settlements or structural distortions.

Figures 2 (a-b) presents the experience of the practitioners in terms of influence zone from the pile and magnitude of the ground deformations, respectively. Most of the ground deformations occur up to a distance of 15 m (50 ft). However, there was no agreement on the expected magnitude of such deformations. Magnitudes less than 25.4 mm (1 in) are considered reasonable for this type of phenomenon. "Others" in Fig. 2b means practitioners' experience was dependent on the site-specific conditions. Respondents stated that the structures and foundations on loose to medium dense sandy soils are susceptible to more settlements and enough to cause cracks and damage to nearby structures. Overall, practitioners stated that they have experienced less than approximately an inch of settlement. Additionally, the practitioners were asked if they consider it important to monitor ground vibrations due to pile driving at different distances away from the pile. Results indicated that a vast 91% of the practitioners do consider this issue important. This highlights the importance of not only being able to predict the ground deformations but also to monitor the effects of pile driving in the surrounding areas. Engineers were also asked to reply to questions regarding their experience on the relationship between peak particle velocity (PPV), pile-driving induced settlement, and distance from the driving source. More than 60% of the participants stated that they have experienced relation between PPV and ground deformations in terms of distance away from the pile.

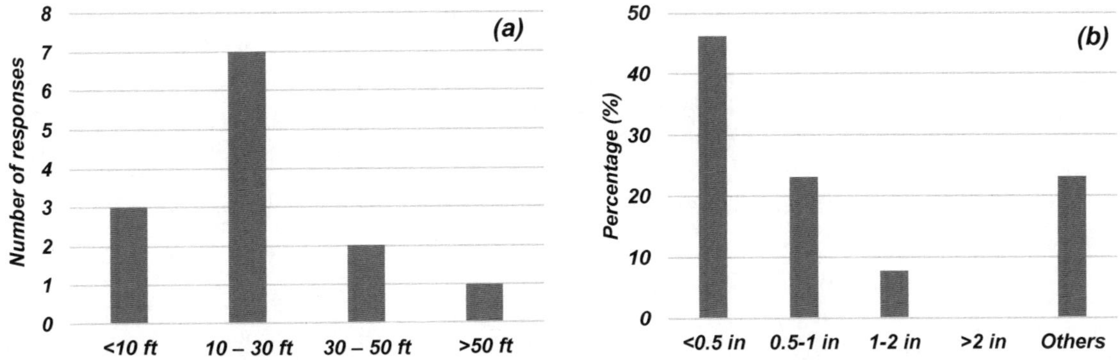

Fig. 2. Responses regarding: (a) distance from the pile driving source of reported settlement and b) the approximate level of ground settlements experienced in the project.

SUSCEPTIBLE CONDITIONS

Figure 3 presents the responses regarding average susceptibility to develop the most amount of settlement by different types of soil conditions around deep foundation installations. Responses are rated from zero to five. Five represents the soils that are the most susceptible to developing large settlements and zero represents the susceptibility to develop the least settlement. Practitioners stated that they have experienced the largest settlement under subsurface conditions constituted of sands regardless of the groundwater table level. Silty and clayey soils caused less settlement than sandy soil conditions around deep foundation activities. Some of the engineers stated specific soil conditions that are susceptible to developing settlements in Florida and they are given as "Others" in Fig. 3. Heave was experienced when Miami Limestone was at shallower depths and most of the ground deformations (i.e., settlement and heave) were experienced under sandy soils with loose to medium relative densities with intermittent clay layers over limestone.

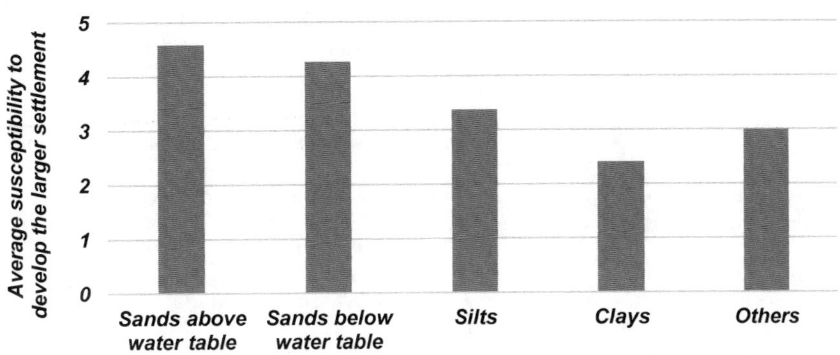

Fig. 3. Responses regarding average susceptibility to develop the most amount of settlement by different types of soil conditions (Rating from 0 to 5, five means the soil with the largest susceptibility to develop settlement).

TRIGGERING FACTORS

Another important objective of this study was to define the main triggering factors of large ground deformations induced by pile driving. Variables such as type of pile, soil mechanics phenomena, and transferred energy to the pile were surveyed. Eighty-six percent (86%) of the engineers agreed that it is important to measure the impact characteristics of the pile driving source and more than 70% of them stated that damage to adjacent urban infrastructure occurs due to the installation of large displacement piles. Figure 4 presents the most common driven piles according to the respondents. Displacement piles such as prestressed concrete piles (PCP) as well as pipe piles were the most voted types of piles with a combined 57% of all the answers. Therefore, displacement piles should be the focus of future studies in order to define proper relationships for pile driving induced ground deformations. Additionally, the practitioners answered with approximately 82% of agreement that it is important to measure the impact characteristics of the driving hammers in order to consider these variables in the definition of the problem.

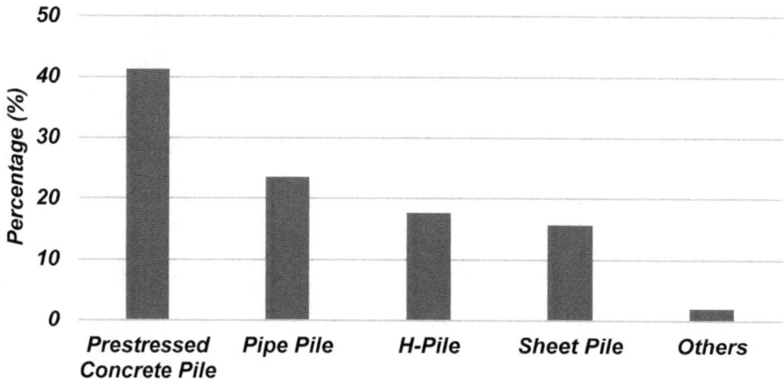

Fig. 4. Responses regarding the most common types of driven piles used for projects in Florida.

Figure 5 presents the main triggering factors, according to the practitioners, of pile driving induced ground deformations. As it can be noted, several variables must be considered to understand this phenomenon. Impact characteristics of the pile driving source should be considered. These characteristics include transmitted energy, force, and frequency content. Soil consolidation due to dissipation of excess pore water pressure was also deemed as the main triggering factor. This phenomenon has also been observed by Turkel et al. (2021). It is worth mentioning that the rest of the variables might also be highly important, but this study only measures the perception of the respondent and not the actual phenomenon.

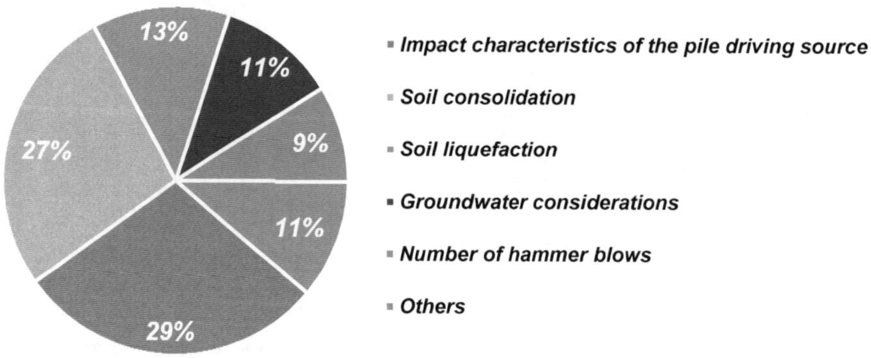

Fig. 5. Responses regarding main sources of pile-driving induced settlements.

MONITORING APPROACHES

Figure 6 presents the responses regarding the time necessary to monitor ground vibrations and soil settlements induced by pile driving. Responses showed the importance of considering ground vibration monitoring during the design phase of deep foundations which was also shown by Orozco-Herrera et al.

(2022). The majority of the respondents indicated that ground vibrations and deformations should be strictly monitored during pile driving operations and "some time" after the pile driving operation. "Others" in Fig. 6 included those who think there should be project site-specific monitoring programs developed based on site conditions, subsurface profile, the scope of work, duration of driving, and proximity to settlement-sensitive structures.

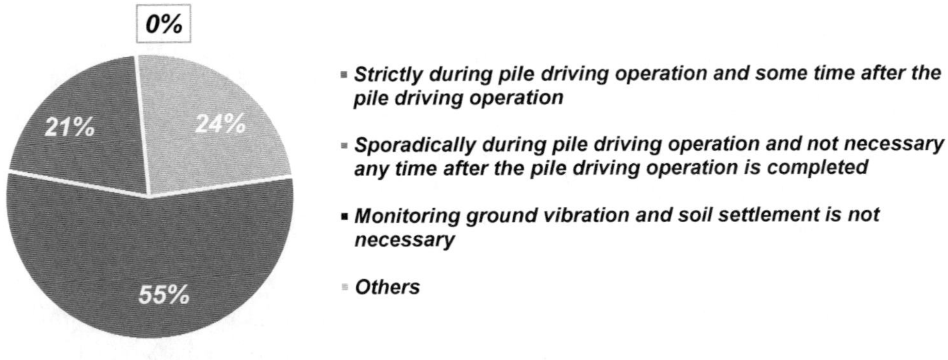

Fig. 6. Responses regarding the monitoring duration necessary to control ground vibrations and soil settlements induced by pile driving.

Figure 7 shows that the farthest sensor should be placed more than 40 ft away from the source based on more than 40% of engineers' responses. Engineers stated many types of sensors to monitor ground vibrations and deformations. "Others" in Fig. 7 included those who think the farthest sensor should be placed based on site-specific conditions, proximity to existing structures, and based on the closest structure of concern.

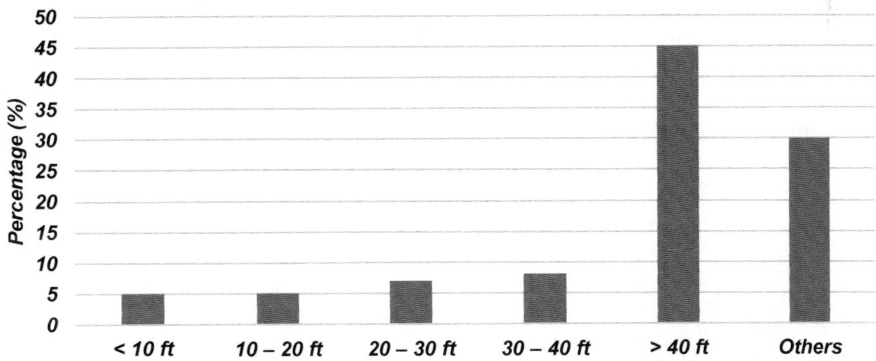

Fig. 7. Responses regarding the location from the pile driving source of the farthest sensor.

Figure 8 presents practitioners' experience in the analysis, design, interpretation, or installation of field sensors. Responses showed that engineers are well experienced with piezometers, inclinometers, and surveyor levels. Some of the respondents indicated that they have experience with additional field instrumentation such as tiltmeters, crack meters, shape array, fiber optics strain measurements (e.g., Bragg grating), and laser scanning.

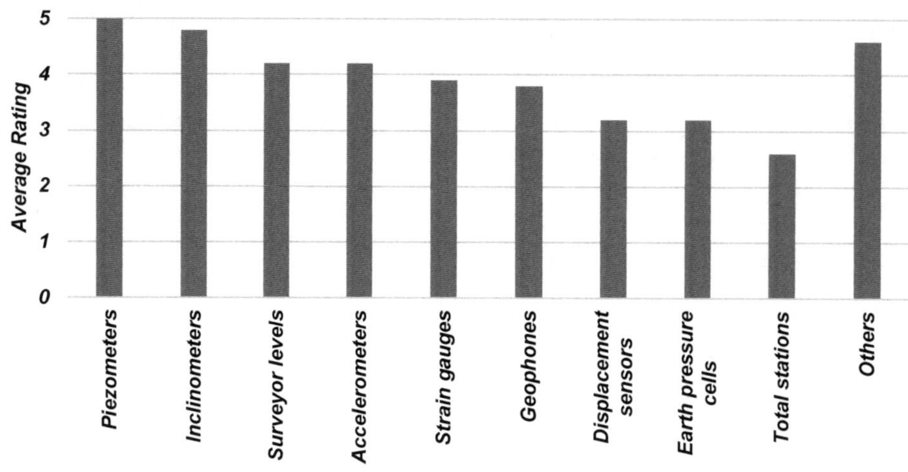

Fig. 8. Responses regarding experience in the analysis, design, interpretation, or installation of field sensors (Rating from 0 to 5, zero means no experience).

DESIGN APPROACHES

Whiffen and Leonard (1971) studied the ground vibrations and their effect on the surrounding structures and the reaction of people to vibration levels. It has been reported that PPV of 0.4 to 0.6 in/s (10 to 15 mm/s) was considered unpleasant by people and it was reported as a level that can cause architectural damage and possible minor structural damage. The Florida Department of Transportation (FDOT) Standard Specifications for Road and Bridge Construction (FDOT 2021) requires the use of vibration monitoring equipment and instrumentation to be installed when foundations are constructed in close proximity to existing infrastructure. Currently, the general limit of peak particle velocity in FDOT projects is 12.7 mm/s (0.5 in/s). Engineers were asked to investigate the common practice and importance shown in the modeling of ground vibrations due to pile driving during the design phase of any deep foundation installation. Figure 9 presents the responses regarding the methods and/or models used to estimate dynamic soil displacement due to pile driving and/or to determine the impact of construction vibrations. "Others" in Fig. 9 constitute those respondents who specifically stated that they simply use their experience or think that such analyses are not common practice or deemed unnecessary.

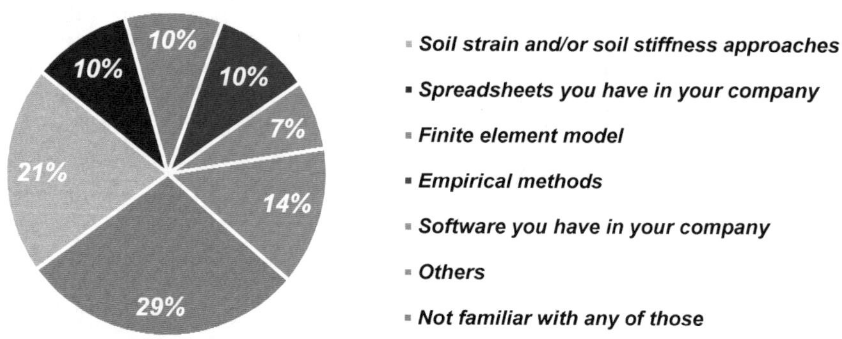

Fig. 9. Responses regarding the methods and/or models used to estimate dynamic soil displacement due to pile driving and/or to determine the impact of construction vibrations.

CONCLUSIONS

A web-based survey was performed to understand the common practice and experiences of geotechnical engineers in Florida regarding the effects of deep foundation installations on surrounding soils and adjacent structures. A total of 44 consultants participated in the survey. The survey results are presented and discussed in this paper. The engineers who have experienced ground vibrations and deformations were further asked to define the susceptible soil conditions and main triggering factors of large ground deformations induced by pile driving. The following conclusions were drawn from this paper:

1. The majority of the respondents have experienced ground deformations induced by pile driving as well as damages in nearby structures associated with these deformations.
2. Most of the ground deformations were experienced up to a distance of 15.0 m (50.0 ft) and the magnitude of settlement was considered to be less than 25.4 mm (1.0 in). More than half of the practitioners experienced the largest ground deformations while installing prestressed concrete piles and pipe piles.
3. Eighty-two percent (82%) of the practitioners consider measuring the impact characteristics of the driving hammers as an important triggering factor of large ground deformations induced by pile driving. Impact characteristics of the pile driving source include transmitted energy, force, and frequency content. Soil consolidation due to dissipation of excess pore water pressure was also deemed as the main triggering factor.
4. Sandy soils with loose to medium relative densities were found to be the most susceptible soil condition to generate ground deformations. 91% of the practitioners stated the importance of monitoring ground vibrations due to pile driving at different locations.
5. The majority of the practitioners stated the importance of considering ground vibration monitoring during the design phase of deep foundations. Approximately 50% of the practitioners stated the location from the pile driving source of the farthest sensor should be at least more than 12.2 m (40.0 ft) away from the pile.

ACKNOWLEDGEMENTS

Financial support was provided by the Florida Department of Transportation, Project No. BDV24 TWO 977-33. The opinions, findings, and conclusions expressed in this publication are those of the authors and not necessarily those of the Florida Department of Transportation or the U.S. Department of Transportation. The authors would like to thank the valuable input of the 44 geotechnical consultants and firms who contributed to this study.

REFERENCES

Athanasopoulos, G. A., and Pelekis, P. C. (2000). “Ground Vibrations from Sheet Pile Driving in Urban Environment: Measurements, Analysis and Effects on Buildings and Occupants.” *Soil Dynamics and Earthquake Engineering*, 19(5), 371–387.

Brandenberg, S. J., Coe, J., Nigbor, R. L., and Tanksley, K. (2009). “Different Approaches for Estimating Ground Strains from Pile Driving Vibrations at a Buried Archeological Site.” *Journal of Geotechnical and Geoenvironmental Engineering*, 135(8), 1101–1112.

Brenner, R. P., and Viranuvut, S. (1977). “Measurement and Prediction of Vibrations Generated by Drop Hammer Piling in Bangkok Subsoils.” *5th Southeast Asian Conference on Soil Engineering*, Southeast Asian Geotechnical Society, Khlong Luang, Thailand, 105–119.

Drabkin, S., Lacy, H., and Kim, D. S. (1996). "Estimating Settlement of Sand Caused by Construction Vibration." *Journal of Geotechnical Engineering*, 122(11), 920–928.

FDOT, 2021. Section 455: Structures Foundations, in: Standard Specifications for Road and Bridge Construction. Florida Department of Transportation (FDOT), Tallahassee, FL, 552–619.

Kim, D.-S., and Lee, J.-S. (2000). "Propagation and Attenuation Characteristics of Various Ground Vibrations." *Soil Dynamics and Earthquake Engineering*, 19(2), 115–126.

Lambe, T. W., and Horn, H. M. (1965). "The Influence on an Adjacent Building of Pile Driving for the MIT Materials Center." *VI International Conference on Soil Mechanics and Foundation Engineering*, 280–284.

Mallard, D. J., and Bastow, P. (1980). "Some Observations on The Vibrations Caused by Pile Driving." *Recent Developments in the Design and Construction of Piles*, Thomas Telford Publishing, 261–284.

Massarsch, K.R. (2004). Vibrations Caused by Pile Driving. The Magazine of the Deep Foundations Institute Summer and Fall Articles.

Massarsch, K. R., and Fellenius, B. H. (2008). "Ground Vibrations Induced by Impact Pile Driving." *International Conference on Case Histories in Geotechnical Engineering*, Arlington, VA, 39.

Massarsch, K. R., and Fellenius, B. H. (2014). "Ground Vibrations from Pile and Sheet Pile Driving. Part 1 Building Damage." *Proceedings of the DFI-EFFC International Conference on Piling and Deep Foundations*, Stockholm, 131–138.

Massarsch, K. R., and Fellenius, B. H. (2015). "Engineering Assessment of Ground Vibrations Caused by Impact Pile Driving." *Geotechnical Engineering Journal*, 46(2), 54–63.

Moore, P. J., Styles, J. R., and Ho, W.-H. (1995). "Vibrations Caused by Pile Driving." *International Conference on Recent Advances in Geotechnical Earthquake Engineering and Soil Dynamics*, St. Louis, MO, 737–741.

Orozco-Herrera, J. E., Turkel, B., Arboleda-Monsalve, L. G., Nam, B. H., & Jones, L. (2022). Continuous Impact Pile Driving Modeling to Elucidate Settlement-PPV-Soil Density-Input Energy Relationships. In *Geo-Congress 2022* (pp. 113-122).

Seo, H., Safaqah, O., and Gudavalli, S. R. (2014). "Ground Vibration Levels Due to Impact Pile Driving in Sands." *Tunneling and Underground Construction*, American Society of Civil Engineers, Shanghai, China, 25–34.

Thandavamoorthy, T. S. (2004). "Piling In Fine and Medium Sand—A Case Study of Ground and Pile Vibration." *Soil Dynamics and Earthquake Engineering*, 24(4), 295–304.

Turkel, B., Orozco-Herrera, J. E., Arboleda-Monsalve, L. G., Nam, B. H., and Jones, L. (2021). "Comparative Analysis of Pile Driving Numerical Modeling Approaches." *International Foundations Congress and Equipment Expo 2021*, American Society of Civil Engineers, Dallas, TX, 484–495.

Whiffen, A. C., Leonard, D. R. (1971). "A Survey of Traffic Induced Vibrations", *RRL report LR 418*, Transportation and Road Research Laboratory, Crowthorne, UK.

Wong, I. H., and Chua, T. S. (1999). "Ground Movements Due to Pile Driving in an Excavation in Soft Soil." *Canadian Geotechnical Journal*, 36, 152–160.

STRUCTURAL ASPECTS OF THE STABILIZATION OF TEN MILE SLIDE

Adrian Gygax, Gygax Engineering Associates Ltd., 103 – 1718 Commercial Drive, Vancouver, Canada, 1 (604) 254 3914, agygax@gea.ca
Rod Kostaschuk, BGC Engineering Inc., 234 St. Paul Street, Kamloops, BC, 1 (250) 374 8600, rkostaschuk@bgcengineering.ca
Sarah Gaib, Ministry of Transportation and Infrastructure, 4C 940 Blanshard Street, Victoria, BC, 1 (778) 974-5371, sarah.gaib@gov.bc.ca

ABSTRACT

The British Columbia Ministry of Transportation and Infrastructure (Ministry) recently completed a slope stabilization project at Ten Mile Slide on Highway 99. The highway is an essential part of the road network connecting the British Columbia (BC) coast with the interior of the province. The landslide was first identified in the mid-1980s and has retrogressed more than 250 m upslope since then, leading to continual and expensive repairs to the highway. The Ministry engaged BGC Engineering Ltd. (BGC) and their structural sub-consultant Gygax Engineering Associates Ltd. (GEA) to develop a stabilization scheme that would halt the slide with an acceptable level of residual risk.

The design required careful consideration of construction stages and the stresses induced in the structural components as the slide movement was progressively stopped as construction advanced. Four stabilization design options were shortlisted with the preferred option selected in Fall 2016. The selected solution comprised the construction of five rows of post-tensioned soil anchors upslope of the highway and a tied-back concrete tangent pile retaining wall for the roadway itself as well as reconstruction of the highway alignment, grade and drainage through the slide. Works were installed by top-down construction within defined time allowances stipulated in the contract. Significant engineering was undertaken beyond standard design processes for this complex project, such as comprehensive geotechnical investigations, aerial and terrestrial laser scanning and change detection analyses, pre-production test anchors, detailed geo-structural numerical modelling and due diligence activities to validate and optimize the design.

The stabilization works have been very successful with the slide response being as predicted and the highway operating again as full two-lane service.

Keywords: slope stabilization, landslide stabilization, ground anchors, bored piles, tangent piles, permanent tie-back anchors

INTRODUCTION

BC Ministry of Transportation and Infrastructure (Ministry) recently completed a slope stabilization project at the Ten Mile Slide on Highway 99, which is an essential part of the road network connecting the British Columbia coast with the interior of the province and which connects the communities of Lillooet and Xaxli'p to the regional services in the City of Kamloops. Approximately 1,600 vehicles travel this section of highway every day, about 20 per cent of the traffic comprises heavy vehicles (trucks and buses). A Canadian National Railways (CN) line crosses the top of the slide and was partially stabilized by the railway in an earlier project, as reported in Piling Canada (2017).

The landslide was first detected in 1988 and is estimated to have regressed more than 250 m (800 ft) upslope in the last 40 years. Historical monitoring and remote sensing data shows that the landslide has moved at average rates ranging from 1.1 m to 3.7 m (3.6 to 12.1 ft) per year. By late 2016, this section of highway was the most expensive, challenging and technically complex site for the Ministry to maintain in the province and was subject to 50% load restrictions with 24-hour flagging to maintain single lane alternating traffic. Slide width has remained stagnant at the highway.

Ten Mile Slide is the active portion of the Tunnel Earth Flow, a geological feature that is 1.5 km (1 mile) long and contains about 13 million cubic metres (460 million ft^3) of soil materials. The active slide is about 300 m (1000 ft) long and 200 m (700 ft) wide and has an average depth to the shear zone of 18 m (60 ft). Prior to stabilization, the moving mass comprised about 1 million cubic metres of soil material, see Fig. 1. Prior to stabilization, the slide was moving continuously at an average rate of 10 mm (0.4 in) per day, accelerating to 50 mm (2 in) per day following precipitation events. Sliding movement is translational, occurring along a 2-metre (7 ft) thick roughly planar basal shear zone. The mobilized body of Ten Mile Slide is mainly dry, has higher strength than the basal shear surface and behaves as a relatively stiff and cohesive mass. Previous efforts to stabilize the highway had only mitigated landslide movement temporarily.

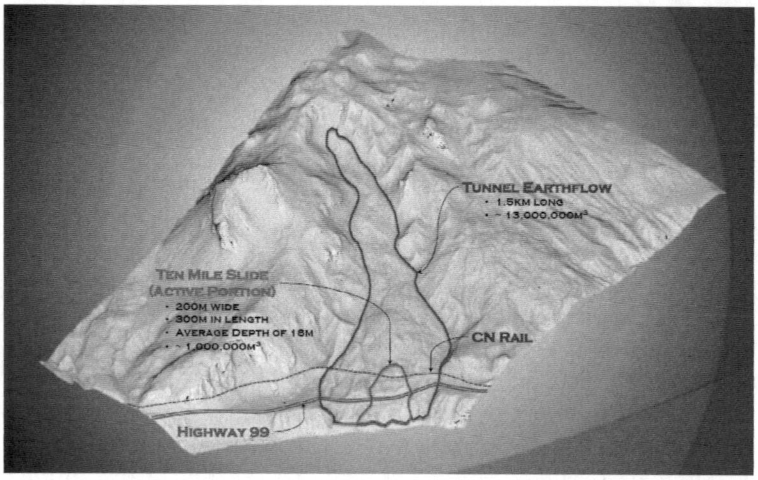

Fig. 1: Overview of the Tunnel Earthflow and Ten Mile Slide

The stratigraphy at the site includes, in order of increasing depth: Tunnel Earthflow, Undifferentiated Glacial Sediment and Jackass Mountain Group Bedrock. The movement of the slide has resulted in a shear zone near the base of the Tunnel Earthflow unit. This shear zone has been characterized spatially and with depth by several boreholes from different investigation efforts over many years. The Tunnel Earthflow is approximately 20 m (65 ft) thick – the texture is heterogeneous and unsorted, and typically comprises plastic, well graded clayey silt, sand and gravel with cobbles and boulders. Except in a few instances, water seepage or wet materials were not encountered. The Undifferentiated Glacial Sediment directly below is comprised of permeable, dense, drained, coarse granular soil. Cobbles and boulders occur at any depth in this soil, as do seams and lenses of low permeability, clayey sediments. Bedrock is located at depth and is not relevant to the stabilization measures.

STABILIZATION CONCEPT

Design Criteria

An important step in the design process was determining an appropriate factor of safety for slope stabilization. Ongoing movement of the landslide suggested the limit equilibrium factor of safety prior to stabilization was near 1.0. After completing a rigorous site investigation, design sensitivity analysis, geo-structural numerical modelling, two test soil anchors and consulting with a technical panel comprised of independent experts, the criteria adopted by the Ministry was to design of the stabilization measures to increase the factor of safety (FoS) on each cross-section analyzed to a minimum of 1.2. To manage uncertainties and risks associated with the work it was further decided to adopt an observational method, comprising significant geotechnical and structural monitoring before, during and after construction for the implementation of the stabilization program.

The minimum design life of the stabilization measures is 30 years based on the rate of slide retrogression. If the combined effect of the stabilization measures constructed by the Ministry and CN prevents future retrogression of the landslide, the design life of the stabilization measures is expected to be longer and largely determined by the life of the structures. The designs for all structural components of the stabilization measures were compliant with the Canadian Highway Bridge Code, (CSA 2014), a 475 year return-period earthquake event and meet the retaining wall performance requirements for 75-year design life.

Climate change projections indicate a small increase in annual precipitation over the design life of the stabilization measures. The impact of these wetter conditions was considered with regard to slope stability by checking the sensitivity of the factor of safety of the stabilized slide to a permanent and sustained increase in groundwater pressure to confirm the resulting factor of safety remained sufficiently high.

Selected Solution

Typical solutions such as re-routing the highway, unloading the crest or buttressing the toe of the slide were concluded not to be feasible due to the steep slope and proximity of the Fraser River at the toe of the slide and the railway at the crest. Groundwater drainage was expected to be ineffective because groundwater pressure is near-zero in both the coarse-grained Undifferentiated Glacial Sediments below the slide and in the overlying Tunnel Earthflow soil, except for occasional perched aquifers. A structural solution which increased the shear resistance along the failure plane was therefore pursued. The design solution needed to consider ongoing slide movement during construction so that components installed in early construction phases could tolerate daily slide movement until enough additional structural support was installed to stop movement. Figure 2 shows the arrangement of the stabilization works along the design section.

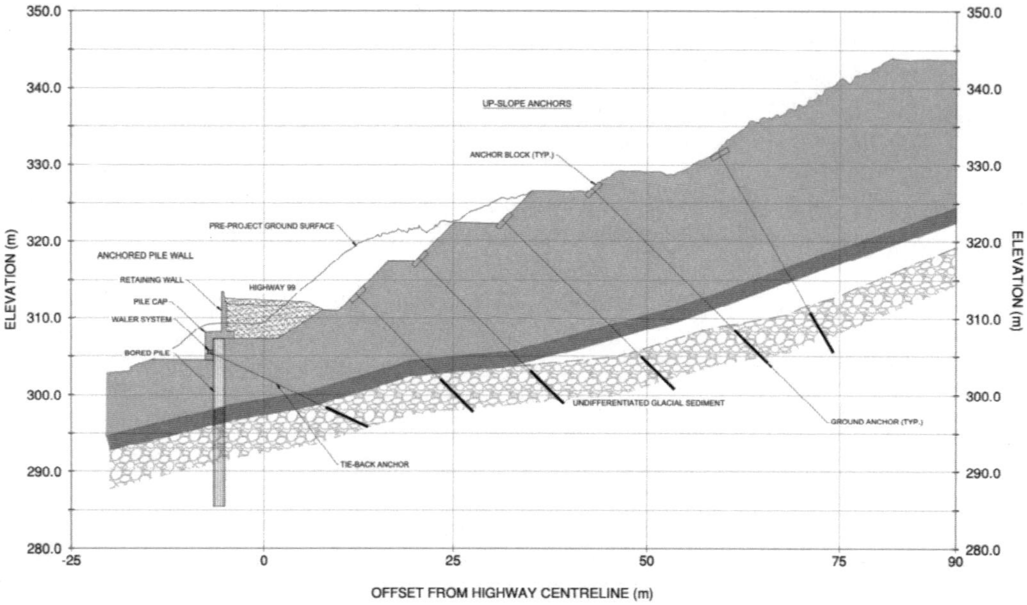

Fig. 2: Interpreted geological profile and selected stabilization solution.
Basal shear zone shown grey, Undifferentiated Glacial Sediments shown yellow

The solution adopted included 276 post-tensioned double corrosion protected (DCP) soil anchors with 2.5 x 2.5 x 0.6 m (8.5x8.5x2 ft) precast concrete bearing blocks installed in five rows above the highway, a tied-back concrete tangent pile wall below the highway and reconstruction of the highway to original alignment and grade. The pile wall comprises 148 bored piles of 1524 mm (60 in) diameter, ranging in length from 20 m to 26 m (65 to 85 ft), 125 tie-back soil anchors, a waler assembly and a cap beam with a cantilever wall stem that extends to 5 m (16 ft) in height. The piles are embedded 10 m (33 ft) into the dense, free-draining, coarse-grained, Undifferentiated Glacial Sediments.

Working top-down, each row of anchored blocks was sized to slow, or even stop, the movement of the slide mass above it. After installation and testing, the anchors of a given row could be locked-off within a fairly short time-frame. However, the pile wall required a large number of piles to be installed and during the time required, the soil mass below the lowermost row of anchored blocks continued to move. The design therefore required a pile solution that would exhibit considerable early shear strength and be tolerant of displacement until the tie-back anchors could be installed, tested and locked-off.

The solution selected was to use steel shell piles with concrete infill (composite piles), with the steel shell providing initial shear stiffness across the slide plane while the concrete infill cured. Installing tie-back anchors through steel shafts would have been prohibitively expensive, so only every third pile was a steel shell composite pile with the remainder being reinforced concrete bored piles. Drilling anchors through the concrete pile shafts is a common construction task. The composite piles were installed ahead of the concrete piles and successfully slowed slide movements to allow the concrete piles to reach adequate concrete strengths without being damaged.

To share the soil loading among both the concrete and the composite piles after the tie-back anchors were installed, a steel waler system was installed to link each composite pile with its adjacent two concrete ones. The waler system involved first attaching a waler stool to each concrete pile, the stool allowing position adjustments to ensure a flat vertical plane for the waler beam was achieved. The stools also incorporated a steel pipe shear stub that was inserted into the hole that was cored through the concrete pile for the tie-back anchor. The shear stub directly transferred the vertical component of the anchor testing and stressing force to the concrete pile shaft. The steel walers, comprising to C380x50 (C15x33.9) channels back-to-back, were then attached to the stools and the tie-back anchors stressed. Subsequently, a concrete pile cap was constructed that encapsulated the pile tops and extended down to cover the walers, with pockets provided for permanent access to the tie-back anchors. By the time the pile wall could be constructed, the design road alignment was between 1 and 6 m (3 and 18') above the pile tops, so the pile cap was topped with a cantilever wall stem that acts integrally with the pile wall system. Figure 3 illustrates this arrangement.

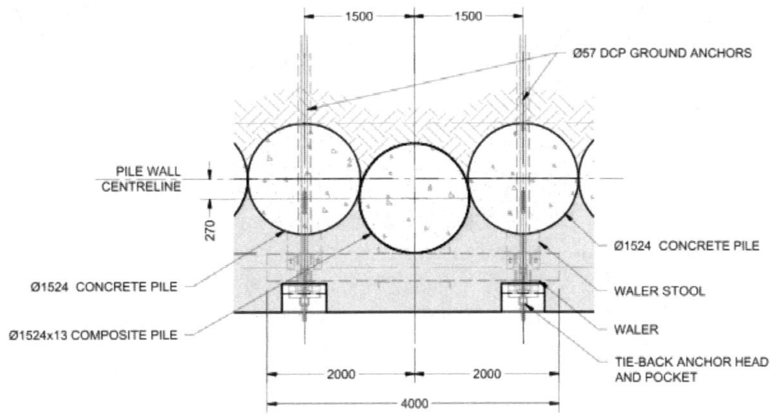

Fig. 3: Pile wall details

DESIGN METHODOLOGY

Detailed 2D and 3D numerical time-step modelling of the slope movements and soil-structure interaction was used to predict stresses on the structural elements throughout construction, and to define time restrictions for the construction contracts. A combination of limit-equilibrium, finite difference and finite element methods were used for both the geotechnical analyses of the slope movements, upslope anchor stresses and the soil loads on the pile wall and for the structural design of the pile wall components.

Geotechnical Analyses

Two-dimensional limit equilibrium stability analyses were used to evaluate and compare various landslide stabilization solutions. For the selected solution, the limit equilibrium stability analyses were then used for defining anchor spacing and loading and for establishing the pile configuration and capacity. The limit equilibrium stability analyses were also used for local stability assessments of the construction excavation and benches required for the installation of the stabilization measures.

More detailed analyses of the slide movement and impact of the stabilizing measures was then carried out on a cross-section through the center of the landslide, which corresponded to one of the limit equilibrium models thus allowing for verification of results. These analyses were carried out on two-dimensional finite difference numerical models using the computer program FLAC (Version 7.0, Itasca, 2011), Figure 4.

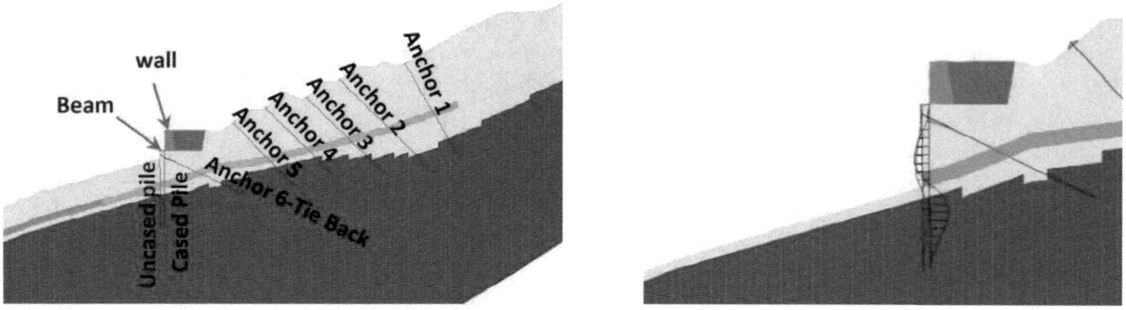

Fig. 4: FLAC model, overall model (left) and detailed view of pile wall with moment diagram (right)

The staging of the stabilization work, and the coincident continual movement of the slide was analyzed in discrete steps. In general terms, each FLAC Step represented a stage of construction, with sub-steps used to calibrate the model to yield results that matched the known ground movements, see Table 1. the Strength Reduction Method (SRM) was used at each step wherein the friction angle of the basal shear zone was reduced slightly below the calibrated value for existing conditions to initiate movement in the model, i.e. achieve a Serviceability Limit State (SLS) with a FoS against sliding of just under 1.0. The SRM was then used again to increase the FoS to 1.2 to determine the loads on the structures for the Ultimate Limit State (ULS). The models were run under SLS conditions to achieve the slide movements anticipated between each phase of construction, e.g., 50 mm movement at the road after installation of Anchor Row 1, etc. By FLAC Step 17, the model included the stiffness of the steel shell of the composite piles, Steps 18 and 19 accounted for the increasing stiffness of the composite pile concrete and by Step 20, the concrete piles were also included. By FLAC Step 26, the pile wall tie-back anchors and anchor pre-stress were included in model. FLAC Step 27 assumed all earthflow soil downhill of the pile wall and above the bottom of the basal shear zone will slide away and cease to support the pile wall.

Table 1. Description of FLAC analysis steps

FLAC Steps	Description
1 to 4	Model development (boundary conditions, soil properties, initial conditions
5 to 7	Installation of Anchor Row 1 and subsequent slide movement
8 to 15	Installation of Anchor Rows 2 to 5
16	Installation of composite pile steel shell
17	Increase composite pile stiffness to account for hardening concrete fill
18 to 20	Installation of concrete piles
21 to 25	Excavation of tie-back and waler construction bench
26	Tensioning of tie-back anchors
27	Loss of soil downstream of wall and above basal shear zone, add earthquake
28	Check case of resetting and re-stressing Anchor Rows 1 to 5

For both the limit equilibrium and finite difference models, a sensitivity analysis was completed to evaluate changes in loading on structures to be expected from a transient rise in pore pressure. Furthermore, the predicted seismic response of the stabilization system was evaluated for the design earthquake by applying eight earthquake time histories to the long-term SLS case where it is assumed all soil downhill of the pile wall and above the basal shear zone moves downhill and ceases to support the pile wall. Each time history was synthesized to match the site-specific target spectrum as defined by CSA 2014.

Structural Analyses and Design of the Pile Wall

The FLAC numerical model, being two-dimensional, incorporated both the composite and the concrete piles as tied-back elements. Thus, the FLAC results do not provide a true representation of the load-sharing between the directly tied-back concrete piles and the waler-restrained composite pile in between. For the detailed structural design, GEA conducted three-dimensional structural analyses of the pile wall system incorporating the three triplets of one composite pile and two concrete piles each and the waler, tie-back anchors, pile cap cantilever wall stem and roadbed. Three models were developed to assess variations in pile lengths and location of the basal shear zone. This wall system was modeled and analyzed using the finite element program RFEM (Dlubal, 2019). The RFEM structural analyses results in the same load demand on tie-back system as the FLAC analyses but provides for a more accurate local distribution of load among the different pile types and the anchors.

For the composite and concrete piles, the effect of the landslide loading was transferred from the 2D FLAC model to the 3D RFEM model as follows:

1. The profile of shear force along the length of each pile type, as determined by FLAC, was used to calculate the profile of soil reaction loads acting on each pile for each relevant analytical FLAC Step.
2. The incremental loads on the piles were then derived for each construction stage. FLAC Step 20 was taken as the starting point. Step 26 minus Step 20 represents the effect of anchor stressing, Step 27 minus Step 26 that of SLS equilibrium being reached, etc.
3. For the pile embedment below the shear plane, the RFEM model used soil-springs determined by P-Y methods.
4. The FLAC analysis had all piles anchored, and as noted above, the anchor load is implicit in the pile soil loads obtained from the FLAC modelling. Therefore, in order to model the wall with only the concrete piles anchored, half of the composite pile soil loads above the plane of slip, for the incremental Step 26 minus Step 20, was added to that of each of the concrete piles.

Highway traffic loads per CSA 2014 were added as surcharge with lateral active earth pressures acting on the pile wall stem. Load and material resistance factors, again per CSA 2014, were applied to the non-geotechnical loads as appropriate to combine with the FLAC ULS loads.

The RFEM load increment analysis results were combined by superposition to yield the shears and moments in each type of pile with due account for load sharing. By way of a check, the total section forces of a concrete-composite-concrete triplet, divided by the wall length of the triplet were compared with the FLAC results and found to be in good agreement.

Figure 4 presents an example of a moment diagram for a concrete pile obtained using the above approach. The moment diagrams for FLAC Steps 20 to 28b are shown in increasing intensity of brown and the seismic case is shown in red. The effect of anchor installation and stressing is clearly visible (Step 26b) as is the increase in anchor load due to seismic effects (Step 27). FLAC Step 28b modelled the effects of de-stressing the upslope anchors, which of course would initiate some slide movement and commensurate increases in soil pressures on the pile wall. The typical concrete pile reinforcement is shown on the right of Figure 4. Meticulous readers will note the extra longitudinal reinforcement provided in the upper portion of the pile. This compensates for bars that were cut during the coring for tie-back anchor installation.

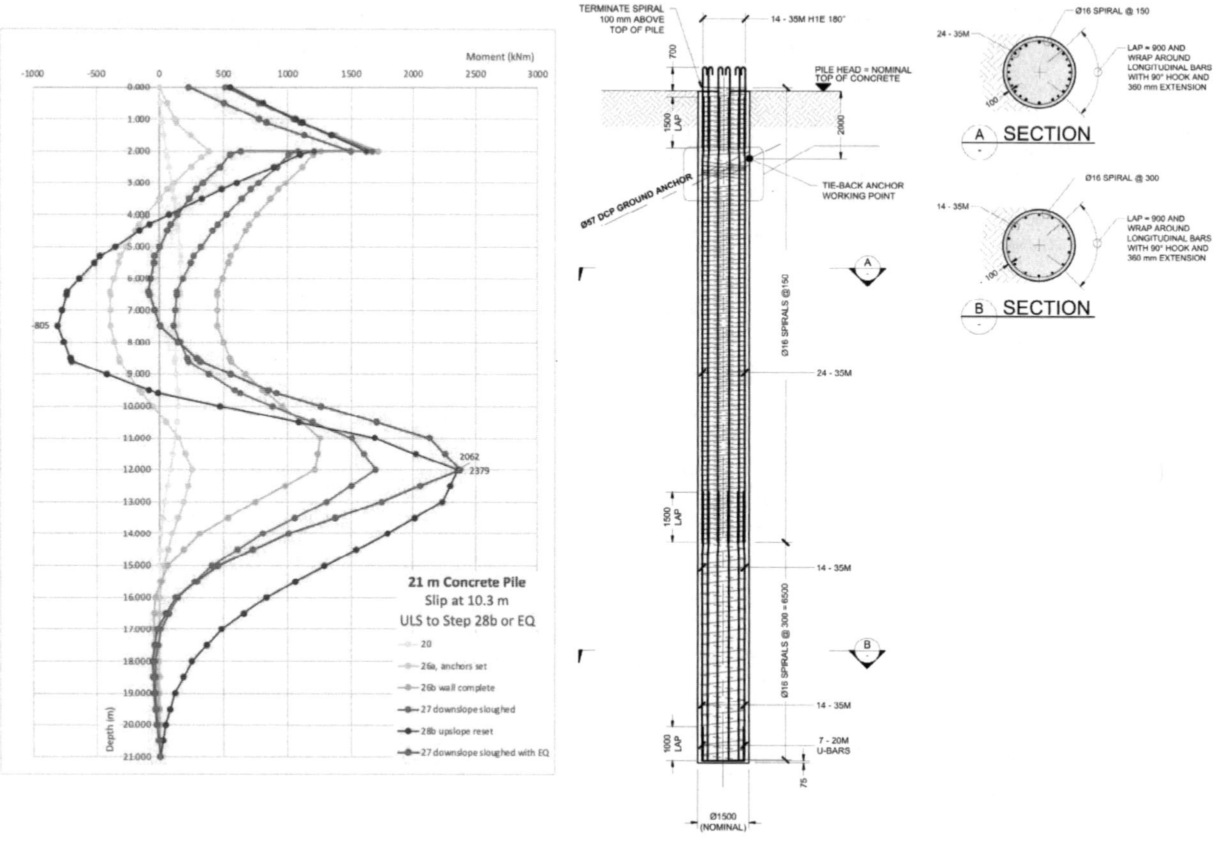

Fig. 5: Concrete pile: example moment development and provided reinforcement

Observational method during construction

The 2D soil-structure interaction modelling in FLAC to estimate loading and response of the stabilization structures to construction and potential future conditions is complex and several assumptions were made that are based on estimated movement rates and soil parameters. While the analyses were considered representative and were the basis of engineering decisions and design, application of the observational method during construction was considered a key component to the project execution. Based on monitoring results, the design could be adapted to suit changing conditions as construction proceeded. The monitoring and instrumentation program included:

- Electronic vibrating wire load cells on selected soil anchors and the tie-back anchors,
- Slope inclinometer casings cast into selected composite piles of the pile wall and distributed on the slope above the highway,
- Ground surface displacement measurements by the Contractor during construction using survey monuments, together with ground-based, terrestrial laser scanning and change detection analysis by BGC.

If monitoring showed that the predicted results were not being achieved, several contingency measures were available, including reducing tension in overstressed anchors and adding anchors. Monitoring results were retrieved daily via satellite upload and web-based plotting. Real-time monitoring was vital to determine when anchors required re-tensioning, which occurred several times for the initial, upper row of anchors until there was sufficient support to stabilize the slide. No additional anchors were required.

Monitoring has shown the slide response being as predicted and movement above the pile wall is also below detection limits. There is some localized, shallow movement and anchor load reduction associated with compression of tension cracks on the extreme eastern edge of the slide, which will be monitored and addressed if required. Displacement below the tangent pile wall continues to occur as anticipated in the design. Currently, anchor and tie-back loads are stable or decreasing very slowly.

Construction

Works were installed top-down starting at the upper row of soil anchors and defined time allowances were specified in the construction contracts for each stage to limit ground movements (Figures 6 and 7). The stabilization work was completed under three construction contracts. In early 2017, 30 soil anchors were installed, an additional 44 soil anchors installed in 2018, and the remaining 202 soil anchors and the tied-back pile wall was completed from 2019 to 2021.

Fig. 6: Top-down construction commenced with up-slope rows of soil anchors and terminated with the pile wall and road re-build

Fig. 7: Pile wall construction: bored pile drilling (left), waler and tie-back installation (right)

The design solution remained the same throughout all construction phases, which speaks for its validity. Even though the project duration was longer than anticipated, much of the assignment was completed at an accelerated pace. Early phases of construction provided the design team with an understanding of the geotechnical conditions, soil anchor installation challenges, construction costs/productivity and validated the design approach, which allowed the team to incorporate lessons learned in subsequent construction contracts. Because of the importance of shear capacity, pile integrity testing was completed using the Thermal Integrity Profiling (TIP) method, which proved very successful. TIP identified the need for smaller diameter reinforcing cages on the few occasions where it was necessary to complete the lower portion of the bored pile within the undifferentiated glacial sediments using open hole methods, resulting in a smaller diameter pile shaft with a more irregular exterior surface than the usual cased shafts.

Conclusions

Ten Mile Slide has been successfully stabilized, significantly reducing the operating costs of a kilometer-long stretch of BC Highway 99 (see Fig. 8 overleaf). The stabilizing system comprises five rows of ground anchors above the highway and a tied-back tangent pile wall on the down-slope side of the highway. The design was based on an extensive monitoring and field-testing program that characterized the slide mass properties and rates of movement. The analytical approach was focused on providing as accurate a picture as possible of the influence of phased construction and the duration of each construction phase. A combination of limit equilibrium, finite difference and finite element techniques allowed for practical numerical model sizes while providing for analytical detail where important, e.g. in determining the load-sharing effects of using tie-backs on two out of every three piles.

A solution was found that addressed stabilization for the unique landslide and offered many firsts for the Ministry in terms of innovative design and monitoring during the construction phase. As a result, the Ministry has contemplated this solution at other slide sites, including those impacted by recent flood events. The project has not only resulted in a safe and reliable highway alignment and minimized maintenance costs for this important corridor to national transportation routes, but it has been essential for Xaxli'p: by maintain community connectivity and cohesiveness, by providing the most direct connection between Lillooet and Kamloops, and by providing community benefits and economic reliability.

Fig. 8: The completed work, autumn 2021. View looking south towards Lillooet, the Fraser River is on the right

ACKNOWLEDGEMENTS

The success of this project is due to many parties. The Xaxli'p Nation worked for many years with the Ministry to develop a viable long-term stabilization solution and contributed labour to its construction. Contractors were Peter Kiewit and Sons (initial upslope work); Flatiron Construction, with subcontractors KanArm (upslope anchors) and Henry Drilling (pile wall). The authors were supported by Scott McKenzie and Brent McAfee (Ministry of Transportation and Infrastructure); Hamid Karimian, Andrew Mitchell, Michael Porter, and Mark Pritchard (BGC); Ernie Naesgaard, Masoud Yazdi (Naesgaard-Amini Geotechnical Ltd.); and Adam Williams (GEA).

REFERENCES

Canadian Standards Association (CSA), 2014. S6: Canadian Highway Bridge Design Code, Canadian Standards Association, Toronto, ON.

Dlubal, 2019. RFEM Ver 5.0. Dlubal Software, Inc., Philadelphia, PA. https://www.dlubal.com/.

Itasca, 2011. FLAC [computer program]. Ver. 7.0. Itasca Consulting Group Inc., Minneapolis, MN. https://www.itascacg.com/.

Piling Canada, Q1 2017. Project Spotlight: Fountain Slide Design-Build Track Mitigation Project. Lester Communications Inc., Winnipeg, MB.

Post-Tensioning Institute (PTI). 2014. Recommendations for Prestressed Rock and Soil Anchors, 5th Edition. Post-Tensioning Institute, Farmington Hills, MI.

THE BENEFITS OF AN EARLY CONTRACTOR INVOLVEMENT FOR A MEGA PROJECT IN SAN FRANCISCO

Robert Small, Malcolm Drilling Co., Inc., Hayward, CA, USA, 510-780-9181, rsmall@malcolmdrilling.com
Ihab Allam, Malcolm Drilling Co., Inc., Hayward, CA, USA, 510-780-9181, iallam@malcolmdrilling.com
Amshu Chappa, Brierley Associates, Moraga, CA, USA, 925-247-9000, achappa@brierleyassociates.com

ABSTRACT

San Francisco Public Utilities Commission (SFPUC) is a department of the city that provides drinking water and wastewater services to San Francisco. The $2 Billion Biosolids Digester Facilities Project (BDFP) will upgrade and modernize the existing facilities at their Southeast Treatment Plant (SEP), which services over 80% of the city's wastewater, by reducing odors in the surrounding neighborhoods, designing for future earthquakes and sea level rise, and ensuring operational redundancy and efficiency for decades to come.

The project required participation of Local Business Enterprises (LBE) and Disadvantage Business Enterprises (DBE) therefore helping the local economy and its residents.

SFPUC decided to use an alternate delivery method to leverage the expertise of the entire construction industry by engaging Trade Core partners at a very early stage in the design process. Malcolm Drilling was chosen to be the Foundation Core Trade partner after an intense pre-qualification process. Our expertise for various deep foundation and support of excavation methods helped the designer to optimize two large excavations in difficult ground conditions and in close vicinity to an active railroad. Our scope of work included the installation of Cased Drilled Shafts, Continuous Flight Auger Piles (CFA), Cutter Soil Mixing (CSM), Diaphragm Walls, Tie-Downs, Tiebacks, Internal Bracing, Mass Excavation, Shotcrete, Demolition and Dewatering. We were also responsible for site access and the required logistic to handle up to 80 concrete trucks coming into the site and the same amount of spoil trucks leaving the site on a daily basis with only one entry and one exit point.

This paper highlights the benefits of early involvement of a specialty contractor in a very complex project. The interaction with the owner's design and management team during the pre-construction phase enabled the project team to apply the most modern and efficient construction techniques proposed by the contractor. Significant cost and schedule savings made this a very successful project for all stakeholders.

Keywords: core trade partner, shoring design, diaphragm walls, tiebacks, construction, young bay mud

INTRODUCTION

The Biosolids Digester Facilities Project (BDFP) is located in the southeast part of San Francisco, CA, at 1800 Jerrold Avenue. The jobsite is bounded by the active Caltrain and Union Pacific Railroad tracks on the west side and the existing Southeast Treatment Plant (SEP) on the north, south and east sides, as shown in Fig. 1 below. The existing SEP was constructed in the 1950s and portions of the facility have reached their operational life. The $2 Billion BDFP will replace and relocate the outdated existing solids treatment facilities with more reliable, efficient, and modern technologies and facilities. In addition, the nearby Bayview and Hunters Point neighborhoods will benefit from the improved odor control.

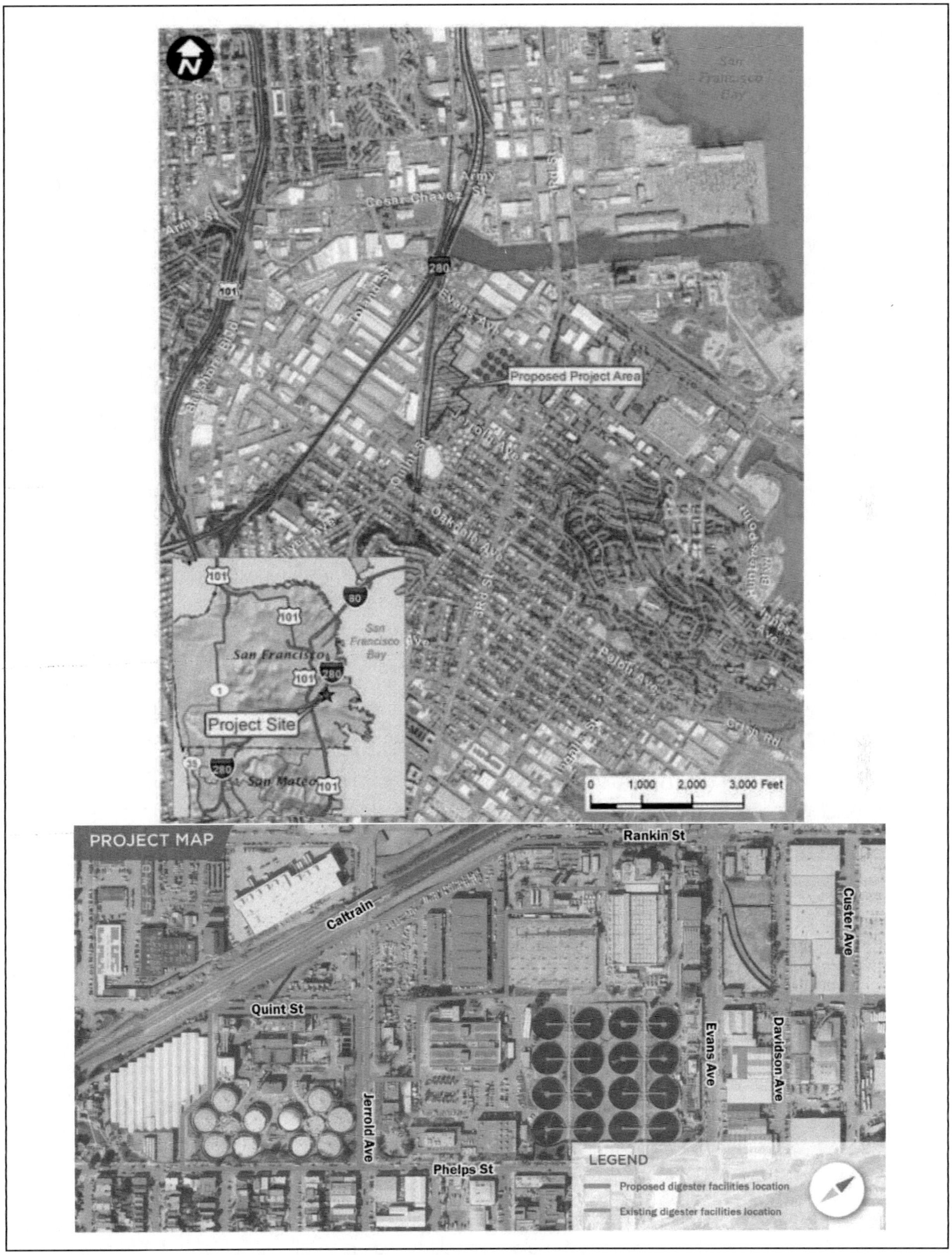

Fig. 1. Jobsite Location Maps

CORE TRADE PARTNER

In 2017, the San Francisco Public Utilities Commission (SFPUC), the owner of the project, selected MWH/Webcor, a Joint Venture, as the Construction Manager General Contractor (CMGC) for the project. As part of an alternate delivery method set forth in the prime contract, the CMGC was permitted to propose the use of a Core Trade Subcontractor Partner for the SFPUC's approval to provide pre-construction services. The CMGC identified three major scopes (Electrical, Mechanical & Foundation) where a Core Trade Partner would be able to assist the project in design-build, design-assist, value engineering, and/or other necessary pre-construction activities. In 2018, Malcolm Drilling Co., Inc. (Malcolm), was approved by the SFPUC and brought on board as the Foundation Core Trade Partner.

Malcolm participated in weekly meetings with the CMGC, SFPUC, and the SFPUC's Design Team to provide value engineering on several major scopes of work. Malcolm proposed Continuous Flight Auger (CFA) piles at certain structures, in lieu of the more traditional drilled shafts, which brought immediate schedule and cost savings to the project. In order to verify the capacities of the CFA piles and drilled shafts, Malcolm performed an early-work onsite load test program, which gave the SFPUC Design Team confidence in the pile foundation selection.

Malcolm also provided guidance on the selection of temporary shoring systems to be utilized at the two major excavations (Facility 610 and Facility 600). Several Cutter Soil Mix (CSM) shoring wall options utilizing Deep Soil Mixing or even Jet Grouting Plugs as well as deeper Secant Piles walls were considered. Initially, a 70 ft deep CSM shoring wall was envisioned for both excavations, however after reviewing the available geotechnical information it was uncertain if the wall would be deep enough at the Facility 610 to provide effective groundwater cut-off. As part of another early-work onsite test program, Malcolm installed and performed a pump test program to provide the Design Team with more information about the groundwater conditions. It was discovered that two underground aquifers (one at 40 to 60 ft below grade and another at 80 to 100 ft below grade) were connected and that a 70 ft deep CSM wall would not provide an effective cut-off to control groundwater drawdown outside the excavation.

Due to the close proximity of two nearby active railroad lines, the project could not utilize a shoring system that might allow uncontrollable drawdown outside the shored excavation. Options for a drilled secant shoring wall and a temporary diaphragm shoring wall were evaluated since they could both reach greater depths than the CSM wall option. Ultimately, a temporary diaphragm shoring wall was selected for the shoring system at Facility 610 due to the faster install time, greater depths it could penetrate, and its ability to meet strict deformation criteria.

EXISTING SITE AND SUBSURFACE CONDITIONS

The project site is located within the Hunters Point Shear Zone. The project site is generally level at about elevation +3 ft and the ground water level was observed at a depth varying between 7 to 12 ft below existing grade, although the piezometric head in the deeper soil strata is generally higher than the unconfined near surface groundwater level. In general, the geologic units encountered at the project site consisted of Artificial Fill (af) extending to a depth ranging from 10 to 18 ft. The fill includes boulders, cobbles, rubble, concrete and existing foundation piles extending to a depth of approximately 40 ft.

The fill is underlain by Young Bay Mud (Qybm) that extends up to 55 ft below ground surface. The Young Bay Mud is a primarily soft to very soft clay with scattered organics and is highly compressible.

Upper Layer Sediments (Quls) are located below the Young Bay Mud and extend to a depth of up to 180 ft and consist of interbedded sands and clays.

Old Bay Clay (Qobc) is stiff to very hard fine-grained soil that underlies the Upper Layer Sediments and extends to as much as 230 ft below ground surface.

Older Colluvium (Qoc) and Franciscan Complex (Kjf) are found at great depth below the site. Fig. 2 is a representative subsurface profile.

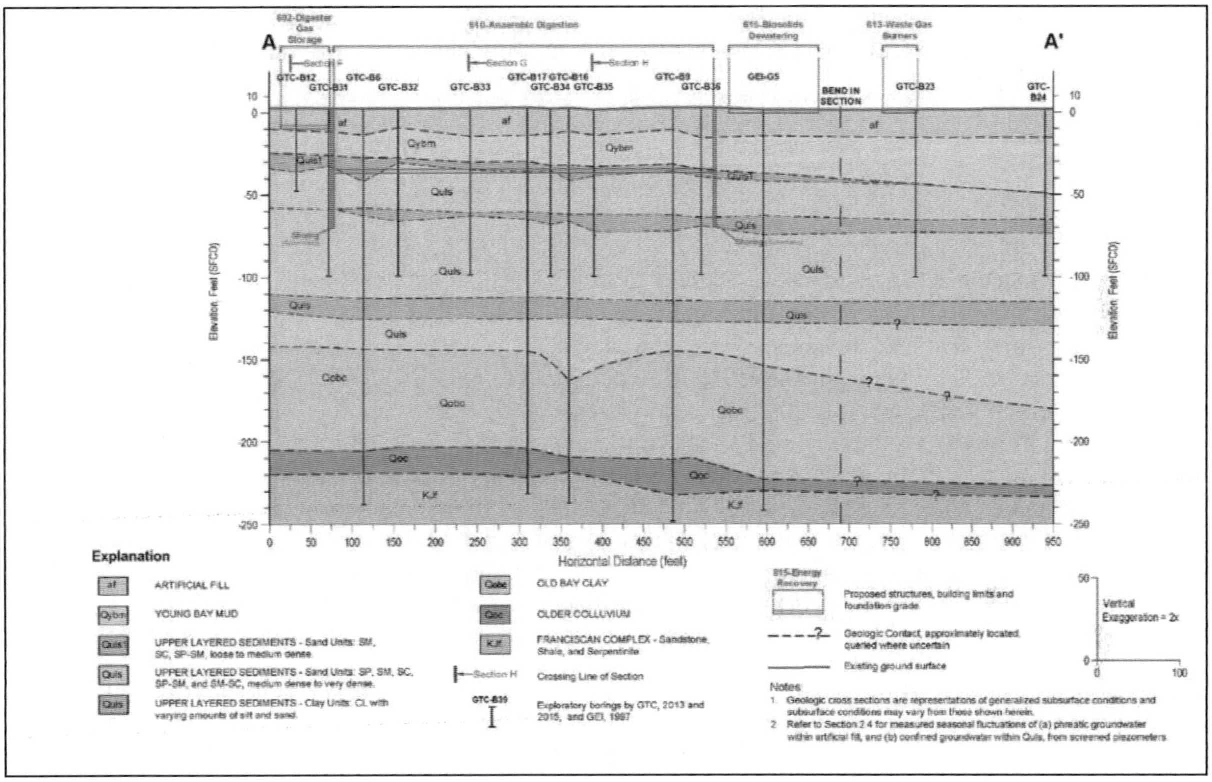

Fig. 2. Subsurface Profile

DESIGN AND CONSTRUCTION CONSIDERATIONS

Facility 610 Shoring and Excavation

The Facility 610 Anaerobic Digestion structure consists of five digestion tanks which extend from a lower basement level that is approximately 34 ft below grade to maximum height of approximately 65 ft above grade (see Fig. 3). Because the excavation is approximately 40 ft deep with numerous site constraints, including two adjacent railroads that could not be impacted by the construction, a rigid shoring system was required. Excavation extended through the artificial fill and Bay Mud and bottomed out in the Upper-Layered Sediments. In order to ensure bottom cutoff, the perimeter shoring wall was extended 160 ft below grade to penetrate into the relatively impermeable Old Bay Clay for a bottom seal.

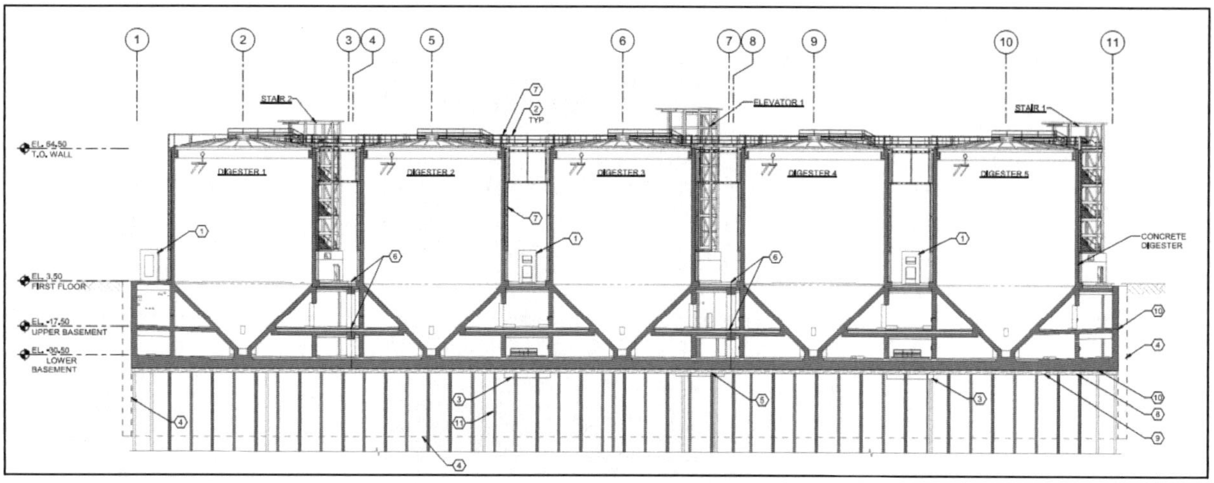

Fig. 3. Profile view of Facility 610

Design and Construction of Excavation Support System

The design and construction of the shoring system navigated several site constraints to ensure the constructability of the tieback/bracing system and the ability to perform the excavation efficiently. The geometry of the excavation would have been ideal for an internally braced shoring system to avoid possible tieback conflicts beyond the structure. However, block-outs through the cast in place digester tanks were not allowed, and the size and shape of the tanks would have required excessively large spans for the internal bracing that could not meet the tight deformation criteria. Other challenges included: high groundwater levels (excessive drawdown outside the excavation was not permitted), very soft and compressible Young Bay Mud, and protecting the active rail lines to the west. The support of excavation system had to be coordinated with existing and new adjacent structures including the existing sewer vault and a forest of piles for new structures and improvements surrounding the Facility 610 excavation. The tieback locations and orientations had to be coordinated to address these constraints.

The constructability at the interfaces of Facility 610 and adjacent facilities (661and 615) and conflict with the existing sewer vault on the western boundary was challenging and special bracing details were developed at these select locations. A supplementary waler detail was added so that tiebacks could be lowered below the deep sewer vault and walers were also added to span the opening at Facility 661.

Fig. 4 shows the shoring system with the excavation at full depth at Facility 610. The 41.5 ft deep excavation is supported by a 3 ft thick, rebar-reinforced diaphragm wall (D-Wall) which is typically restrained by three levels of tiebacks, although localized bracing was used to work around conflicts as noted above. Fig. 5 presents a typical section view of the shoring system at the western side of the site.

Fig. 4. Shored Excavation at Full Depth

The primary wall panels were excavated by mechanical and hydraulic grab as well as hydro cutters mounted on a Bauer base crane MC96. To ensure verticality and watertight joints, the hydro cutter was used to excavate each secondary panel with real time verticality monitoring. A support fluid processing system maintained the fluid properties required to stabilize the trench excavation and recycle the fluid as concrete was placed. The primary panels were typically 22 ft-4 in long and were excavated in three bites. The one bite secondary panels were cut into the previously constructed primary panels. The diaphragm wall was designed to act both as a structural wall and to provide groundwater cutoff. The wall extended to a depth of 158.5 ft below existing grade.

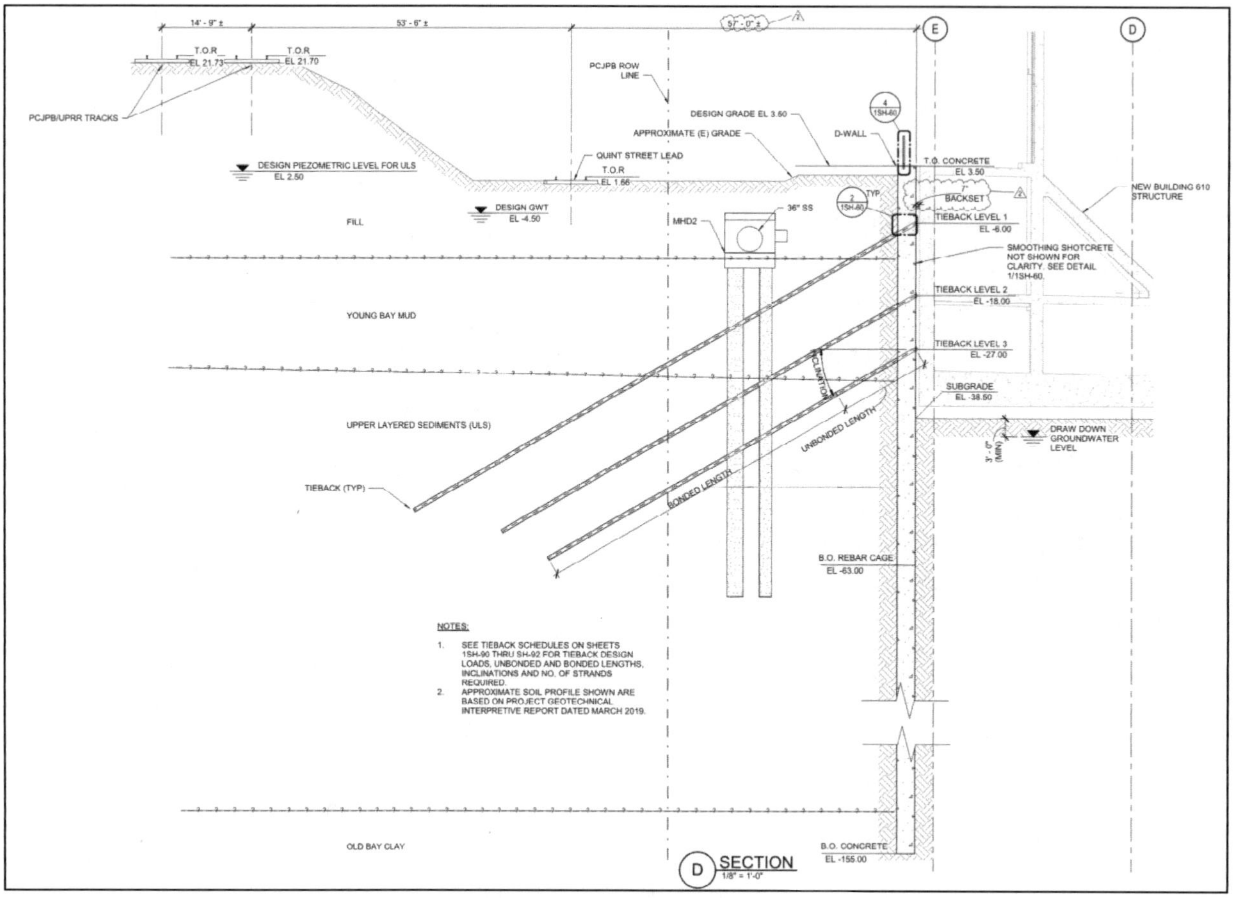

Fig. 5. Typical west wall section of D-Wall Shoring System with tiebacks

Tiebacks were typically installed at 5 ft on center and had design loads of 180 kips at Level 1, 205 kips at Level 2 and 245 kips at Level 3. The tiebacks were installed at 30-degree angles with sufficient unbonded length such that bond zone reached the competent Upper Layered Sediments that are below the Young Bay Mud. Bond capacity was confirmed with field proof tests. The tieback tendons varied from 5 to 10 high-strength strands depending on the tieback design load. The tiebacks were installed through pipe block-outs that were pre-tied in with the diaphragm wall rebar cages. The north-east corner bracing at Level 3 was required to avoid conflict with future permanent piles at the adjacent structure. The typical bracing consisted of HSS20 pipe struts at this location.

Drilling the Level 2 and Level 3 tiebacks (see Fig. 6) below the ground water table and risking excessive water and soil inflow to the excavation was a challenging part of the construction. However, Malcolm used a wall mounted preventer system and developed cased drilling and grouting techniques to minimize the amount of soils and water that entered the excavation during the tieback install. Ultimately, hydroactive grout was pumped into the locked-off tieback wedge plate to stop the leaking water.

Fig. 6. Level 3 Tieback Drilling

All the tieback anchors were designed to avoid the future 36-inch sewer line foundation piles on the west side and drilled shaft permanent foundation piles of adjacent buildings around the entire perimeter of the excavation. It was very important that the drilled shaft foundation piles were installed first in order to avoid the risk of drilling through tensioned tiebacks. Fig. 5 above shows a section view of the D-Wall along with west side with tiebacks navigating existing sewer line pile foundations and being drilled below the adjacent railroad embankments.

The process of locating the tiebacks had to be iterated several times to yield an efficient design in conjunction with the ongoing design of the adjacent structure piles. See Fig. 7 below, showing a partial plan view of the diaphragm wall and the 100 ft long tiebacks that had to be drilled straight in order to miss the already existing pile foundations.

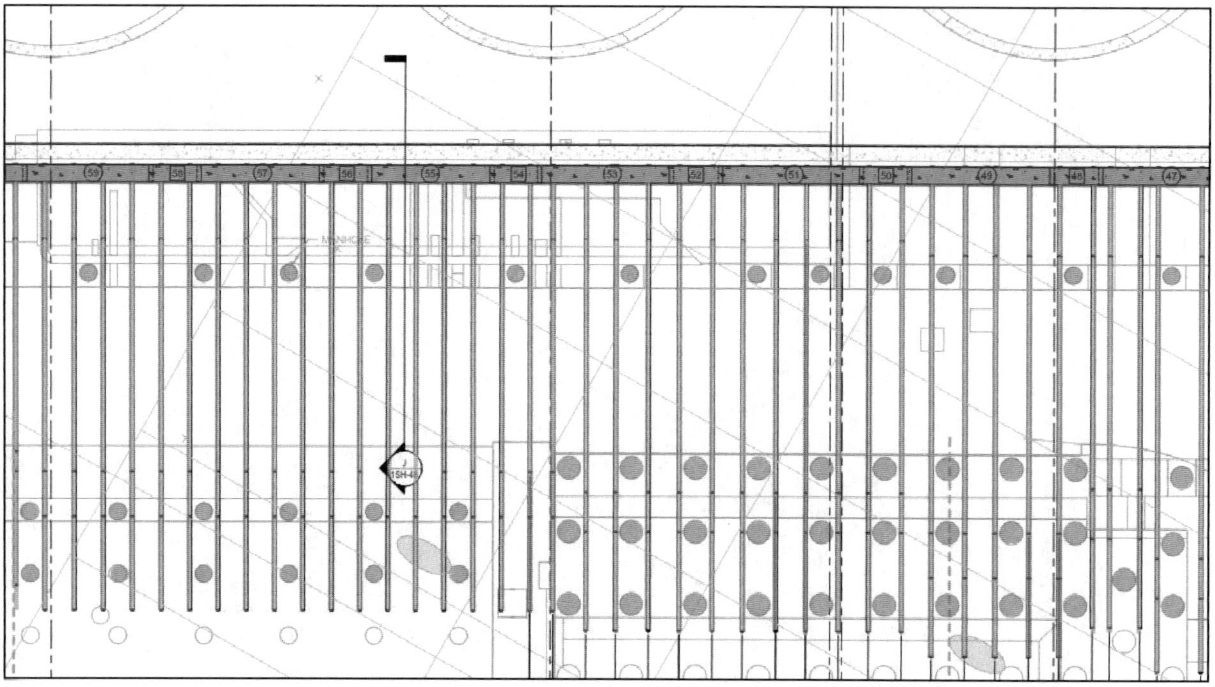

Fig. 7. Plan view of diaphragm wall and designed tieback locations relative to existing pile foundations.

Because the construction sequence for the adjacent buildings was uncertain at the initial phase of the design, the Facility 610 system had to be designed such that all structures could be excavated at the same time, which was extremely challenging due to the complex interaction between the adjacent excavations.

Monitoring and Performance criteria

Movement of the top of slurry wall was monitored on a weekly basis using optical survey techniques. Approximately 40 survey monuments were mounted on top of the diaphragm walls at 40 ft maximum spacing. Six inclinometers were installed around the perimeter of the Facility 610 excavation as shown in Fig. 6. The inclinometers were read manually every week and interpretations of the data was calibrated against deflections measured at the nearby survey monuments.

The contract specifications set threshold horizontal deflections limits of the diaphragm wall at 0.5 inches with a maximum movement limit of 1.0 inches. The baseline measurements were established prior to excavation and monitoring will be continued until the structure is constructed to grade. The survey data was made accessible to the contractors, engineers and other stakeholders.

Groundwater outside the excavation was monitored via manual reading of four multi-level piezometers installed in four boreholes around the entire site with two boreholes closer to Facility 610 that monitored groundwater levels within the Young Bay Mud and Upper-layered Sediments. The dewatering within the excavation footprint did not lower the groundwater more than 5 to 6 ft outside the excavation.

At the time of preparation of this paper, build-out of the basement was underway and lateral deflections had been limited to approximately 0.5 inches or less. The as-built deflection numbers were comparable to estimates made during the design process.

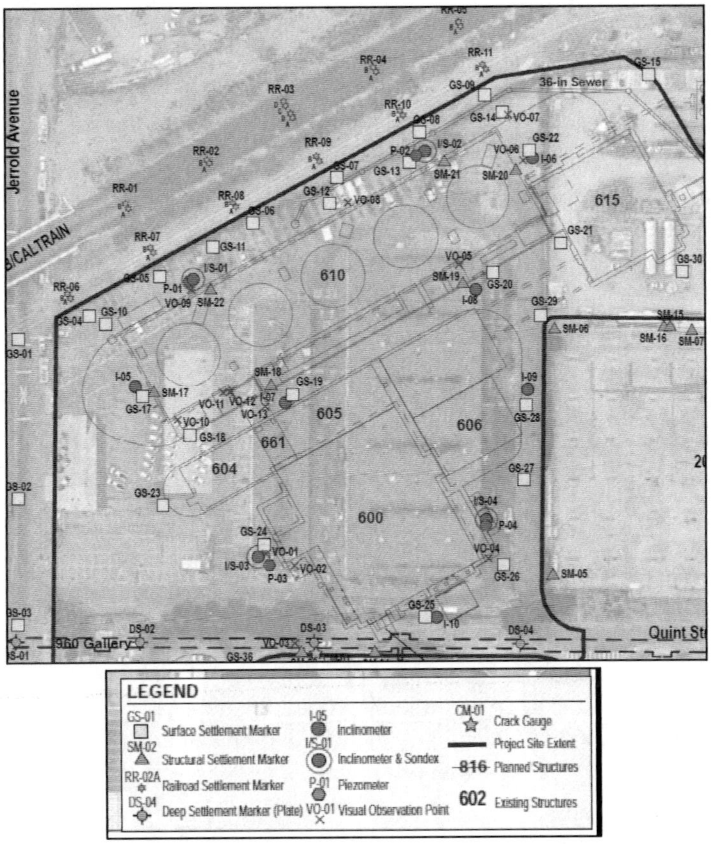

Fig. 8. Instrumentation Plan (Terra Engineers, Inc.)

Construction Verification

The diaphragm wall concrete was specified to have a minimum unconfined compressive strength of 5,000 psi at 28 days. Cylinders were taken from each concrete panel for UCS testing from both pre-production and production wall panels to confirm that the design strength was achieved.

To check the verticality of the excavated wall trench, survey of all the panel excavations was performed using the Koden test. The support fluid properties were monitored during excavation and prior to concrete placement by the field engineers using American Petroleum Institute (API) standard procedures. The concrete level was measured by soundings using weighted tapes after every three trucks discharged (one truck for secondary panels) to monitor concrete volume.

For the west side of the project, where tiebacks encroached on the railroad easement, the first three tiebacks and ten percent remaining tiebacks were performance tested and all other tiebacks were proof tested. For all other sides of the excavation a minimum of the first two tiebacks and two percent remaining tiebacks were performance tested and all other tiebacks proof tested. All tiebacks were required to be de-tensioned during the Facility 610 build-out, so future construction around the excavation would not encounter a locked-off tieback.

CONSTRUCTION

Mobilization and Site Utilization

Logistics was a primary challenge for the construction of this project due to multiple concurrent activities to meet the project schedule. In addition to the diaphragm wall construction on this project, Malcolm also installed 290 each 3 ft and 4 ft diameter auger cast piles, 233 each 4 ft diameter cased drilled shafts and 526 each 1 ft diameter tie down anchors. For 3 shallower structure excavations (up to 26ft in depth), Malcolm installed a Brierley Associates designed, temporary CSM shoring wall which was supported by 1 level of internal bracing. Over 50 dewatering wells were installed by Malcolm to manage and treat the groundwater at the various excavations.

The site configuration including access/egress, installation sequence, and support equipment layout was carefully planned prior to mobilization. The general site configuration is shown below in Fig. 9. Onsite daily meetings included all drilling superintendents and subcontractors to plan the next day's activities, and a unique coordination map was created each day to reflect the new work areas. This level of coordination was required to manage the large site and ensure that spoils were efficiently off-hauled and ready-mix concrete made on time deliveries to drilling locations. At the peak, 80 trucks of spoil were being loaded out while 80 trucks of concrete were being delivered to the jobsite daily.

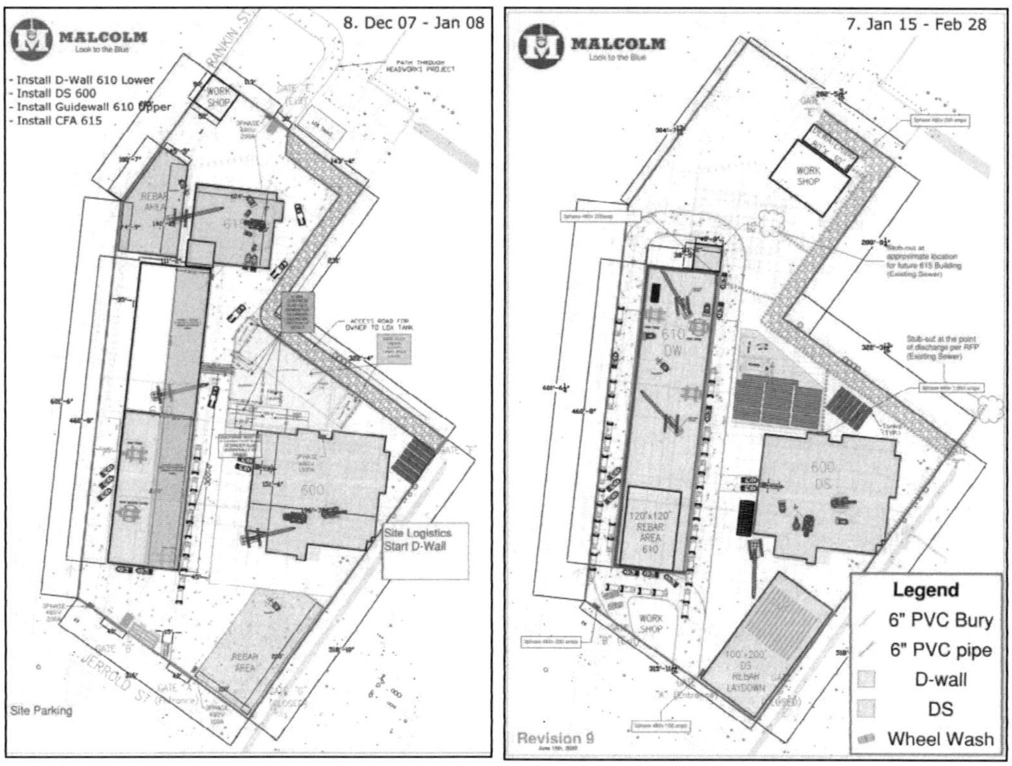

Fig. 9. Malcolm Logistics Plans to manage work flow

Fig. 10. Photo of actual site construction during D-Wall installation

The equipment used to install the diaphragm wall included a Bauer BC40 Hydro cutter mounted on a Bauer MC96 crane, one hydraulic grab mounted on a MC64 crane, one mechanical grab mounted on Liebherr HS 885, one 300T class support crane, one 150T class support crane, a Bauer BE550 Desander unit, a centrifuge, and (22) 21,000-gallon open top mixing tanks (see Fig. 10 above).

Diaphragm Wall Construction

Temporary reinforced concrete guide walls were constructed along the alignment of the diaphragm wall to be utilized as a guide for the excavation equipment and the setting of the rebar cages. The guide walls consisted of two parallel reinforced concrete beams that were 1 to 2 ft wide by 3 to 4 ft deep. Extremely high levels of accuracy and quality control for the guide wall construction are critical as they are used to maintain diaphragm wall panel verticality, location, and rebar cage elevation control. The top of the guide wall was also used for the installation of hard survey control points for the diaphragm wall construction.

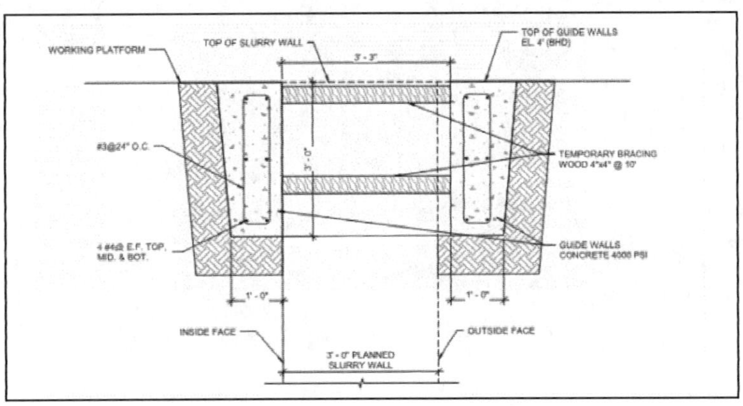

Fig. 11. Guide Wall Construction

The BC40 Hydro cutter is an extremely powerful, versatile, and proven machine with a 41 ft guide frame equipped with steering flaps and real time verticality monitoring to control the plumbness of the excavation. At the bottom of the large, rigid steel frame are two cutting wheels that can swapped out to match the precise width of the required diaphragm wall design and cutting teeth that can be changed to accommodate the type of soil or rock being excavated.

Fig. 12. BC40 Hydro cutter mounted on MC96 crane.

Fig. 13. Two Hydro cutters and One Hydraulic grab

Quality control of the panel excavation is critical for the water tightness of the diaphragm wall system. Prior to concrete and rebar placement, the panel excavation is again independently checked via a KODEN drilling monitor, which uses ultrasonic waves to measure a precise profile of the panel excavation to confirm it meets dimensional tolerances.

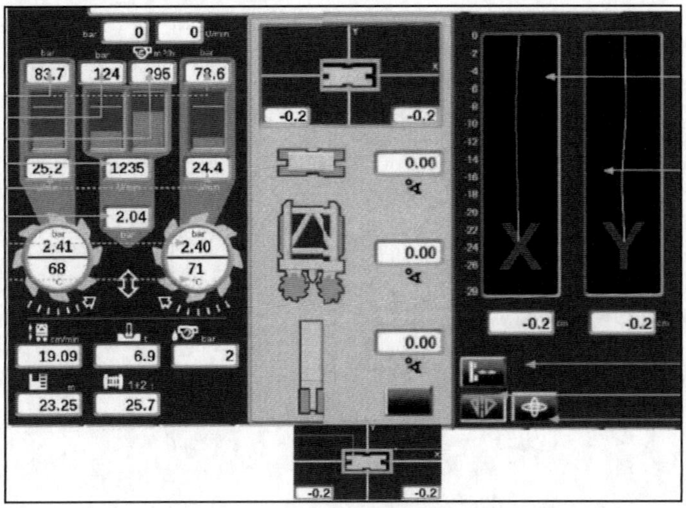

Fig. 14. Real-time monitoring by B-Tronics of panel excavation

Panel reinforcing cages were assembled horizontally on the ground at the project site and later uplifted to a vertical position for installation. Bracing embeds with shear studs, and tieback blockout pipe sleeves, were all installed in the rebar cage prior to lifting and installation. Right angle corner panel cages were assembled, lifted, and installed monolithically as well. All of the cages were lifted with the single 300 ton support crane and brought to the panel for installation, with the heaviest cage weighing approximately 20 tons.

Fig. 15. Rebar cage installation in tandem with slurry wall excavation.

Placement of the 5,000 psi design strength concrete was achieved via tailgate placement into gravity tremies. Each primary panel pour was approximately 400 CY and required the use of three simultaneous tremies. Each secondary panel pour was approximately 160 CY and required the use of two simultaneous tremies. In total approximately 25,000 CY of structural concrete were placed for the slurry diaphragm wall.

Fig. 16. Primary Panel concrete placement with simultaneous excavation in background.

Quality control of the concrete mix design is a critical factor for the slurry diaphragm wall system. An extensive pre-production trial batch program was implemented prior to mobilization for this project to develop a mix with local suppliers that met the required design strengths and workability parameters. During construction, continuous testing of the delivered concrete for flow, flow retention, and segregation following the EFFC-DFI Guide for Tremie Concrete was implemented prior to the concrete going in the ground.

CONCLUSIONS

The early involvement of a core trade subcontractor proves successful to the design and construction of mega infrastructures that are complex and require intensive coordination between owner, CMGC, Design Team and contractors to overcome the many challenges a project like this presents.

In 2018 and 2019, Malcolm spent many hours traveling to, and attending, in-person meetings with the Design Team and the CMGC, which were indeed necessary, but now with online meetings so commonplace, this collaboration process only became more efficient as the project progressed.

For this mega project, not only did Malcolm deploy nearly every type of drilling method in our arsenal, but we undertook many scopes that normally the General Contractor handles. In addition, Malcolm had to meet a local small business hiring goal for the project which essentially meant that any subcontractor

Malcolm wanted to hire had to be a local small business San Francisco company. Through community outreach meetings and partnering with local companies, Malcolm was successful at finding and managing multiple local San Francisco subcontractors for the following scopes of work: excavation, spoil off-haul/disposal, rebar cage fabrication, surveying, site security, and dust control.

Many valuable lessons were learned on this project and while they don't come around often, Malcolm and Brierley are well suited to take on the next challenging mega project in the future.

Fig. 17. Overview of project site with completed diaphragm wall excavation at Facility 610 (left) and internally braced CSM shoring wall at Facility 600 (right)

Acknowledgements

The authors would like to acknowledge the following organizations who were involved with the project and whose collaboration was valuable throughout design and construction: SFPUC, The SFPUC Design Team, MWH/Webcor JV, Silverado Contractors Inc., Bertco Inc., Central Concrete Supply, Cemex, CMC Rebar, Alta Group Inc., and Mike O'Dell Surveys.

SEISMIC ANALYSIS OF DRILLED SHAFTS AS EARTH RETAINING STRUCTURE BASED ON SITE CHARACTERIZATION

Bernardo J. Jiménez Vega, H Solís, Costa Rica, bjimenez@hsolis.com
Edy L. T. Montalvan, PhD, edytemon@gmail.com

ABSTRACT

During the last phase of the *Circunvalacion Norte* project in San Jose, Costa Rica, preliminary designs for an underpass structure of over 16 km (10 miles) considered drilled shafts as an earth retaining system. Since the project's execution was delayed and close to the deadline, it was necessary to optimize this structure, with the drilled shafts being the most critical aspect. The key element in order to obtain an improved design was characterizing the job site with a specific site study for seismic demand, and the use of advanced numerical modelling, considering soil-structure interaction. The site investigation comprised the CPT, PMT, MASW geophysical tests, deep exploration drilling, and triaxial compression tests. The terrain was modeled using an advanced constitutive model with hardening plasticity calibrated from the triaxial tests. Plaxis software was used to calibrate the seismic model for the terrain and for the dynamic analysis. Two conditions were considered for the retaining structure: free-head and fixed head. A parametric study with drilled shafts of 1.0 m (3.3 ft) and 1.2 m (3.9 ft) in diameter, and center-to-center spacing of 1.3 m (4.3 ft) and 3.0 m (9.8 ft) was conducted in order to obtain an optimized structure. For each condition of the drilled shaft wall, the computed deformations were compared for two spectrum designs: site-specific spectrum and seismic code generic spectrum. Finally, a dynamic analysis was performed to check the performance of the drilled shafts with optimized spacing using a control earthquake. Based upon seismic performance curves, the analysis showed the possibility of greater spacing for the drilled shafts compared to the original design, obtaining a 32% reduction in the total length of the drilled shafts in the project.

Keywords: drilled shafts, seismic hazard, geotechnical site study, earthquake, dynamic analysis

INTRODUCTION

In Costa Rica, for structures like bridges and viaducts, there is a guideline code for seismic resistant design (known as LDSP in Spanish). In the absence of seismic hazard studies with site-specific characterization, the LDSP code stipulates a procedure to create a generic design response spectra, based on seismic zoning and soil type.

The *Circunvalacion Norte* project is the last segment of an urban peripheral highway around San Jose, the capital city of Costa Rica. The initial design considered using the generic design spectrum from the LDSP guideline. During the redesign stage of its last segment, a site-specific seismic hazard study was conducted. This study allowed design optimization of drilled shafts for the underpass structure, by increasing spacing and decreasing the length of the shafts.

This paper presents a summary of the seismic hazard characterization and a comparison of the two design response spectra (the generic versus the site-specific). Finally, the optimization of the design through pseudo-static analysis and dynamic analysis using control earthquakes is presented.

One intent of this paper is to demonstrate the importance of site-specific seismic studies, especially on large projects. In the case herein presented, a site-specific spectrum allowed designing with lower seismic demands (e.g., lower resistance factors) than the LDSP's generic response spectra, while maintaining the same return period (2500 years), and providing a lower degree of uncertainty. Hence, using a lower seismic demand results in money saving and reduced execution time, which validates and justifies the use of this type of seismic study.

SEISMIC CHARACTERIZATION

The objective of seismic characterization is to determine the seismic hazard associated with a given return period for a specific site. In this study, the return period was 2500 years. Furthermore, other parameters considered were possible earthquake sources, propagation of waves between the source and the study site, and geotechnical characteristics of the subsoil. A hazard calculation was carried out from three perspectives: deterministic, probabilistic, and hybrid, integrating the results of the modeling of the source, propagation, and local ground effects in all cases. Finally, the critical case for the analyzed structures has been presented, corresponding to a drilled shaft wall, which can be related to a structure with a short period (T = 0.1 s), close to peak ground acceleration (PGA).

Source model

A survey was conducted to map faults near the study area. A total of 13 active or potentially active faults were located within the study area. The faults were divided into two groups: near faults, which could lead to the highest accelerations, and remote faults within 40 km from the study site, that must be accounted for displacements during long lasting earthquakes. Table 1 presents information of the faults considered in this study.

Potential hazard

Estimation of a potential maximum earthquake, with its source on a geological fault, consists in using empirical regressions that relate the area or length of a rupture with the magnitude of large earthquakes that can occur. The energy released by an earthquake is directly related to the instantaneous stress drop that occurs in the rupture plane, and this is related to the area of the fault surface that slides.

One important challenge is the selection of a suitable regression model for the region, according to the relationship between fault length and maximum moment magnitude (Length/M_{max}). These regressions are fundamentally based on a database of historical earthquakes in which the rupture dimensions have been determined. The mixture of earthquakes in different tectonic environments are associated with different types of faults, which results in large differences in the existing Length/M_{max} relationships. This study uses the Stirling et al. (2008) equation (Eq. 1) for the investigated site.

$$M_w = 4.18 + 2/3 \log W + 4/3 \log L \qquad [1]$$

Where, M_w is the moment magnitude, W is the fault width and L is the fault length. This regression equation was developed for obliquely slipped New Zealand faults, which are applicable to volcanic tectonic environments such as this study case. Table 1 shows the results of the potential hazard, while M_w values higher than 7.0 can be found for faults in less of 40 km distance, all the faults nearest to the specific site (less than 10 km), have values of Mw varying from 6.44 to 6.76.

Table 1. Potential hazard for faults near the specific site.

Fault name	Distance (km)	Style	Length (km)	Width (km)	Slip rate (mm/yr)	M_w (±0.18)
Navarro 3	<40	Left	45.0	12	3±2	7.10
Atirro	<40	Dextral	47.3	12	2±1	7.13
Rio Sucio	<40	Dextral	17.7	12	2±1	6.56
Alajuela	<10	Reverse	20.9	17	10	6.76
Aguacaliente	<10	Left	16.6	12	2±1	6.53±0.06
Guapiles	<40	Reverse	13.2	17	0.6±0.4	6.49
Siquirres	<40	Reverse	56.1	17	<1	7.33
Cipreses	<10	Reverse	12.0	17	2±1	6.44
Escazú	<10	Normal	16.0	17	2±1	6.61
Lara	<40	Dextral	29.7	13	1-5	6.89

Ground Motion Prediction

The seismic potential of a fault is determined by its geometric characteristics as shown in the previous section, but this potential release energy is distanced from the study site. One of the greatest challenges in seismic site studies is to estimate the wave propagation effects from the fault on to the study site. For this determination, ground motion prediction equations (GMPE) are used to calculate earthquake attenuation.

To estimate earthquake attenuation, it was decided to use the second version of the Next Generation Attenuation models (NGA-West2). This study implemented the best model from the NGA-West2, and the results were compared with data from historic accelerograms using the criteria proposed by Kale & Akkar (2012), specifically, the CB-13 model (Campbell & Bozorgnia, 2013);, this model uses analysis of residuals from the ground motion data to create response spectra for periods ranging from 0.01 to 10 seconds using the following terms: magnitude moment, distance of rupture, sediments depth, hypo-central depths, shear wave velocities V_{s30}, and rupture dips. For further details of the model check Campbell & Bozorgnia (2014).

Site response

Seismic analyses generally use average shear wave velocity for 30 m depth (V_{s30}) for subsoil characterization and evaluation of site response effects. The value of V_{s30} is calculated as follows:

$$V_{s30} = \frac{30}{\sum_{i=1}^{n} \frac{d_i}{Vs_i}} \qquad [2]$$

For the study site, Multi-channel Analysis of Surface Waves (MASW) tests were performed to characterize the subsoil. Fig 1a shows the results for one of the profiles, with a V_{s30} value of 284 m/s. In all the tests, V_{s30} values varied between 230 and 290 m/s. Figure 1 b shows a typical profile of the study site, where a layer of more resistant material stands out with values greater than 360 m/s and a thickness of 5.0 m.

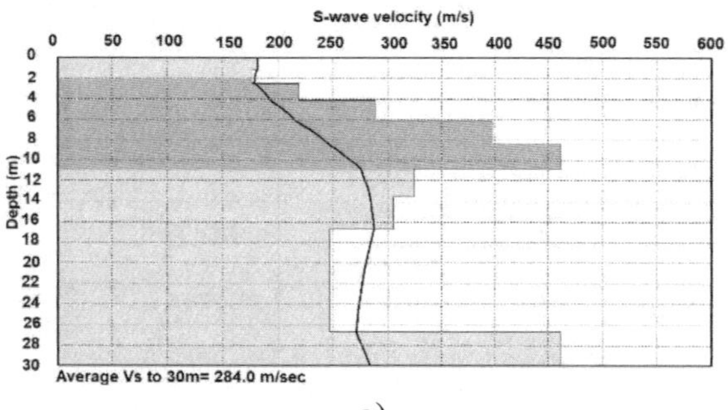

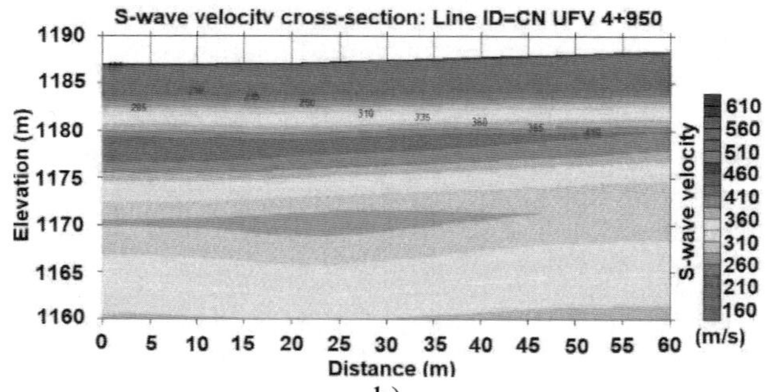

Fig 1. Site specific characterization: a) V_{s30} profile; b) Shear wave site map for station 4+950.

Amplification factors

Once the site response was classified, the degree of amplification relative to the seismic movement was estimated. It is understood that the softer the material (lower V_{s30} value), the higher the amplification effects. Fig 2 presents the amplification factors (FSA) for this study. These factors were estimated based on the NEHRP (2015) guidelines.

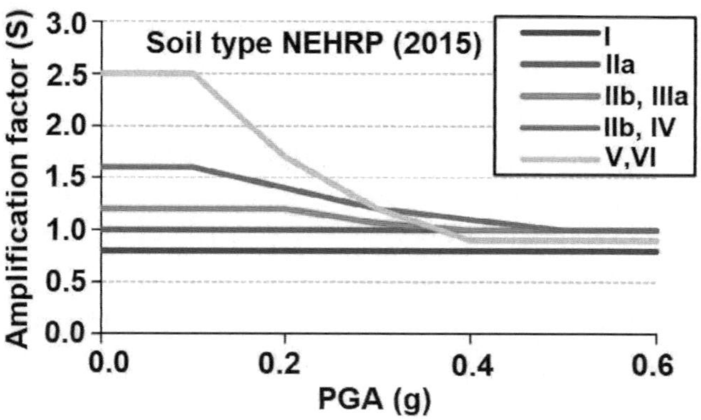

Fig 2. Amplification factor based on V_{s30}.

The site zone corresponds to NEHRP's IIIa soil type, where amplification values varied between 1.0 and 1.2 depending on the PGA obtained without considering local effect. These amplification factors are applied to the acceleration corresponding to the rock material (Eq. 3).

$$\vec{A}(site - specific) = FSA \times \vec{A}(rock) \qquad [3]$$

Seismic hazard

The seismic hazard considered three approaches: deterministic, probabilistic, and a hybrid scenario using control earthquakes. The probabilistic approach implemented a catalog of earthquakes within a radius influence of 300 km, this catalog has been updated with data from 20 years of national records registered by the Earthquake Engineering Laboratory of the University of Costa Rica (Moya-Fernandez et al., 2020). Fig 3 presents the disaggregation of earthquakes through probabilistic analysis. It is evident that the earthquakes with the highest probability of occurrence are those with distances of less than 10 km and moments of magnitude M_w close to 6.5.

Based on the most possible earthquakes in the area, it was decided that the critical scenario, for short periods (close to the PGA), would correspond to a hybrid analysis using the characteristics of the *Escazu* fault. This fault is located at a distance of less than 10 km from the study site and has a M_w value of 6.61, coinciding with the results of probabilistic analysis. Fig 4 shows the spectrum results for the case considering the *Escazu* fault. Fig. 4 presents the computed response spectrum for the mean, the mean plus one standard deviation (std), and the mean minus one std, for 84% confidence intervals. In the same figure, the generic spectrum (LDSP) used in the first design of the project is also presented. In general, all the spectrum from the site-specific study indicates lower values of spectral acceleration than the LDSP spectrum. Regarding the PGA, the original study adopted 0.61 g, while the site-specific study presented in this paper estimated a maximum value of 0.45g. This reduction of more than 25% of the seismic forces allows for the use of sets of less rigid elements in the drilled shaft wall, maintaining the same expected level of performance.

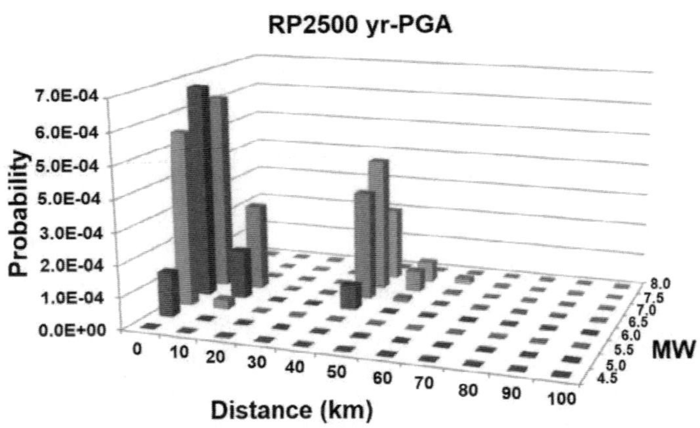

Fig 3. Disaggregation of the probabilistic calculation

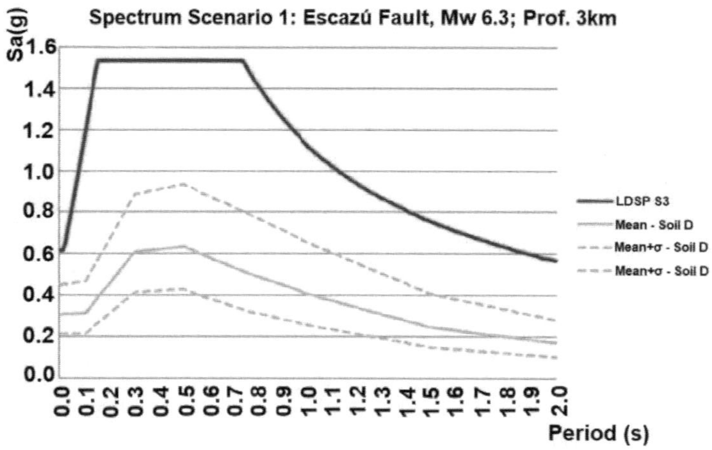

Fig 4. Response spectrum for the critical scenario.

GEOMECHANICAL CONSTITUTIVE MODEL

Drilled shaft wall optimization was performed using a suitable constitutive model to describe deformations and ground-structure interaction. Historically, elastic perfectly-plastic models have been used for this kind of problems, however, it is known that these models fail to realistically simulate deformations, since they describe exaggerated dilatancies, and adopt constant deformation modules as well as constant yield surfaces. For soil-structure interaction problems, and specifically for the analyzed problem, it is recommended to use hardening plasticity models. In this case, the Hardening Soil Model with Small Strain Stiffness (HSSmall) was employed, as it allows for the variation of the modulus of deformation and adjustment of the yield surface depending on the stress-strain state of the material. Additionally, for small deformations such as those presented in earthquakes, it allows for a degradation of the shear modulus.

The HSSmall-model parameters were calibrated based on MASW, CPT, Pressuremeter and CD triaxial compression tests. Table 2 presents the parameter's values for the adopted constitutive model. Cohesion values presented in Table 2, are about 1 to 2 times smaller than those obtained from the triaxial tests. It was decided to conservatively reduce cohesion values for design.

Table 2. HSSmall calibrated parameters.

Parameter	UG 1	UG 2	UG 3
V_{s30} (m/s)	<220	220-360	>360
Friction angle φ'	28	30	28
Cohesion c' (kPa)	10	20	40
E_{50} (MPa)	20	45	80
E_{oed} (MPa)	20	45	80
E_{ur} (MPa)	60	100	200
G_0 (MPa)	550	951	1200
Poisson's coef. υ	0.3	0.3	0.3
$\gamma_{0.7}$	0.1%	0.1%	0.1%

FINITE ELEMENT ANALYSIS

Design optimization was performed using the PLAXIS 2D software. For this case study, it is appropriate to use a plane strain model, since the alignment of the wall does not present abrupt changes that would cause stress concentration on the analyzed plane. Due to the limits of the software, the arching effect between drilled shafts could not be considered, however, this effect is believed to give higher stability; thus, not considering this effect is on the conservative side.

Two analysis models were made for the drilled shaft wall. The first model, presented in Fig 5 a, corresponds to free head drilled shaft, with the shafts only being a retaining wall. The second model, presented in Fig 5 b, corresponds to braced drilled shaft with a fixed head, with the placement of beams and concrete slabs in the upper part. In this case, the drilled shafts work as retaining walls and as foundation for the overpass structure.

Drilled shafts were modeled using two diameters, 1.2 m and 1.0 m. Center-to-center spacing of drilled shafts was varied adopting the following values: 1.3, 1.6, 2.0, 2.6, and 3.0 m. Shafts depth have values between 10 and 19 m, and were kept constant in the numerical analysis. PLAXIS considers an equivalent spacing d_{eq} (eq. 4) between drilled shafts, using the normal stiffness (EA) and flexural rigidity (EI) of the drilled shaft. The idea is to create plates of equivalent stiffness for each case.

$$d_{eq} = \sqrt{\frac{12EI}{EA}} \qquad [4]$$

For each model, construction stages were simulated. Additionally, pseudo-static analyses were modeled considering a horizontal seismic coefficient (k_h) estimated from the LDSP response spectrum (0.34) and from the site-specific spectrum (0.25). This pseudo-static model was placed after the service stage. Finally, for the adopted spacings, two control earthquakes were modeled by using the compliant base and free field dynamic boundary options, to verify the performance of the wall.

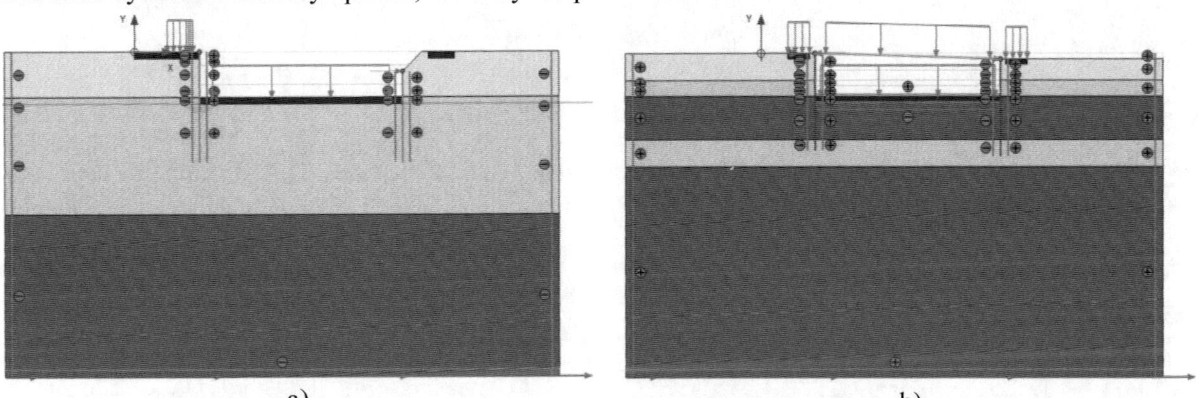

a) b)

Fig 5. Plaxis model a) Free head drilled shaft; b) Fixed head drilled shaft.

Comparison between LDSP and site-specific

The results from Plaxis analyses are presented in Fig 6 and 7, showing a comparison of lateral deformations (Ux) of the fix and free head drilled shaft with varying diameter and spacing, and considering the two response spectrum (generic LDSP and site-specific). The purpose of this analysis is to achieve the optimization of drilled shafts spacing. The original project design proposed a 1.2 m diameter and 1.5 m spacing for free head conditions, and drilled shafts with 1.0 m diameter and 1.3 m spacing for fixed head conditions.

For 1.0 m diameter drilled shafts, head lateral displacements (Ux) obtained for the LDSP generic spectrum were around 50% of those obtained for the site-specific spectrum. Analyses with the site-specific spectrum computed horizontal displacements between 15 and 20 cm for free head conditions (Fig 6 a). For fixed head conditions, the displacements were around 6 cm (Fig 6 b). Drilled shaft spacing has higher influence on displacements for free head conditions, increasing up to 25% between the smallest and the largest analyzed spacing. For fixed head conditions, this difference is in the order of 10%. Results for the case of drilled shafts with spacing of 3.0 m and free head conditions is not presented, since the displacements are excessive and the model collapses.

For 1.2 m diameter drilled shafts, the same magnitudes of displacements were observed as well as a significant reduction in displacement when using the site-specific spectrum. Regarding the influence of diameter, 1.2 m diameter drilled shafts with free head conditions presented displacements up to 20% lower than those of 1.0 diameter drilled shaft, this tendency increases as the spacing between the shafts increases. The same behavior was observed for fixed head conditions, however, displacements for 1.0 m diameter shafts were 15% lower than those of 2.0 m diameter shafts.

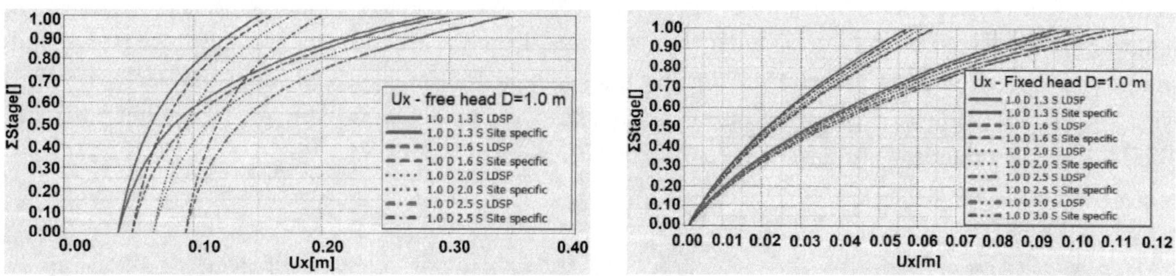

a) b)

Fig 6. Deformations at head for 1.0 m diameter drilled shafts a) Free head drilled shaft b) Fixed head drilled shaft.

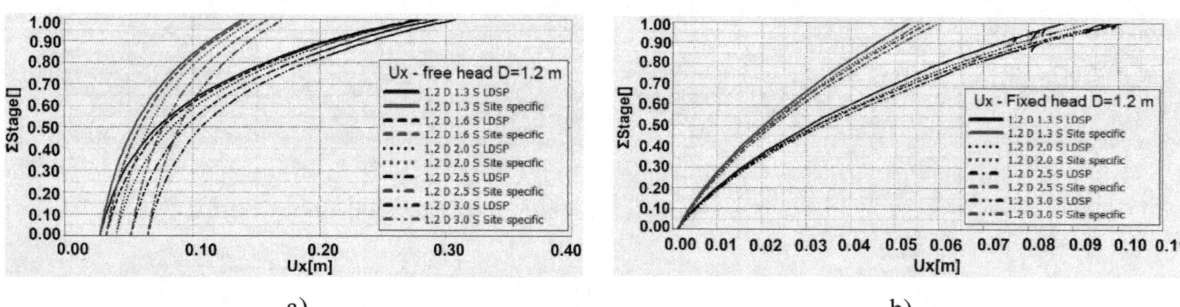

a) b)

Fig 7. Deformations at head for 1.2 m diameter drilled shafts a) Free head b) Fixed head.

In general, all tested conditions present deformation values within the same range and they meet global stability requirements (detailed discussion of stability analysis is outside the scope of this paper). Based on these results, which are considered to have lower uncertainty since they come from a site-specific study, and knowing that the pseudo-static analysis is usually conservative compared to a dynamic analysis, a decision was made on the diameter to be used in the optimization. For the free head condition, it was decided to use drilled shafts with 1.0 m diameter and 2.0 m center-to-center spacing, and for the fixed head condition, it was decided to use drilled shafts with 1.2 m diameter and 2.4 m center-to-center spacing. This new shaft spacing results in a reduction of 32% of the total length of drilled shafts in the project. Depth of the shafts kept constant in all analysis.

Control Earthquake Performance

The results of the dynamic analysis using the control earthquakes are shown in Figs 8, 9 and 10. These earthquakes come from local records of previous events. Fig 8 a shows the pseudo-spectral acceleration graph: earthquake 1 has a PGA value of 0.24g and earthquake 2 a value of 0.30g. Fig 8 b shows the accelerograms of the earthquakes: earthquake 1 is characterized by being a long-duration event with distorted large amplitude values, and earthquake 2 is a short event with more frequent large amplitudes. The analysis was carried out for both conditions: free-head drilled shaft (1.0 m diameter and 2.0 m spacing) and fixed-head drilled shaft (1.2 m diameter and 2.4 m spacing). The objective of this analysis was to evaluate the behavior of the structures using the optimized spacing.

Fig 9 shows the Fourier spectrum of the seismic waves at three depth points of the drilled shaft: head, middle and tip. In the case of the free head drilled shaft, a signal amplification between 2 and 6 Hz is observed at the head point, which can cause amplification of the accelerations in the upper part, since it coincides with the fundamental period of the structure (0.1 s), modifying the vibration mode at the top of the drilled shaft. In the case of the drilled shaft with fixed head, the signal amplification at the head point occurs in a smaller range, between 2 and 3.5 Hz, moving further away from the fundamental period of the structure. In this case, signal amplification was observed between the bottom of the drilled shaft and the middle, for waves less than 2 Hz.

The expected performance of the drilled shafts during the seismic event at the field is presented in Fig 10. Control points were placed on the drilled shaft to evaluate lateral displacements of the whole structure (block movement) and deflections in the upper part of the drilled shaft. Block movements are less damaging, they prevent cracking fissures and loss of section, as they can occur in large deflections. In the case of the free head drilled shaft, it is observed that for both earthquakes the displacements along the drilled shaft are similar. In comparison to deflections before the earthquake, deflections after both events increased by 1.0 cm between the lower and middle part and by 1.0 cm between the middle part and above. This increase in deflection is due to a difference in vibration modes in the upper and lower part of the drilled shaft. The computed deflection at drilled shaft head corresponds to approximately 1% of the excavated height, which is considered an acceptable limit after a control earthquake. In the case of fixed head drilled shaft, for earthquake 1 there is a minimum increase (2 mm) in the deflection of the drilled shaft. For earthquake 2, no increase in deflection was observed. In this case, the deformation at all control points coincides for both earthquakes, showing that for fixed head drilled shafts, the bracing allows the entire structure to vibrate with the same period, creating block displacements and not allowing deflections to increase. In this case, the deflection found corresponds to less than 0.1% of the height. These results show that elements meeting minimum safety requirements will be used in the spacing of drilled shafts for an optimized design.

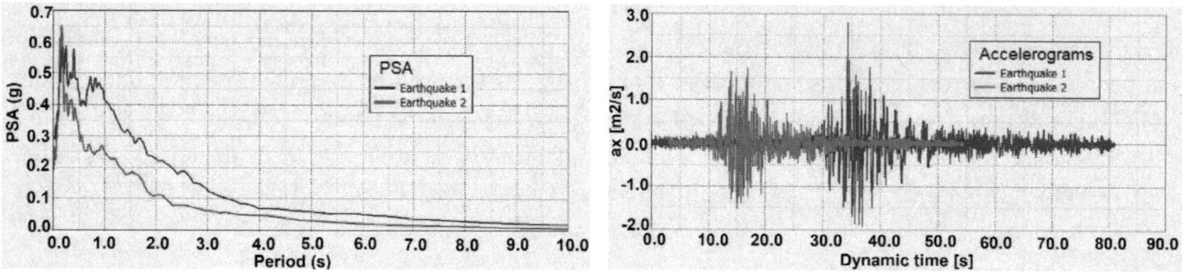

a) b)

Fig 8. Earthquake events in dynamic analysis a) pseudoespectral acceleration b) accelerograms.

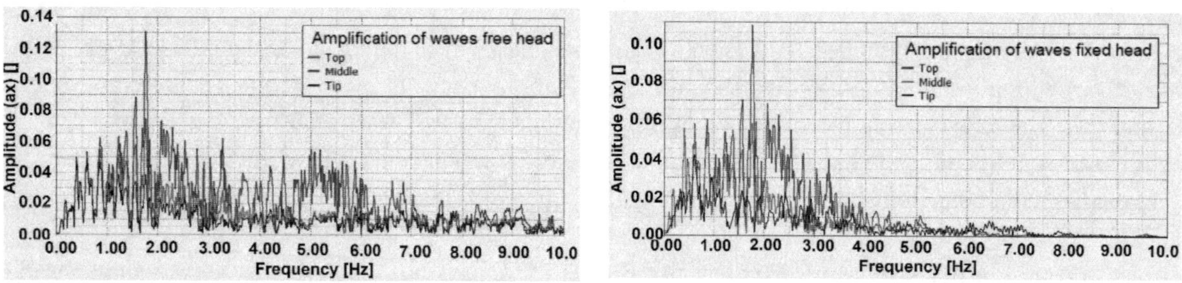

a) b)

Fig 9. Amplification of seismic waves a) free head condition b) fixed head condition.

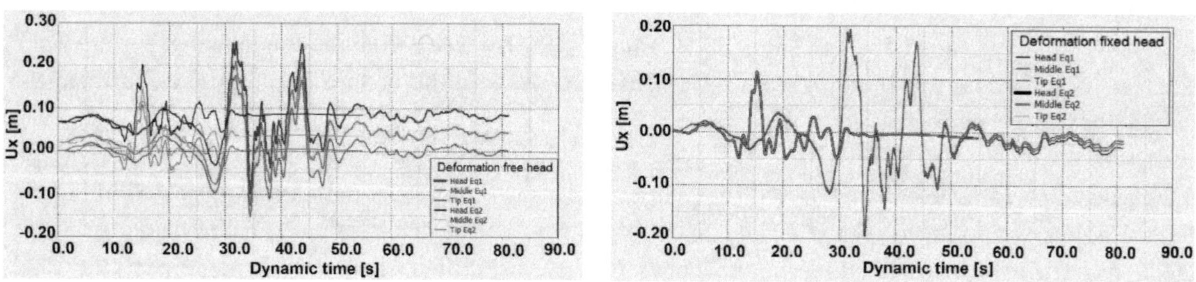

a) b)

Fig 10. Dynamic horizontal deformation for earthquakes 1 and 2: a) free head; b) fixed head.

CONCLUSIONS

In this paper, conditions and seismic behavior of drilled shafts on the last stage of the *Circunvalacion Norte* project in Costa Rica were analyzed. The possibility of optimizing the center-to-center spacing of the drilled shafts through implementation of a site-specific study to characterize the seismic demand was analyzed. The spectrum from the site-specific study was compared to that of the generic spectrum stipulated by national regulations.

During the site study, for each fault, the associated potential was presented according to its physical characteristics. The GMPE model selected as most appropriate was the Campbell & Bozorgnia, (2013) model. Using MASW tests, the site was characterized to determine site amplification factors. The seismic hazard was evaluated by three approaches, namely deterministic, probabilistic and hybrid. This made it possible to determine that, for the analyzed structure, the Escazu fault presented the most important hazard. The earthquake spectrum from the site-specific study was presented and compared with the generic spectrum, obtaining lower spectral accelerations and less uncertainty.

Finite element analysis was performed using the Plaxis software for various shafts spacings and diameters, in order to determine an optimized spacing for the drilled shaft wall. The results were presented comparing them with the results of the generic study. In all cases, the deformations from the generic study were about two times greater than those from the site-specific study. The optimized design was 2.0 m spacing with 1.0 m diameter for free head drilled shafts and 2.4 m with 1.2 m diameter for fixed head drilled shafts. Control earthquakes were used to verify the earthquake performance of the optimized spacing. An amplification of the signal was evidenced at the head of the free head drilled shafts, increasing its horizontal deflection. For the case of the fixed head drilled shaft, the structure behaves as rigid body, without relative deflections during the earthquake. The computed deflections for both cases are found within allowed tolerances.

Finally, the importance of site-specific studies is highlighted which, with characteristic information of the place, can allow for reductions in the length of the drilled shafts; in this case, a reduction of 32% of total length of drilled shafts in the project. It should be noted that there may also be a case in which a site-specific study gives greater seismic demands, however, the use of site-specific data with a lower degree of uncertainty should be rewarded.

REFERENCES

Campbell, K. W., and Bozorgnia, Y., 2013. NGA-West2 Campbell-Bozorgnia Ground Motion Model for the Horizontal Components of PGA, PGV, and 5%-Damped Elastic Pseudo Acceleration Response Spectra for Periods Ranging from 0.01 to 10 s, PEER Report No. 2013/06, *Pacific Earthquake Engineering Research Center*, University of California, Berkeley, CA, 238 pp

Kale, O., & Akkar, S. (2012). A novel procedure for selecting and ranking candidate ground-motion prediction equations (GMPEs) for seismic hazard analysis: Euclidean distance-based ranking (EDR) method. *Bulletin of the Seismological Society of America*, *103*(2A), 1069-84.

Moya-Fernández, A., Pinzón, L. A., Schmidt-Díaz, V., Hidalgo-Leiva, D. A., & Pujades, L. G. (2020). A Strong-Motion Database of Costa Rica: 20 Yr of Digital Records. *Seismological Research Letters*, *91*(6), 3407-3416.

NEHRP (2015). Council, B. S. S. (2009). NEHRP recommended seismic provisions for new buildings and other structures. *Federal Emergency Management Agency*, FEMA P-1050-1/2015

Stirling, M., Gerstenberger, M., Litchfield, N., McVerry, G., Smith, W., Pettinga, J., & Barnes, P. (2008). Seismic hazard of the Canterbury region, New Zealand: New earthquake source model and methodology. *Bulletin of the New Zealand Society for Earthquake Engineering*, *41*(2), 51-67.

ANALYTICALLY MODELLING DST ARRIVAL TIME DATABASES WITH HIGH ORDER POLYNOMIALS FOR OPTIMAL HIGH RESOLUTION IMAGING

Erick Baziw
Baziw Consulting Engineers Ltd., Vancouver, Canada
Gerald Verbeek
Baziw Consulting Engineers Ltd., Vancouver, Canada

ABSTRACT:

Downhole Seismic Testing (DST) is an extensively utilized geotechnical tool for site characterization. DST provides low strain ($<10^{-5}$) *in-situ* interval shear and compression wave velocity estimates. These velocities are determined by obtaining relative arrival times of source waves as they travel through the stratigraphy and are recorded by one or more vertically offset seismic sensors. The DST arrival times have associated measurements errors and resolution limitations, which become more pronounced as the depth interval of analysis is reduced. Baziw Consulting Engineers (BCE) has developed a new DST analysis technique, the so-called *DSTPolyKF* algorithm, whereby the arrival times of DST data set are used to define a high order polynomial. The main advantages of this new technique are four-fold: 1) the ability to utilize all arrival time estimates irrespective of measurement errors. 2) the ability to analyze the data with user specified and even small ($\leq$ 0.5m) depth intervals. 3) the ability to apply sophisticated data fusion for significantly more accurate DST interval velocity estimation. 4) the use of polynomial regression accuracy parameters to quantify how well the "best fit" polynomial fits the acquired arrival time data sets. This paper outlines the mathematical details of the best fit polynomial where a Kalman filter formulation is implemented. The performance of the *DSTPolyKF* algorithm is demonstrated by processing real DST data sets.

Keywords: downhole seismic testing, interval velocity estimation, interval velocity profile

1. INTRODUCTION

Downhole Seismic Testing (DST) is utilized extensively in both onshore and offshore geotechnical site investigations for the estimation of *in-situ* shear and compression wave velocities (ASTM, 2017). The shear and compression wave velocities allow for the calculation of various important geotechnical parameters, such as the shear modulus, Poisson's ratio and Young's modulus. Figure 1 shows a schematic of the typical DST configuration: a seismic source is used to generate a seismic wave train at the surface. Downhole seismic receivers are used to record the seismic wave train at predefined depth

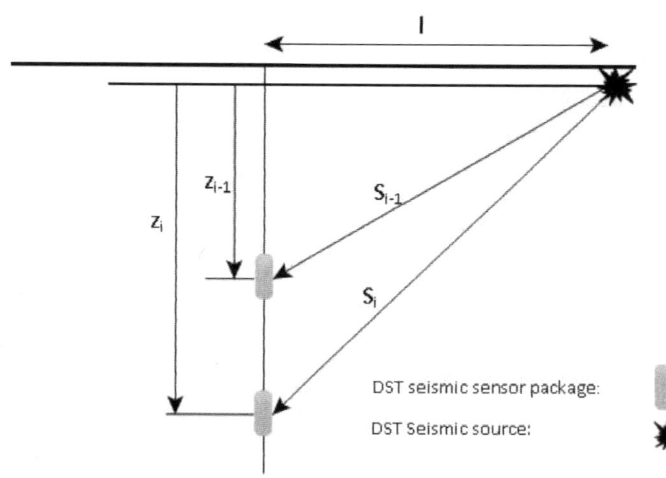

Figure 1. Schematic of the typical DST configuration.

increments. When triggered by the seismic source a data recording system records the response of the downhole receiver(s). Interval velocity estimates are obtained by measuring the relative and true travel times between the source waves recorded at subsequently greater depths.

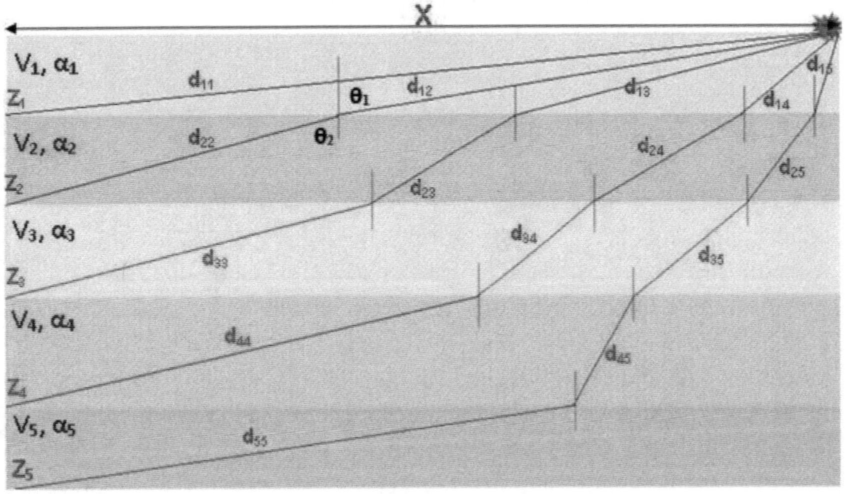

Figure 2. Schematic of thin layer effect for a sand layer embedded in a clay layer (Baziw and Verbeek, 2021).

In DST source wave trajectories adhere to *Fermat's principle*, which means the raypath travels along the trajectory that minimizes the travel time between points, and at every depth the raypath will be different as shown in Fig.2. In Fig. 2, V_i and α_i denote the soil interval velocity and absorption, respectively. Angle θ_1 is the angle of incidence and θ_2 is called the angle of refraction. Equation (1), which is generally referred to as Snell's Law, defines the relation between θ_1, θ_2, V_1 and V_2.

$$Sin\ \theta_1 / V_1 = Sin\ \theta_2 / V_2 \quad (1)$$

Baziw and Verbeek (Baziw, 2002; Baziw and Verbeek, 2012, 2019, 2021a) have developed a methodology, the so called Forward Modelling Downhill Simplex Method (FMDSM), which implements both *Fermat's Principle* and *Snell's Law* when processing DST data sets.

There are numerous factors that can lead to DST arrival time measurement errors, but these errors are predominantly due to poor data quality, low resolution of the seismic sensor and corresponding small DST depth increments, and the type of data trigger utilized. Poor quality DST seismic data recordings are generally caused by "dirty" sources (e.g., poor coupling between seismic hammer plate and soil), poor sensor-stratigraphy coupling, complicated stratigraphy resulting in source wave reflections, refraction etc., insufficient energy output of the seismic source, or a combination of them.

To address the outlined DST measurement errors, a new DST analysis technique has been developed, the so-called *DSTPolyKF* algorithm, which "best fits" a high order polynomial to arrival times established for a particular DST data set. This technique implements a Kalman filter (KF) formulation so that arrival time data sets can be processed sequentially with measurement errors assigned for each estimated arrival time (allowing for variable weight on arrival time estimates); Furthermore, the KF is an ideal mathematical tool for data fusion. In this case numerous and relevant data sets are utilized for optimal estimation of desired parameters.

2. MATHEMATICAL BACKGROUND AND *DSTPolyKF* ALGORITHM FORMULATION

2.1 Kalman Filter

The Kalman Filter (KF) is a Bayesian Recursive Estimation (BRE) filtering technique based on state-space, time-domain formulations of physical problems (Gelb, 1974; Arulampalam et al., 2002; Baziw, 2007; Baziw and Verbeek, 2021b). Application of this filter type requires that the dynamics of the system and measurement model, which relates the noisy measurements to the system state equations, be describable in a mathematical representation and probabilistic form that uniquely define the system behaviour. Baziw and Verbeek (2021b) give a detailed outline of the KF governing equations and the implementation of these equations. The KF equations are outlined in Table 1.

Table 1
KF Governing Equations

DESCRIPTION	Mathematical Representation	Eq.
System equation	$x_k = F_{k-1}x_{k-1} + G_{k-1}u_{k-1}$	2
Measurement equation	$z_k = H_k x_k + v_k$	3
State estimate extrapolation	$\hat{x}_{k\|k-1} = F_{k-1}\hat{x}_{k-1\|k-1}$	4
Error covariance extrapolation	$P_{k\|k-1} = F_{k-1}P_{k-1\|k-1}F_{k-1}{}^T + G_{k-1}Q_{k-1\|k-1}G_{k-1}$	5
Measurement extrapolation	$\hat{z}_k = H_{k-1}\hat{x}_{k\|k-1}$	6
Innovation	$\Delta_k = z_k - \hat{z}_k$	7
Variance of innovation	$S_k = H_k P_{k\|k-1} H_k^T + R_k$	8
Kalman gain matrix	$K_k = P_{k\|k-1} H_k (S_k)^{-1}$	9
State estimate update	$\hat{x}_{k\|k} = \hat{x}_{k\|k-1} + K_k \Delta_k$	10
Error covariance update	$P_{k\backslash k} = [I - K_k H_k] P_{k\|k-1}$	11

In (2) and (3) $\boldsymbol{v}_k$ and $\boldsymbol{u}_k$ are *i.i.d* Gaussian zero mean white noise processes with variances of $\boldsymbol{Q}_k$ and $\boldsymbol{R}_k$, respectively (i.e., $v_k \sim N(0, R_k)$ and $u_k \sim N(0, Q_k)$).

In Table 1 $\boldsymbol{x_k}$ denotes the state to be estimated, $\boldsymbol{F}_{k-1}$ denotes the state transition matrix which describes the system dynamics, $\boldsymbol{u}_{k-1}$ the process or system noise (model uncertainty), $\boldsymbol{G}_{k-1}$ describes the relationship between $\boldsymbol{x_k}$ and $\boldsymbol{u}_{k-1}$, and $\boldsymbol{H}_k$ the relationship between the state and the available measurement (source wave arrival time). The KF can be applied to problems with linear time-varying systems and with non-stationary system and measurement statistics. Problems with nonlinearities are handled by linearizing the system and measurement equations. The Kalman Filter is readily applied to estimation, smoothing and prediction.

2.2 DSTPolyKF Algorithm Formulation

2.2.1 Polynomial Regression and the Assessment of its Accuracy

Polynomial regression is a commonly utilized technique for analytically modelling data sets with measurement errors. The polynomial regression equation is given as

$$z_k = \hat{z}_k + v_k = a_0 + a_1 d + a_2 d^2 + a_3 d^3 + a_4 d^4 + \cdots + a_n d^n + v_k \quad (12)$$

where $\hat{z}_k$ is the polynomial regression estimator, n is the degree of the polynomial, v_k is the measurement error and $d = k\Delta$, Δ is the sampling rate.

Ostertagová, (2012) outlines four parameters which can be utilized to evaluate the accuracy of the polynomial regression best fit. These four parameters are mean squared error (MSE) of the polynomial estimator, Mean Absolute Percentage Error (MAPE), coefficient of determination (R^2), and adjusted coefficient of determination (R^{*2}). Table 2 outlines the four polynomial regression accuracy parameters by their mathematical representations and important characteristics. In the *DSTPolyKF* algorithm polynomial regressions of order 2 to 7 are derived for the estimated DST source wave arrival times. The polynomial aggression order[1] which results in the "best" accuracy parameter values as defined in Table 2 and lowest polynomial order is utilized. This approach addresses the well-known bias-variance tradeoff of polynomial regression (Emmert-Streib and Dehmer, 2019).

2.2.2 DSTPolyKF algorithm state-space formulation and measurement model

n^{th} Degree Polynomial Measurement Model

The following states need to be defined for estimating the nth degree polynomial coefficients outlined in eq. (12) with a KF

$$\begin{bmatrix} x_1 \\ x_2 \\ x_3 \\ x_4 \\ \vdots \\ x_{n-1} \\ x_n \\ x_{n+1} \end{bmatrix} \equiv \begin{bmatrix} a_0 \\ a_1 \\ a_2 \\ a_3 \\ \vdots \\ a_{n-2} \\ a_{n-1} \\ a_n \end{bmatrix} \tag{17}$$

The discrete system equation (eq. (2)) is given as

$$\begin{bmatrix} x_{1k+1} \\ x_{2k+1} \\ x_{3k+1} \\ x_{4k+1} \\ \vdots \\ x_{(n-1)k+1} \\ x_{(n)k+1} \\ x_{(n+1)k+1} \end{bmatrix} = \begin{bmatrix} 1 & 0 & 0 & 0 & 0 & 0 & 0 & 0 \\ 0 & 1 & 0 & 0 & 0 & 0 & 0 & 0 \\ 0 & 0 & 1 & 0 & 0 & 0 & 0 & 0 \\ 0 & 0 & 0 & 1 & 0 & 0 & 0 & 0 \\ 0 & 0 & 0 & 0 & \ddots & 0 & 0 & 0 \\ 0 & 0 & 0 & 0 & 0 & 1 & 0 & 0 \\ 0 & 0 & 0 & 0 & 0 & 0 & 1 & 0 \\ 0 & 0 & 0 & 0 & 0 & 0 & 0 & 1 \end{bmatrix} \begin{bmatrix} x_{1k} \\ x_{2k} \\ x_{3k} \\ x_{4k} \\ \vdots \\ x_{(n-1)k} \\ x_{nk} \\ x_{(n+1)k} \end{bmatrix} \tag{18}$$

---

[1] The maximum number of turnings of a polynomial function is always one less than the degree

Table 2. Polynomial regression best fit accuracy parameters

<table>
<tr><th>Parameter and Mathematical Equation</th><th></th><th>Notes</th></tr>
<tr><td>$$MSE = \frac{SSR}{df_E} = \frac{\sum_{k=1}^{m}(z_k - \hat{z}_k)^2}{m-(n+1)}$$
SSR is the Sum of Squares Residual. The SSR has χ^2 distribution with $df_E = m-(n+1)$ degrees of freedom.</td><td>(13)</td><td>MSE is an unbiased estimator of the variance of the measurement error. Parameter RMSE is the Root Mean Square Error where RMSE = √MSE. RMSE is a biased estimator of the standard deviation of the measurement error.</td></tr>
<tr><td>$$MAPE = \frac{100}{m}\sum_{k=1}^{m}\left|\frac{z_k - \hat{z}_k}{z_k}\right|$$</td><td>(14)</td><td>MAPE measures relative performance or prediction accuracy of a forecasting method in statistics. MAPE values and the associated prediction accuracy are outline in below.
<table>
<tr><th>MAPE Value</th><th>Prediction Accuracy</th></tr>
<tr><td>MAPE<10%</td><td>excellent</td></tr>
<tr><td>10% ≤ MAPE ≤ 20%</td><td>good</td></tr>
<tr><td>20% < MAPE ≤ 50%</td><td>acceptable</td></tr>
<tr><td>50% < MAPE</td><td>inaccurate</td></tr>
</table></td></tr>
<tr><td>$$R^2 = 1 - \frac{SSR}{TSS} = 1 - \frac{\sum_{k=1}^{m}(z_k - \hat{z}_k)^2}{\sum_{k=1}^{m}(z_k - \bar{z})^2}$$
$\bar{z}$ is the arithmetic mean. TSS is the Total Sum of Squares.</td><td>(15)</td><td>R^2 measures how the percentage of variation in variable z_k can be explained by the explanatory variable d in eq. (12). In general terms, R^2 quantifies what percentage of the measured z_k values reside on the best fit polynomial $\hat{z}_k$. R^2 is similar to the correlation coefficient. The correlation coefficient quantifies how strong of a linear relationship there is between two variables. R^2 values and the associated prediction accuracy are outline below.
<table>
<tr><th>R^2 value</th><th>Prediction Accuracy</th></tr>
<tr><td>$0.9 \leq R^2$</td><td>excellent</td></tr>
<tr><td>$0.8 \leq R^2 < 0.9$</td><td>good</td></tr>
<tr><td>$0.6 \leq R^2 < 0.8$</td><td>acceptable</td></tr>
<tr><td>$R^2 < 0.6$</td><td>inaccurate</td></tr>
</table></td></tr>
<tr><td>$$R^{*2} = R^2 - \frac{(1-R^2)n}{m-(n+1)}$$</td><td>(16)</td><td>R^{*2} is the adjusted value of the coefficient of determination in which the order of the polynomial is taken into account. Prediction accuracy is poor if R^{*2} is significantly lower than R^2.</td></tr>
</table>

The linear measurement matrix, H, is given by the following equation

$$H_k = [1 \quad d \quad d^2 \quad d^3 \quad \cdots \quad d^{n-3} \quad d^{n-2} \quad d^{n-1}] \tag{19}$$

3. *DSTPolyKF* ALGORITHM PERFORMANCE ASSESSMENT

The *DSTPolyKF* algorithm was implemented on both onshore and offshore challenging DST arrival time data sets for performance assessment of the algorithm. The onshore DST data sets had horizontal shear (SH) source waves generated on both the Right Side (RS) and Left Side (LS) of the downhole seismic probe resulting in two source wave traces recorded for each depth increment. The onshore DST instrumentation utilized high precision triaxial accelerometers and contact triggers. The offshore investigation also generated both RS and LS SH source waves. In this case both "top" and "bottom" triaxial sensor arrangement packages were utilized. This resulted in four source wave traces for each depth increment. The Top Sensor (TS) and Bottom Sensor (BS) were vertically offset by 1m. The offshore DST instrumentation utilized geophone sensors and a geophone sensor trigger.

3.1 Onshore Data Analysis

Table 3. Estimated polynomial accuracy parameters for onshore data analysis

Polynomial Order	RMSE	MAPE	R^2	R^{*2}
2	6.0434	7.9976	0.99727	0.996966
3	3.9483	4.9092	0.998899	0.998705
4	1.9703	2.0815	0.999742	0.999678
5	1.8957	1.5455	0.999776	0.999702
6	1.7769	1.4371	0.999816	0.999738
7	1.6805	1.5168	0.999848	0.999765

Figure 3 illustrates the estimated LS and RS estimated arrival times for the onshore DST data set. As is illustrated in Fig. 3, there is significant variability between the LS and RS arrival estimates from 12m to 21m. This results in significant variability in the resulting interval velocity estimates. Table 3 outlines the corresponding polynomial regression estimated accuracy parameters for the averaged arrival times (LS and RS) illustrated in Fig. 3 and orders 2 to 7. From the results outlined in Table 3, the 6th and 7th order polynomial regressions have overall "best" accuracy parameter values as defined in Table 2. Figure 4 illustrates the averaged LS and RS arrivals times with the 7th order polynomial regression best fit line. Figure 5 illustrates the FMDSM output with source wave raypaths illustrated when processing the 7th order best fit polynomial regression arrival times. Table 4 outlines the estimated interval velocities and percent differences for polynomial regressions of orders 6 and 7. The 6th and 7th order percent differences are below the desired 10% for all depths except 18.5, 19m, 19.5m (just exceeding 10%). The estimate at 21m should be dropped due to the large percent difference. In general terms, when comparing polynomial regression of varying orders with similar "best" accuracy parameters the discrepancies of velocity estimates occurred at the end of the polynomial regression best fit lines. Figure 6 illustrates the 7th order best fit polynomial regression interval velocity plot with an interpolated 0.25m depth increment and the interval velocity estimate at 21m dropped.

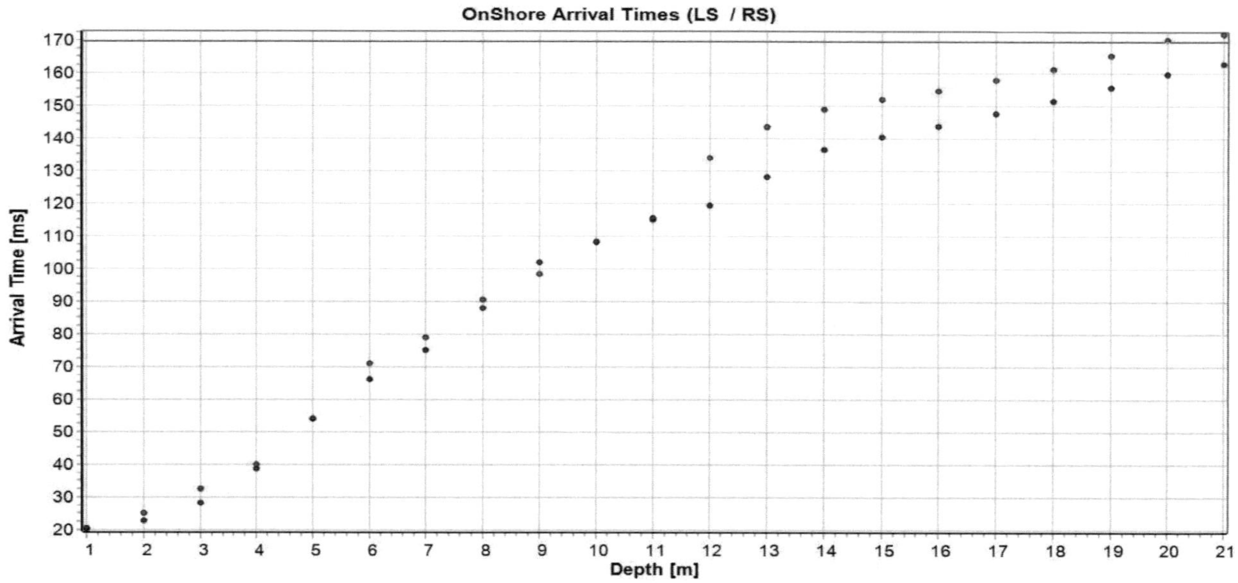

Figure 3. LS and RS arrival times for the onshore real DST analysis.

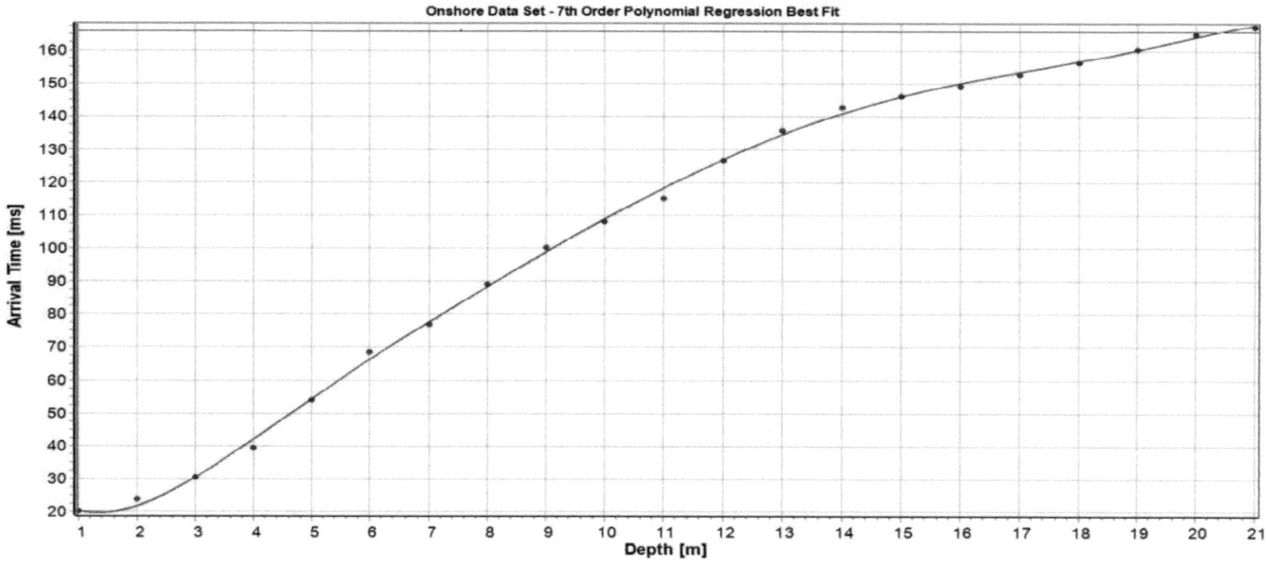

Figure 4. Averaged arrival times (red dots) and 7th order polynomial regression best fit (blue line) for the onshore real DST analysis.

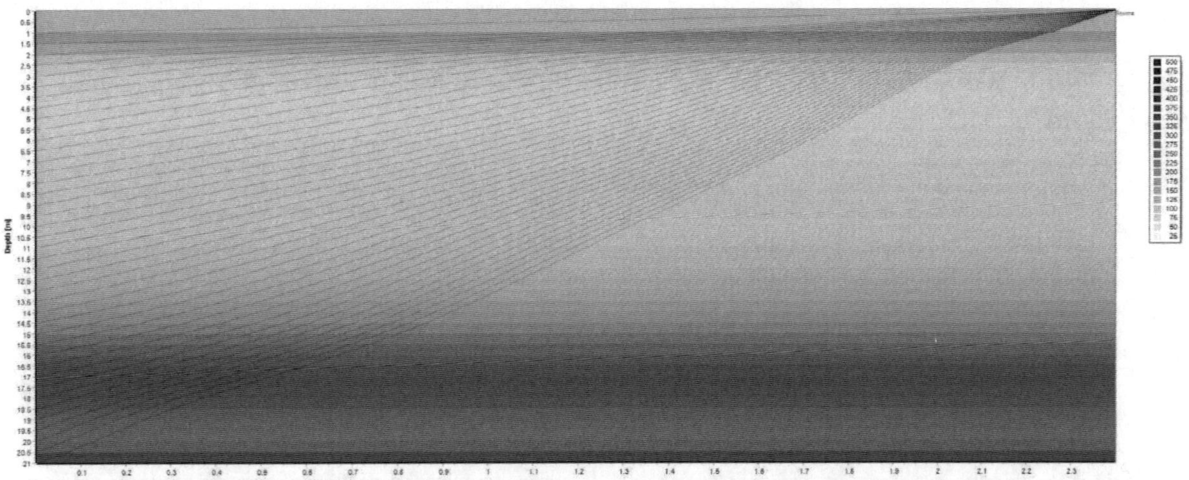

Figure 5. FMDSM output with source wave raypaths (black lines) illustrated when processing the 7th order best fit polynomial regression arrival times shown in Fig. 4.

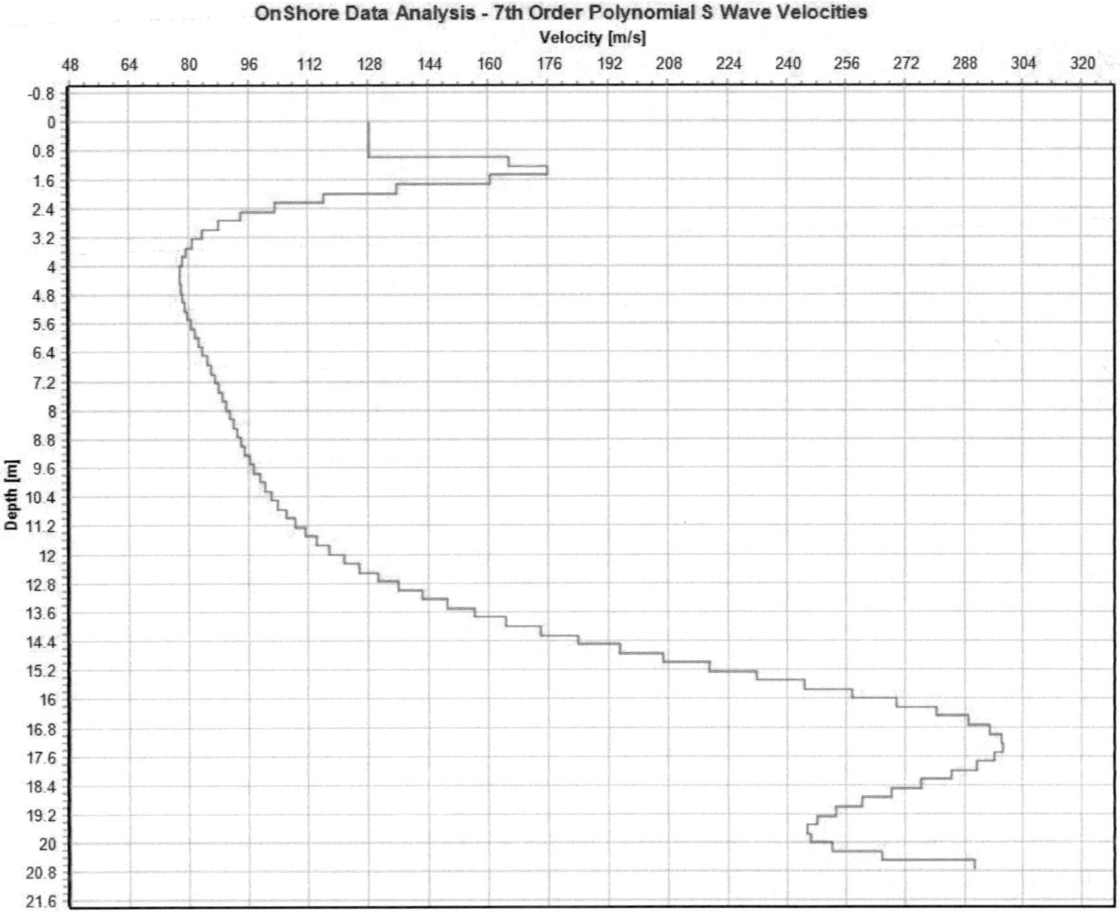

Figure 6. 7th order best fit polynomial regression interval velocity plot with an interpolated 0.25m depth increment and the estimate at 21m dropped for the onshore data set.

Table 4. Estimated interval velocities and percent differences for polynomial regressions of orders 6 and 7.

Depth [m]	6th Order Interval Velocity [m/s]	7th Order Interval Velocity [m/s]	6th and 7th Order Precent difference
1	128.3	128.3	0
1.5	151.2	171.1	6.2
2	136.9	147.8	3.8
2.5	113.8	109.3	2
3	98.1	91	3.8
3.5	88.5	82.4	3.6
4	82.7	78.5	2.6
4.5	79.5	77.3	1.4
5	77.9	77.8	0.1
5.5	77.6	79.2	1
6	78.2	81.2	1.9
6.5	79.6	83.4	2.3
7	81.6	85.7	2.5
7.5	84	87.9	2.3
8	86.9	89.9	1.7
8.5	90.1	91.9	1
9	93.6	93.9	0.2
9.5	97.5	96	0.8
10	101.6	98.4	1.6
10.5	106	101.4	2.2

Depth [m]	6th Order Interval Velocity [m/s]	7th Order Interval Velocity [m/s]	6th and 7th Order Precent difference
11	110.7	105.1	2.6
11.5	116	109.8	2.7
12	121.7	115.9	2.4
12.5	128.2	123.5	1.9
13	135.6	133.2	0.9
13.5	144.2	145.3	0.4
14	154.4	160.3	1.9
14.5	166.6	178.6	3.5
15	181.2	200.2	5
15.5	199	224.7	6.1
16	220.7	250.4	6.3
16.5	246.5	274.1	5.3
17	276.5	291.3	2.6
17.5	308.9	298.2	1.8
18	338.9	293.7	7.1
18.5	357.9	280.6	12.1
19	355.1	264.5	14.6
19.5	325.5	251.1	12.9
20	275.6	246.4	5.6
20.5	219.3	259	8.3
21	168.1	312.7	30.1

3.2 Offshore Onshore Data Analysis

Table 5. Estimated polynomial accuracy parameters for onshore data analysis

Polynomial Order	RMSE	MAPE	R^2	R^{*2}
2	4.8413	4.2236	0.998327	0.998264
3	2.4921	1.9924	0.999565	0.99954
4	1.2825	1.0099	0.999887	0.999878
5	1.0571	0.7638	0.999925	0.999917
6	1.0357	0.6924	0.999929	0.999921
7	0.9672	0.6404	0.99994	0.999931

Figure 7 illustrates the estimated (Top Sensor LS) TSLS, TSRS (Top Sensor RS), BSLS (Bottom Sensor LS) and BSRS (Bottom Sensor RS) estimated arrival times for the offshore DST data set. Table 5 outlines the corresponding polynomial regression estimated accuracy parameters for the averaged arrival times (TSLS, TSRS, BSLS, and BSRS) illustrated in Fig. 7 and orders 2 to 7. From the results outlined in Table 5, the 6th and 7th order polynomial aggressions have overall "best" accuracy parameter values as defined in Table 2. Figure 8 illustrates the averaged TSLS, TSRS, BSLS and BSRS arrivals times with the 7th order polynomial regression best fit line. Figure 9 illustrates the FMDSM output when processing the 7th order best fit polynomial regression arrival times. Table 6 outlines the estimated interval velocities and percent differences for polynomial regressions of orders 6 and 7. Figure 9 illustrates the 7th order best fit polynomial regression interval velocity plot with an interpolated 0.5m depth.

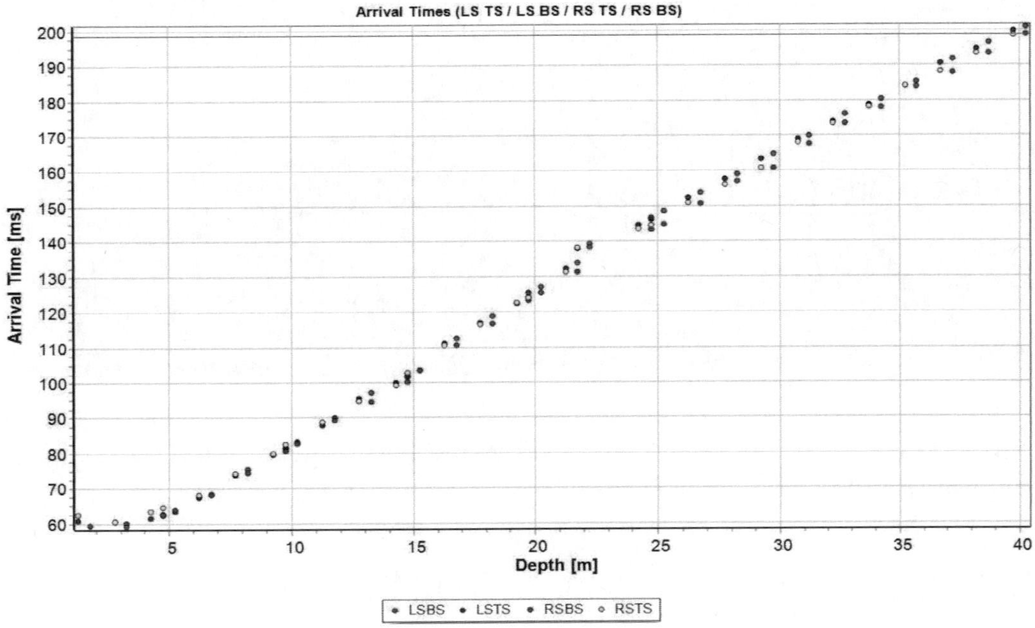

Figure 7. TS LS, TS RS, BS LS and BS RS arrival times for the offshore real DST analysis.

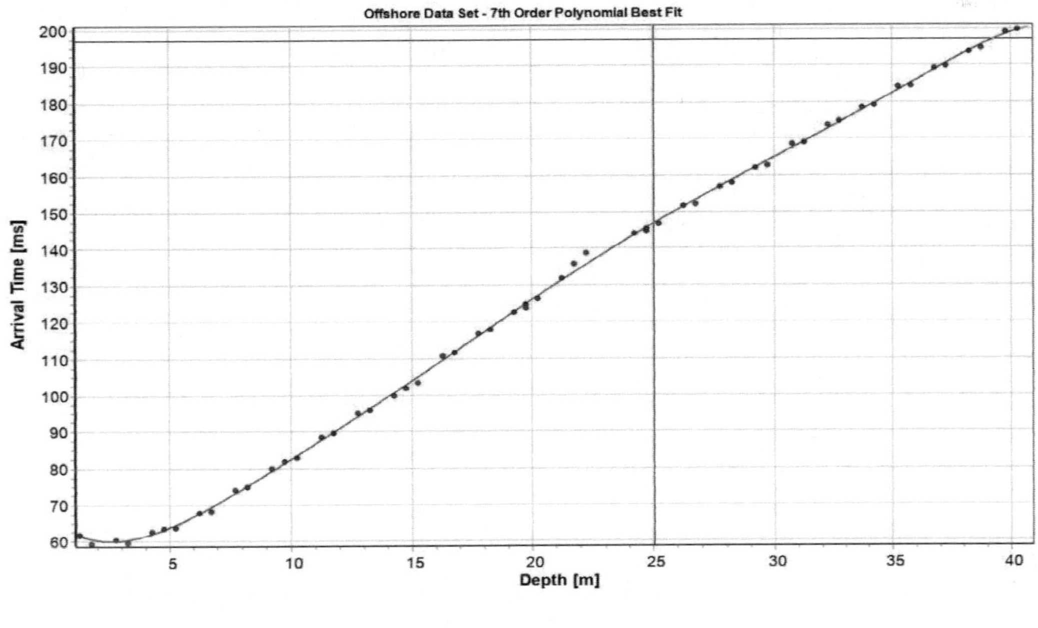

Figure 8. Averaged arrival times (red dots) and 7th order polynomial regression best fit (blue line) for the offshore real DST analysis.

Figure 9. FMDSM output when processing the 7th order best fit polynomial regression arrival times shown in Fig. 7.

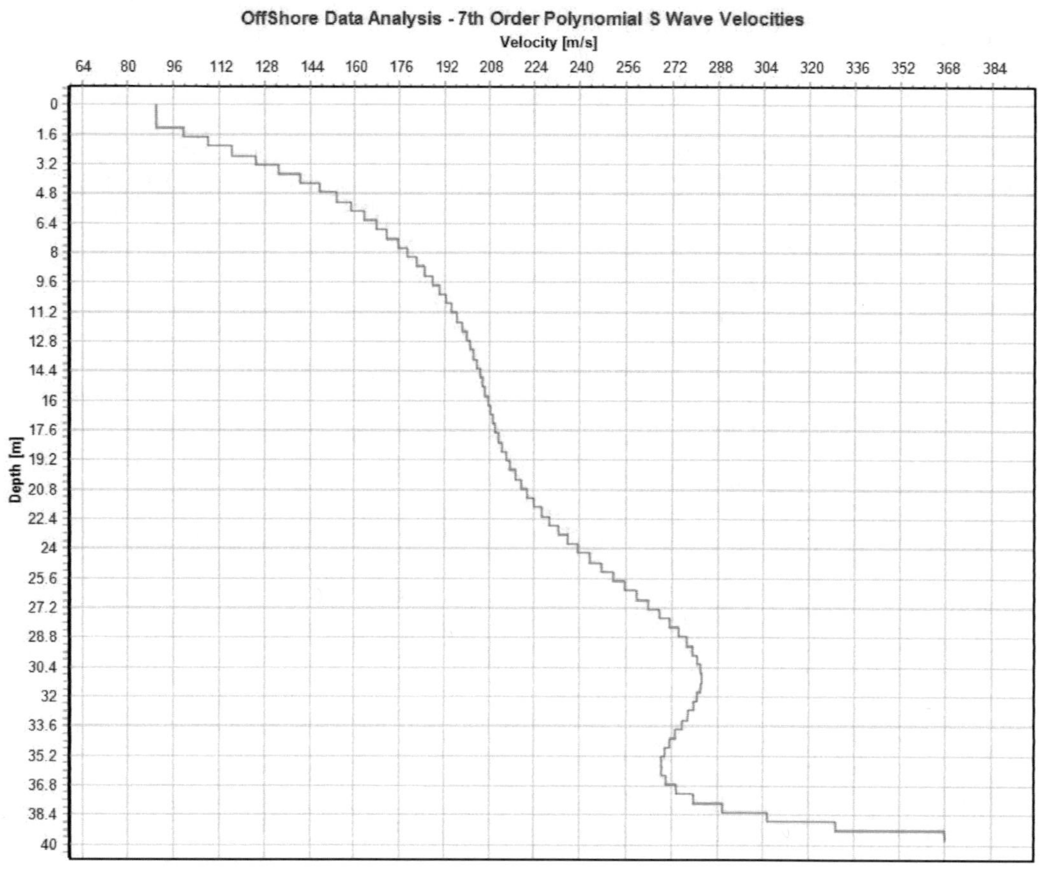

Figure 10. 7th order best fit polynomial regression interval velocity plot with an interpolated 0.5m depth increment for the offshore data set.

Table 6. Estimated interval velocities and percent differences for polynomial regressions of orders 6 and 7.

Depth [m]	6th Order Interval Velocity [m/s]	7th Order Interval Velocity [m/s]	6th and 7th Order Precent difference
1.24	89.7	89.7	0
1.74	98.3	99.6	0.7
2.24	105.9	108	1
2.74	114.2	116.5	1
3.24	123.1	125	0.8
3.74	132.3	133.1	0.3
4.24	141.2	140.6	0.2
4.74	149.6	147.3	0.8
5.24	157.1	153.3	1.2
5.74	163.7	158.7	1.6
6.24	169.1	163.4	1.7
6.74	173.5	167.8	1.7
7.24	177	171.7	1.5
7.74	179.8	175.4	1.2
8.24	181.9	178.8	0.9
8.74	183.6	181.9	0.5
9.24	185	184.8	0.1
9.74	186.2	187.6	0.4
10.24	187.4	190.1	0.7
10.74	188.5	192.4	1
11.24	189.7	194.6	1.3
11.74	190.9	196.6	1.5
12.24	192.2	198.4	1.6
12.74	193.7	200	1.6
13.24	195.3	201.4	1.5
13.74	196.9	202.7	1.5
14.24	198.7	203.8	1.3
14.74	200.6	204.9	1.1
15.24	202.5	205.8	0.8
15.74	204.6	206.7	0.5
16.24	206.7	207.6	0.2
16.74	208.8	208.4	0.1
17.24	211	209.3	0.4
17.74	213.3	210.3	0.7
18.24	215.5	211.4	1
18.74	217.8	212.6	1.2
19.24	220.1	214	1.4
19.74	222.3	215.5	1.6
20.24	224.6	217.2	1.7

Depth [m]	6th Order Interval Velocity [m/s]	7th Order Interval Velocity [m/s]	6th and 7th Order Precent difference
20.74	226.8	219.2	1.7
21.24	229.1	221.4	1.7
21.74	231.3	223.8	1.6
22.24	233.5	226.5	1.5
22.74	235.6	229.4	1.3
23.24	237.8	232.6	1.1
23.74	239.9	236	0.8
24.24	242	239.7	0.5
24.74	244.1	243.5	0.1
25.24	246.2	247.4	0.2
25.74	248.3	251.5	0.6
26.24	250.5	255.7	1
26.74	252.6	259.8	1.4
27.24	254.8	263.8	1.7
27.74	257	267.7	2
28.24	259.2	271.2	2.3
28.74	261.5	274.5	2.4
29.24	263.8	277.2	2.5
29.74	266.1	279.5	2.5
30.24	268.5	281.1	2.3
30.74	270.9	282.1	2
31.24	273.4	282.4	1.6
31.74	275.9	282.1	1.1
32.24	278.3	281.1	0.5
32.74	280.7	279.7	0.2
33.24	283.1	277.8	0.9
33.74	285.4	275.6	1.7
34.24	287.5	273.4	2.5
34.74	289.4	271.3	3.2
35.24	291.1	269.6	3.8
35.74	292.4	268.6	4.2
36.24	293.3	268.6	4.4
36.74	293.8	270.1	4.2
37.24	293.7	273.6	3.5
37.74	292.9	279.8	2.3
38.24	291.3	289.9	0.2
38.74	289	305.5	2.8
39.24	285.7	329.7	7.1
39.74	281.4	368	13.3

CONCLUSIONS

The focus on the arrival times obtained for a Downhole Seismic Test (DST) to generate a mathematical function to describe the wave arrival time as a function of depth seems to be a promising analysis approach to determine the seismic wave velocity profile. In this paper the mathematical details of the *DSTPolyKF* algorithm approach have been outlined and the only major remaining basic question is the number of data points required to generate an interval velocity profile that is reliable. The authors will continue to address that topic and also use this approach to validate the results that are generated with this analysis approach. Meanwhile the approach will be presented at conferences like the DFI annual conference as an alternate analysis approach.

REFERENCES

[1] Aki, K. and Richards, P.G. (2002), Quantitative Seismology, 2nd edition, Sausalito, California: University Science Books. pp. 88-89.

[2] Arulampalam, M.S., Maskell, S., and Clapp, T. (2002), "A tutorial on particle filters for online nonlinear/nonGaussian Bayesian tracking", IEEE Transactions on Signal Processing, vol. 50, no. 2, pp. 174-188, Feb. 2002

[3] ASTM D6067 / D6067M – 17 (2017), "Standard Practice for Using the Electronic Piezocone Penetrometer Tests for Environmental Site Characterization and Estimation of Hydraulic Conductivity", ASTM Vol. 4.09 Soil and Rock (II): D5877-latest.

[4] Baziw, E., Tichy, J, and de Caprona, G (2000),"Data Acquisition in Seismic Cone Penetration Testing". In Proceedings of the 3rd International Symposium on Integrated Technical Approaches to Site Characterization (ITASCE), Argonne, IL, 11 14 Sept. 2000. Argonne National Laboratory. pp. 69-72.

[5] Baziw, E. (2007), Application of Bayesian Recursive Estimation for Seismic Signal Processing, Ph.D. Thesis, Dept. of Earth and Ocean Sciences, University of British Columbia.

[6] Baziw, E. (2002), "Derivation of Seismic Cone Interval Velocities Utilizing Forward Modeling and the Downhill Simplex Method", Can. Geotech. J., 39(5), pp.1181-1192.

[7] Baziw, E., and Verbeek, G. (2012), "Deriving Interval Velocities from Downhole Seismic Data", Geotechnical and Geophysical Site Characterization 4 – Mayne (eds), CRC Press, 1019–1024.

[8] Baziw, E. and Verbeek, G. (2019), " Unique algorithm which takes into account source wave travel paths when estimating absorption values from DST data sets" published in the DFI 44rd Annual Conference on Deep Foundations conference proceedings. October 14-17, 2019 - Chicago, IL.

[9] Baziw, E., and Verbeek, G. (2021a), "Implementation of the FMDSMAA algorithm" published in the conference proceedings of the 6th International Conference on Geotechnical Site Characterization (ISC-6), Budapest – Hungary, September 26-29, 2021.

[10]Baziw, E. and Verbeek, G. (2021b), "Implementation of Kalman Filtering Techniques for Filtering CPT Cone Bearing Measurements", published in the DFI 46th Annual Conference on Deep Foundations conference proceedings. October 12-15, 2021 - Las Vegas, NV.

[11]Emmert-Streib, F. and Dehmer, M. (2019) "Evaluation of Regression Models: Model Assessment, Model Selection and Generalization Error", Mach. Learn. Knowl. Extr. 1, 521–551.

[12]Gelb, A. (1974), Applied Optimal Estimation (4th Edition). Cambridge, Mass: MIT Press.

[13]Ostertagová, E. (2012), "Modelling Using Polynomial Regression", Procedia Engineering, 48 (2012), pp. 500-506.

[14]Shearer, P.M. (1999), Introduction to Seismology, 1st edition Cambridge: Cambridge University Press.

ADVANCED CONTINUOUS SURFACE WAVE TESTING (ACSW): GEOLOGIC AND GEOTECHNICAL SITE CHARACTERIZATION FOR THE 21ST CENTURY

David Wilshaw, BSc., MSc.(Eng.), CGeol, FGS, PG, Ground Stiffness Surveys LLC,
P.O. Box 783547, Winter Garden, FL; e-mail: David@GroundStiffnessSurveys.com

ABSTRACT

Continuous Surface Wave (CSW) testing has been the "next big thing" in ground characterization for the last 25 years. The technique uses surface geophones to measure the speed at which Rayleigh waves (generated by a vibratory source) travel through the ground over a range of frequencies. The velocity at which surface waves travel through the ground is directly related to ground stiffness; the velocity of shorter wavelengths is influenced by the stiffness of shallower soil layers, while the velocity of longer wavelengths is influenced by the stiffness of deeper layers. Measuring the surface wave velocity over a range of frequencies, therefore, allows a stiffness / depth profile to be produced. Using the small-strain stiffness data measured by the CSW technique greatly improves the accuracy of simple limit state analyses; the data are also of sufficient quality for use in Finite Element modelling. Nonetheless, the technique has suffered from a perception of high cost, low productivity, and poor repeatability. An improved Advanced Continuous Surface Wave (ACSW) testing system has been developed by Ground Stiffness Surveys in the United Kingdom, has been deployed there commercially for several years, and is now available in the United States. ACSW uses an automated electronic control system that generates vibrations (from a purpose designed and manufactured portable seismic "shaker") at different frequencies, measured across a 3m (10-foot) array of geophones, linked to a robust data acquisition and analysis software package. The range of applications in geologic and geotechnical site characterization includes rapid preliminary non-intrusive site assessments (utilizing published soil correlations); stiffness profiling to detect discrete layers (such as a variation in rockhead); average stiffness profiling for earthwork or ground improvement control testing, Working Platform stability assessments, foundation settlement analysis, rock quality and rippability assessments, and void detection.

Keywords: Continuous Surface Wave testing, small strain stiffness, Rayleigh waves, in-situ testing

INTRODUCTION

Geotechnical Engineers have become adept at maintaining a safe distance from Ultimate Limit States, which are generally controlled by soil strength. However, as serviceability requirements have become more demanding and construction adjacent to existing structures more commonplace, the focus in design has increasingly shifted to the movements which take place far from the threshold of soil yield and possible failure. The realistic prediction of these ground movements (key to the production of an economical design) requires a sound knowledge of ground stiffness and its stress strain behaviour.

While the simple, traditional investigation techniques employed to date generally permit the ground profile and strength to be estimated with reasonable accuracy, direct assessment of ground stiffness is rarely attempted, either due to the complexity of in-situ measurement or the difficulty that obtaining high quality samples suitable for laboratory testing poses. As a result, ground stiffness values are often determined from simple empirical relationships with either Standard Penetration Test (SPT) N values or basic in-situ measurements of undrained strength. This approach leads to conservatism and results in an uneconomic design.

The Continuous Surface Wave (CSW) methodology presents an attractive opportunity to obtain high quality ground stiffness data rapidly, but historically its use has been restricted due to limitations in equipment ruggedness, ease of use, data processing and available computing power. Over the last 5 years

Ground Stiffness Surveys (GSS) has been developing the Advanced CSW testing system described herein, which addresses these limitations to provide a rapid, portable, and rugged field system for obtaining accurate ground stiffness profiles.

TRADITIONAL METHODS FOR MEASURING GROUND STIFFNESS PROFILES

Since the pioneering days of soil mechanics, there have been various attempts to devise empirical correlations between the deformation modulus of soil and rock and available in-situ and laboratory testing techniques. Correlations with SPT N-value, CPT q_c and laboratory undrained shear strength are plentiful in the literature but tend to be littered with assumptions and qualifiers.

Beyond these empirical correlations, direct field measurements of stiffness range from California Bearing Ratio (CBR) testing and static (plate) load tests, through pressuremeter / dilatometer testing and large-scale instrumented maintained load tests, with a commensurate increase in accuracy (and cost) throughout the range. In the current competitive marketplace, pressuremeter testing has undergone something of a renaissance, as geotechnical engineers are recognizing the commercial advantages of selling a shallow foundation solution to their clients, thus avoiding piles and/or ground improvement. However, despite obvious opportunities to achieve cost savings during construction, there is still a reluctance from clients to apply a realistic budget to the geotechnical exploration. Since pressuremeter testing is relatively slow and ties up drilling equipment and personnel for hours and days, often only a handful of stiffness readings are made.

Stiffness characteristics can also be derived from the results of laboratory testing, such as triaxial and oedometer consolidation tests, but to fully characterize soils in this manner, it is necessary to vary parameters such as shear strain, confining pressure, loading cycles, overconsolidation ratio, etc. to analyse how the soil might respond. Thus, laboratory testing tends to be slow and hence expensive, with a limited number of test results being available for use in design.

Inevitably, against the background of unrealistic project deadlines and budget constraints, even relatively straightforward in-situ or laboratory testing is avoided, and an emphasis placed simply on the SPT N-value as the basis of all ground characterization. Many hours of research and dozens of papers have been devoted to settlement prediction using SPT N-value or CPT q_c converted to an equivalent modulus of elasticity. However, as Das and Sivakugan (2007) succinctly put it, "*(o)ne of the main factors that contribute to the uncertainty in settlement predictions is our inability to quantify the soil stiffness*".

CONTINUOUS SURFACE WAVE (CSW) GROUND STIFFNESS PROFILING

The relationship between seismic shear wave velocity and shear stiffness as a function of soil density has long been understood, being first established in the early part of the 20th Century (Love, 1927). However, the opportunity this presented to determine ground stiffness was little utilized until the early 1990's and the arrival of personal computers combined with an increasing awareness of the non-linear stiffness behaviour of geomaterials.

While the non-linear stiffness behaviour of soils and weak rocks is now well understood, it is not always adequately considered in the selection of geotechnical design parameters and methodology. This is partly explained by the continuing widespread use of empirical methods of stiffness measurement, which do not directly measure stiffness (i.e., SPT N value) and which are routinely used to calculate anticipated settlement. When stiffness is measured directly by commonly utilized methods (i.e., plate load test, oedometer, pressuremeter) further complication is introduced by the range of strains across which testing is undertaken. This makes subsequent strain adjustment for operational levels difficult.

In contrast, ground stiffness values determined from seismic tests are measured in the small strain linear part of the stiffness degradation curve, where changes in stiffness with strain are very small (see Figure

1), corresponding to strains of less than <0.002% (Heymann, 2007). Conveniently, the stiffness degradation (or softening) curve for a wide range of geomaterials is remarkably uniform (Clayton & Heymann, 2001), permitting straightforward adjustment of small strain stiffness values to operational strain levels either prior to, or as part of, the analysis process.

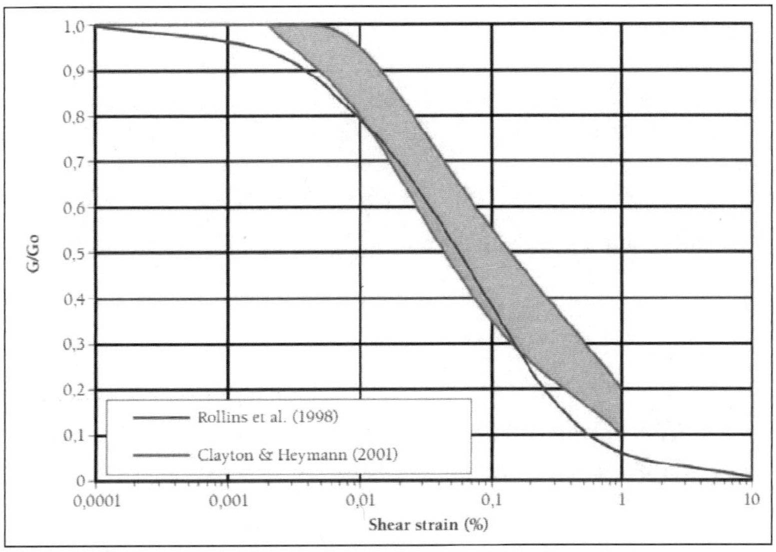

Figure 1 Typical soil softening functions (after Heymann, 2007)

The use of Surface Waves (as opposed to Body Waves) for ground stiffness measurement is particularly attractive for two reasons. Firstly, approximately 2/3 of the energy imparted at the ground surface produces one of the two types of surface wave known as Rayleigh waves. This means that seismic sources can be relatively small and therefore portable. Secondly, Rayleigh waves are dispersive in nature, meaning that longer wavelengths penetrate to greater depth, with a typical effective depth of a Rayleigh wave being the equivalent of the wavelength, $\lambda/2.5$.

This permits a stiffness versus depth profile to be developed for measurements made across a range of frequencies (and therefore wavelengths). Once measured, the Rayleigh Wave velocity (v_r) is simply converted to shear wave velocity (v_s) through a function with Poisson's ratio, ν and hence to shear stiffness as follows (Heymann, 2007):

$$v_r = v_s . f(\nu) \quad [1]$$

where

$$v_r/v_s = (0.874 + 1.117\, \nu) / (1 + \nu) \quad [2]$$

v_s is relatively insensitive to variations in ν. Seismic tests are always undrained, which for most saturated soils implies a ν of 0.5. When soils are partially saturated, or when the soil matrix is stiffer than the porewater, then Poisson’s ratios may be less than 0.5, resulting in the calculation of higher v_s values.

However, ν typically varies between 0.2 - 0.5, meaning that the ratio of v_r/v_s = 0.91 to 0.95, so the maximum variability in v_s over the possible range in Poisson's ratio is about 7%.

Shear wave velocity, v_s can then be converted to small strain shear modulus (G_0) using:

$$G = \rho . v_s^2 \quad [3]$$

Where ρ is the unit weight / bulk density. G is more sensitive to variations in v_s (square function) than unit weight (direct function). Given that ρ generally varies within a relatively small range from 1.6 to 2.1 Mg/m^3 for most ground conditions, derivation of G is relatively insensitive to assumed soil density (if not known), and conservative (i.e., lower bound) if a low soil density (say, 1.8 Mg/m^3) is assumed.

Finally, we can use the relationship between G, Young's Modulus, E, and Poisson's ratio, ν as follows:

$$E = G.(2.(1+\nu)) \quad [4]$$

The choice of Poisson's ratio determines drained or undrained parameters and since we are dealing with G_0, then a strain softening function can be applied to E to give a design value for the specific problem to hand.

ADVANCED CONTINUOUS SURFACE WAVE (ASCW) TEST SYSTEM

Determination of ground stiffness through the measurement of Surface Waves is well established in the USA using the Spectral Analysis of Surface Waves (SASW) methodology and its more recent development the Multi-Spectral Analysis of Surface Waves (MASW) method. However, both techniques rely on the use of impact sources which also produce P & S Waves (i.e., Body Waves) in addition to Surface Waves; this can make the detection of Rayleigh waves difficult. It is also difficult to generate a full frequency spectrum of Surface Waves using an impact source which, especially if combined with cultural noise, can greatly reduce data quality.

The Continuous Surface Wave (CSW) methodology overcomes the problems associated with an impact source by using a vertically polarized surface vibrator (or "shaker"). This source is rich in Rayleigh waves and generates little Body Wave 'noise', resulting in a much higher signal to noise ratio compared to impact sources (Jiang et al. 2015). The continuous nature of the source permits data collection over many wave cycles at a monotonic frequency, permitting signal quality to be assessed in the field and allowing random cultural noise to be largely removed. As a result, accurate velocity measurement is possible over a short geophone array, making the CSW technique the basis for a rapid, robust, and practical stiffness profiling methodology, providing more repeatable results than the SASW technique (Simons et al. 2002). A detailed description of Continuous Surface Wave (CSW) testing theory is provided by Heymann (2007).

Until recently, adoption of CSW testing as a routine ground investigation technique has been constrained by the delicate nature of the equipment, the time-consuming nature of data acquisition, and outdated software systems that were unable to deal with complex or poorer quality data. These issues have all largely been addressed with the advent of high power, low frequency-tuned seismic sources, combined with advanced data acquisition and inversion techniques. This has resulted in the advent of the GSS Advanced CSW (ACSW) testing system, which has several attractive features:

- Ruggedized, weatherproof and easily portable equipment, comprising a single high-power, wide frequency (8-91 Hz) mechanical vibrator and single combined data acquisition and shaker control box.
- Rapid, non-intrusive field technique, minimising the risk to buried utilities and permitting increased spatial coverage and reduced cost; typically, 3-4 tests per hour can be undertaken (compared to typical 3 to 4 tests per shift with early systems, Heymann 2007).
- Custom control and analysis software (C-DAS), providing full test automation and quality assurance with full data review and stacking of multiple tests possible in the field.
- Implementation of advanced search algorithms and forward models within the software, enabling the widest range of field data to be processed.
- Provision of small strain shear stiffness (G_0) profiles to a typical depth of 10m (33 feet) which may be automatically converted to Young's Modulus values and softened to operational strain levels.

A typical site layout for the ACSW system is illustrated below by Figures 2 and 3:

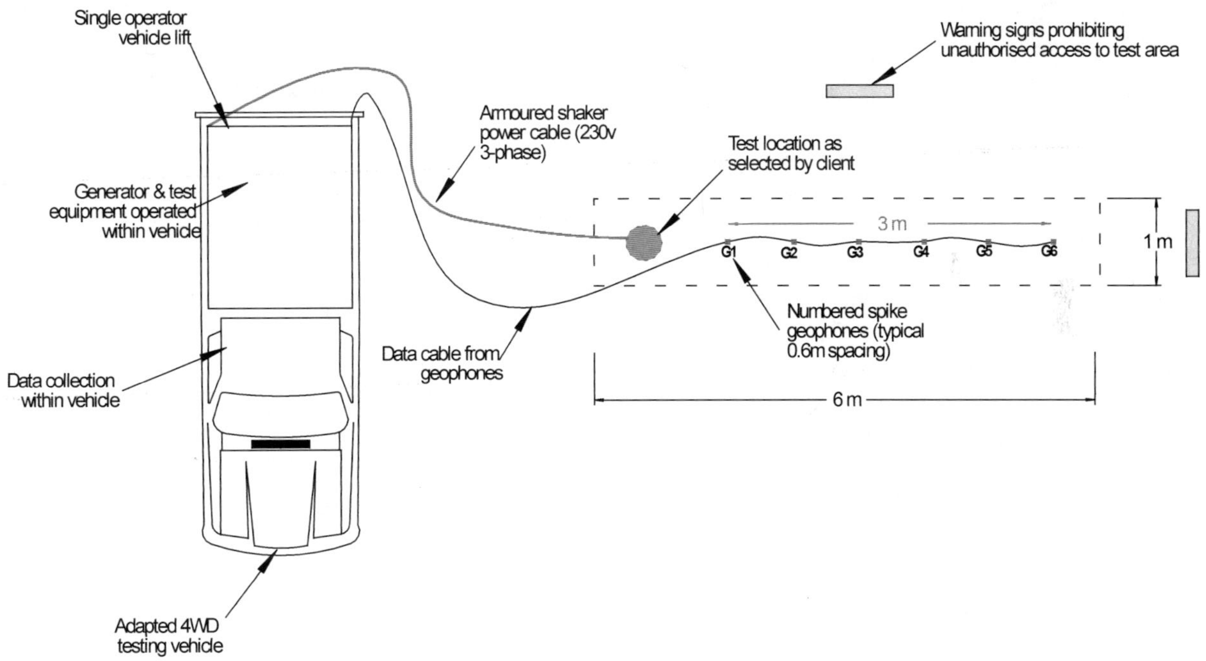

Figure 2 – ACSW Test Layout

Figure 3 ACSW testing being undertaken in Florida

An ACSW test takes less than 10 minutes to complete. The C-DAS software controls the wave frequency generated by the vibration source. The ground response to 100 waves generated at each of 100 different frequencies is measured across a short (3m / 10-foot) array of six (6) surface geophones. Each ACSW test comprises data acquisition during a sweep across a full range of shaker frequencies (Sweep test) and velocity measurements over a series of discrete constant frequencies (Monotonic test). This provides two complete sets of dispersion curve data. C-DAS automatically checks the quality of the data and the calibration of the geophones and shaker by:

1. comparing the frequency being measured by the geophones against that being generated by the shaker and
2. comparing the phase difference across each pair of geophones, to ensure that the assumption of cross-isotropy (constant lateral stiffness) is within acceptable limits (see Figure 4 below).

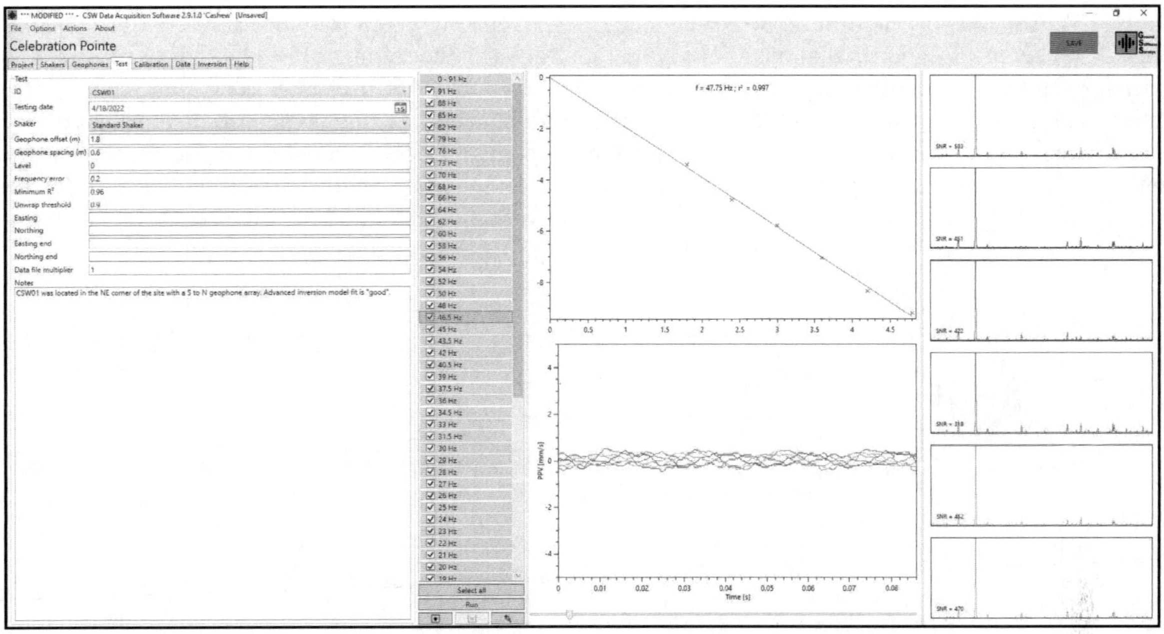

Figure 4 C-DAS software ACSW test control screen

C-DAS also carries out a real-time simple inversion of the field dispersion curve by assigning the measured value of v_r and the equivalent v_s to a fraction of the wavelength ($\lambda/2.5$), creating a velocity versus depth profile as testing proceeds, illustrated in Figure 5 below:

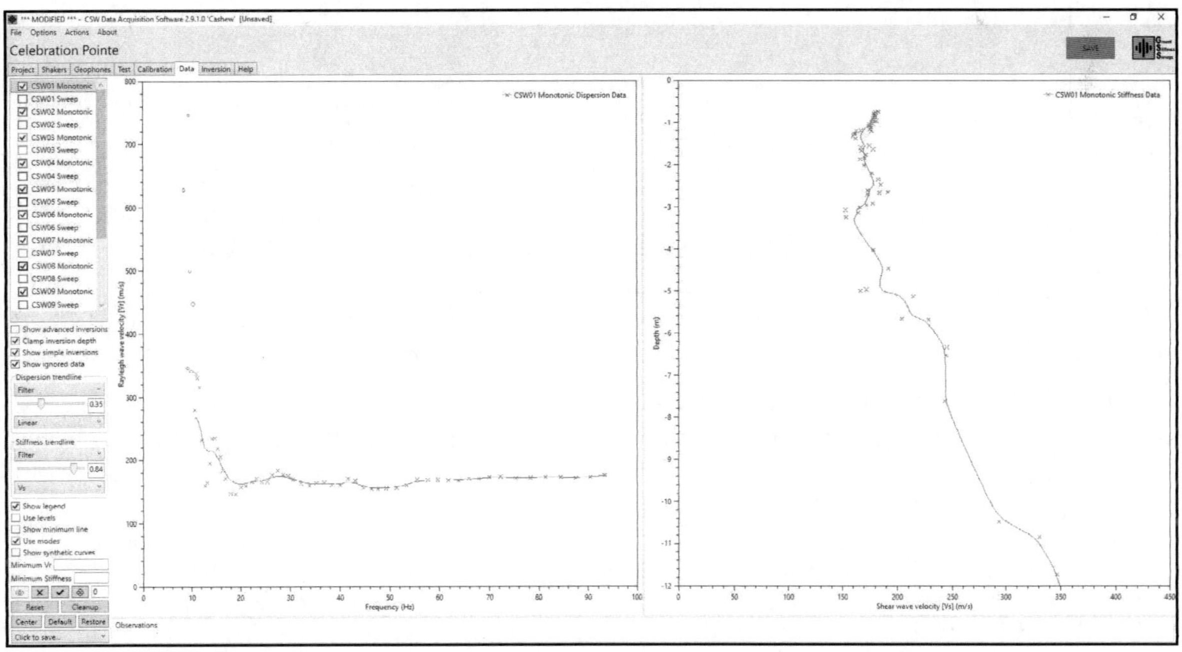

Figure 5 - C-DAS software Dispersion Curve and Simple Inversion screen

Finally, C-DAS offers a function known as an advanced inversion, which provides a layered soil stiffness profile for use in design calculations. Carrying out forward modelling of a layered v_s profile from dispersion curve data is mathematically complex, therefore C-DAS uses the Leong & Aung (2013) published algorithm to generate a synthetic dispersion curve. The fit of the synthetic dispersion curve to the test dispersion curve is used to guide an iterative adjustment of the layered v_s profile using a custom Monte Carlo method until a good fit is obtained. The output is provided to the client as an Excel spreadsheet indicating the different layer thicknesses and v_s values and allowing the user to vary ρ, ν, and design strain using the Rollins et al (1998) strain softening function, as illustrated in Figure 6:

Project:	Celebration Pointe	**Report:**	EFFC-DFI
Shift:	4/18/2022	**Client:**	Keller
Test:	CSW01	**Date:**	4/18/2022

Test notes: CSW01 was located in the NE corner of the site with a S to N geophone array. Advanced inversion model fit is "good".

Default values of density and Poisson's Ratio in the highlighted columns may be adjusted to known values
Strain level of softened value of Young's Modulus using the Rollins equation can be adjusted in the cell below.
See the GSS report ref EFFC-DFI for conditions of use of data

Eastings
Northings ***Level (m)*** Strain level softened to: 0.1 %

Vs (ft/s)	Thickness (ft)	Depth (ft)	Unit weight (lb/ft^3)	Go (tsf)	v	Eo (tsf)	E at x% strain (tsf)
601	4	0.0	112	561	0.26	1413.9	481.6
533	5	3.6	112	442	0.26	1113.6	379.3
557	6	8.6	112	482	0.26	1213.4	413.3
598	6	14.6	112	556	0.26	1401.4	477.3
774	6	20.7	112	931	0.26	2345.4	798.8
898	5	27.1	112	1254	0.26	3160.4	1076.4
980	2	32.0	112	1492	0.26	3760.2	1280.7
1060	3	33.9	112	1747	0.26	4401.8	1499.2
1111	4	36.7	112	1918	0.26	4833.0	1646.1

Notes:

1. Vs values have been determined from advanced inversion of field dispersion data. Vs values derived by inversion rely on assumptions of soil density and Poisson's Ratio. Where no project specific data is provided then default values of Υ=112 lb/ft^3 and ν=0.26 are assumed. *Refer to above GSS report conditions for further information.*
2. Density = ρ = Υ_t / g (where Υ_t = unit weight in lb/ft^3)
3. Stiffness is provided in units of short tons/ft^2 or tsf where 1 short ton = 2,000lbs
4. Shear modulus = G = $\rho . v_s^2$
5. Youngs modulus = E=G.(2.(1+v))
6. Softened values of stiffness are calculated using Rollins equation

$$\frac{G}{G_o} = \frac{1}{\left[1+16\gamma\left(1.2+10^{-20\gamma}\right)\right]}$$

where Υ is shear strain

Rollins et al. (1998)

Figure 6 – Excerpt from Excel report

POTENTIAL APPLICATIONS OF ACSW

The ACSW system was designed and manufactured by practicing geotechnical engineers who wanted more accurate data for use in calculating the anticipated settlement of a shallow foundation. Hence a 3m / 10-foot geophone array was chosen, so that the bulk soil stiffness under a large pad foundation could be measured to the typical 3B depth of influence (9m / 30-feet) and adjusted for the design strain. This remains a powerful application of the ACSW system. A detailed case study comparing anticipated settlements of a foundation using GSS ACSW data with actual settlements measured during a full-scale maintained load test of a 5m by 5m (16-foot by 16-foot) concrete test pad foundation can be found in Heymann et al. (2017).

There are also many other potential applications for a non-destructive, portable, rapid test of ground stiffness. These have included: stiffness profiling to detect discrete layers (such as a variation in rockhead); average stiffness profiling for earthwork or ground improvement control testing; Working Platform stability assessments for piling rigs and cranes; rock quality and rippability assessments; and void detection. Forensic applications have included assessment of grout distribution following sinkhole repair and ground characterization following a major building collapse. Since the test is rapid and the equipment portable, completed projects have also included active railroad embankment assessments during short night-time track possessions and riverbank assessments of retaining wall foundation conditions during short tidal windows. ACSW is non-destructive, so is attractive for the exploration of existing dams and levees, areas with underground or overhead utility conflicts, and within buildings (as illustrated in Figure 7):

Figure 7 – ACSW testing within residential property

CONCLUSIONS

1. ACSW testing is a fast, non-intrusive means of obtaining an in-situ ground stiffness profile in a construction environment.
2. ACSW stiffness values, softened to a conservative operational strain level of 0.1%, are significantly higher than the equivalent empirically derived values. It is noted that there is significant subjectivity involved in the selection of the correct empirical relationship for determining stiffness values from SPT N values, CPT q_c and undrained strength data which can lead to large differences in derived parameters and greater conservatism in design.

3. ACSW ground stiffness profiles have been demonstrated to provide the basis for accurate predictions of settlement which are more realistic when compared with actual foundation performance.
4. In addition to the provision of accurate ground stiffness profiles, ACSW derived seismic velocity profiles may be used to undertake rapid ground profiling, estimation of other geotechnical parameters, Working Platform assessments, earthworks control testing/verification, and forensic analysis of building failures.

ACKNOWLEDGMENTS

The author is indebted to the founders and fellow directors of Ground Stiffness Surveys LLC, the designers, manufacturers, and developers of the ACSW system and accompanying C-DAS software: Chris Milne MSc., EurGeol, CGeol, FGS, EurIng, CEng, FICE; John Rigby-Jones MSc., EurIng, CEng, MICE; and Hamish Milne, MEng, ACGI, MIET.

REFERENCES

Clayton, C.R.I. (1995) The standard penetration test (SPT): methods and use. CIRIA report R143. ISBN: 0-86017-419-0. 144pp.

Clayton, C.R.I., and Heymann, G. (2001). "The stiffness of geomaterials at very small strains." *Géotechnique*, 51(3): 245–256.

Das, B.M. and Sivakugan, N. (2007). "Settlement of shallow foundations on granular soil – an overview." *International Journal of Geotechnical Engineering,* (1): 19-29.

Heymann, G. (2007). "Ground stiffness measurement by the continuous surface wave test." *Journal of the South African Institution of Civil Engineers*, Vol 49, No. 1

Jiang, P., Tran, K.T., Hiltunen, D.R., and Hudyma, N. (2015). "An Appraisal of a New Generation of Surface Wave Techniques at a Test Site in Florida." *International Foundations Congress and Equipment Expo 2015* (IFCEE)

Love, A.E.H. (1927). "A treatise on the mathematical theory of elasticity." 4th ed. Cambridge University Press

Mair, R.J. (1993). "Developments in geotechnical engineering research: application to tunnels and deep excavation." *Unwin Memorial Lecture. Proceedings of the Institution of Civil Engineers*, Civil Engineering, Vol.93, 27-41.

Rollins, K.M., Evans, M.D., Diehl, N.B., and Daily, W.D. III (1998). "Shear modulus and damping relationships for gravels." *Journal of Geotechnical and Geoenvironmental Engineering*, ASCE, Vol. 124, No. 5, 396-405.

Leong, E. and Aung, A. (2013). "Global inversion of surface waves dispersion curves based on improved Weighted Average Velocity (WAVe) method." *Journal of Geotechnical and Geoenvironmental Engineering*, ASCE, Vol. 139 No.12.

Heymann, G., Rigby-Jones, J., and Milne, C.A. (2017). "The application of continuous surface wave testing for settlement analysis with reference to a full-scale load test for a bridge at Pont Melin, Wales, UK." *Journal of the South African Institution of Civil Engineering,* Vol 59, No. 2

THE ADVANTAGE OF MEASUREMENT WHILE DRILLING (MWD) FOR GEOTECHNICAL SITE CHARACTERIZATION

Anahita Modiriasari, WSP USA, New York, NY, USA, (646) 467-6206, anahita.modiriasari@wsp.com
Benjamin S. Rivers, FHWA, Atlanta, GA, USA, (404) 562-3926, benjamin.rivers@dot.gov

ABSTRACT

Measurement while drilling (MWD) is a subsurface exploration technique associated with real-time and continuous monitoring and recording of drilling data during the drilling process. This technique can be used for underground investigations to differentiate ground strata, detect the presence of fractures and voids, and evaluate geological features and mechanical properties of geomaterial. When MWD is used for geotechnical investigations, it can improve site characterization, identify variable conditions, improve design and construction, lead to shorter project delivery times, reduce the cost and risks associated with limited subsurface data and geotechnical uncertainties, and maximum return-on-investment. Given these benefits, MWD is gaining attention among transportation agencies and the geotechnical industry. Federal Highway Administration's (FHWA's) innovation deployment program "Every Day Counts (EDC-5)" includes the implementation of Advanced Geotechnical Methods in Exploration (A-GaME), which aims to implement effective, underutilized subsurface exploration methods, including MWD, in U.S. practice. This paper provides an overview of the MWD technology for geotechnical site investigations and presents a case history where MWD was employed to design deep foundation for the construction of a building in Massachusetts. The advantages of this technology for risk reduction and project cost and time saving are also discussed.

Keywords: subsurface investigation, measurement while drilling, MWD, geotechnical site characterization

INTRODUCTION

Geotechnical site characterization is essential for transportation projects. Traditionally, borings are commonly advanced to perform standard penetration tests (SPTs) and obtain representative soil samples at discrete locations, often at depth intervals of 5- to 10-feet. Conventional borings conducted to minimum standards often provide limited data for project design and construction, which can result in constructability issues and increased cost and construction delays. Under the Federal Highway Administration's (FHWA's) innovation deployment program "Every Day Counts (EDC-5)", the initiative on Advanced Geotechnical Methods in Exploration (A-GaME) aims to improve U.S. geotechnical site characterization practices by promoting the effective implementation of underutilized subsurface exploration technologies. By strategically planning geotechnical investigations that consider project risks and incorporate A-GaME technologies, the A-GaME initiative intends to reduce uncertainties in site conditions, improve reliability within design and construction and optimize subsurface exploration programs while maximizing the return-on-investment from improved site characterization during project delivery to provide time and cost savings for project stakeholders.

Measurement While Drilling (MWD) is one of the A-GaME methods. MWD provides real-time and continuous measurement and monitoring of drilling parameters such as rotation speed, penetration rate, torque, thrust, drilling fluid pressure and flow rate, and allows for examination of the ground-machine interactions responsible for the measured drilling characteristics. MWD data is acquired electronically from sensors placed on the drill rig. The sensors continuously collect data for each monitored parameter, typically displayed in real-time, without interfering with the drilling process. The continuous collection of this data produces practically continuous profiles of measured and calculated drilling parameters that can be used to identify changes in subsurface conditions.

This paper focuses on the application of MWD technology as a field test technique to characterize subsurface conditions during drilling and to investigate the influence of in-situ materials, drilling techniques and drilling tools on the recorded drilling parameters in both qualitative and quantitative terms. An example of the implementation of the MWD technology for subsurface investigation in the design of deep foundation for the construction of a building in Massachusetts is also described in the paper. The purpose of this paper is to provide an overview on the MWD technology, its application and current efforts to get this technique into routine practice for geotechnical site characterization.

MWD APPLICATIONS IN GEOTECHNICAL SITE CHARACTERIZATION

MWD is relatively new to U.S. geotechnical site characterization practice, although it has been used for decades for this purpose in Europe and was recently standardized under the International Organization of Standardization (ISO) as ISO 22476-15 in 2016. For geotechnical investigation purposes, the drilling parameters are used to distinguish material changes during drilling. The parameters from direct measurements or compound relationships describing the mechanical response (e.g., rate of penetration with respect to applied energy) during the advancement of a borehole can be displayed graphically on a dialog monitor or computer, as seen in Fig. 1. In geotechnical engineering, MWD was first applied using ENPASOL device in the early 1970s to measure drilling parameters of typically continuous non-coring rotary rigs or to complete the core description in case of poor recovery on coring rigs (Hamelin et al. 1982). Similar systems have been used for determining geological strata, assessing ground improvements, evaluating bridge foundations (Pfister 1985; Pazuki and Doran 1995), identifying soil and rock for dredging operations (Smith 1994), investigating landslides (Anagnostopoulos and Papadopoulos 1995), and assessing the condition of tunnel lining (Gui et al. 2002). Yue et al. (2004) developed an in-situ device, the Drilling Process Monitoring (DPM) system, that could be mounted onto existing drilling machines with minor modifications to monitor and record the full drilling process during soil nailing for slope stabilization.

Fig. 1. MWD system displaying the real-time monitoring of drilling parameters (Rodgers et al. 2020)

To date, MWD technology has been mostly used in the U.S. for drill-and-blast and grouting projects. For instance, Lindenbach (2016) used MWD technology for a ground improvement grouting project in Florida to monitor the grouting operation directly, provide real-time data generation from the grouting operation for daily transmittal to the client, compare with the grouting logs to confirm that identified voids had been grouted effectively, and find regions where grout takes should have been higher based on the drilling logs (quality control check). In addition, Charlton et al. (2010) successfully used MWD technology at Bear Creek Dam in Tennessee during a grouting program to address seepage through the karstic limestone foundation to create a detailed geologic model by identifying areas of fractured rock, clay infill, and changes in stratigraphy with higher conductivity to be grouted appropriately.

In addition to being used on grouting and drill-and-blast projects worldwide (Rai et al. 2016; Høien and Nilsen 2014; Leung and Scheding 2015; Hatherly et al. 2015; Shi 2013; Zhao et al. 2011; Zetterlund et al. 2017), this technology has been successfully used in highway and transportation projects in many European countries to support borehole permeability tests (e.g., Lefranc and Lugeon tests), prebored pressuremeter testing, installation of instrumentation (e.g., piezometer and inclinometer), slope stability analysis (e.g. during soil nailing), deep foundation installations to estimate grout take or depth to sound rock, detection of cavities (Garassino and Schinelli 1998), and bearing capacity estimations. The Florida Department of Transportation (FDOT) took the first steps in the U.S. to understand and develop MWD practices for drilled-shaft construction by correlating drilling parameters to in-situ rock strength for bridge foundations. FDOT subsequently studied and related drilling operations using equipment and tooling common to geotechnical site investigations. MWD performed prior to design provided a significant increase in design data, and improvement of sample recoveries, drilling practices and equipment selection. MWD performed during drilled-shaft construction provided a construction monitoring technique (Rodgers et al. 2018b). Outside of the U.S., MWD has been extensively implemented for exploration of subsurface stratigraphy (Gui et al. 2002; Gui and Hamelin 2004; Gui 2008; Sadkowski 2010b; Reiffsteck et al. 2018; Laudanski et al. 2013). Additionally, it has been extensively used during geotechnical investigations for rock classification (Vezhapparambu 2018) and for determining schistosity orientation, degree of weathering (Yue et al. 2004), the detection of cavities (Patel 2013), frequency of rock jointing (Sadkowski et al. 2010b; Schunnesson 1996), and infilling (Babaei Khorzouhi et al 2018).

MWD has been used to infer the mechanical properties of rock including the uniaxial compressive strength (UCS), Young's modulus, and tensile strength (Smith 1994; Rodgers et al. 2020; Rodgers et al. 2018a; Rodgers et al. 2019; Taheri et al. 2016; Rodgers et al. 2018b; Celada et al. 2009; Eldert et al. 2019; Finfinger et al. 2000; Hoberock and Batcher 1996; Kalantari et al. 2018; Li and Itakura 2012; Lonstein et al. 2015). MWD has also provided a means to assess general rock quality characteristics (Rodgers et al 2020; Rai et al. 2016; Schunnesson 1996; Naeimipour et al. 2014).

BENEFITS OF MWD APPLICATION

MWD provides continuous drilling performance information with depth, as previously mentioned, which can be correlated to the drilling ease/difficulty and is representative of subsurface characteristics. MWD can be cost effective since it is simply measuring and recording parameters during drilling operations. Additionally, the in-situ MWD data is extremely valuable when samples are difficult to obtain or when samples might rapidly deteriorate when exposed to the atmosphere and are not representative of the actual ground conditions (Lindenbach 2016). MWD can also be used for intermediate/regional geomaterials where more common methods are not as well suited (residuum, till, conglomerate, cobbles, karst, etc.). MWD is well suited as a supplemental technique to acquire additional information during conventional operations (minimal added effort for more data). MWD systems can be installed on a drill rig at modest cost without being disruptive to operators, who require some training. Furthermore, MWD provides real-time feedback to drillers so that drilling operations can be adjusted to optimize data and sampling quality, production and operational cost. MWD equipment can also be installed on rotary drilling equipment capable of oriented drilling. The continuous record of parameters with depth can relate drilling resistance and penetration rates at specific anchor or foundation locations for improved reliability of anticipated drilling conditions during construction. The increased frequency of measurements provides a means to assess variability with depth for design purposes, too. When integrated within a well-planned and -executed subsurface exploration program, MWD can reduce uncertainties and risks associated with site variability and limited subsurface data that could lead to construction delays and cost escalations (Rodgers et al. 2020; Lonstein et al. 2015).

MWD EQUIPMENT REQUIREMENTS

Drilling Equipment Requirements

Most geotechnical drill rigs can be instrumented to achieve high quality data and resolution. Any drill rig that can satisfactorily obtain quality testing and sampling in soil and rock for geotechnical investigation using rotary-wash drilling and coring can be instrumented for MWD and doing so provides the greatest flexibility for applications. For now, per ISO 22476-15, the only drilling methods standardized include rotary and rotary-percussive drilling. Several State Departments of Transportation are currently studying the application of hollow-stem auger (HSA) drilling with MWD, however, MWD practice with HSA has not yet been standardized. Both mechanical and hydraulic driven and operated drill rigs can be outfitted to allow monitoring and data acquisition of drilling parameters; mechanical systems are more widely used in the U.S. The drilling system should always have ample power, stability, and the appropriate energy transfer (drill-rods), bit-types, and drilling fluid to penetrate the various geologic layers efficiently to required depths for the investigation objectives while maintaining borehole stability and effectively removing cuttings.

The drilling technique should be governed by the test objectives (subsurface profiling and/or estimation of soil and rock properties), necessary resolution and application. Methods without percussion tend to provide higher resolution. If MWD is used with coring operations, MWD can help to complete the core description in case of poor recovery. More preferably, it can be used to control drilling to optimize core quality (Rodgers et al. 2020). For applications where coring is not required, such as for soundings used to probe for boulders or karst features, MWD provides an effective means to acquire a record of useful continuous drilling data using destructive roto-percussion drilling often at a one third of the cost of coring or sampling methods (Laudanski et al. 2013). Drilling operations should remain as constant as possible in all cases to obtain comparable consistent drilling results and correlations with soil and rock information. The reasoning for this is to minimize variations in the drilling data due to the influences of different drilling tools, drilling rig types, and drilling techniques (Gui et al. 2002).

Measuring System Requirements

Modern systems acquire measurements digitally. A typical system includes a set of transducers and sensors, a digital data recorder with a data display, and a data acquisition system including a signal converter (sometimes part of a data recorder). The real-time display allows adjustments to the drilling process. Figure 2 shows an example of the MWD sensors mounted on a hydraulic drill rig. ISO 22476-15 specifies an allowable minimum accuracy of measured values for three quality class systems. A Quality Class 1 system provides the highest quality and requires that at least six parameters are measured. These may include penetration length, down-thrust pressure, holdback pressure (not common on U.S. geotechnical rigs), drill-fluid pressure, drill-head rotational speed, drill-fluid circulation rate, and torque. The measuring equipment should be checked and calibrated regularly according to the manufacturers' specification and calibration frequency requirements. Calibration records should be documented, and a copy of the records should be kept with the maintenance log.

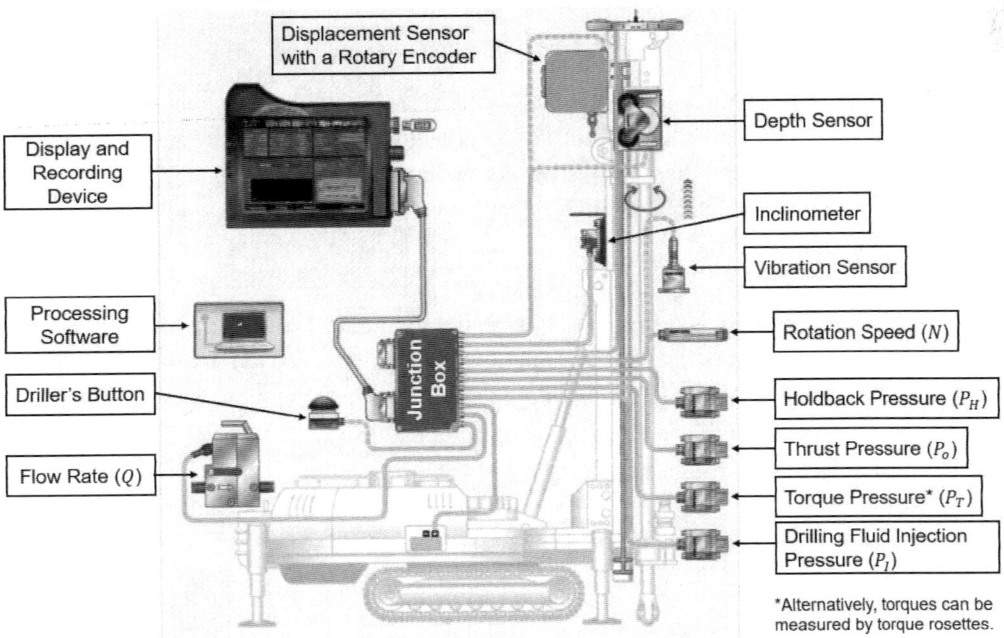

Fig. 2 MWD sensors and transducers (modified after Teale 1965)

MWD TEST PROCEDURE

Disturbance from nearby previously drilled locations can influence in-situ conditions and MWD measured values. Therefore, locations of new MWD borings should avoid close proximity to previously executed boring or coring locations. Consequently, any selected locations where core-only borings may be used for general rock investigations or for correlating MWD test data to rock properties should be performed an adequate distance away from comparative MWD test locations. According to McVay, Rodgers, and Horhota (2020), a distance of at least three to five times the borehole diameter between MWD tests is suggested. Some drilling techniques, such as air-flush drilling, may require larger distances (ISO 22476-15 2016).

Current MWD international practice does not provide guidance on oriented borings, although performing MWD for oriented borings is possible (e.g., oriented exploratory drilling for soil nailing in cut slopes). The drill rig mast and rotary head track should be oriented to apply a thrust vector to the drilling rods as close to vertical or the desired axis of inclination as possible. ISO 22476-15 (2016) recommends that the deviation from the intended axis should be less than 2° or 1.3 inches per 3.3 feet. The inclination can be measured and controlled in real-time with a biaxial inclinometer or tilt sensor attached to the drilling mast.

Since it is possible that two different soil or rock units show the same range of drilling parameter measurements, it is essential to do an initial calibration with the execution of at least one logged MWD borehole near a sampled one to compare variations with observed conditions. Calibration should also aim to optimize drilling parameters that result in achieving maximum mechanical efficiency of the drilling tool and minimum energy required for drilling. Note that the measured responses are specific to the drilling system, bit-type and -wear. As such, site calibrations and production-drilling should document the drilling system, the size and type of bit used, and any observable bit-wear before and after each hole (or more frequently, as needed).

The frequency of comparative borings and corings depends on the application and variability of the site. The MWD system should allow recording of all parameters as a function of penetration depth intervals. If data are recorded in time-series, they are inverted to depth-serial data at regular intervals. With appropriate filtering or improvements to acquisition systems, MWD between sampling and in-situ testing intervals are possible within the same hole.

Upon completion of the MWD boring, the total boring depth should be verified and recorded. The total length of the drilling rods used to reach target depth should be cross-checked against the depth measured during MWD. The change or replacement of any equipment should be reported as well. The final wear of the drilling bit should be assessed for comparison with the pre-drilling condition. These can be used later for assessing the influence of wear in the results (ISO 22476-16 2016).

INTERPRETATION OF MWD RESULTS

Drilling Parameters

Drilling parameters are measured directly and recorded during drilling. Throughout the drilling process, a relatively constant drilling-fluid pressure (flow rate), rotation speed, and thrust on the bit (or crowd) should be maintained to obtain comparable data from other parameters (Gui et al. 2002). The standard values for each of these three parameters should be determined on each site to suit the drilling and ground conditions (Gui and Hamelin 2004). The values are generally first adjusted at the beginning of the drilling to obtain the best compromise between production and the instrumented drilling investigation, and to optimize the drilling operation (minimize the drilling energy and core-sample disturbance).

The data recorded from the various drilling parameter sensors require some processing and filtering for conversion into useful engineering units and to yield data during actual drilling. Data produced during idle running, rod changes, ceasing, bit lifting, or from signal noise is removed (Li et al. 2014). It should also be noted that some research results showed that it might be possible to interpret the noise in the raw data and characterize the mechanical vibrations to distinguish soil formation quantitatively. This is because the noise can be related to the material and drilling bit interaction and each material type creates noise that can have a "signature" in specific unfiltered drilling parameters (Gui et al. 2002; Gui 2008).

Compound Parameters

Compound parameters are derived from the combination of various drilling parameters. Since the early 1980s, compound parameters have been developed to limit the influence of varying drilling techniques and assist the understanding and characterization of the ground conditions (Somerton 1959; Pfister 1985; Gui et al 2002; Teale 1965; Moller et al. 2004; Nishi et al. 1998; Girard et al. 1986). Compared to measured drilling parameters, compound parameters are less dependent on the borehole advancement imposed by the driller and drilling procedures, and they represent physical characteristics of the ground.

In order to interpret the MWD results, a standardized drilling system, in terms of hardware and drilling method, should be used along with a consistent range of drilling-fluid flow rate, rotation speed, thrust and torque provided during drilling. Differences and changes in drilling method, drilling tool, bit-wear, flow rate, applied thrust, and operator will influence measures. The measured drilling- and compound-parameters are used to identify geological features and transitions between soil and rock layers, evaluate changes in relative density or weathering, evaluate discontinuities, predict soil and rock properties, or to optimize operating parameters. This helps to improve the geotechnical model in combination with other site investigation and sampling methods.

USE OF MWD AS A GROUND INVESTIGATION TOOL

To date, MWD has been mostly used qualitatively to describe geological profiles and has not yet become a standard ground investigation tool in civil and geotechnical engineering projects in the U.S. The assessment of rock strength and geospatial variability from MWD are new applications with limited work completed. The following section summarizes the implementation of MWD for a deep foundation project. This case study presents how MWD technology can be successfully used in the U.S. practice to minimize the uncertainties of subsurface investigations, improve drilling operations, provide construction monitoring, reduce project cost and reduce drilling time.

Detection of Stratigraphy Changes and Fracture Zones for a Deep Foundation Project in Boston, MA (Sadkowski et al. 2010b)

Project Overview

A subsurface investigation was conducted using the MWD technology to provide geotechnical recommendations for the construction of a nine-story mixed-use (clinical, retail, office) building with a footprint of approximately 57,372 square feet and with four to six levels of below-ground parking. The project is located in the Back Bay area of Boston, Massachusetts. The building was constructed with excavation support provided by a 3.3-ft thick reinforced concrete diaphragm wall, which provided the permanent exterior walls for the below-grade portion of the structure. The interior building column loads were supported by spread footings (49 to 59 ft below proposed finished grade) to be constructed below a pressure relief slab.

Sadkowski et al. (2010b) identified the stratigraphic changes from coarse to fine-grained soils and fractured zones within the bedrock. The results of this investigation were used to determine the design depth of the reinforced concrete diaphragm wall to allow for an adequate cutoff from groundwater during the deep excavation. The MWD drilling system was used on eight 4-in (101.6 mm) diameter test borings with depths ranging from 123 to 159 ft (37.5 to 48.5 m) below the ground surface, and with approximately 20 to 40 ft (6.1 to 12.2 m) of rock coring at each location. The drilling parameters were measured during the advancement of a tri-cone roller bit using rotary technique through the overburden glacial soils. The drilling parameters that were monitored and recorded included: depth, penetration rate, rotary speed, thrust pressure, torque, drilling fluid pressure, and drilling fluid inflow and outflow parameters at a 0.2 in (5 mm) frequency (Fig. 3). This allowed for identification of very thin lenses of stratigraphic variation.

Implementation of MWD Technology in Soil Drilling

The trends in the drilling parameters were correlated with the soil samples obtained from 2-in split spoons (Fig. 3). The ratio of drilling fluid (water) pressure to the thrust pressure best identified changes from granular to cohesive soil types since the fluid pressure on the bit is a function of the soil type and the pressure applied to the bit by the drill rig (with a relatively constant rotation rate). The correlations between collected soil samples and the normalized ratio of water pressure over thrust pressure allowed drilling parameters to characterize subsurface conditions with respect to soil composition. This resulted in reduction of the frequency and number of split spoon samples taken and increased the boring completion efficiency. Figure 3 also shows the decrease of the penetration rate (by maintaining a relatively consistent rotation and thrust pressure) observed upon encountering the glacial till layer which was confirmed by the split-spoon samples collected in the glacial till layer. The glacial till was evaluated as a potential alternative layer to provide groundwater cutoff since the clay and silt layer were discontinuous.

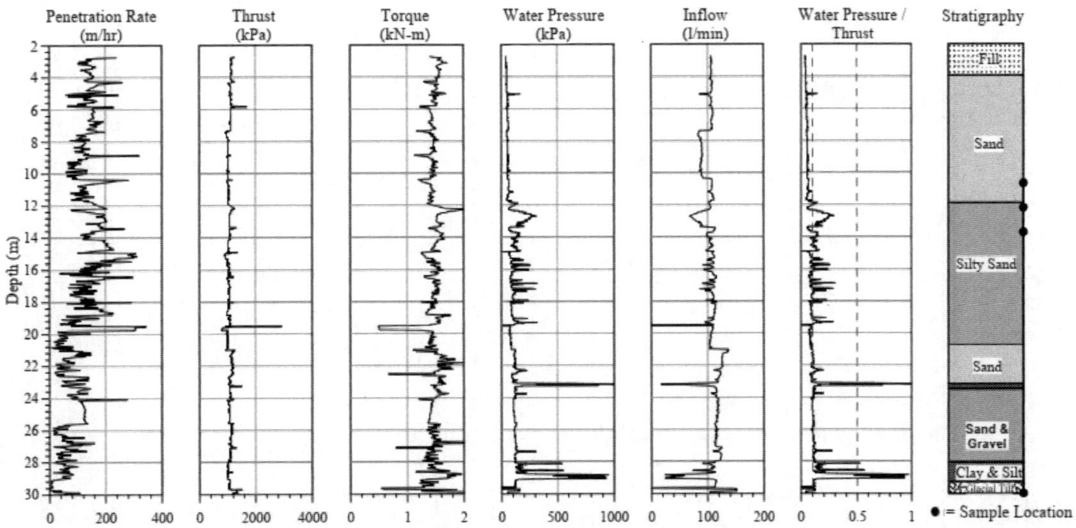

Fig. 3. Measured drilling parameters, water to thrust pressure ratio, and soil sample depths stratigraphy (Sadkowski et al. 2010b)

Implementation of MWD Technology in Rock Drilling

Sadkowski et al. (2010b) also used the MWD method to record the drilling parameters during rock coring to evaluate the rock mass quality and competency, presence of fractures, and changes in lithology. This provided input for selection of depth intervals to perform packer tests to measure the hydraulic conductivity of the fractured bedrock formation. The correlation between the trends of the drilling parameters with the physical characteristics of the bedrock core specimens helped to discern between natural fractures and drilling-induced fractures. In addition, bit-wear was minimized, and higher quality core-samples were obtained by monitoring water loss or gain.

Figure 4 shows the drilling parameters during bedrock coring where bedrock was encountered between 105 and 144 ft (32 and 44 m) below the ground surface. The approximate depth of natural fractures – based on the drilling data and core specimens – are also shown in the figure along with measured core-run Rock Quality Designation (RQD) indices. Two flowmeters were used to measure flow rate into the borehole (inflow) and the return flow rate out of the borehole (outflow). The purpose of using two flowmeters was to measure fluid loss from fractures. Fluid pressure and penetration rate data also indicate the approximate depth of the natural fractures. These measures were also used to estimate RQD index values. Using a constant thrust and rotation rate, a variation in other measured drilling parameters such as the penetration rate, drill-fluid pressure, or flow would suggest either a change in bedrock lithology or the presence of an anomaly such as a fracture or a cavity if the variation was over a larger depth. The drill-fluid pressure decreased significantly at a depth of about 110 ft (33.5 m). This was attributed to drilling-fluid consumption by a weathered Argillite seam within the Quartzite bedrock formation. This was also observed in the outflow data as it approached zero return flow at this depth. The recorded drilling data was used to select zones within the fractured rock to perform packer tests to measure the permeability of the fractured bedrock. This helped to evaluate a suitable design depth for the concrete diaphragm wall where adequate groundwater-cutoff could be obtained during excavation.

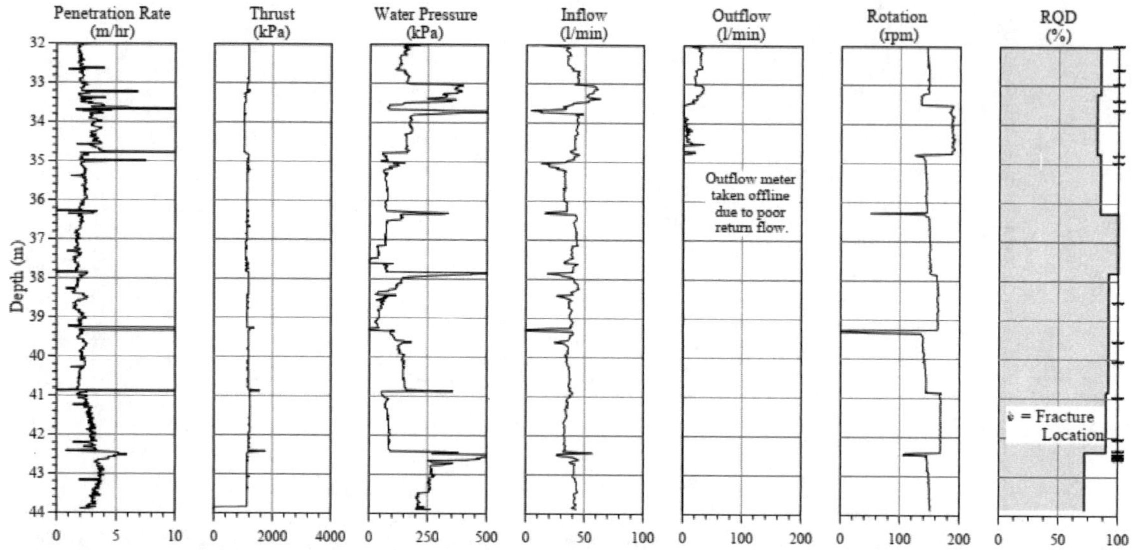

Fig. 4. Measured drilling parameters, fracture locations and RQD in a bedrock profile (Sadkowski et al. 2010b)

Project Observations

Use of MWD technology led to reduction of the number of required samples while also gaining valuable subsurface data for the entire borehole profile. In addition, the data resolution obtained through MWD technology is greater than that available from typical sampling protocols since drilling parameters are monitored and recorded at a near-continuous very high frequency. Ultimately, the use of drilling parameters from MWD technology increased the efficiency in which boreholes were drilled, which led to estimated savings of $40,000 to $50,000 (associated with approximately 12 to 16 days of drilling) in drilling related expenses.

SUMMARY AND CONCLUSIONS

The overview, case history and results from MWD shown in this paper highlight its potential value to improve subsurface models and supplement conventional site investigation and sampling methods. MWD is a viable method that provides real-time, near-continuous drilling parameter measurements related to subsurface conditions, including stratigraphic and lithologic changes, geologic structure and features (discontinuities, voids, fracture zones), and material properties and strength assessment of soil and rock. The benefits (similar to cone penetration test, CPT) of being nearly continuous and repeatable provide for reliable measurement with depth. The automated nature of digital acquisition potentially makes for easy data transfer and use in evolving digital areas of the profession (e.g. visualization, computational modeling, artificial intelligence (AI) and machine learning) that require robust datasets. In addition, MWD helps to quantify drilling procedures that improve core recovery, sample quality, and drilling efficiency. Implementing MWD only requires a relatively small investment to procure equipment, instrument a drill-rig and train a drill-crew. The added value from the additional data can ultimately provide better site characterization, reducing geotechnical uncertainties and project risks for more effective project delivery. MWD does not provide a direct measurement of ground properties nor a direct method to discern material. However, the parameters can be used relationally and site-specifically to discern changes in strata/lithology and to make correlation to material types and properties. Finally, MWD can potentially reduce the number of required physical samples needed during site characterization, but it cannot substitute for sampling. Representative and undisturbed samples will be required to calibrate and validate correlations between the MWD drilling parameters and subsurface properties; correlations are specific to the existing soil or rock in each site.

To date, MWD has been used primarily to support other in-situ tests and to describe geological profiles, and it has not yet become a standard ground investigation tool for civil and transportation projects in the U.S. Future studies and continuing field-trials are anticipated to help develop strategies to select appropriate tools for different material types, maximize the benefits of drilling parameters, and behaviorally classify soils and rocks by interpreting MWD data in a systematic approach similar to the use of CPT and SPT data. Overall, this paper shows that a few key drilling parameters are sufficient to develop empirical relationships to assist in soil and rock identification, provide an indication of general soil and rock behavior, and to investigate the influence of in-situ materials, drilling techniques, and drilling tools on the recorded drilling parameters in both qualitative and quantitative terms. For each site with similar ground conditions, correlation between the MWD data and sampling results can be determined during the initial site investigation phase. The MWD data can then be used to assess the subsurface properties in real-time during the construction phase.

ACKNOWLEDGMENTS

This material is based upon work supported by Professor Jean Benoît, Ph.D., Department of Civil and Environmental Engineering, University of New Hampshire, Professor Michael McVay, Ph.D. and Michael Rodgers, Ph.D., P.E., Research Assistant Scientist, Department of Civil and Coastal Engineering, University of Florida and David Horhota, Ph.D., P.E., State Geotechnical Materials Engineer, Florida Department of Transportation. The authors are grateful for this support.

No content within this paper has the force and effect of law and is not meant to bind the public in any way. The U.S. Government, including the Federal Highway Administration (FHWA), does not endorse products or manufacturers. Any references to products or manufacturers are included for informational purposes only and are not intended to reflect a preference, approval, or endorsement of any one product or entity.

REFERENCES

Anagnostopoulos, A. and Papadopoulos, B. 1995. Restraint of An Active Landslide by Bored Piles. Proceedings of the Bengt B. Broms Symposium on Geotechncial Engineering, 187–193.

Babaei Khorzoughi, M., Hall, R., and Apel, D. 2018. Rock fracture density characterization using measurement while drilling (MWD) techniques. International Journal of Mining Science and Technology, 28(6), 859–864.

Celada, B., Galera, J., Muñoz, C., and Tardáguila, I. 2009. The use of the Specific Drilling Energy for Rock Mass Characterisation and TBM Driving During Tunnel Construction. Proceedings of the ITA-AITES World Tunnel Congress, (V), 1–13.

Charlton, J. E., Ginther, C. H., and Bruce, D. A. 2010. Comprehensive foundation rehabilitation AT BEAR CREEK DAM. Proceddings of the Association of State Dam Safety Officials Annual Conference 2009, Dam Safety 2009, 3(3), 1452–1476.

Eldert, J., Schunnesson, H., Johansson, D., and Saiang, D. 2019. Application of Measurement While Drilling Technology to Predict Rock Mass Quality and Rock Support for Tunnelling. Rock Mechanics and Rock Engineering, 53(3), 1349–1358.

Finfinger, G., Peng, S., Gu, Q., Wilson, G., and Thomas, B. 2000. An Approach to Identifying Geological Properties from Roof Bolter Drilling Parameters. Proceedings of the 19th Conference on Ground Control in Mining, 1–12.

Garassino, A. ., and Schinelli, M. L. 1998. Detection of Cavities by Monitored Borehole Drilling (TMD). Proceedings of the 1st International Conference on Site Characterization, 1, 365–370. London: Geological Society of London.

Gui, Soga, K., Bolton, M. D., and Hamelin, J. P. 2002. Instrumented borehole drilling for subsurface investigation. Journal of Geotechnical and Geoenvironmental Engineering, 128(4), 283–291.

Gui, M. W., and Hamelin, J. P. 2004. Development of an instrumented borehole drilling system for ground investigation. Electronic Journal of Geotechnical Engineering, 9 A, 1–11.

Gui. 2008. The Basics of Noise Detection and Filtering for Borehole Drilling Data. The Open Civil Engineering Journal, 2(1), 113–120.

Hamelin, J. P., Levallois, J., and Pfister, P. 1982. Enregistrement des parametres de forage: Nouveaux développements. Bulletin of the International Association of Engineering Geology - Bulletin de l'Association Internationale de Géologie de l'Ingénieur, 26(1), 83–88.

Hatherly, P., Leung, R., Scheding, S., and Robinson, D. 2015. Drill monitoring results reveal geological conditions in blasthole drilling. International Journal of Rock Mechanics and Mining Sciences, 78, 144–154.

Hoberock, L. L., and Bratcher, G. J. 1996. A New Approach for Determining In-Situ Rock Strength While Drilling. Journal of Energy Resources Technology, 118(4), 240–255.

Høien, A. H., and Nilsen, B. 2014. Rock mass grouting in the Løren tunnel: Case study with the main focus on the groutability and feasibility of drill parameter interpretation. Rock Mechanics and Rock Engineering, 47(3), 967–983.

Kalantari, S., Hashemolhosseini, H., and Baghbanan, A. 2018. Estimating rock strength parameters using drilling data. International Journal of Rock Mechanics and Mining Sciences, 104, 45–52.

Laudanski, G., Reiffsteck, P., Tacita, J. L., Desanneaux, G., and Benoît, J. 2013. Experimental study of drilling parameters using a test embankment. Geotechnical and Geophysical Site Characterization 4 - Proceedings of the 4th International Conference on Site Characterization 4, ISC-4, 1, 435–440.

Leung, R., and Scheding, S. 2015. Automated coal seam detection using a modulated specific energy measure in a monitor-while-drilling context. International Journal of Rock Mechanics and Mining Sciences, 75, 196–209.

Li, Z., and Itakura, K. I. 2012. An analytical drilling model of drag bits for evaluation of rock strength. Soils and Foundations, 52(2), 216–227.

Li, Z., Itakura, K. I., and Ma, Y. 2014. Survey of measurement-while-drilling technology for small-diameter drilling machines. Electronic Journal of Geotechnical Engineering, 19(Z2), 10267–10282.

Lindenbach, E. J. 2016. Improving investigations with drill parameter recorder technology. Geotechnical Practice Publication, (GPP 10), 51–59.

Lonstein, E., Benoît, J., Sadkowski, S., and Stetson, K. 2015. Estimation of Cambridge Argillite Strength Based on Drilling Parameters. Proceedings of the 66th Highway Geology Symposium, 74–102.

Naeimipour, A., Bahrampour, S., Rostami, J., and Dogruoz, C. 2014. Ground characterization while drilling roofbolters in tunneling operations. Proceedings of the North American Tunneling - 2014 Proceedings, NAT 2014, 51–59.

Patel, A. 2013. Characterization of cavities in rocks using drilling parameters. International Journal of Rock Mechanics and Mining Sciences, 63, 122–130.

Pazuki, A., and Doran, S. 1995. Soil Investigation for Cross Passages. Proceedings of the XI European Conf. on Soil Mechanics and Foundation Engineering, 63–72.

Pfister, P. 1985. Recording Drilling Parameters in Ground Engineering. Jopurnal of Ground Engineering, 18(3), 16–21.

Rai, P., Schunnesson, H., Lindqvist, P. A., and Kumar, U. 2016. Measurement-while-drilling technique and its scope in design and prediction of rock blasting. International Journal of Mining Science and Technology, 26(4), 711–719.

Reiffsteck, P., Benoît, J., Bourdeau, C., and Dessanneaux, G. 2018. Enhancing geotechnical investigations using drilling parameters. Journal of Geotechnical and Geoenvironmental Engineering, 144(3), 1–11.

Rodgers, M., McVay, M., Ferraro, C., Horhota, D., Tibbetts, C., and Crawford, S. 2018a. Measuring rock strength while drilling shafts socketed into Florida limestone. Journal of Geotechnical and Geoenvironmental Engineering, 144(3), 1–8.

Rodgers, M., McVay, M., Horhota, D., and Hernando, J. 2018b. Assessment of rock strength from measuring while drilling shafts in Florida limestone. Canadian Geotechnical Journal, 56(5), 662–674.
Rodgers, M., McVay, M., Horhota, D., Sinnreich, J., and Hernando, J. 2019. Assessment of shear strength from measuring while drilling shafts in Florida limestone. Canadian Geotechnical Journal, 56(5), 662–674.

Rodgers M., McVay M., Horhota D., Hernando J., Paris J. 2020. Measuring While Drilling in Florida Limestone for Geotechnical Site Investigation. Canadian Geotechnical Journal, 57(11), 1733-1744.

Sadkowski, S. S., Stetson, K. P., Benoît, J., and Roche, J. T. 2010b. Characterizing Subsurface Conditions Using Drilling Parameters for a Deep Foundation Project in Boston, MA, USA. GeoFlorida 2010: Advances in Analysis, Modeling and Design. American Society of Civil Engineers.

Schunnesson, H. 1996. RQD predictions based on drill performance parameters. Tunnelling and

Underground Space Technology, 11(3), 345–351.

Shi, H. 2013. Application Prospect of MWD Drlling Technology in Water Prevention and Control of Coal Seam Floor. ICPTT 2013: Trenchless Technology - The Best Choice for Underground Pipeline Construction and Renewal, Proceedings of the International Conference on Pipelines and Trenchless Technology, 554–563.

Smith. 1994. New Approach for Determination of Rock and Rock Mass Properties at Dredging Sites. Development and Change, 25, 497–526.

Somerton, W. 1959. A Laboratory Study of Rock Breakage by Rotary Drilling. Society of Petroleum Engineers, 216(1), 92–97.

Taheri, A., Qao, Q., and Chanda, E. 2016. Drilling Penetration Rate Estimation using Rock Drillability Characterization Index. Journal of The Institution of Engineers (India): Series D, 97(2), 159–170.

Vezhapparambu, V. S., Eidsvik, J., and Ellefmo, S. L. 2018. Rock classification using multivariate analysis of measurement while drilling data: Towards a better sampling strategy. Minerals, 8(9).

Yue, Z. Q., Lee, C. F., Law, K. T., and Tham, L. G. 2004. Automatic monitoring of rotary-percussive drilling for ground characterization-illustrated by a case example in Hong Kong. International Journal of Rock Mechanics and Mining Sciences, 41(4), 573–612.

Zetterlund, M., Martinsson, L., and Dalmalm, T. 2017. Implementation of MWD-Data for Grouting Purposes in a Large Infrastructure Project - The Stockholm Bypass. Geotechnical Special Publication, (288 GSP), 61–70.

Zhao, Y., Shi, Z., and Liu, Q. 2011. Research and Application of Down-hole Motor Borehole Technology for Directional Drilling in Coal Mine. ICPTT 2011: Sustainable Solutions for Water, Sewer, Gas, and Oil Pipelines - Proceedings of the International Conference on Pipelines and Trenchless Technology 2011, 2032–2038.

CORROSION PROTECTION OF MICROPILES

Racquel Nottingham, Friedr. Ischebeck GmbH, Ennepetal, GERMANY

Marc Mastrantuono, Ischebeck USA Inc., Naples FL, USA

ABSTRACT

Micropiles often provide a favourable solution for many deep foundation problems especially when significant site constraints such as confined construction sites or vibration limitations have been identified. Often the durability of buried steel elements are heavily scrutinized in order to effectively determine the rates of corrosion and methods to reduce this while still maintaining the structural integrity of the engineered solution. Practices such as sacrificial corrosion loss and the introduction of plastic sheaths have become the go-to methods of corrosion protection. Since the introduction of the micropile, other solutions have been developed, namely protection with an adequate grout encapsulation.

This paper will explore corrosion protection of micropiles and the guidelines provided in the published codes and norms. Since few codes define the exact way to verify the durability of these micropiled structures, the existing data is presented and the analysis of grout encapsulation as an effective corrosion protection method is evaluated.

Keywords: micropiles, grout encapsulation, thread patterns

INTRODUCTION

Micropiles are drilled and grouted, reinforced, non-displacement piles which have minimal diameters of less than 300mm (12in) (Federal Highway Administration (FHWA) 2012). These small elements can transfer very high axial loads to the ground but only very minimal lateral loads. These slender piles transfer loads to the ground through skin friction at the interface grout-ground and the end bearing is often neglected in soft soils and weak rock.

Micropiles are categorized by the method of grout placement and five categories are given in the FHWA's Micropile Design Guideline. In this research, Type E micropiles, highlighted in Figure 1, would be explored. These refer to micropiles that are installed using a continuously threaded hollow bar which functions as an injection tube, drilling rod and reinforcing element.

Of all the micropile variants, the self-drilling hollow bar system has proven to be versatile and often very efficient in urban interventions including locations with significant site constraints. The self-drilling system comprises a hollow threaded bar and sacrificial drill bit. The elements are manufactured in standard lengths usually 3m and 6m bars and can be extended via coupled connections.

Corrosion protection of steel elements often plays a distinctive role in the initial engineering and design considerations. The durability of a certain element often influences the use of the element to perform the assigned task. Countless articles and research have been done on exposed structures and codes but not much effort is placed on highlighting the specific corrosion protection measures of buried steel elements.

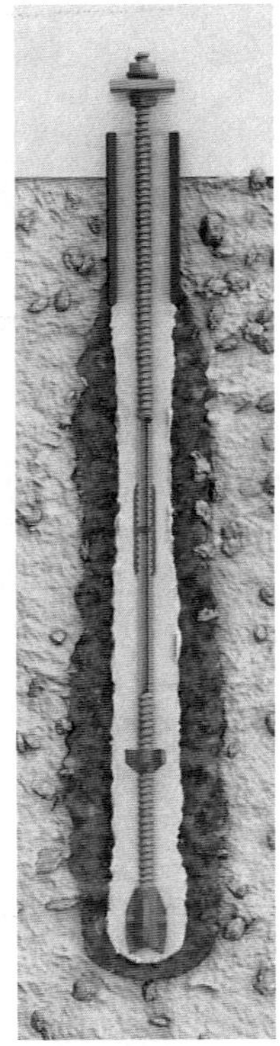

Figure 1 - Cross-section through typical Type E micropile.

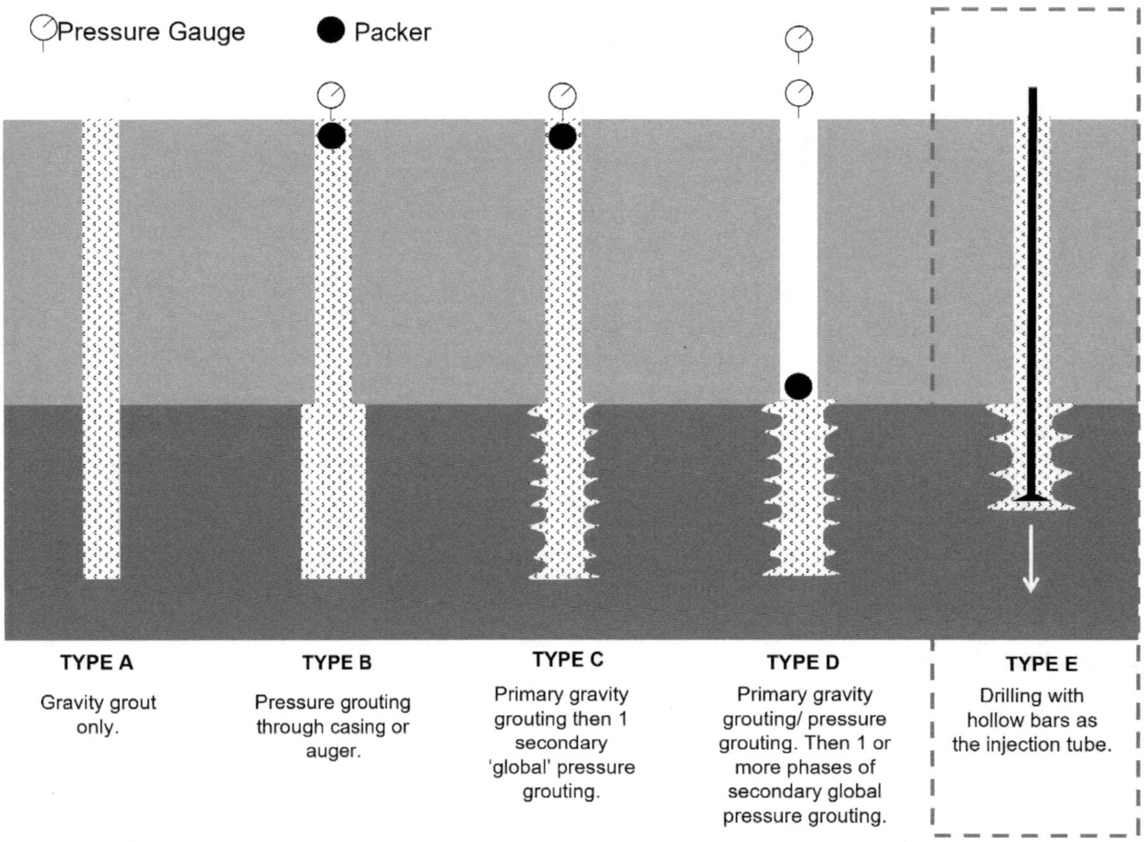

Figure 2: Micropile Classification recognized in the United States *(Federal Highway Administration (FHWA) 2012)*

In the past, many design engineers have simply adopted the concept for corrosion protection of exposed elements to buried elements. In sheet piling, the metal is in direct contact with the elements such as air and water which encourages the development of oxides. Extensive research has been done to estimate the corrosion rate of these metals and the effect of atmospheric conditions such as humidity, rainfall, pollution, salts etc. on the corrosion rate (American Galvanizers Association n.d.).

In the ground however, the behaviour of steel is quite different. In addition to the elements being difficult to observe because they are buried, the wide range of soil types makes it extremely difficult to predict the corrosion of metals in every soil type. Furthermore, when the elements are encapsulated in an adequate and effective grout body, as is the case in some micropile installations, the research is limited. An adequate grout body can provide a further corrosion protection barrier to the steel and is often overlooked. It is also substituted for those known solutions for exposed pipes and/or exposed steel structures such as sheet piles.

The AASHTO and FHWA have outlined different types of micropiles based on the method of grout placement. The different corrosion protection methods in the code is presented and analysed in terms of the appropriateness and useability in practice.

THE CODES AND NORMS

In Europe the EN 1997-1 (EN 1997-1 2004), governs geotechnical designs and for each application an execution norm is produced. For micropiles the EN 14199 is specifically for the execution of micropiles. Each country however has its own norms that may form part of the common practice.

In the United States, the AASHTO and FHWA design guides are often used in the design and implementation of micropiles.

These would be further used to argue the case of durability later in this paper

CORROSION

Corrosion is a natural process in which refined metal is transformed into a chemically stable oxide. For embedded steel the corrosion occurs through an electrochemical process whereby the metal is transformed into free ions and electrons due to the interaction of an electrolyte, which in this case is the soil. The process is both anodic – ions pass from the metal to the solution as free hydrated ions – and cathodic – metals separate from the solution and recombine with the metal – reactions (Hobst und Zajic 1983).

Specific soil characteristics can present an environment which is conducive to corrosion and thus it is imperative to evaluate the physical and chemical properties in the design phase. In the case of the physical parameters, these can welcome the egress of the elements, increasing the corrosion potential of the steel element. In soils such as clays, which have a generally higher plasticity, is very impermeable when saturated but upon drying the particles shrink. Cracks then form providing the prefect environment for moisture and oxygen to make its way toward the buried element (Wasim, et al. 2018).

Resistivity refers to the concentration of ions from dissolved salts and minerals that flow through the soil electrolyte and is an indicator of the soils aggressivity and susceptibility to corrosion (Merzah, Hafiz und Mohammed 2014). Resistivity is a function of moisture content and concentration of ionic soluble salt. As it decreases, such as an increase in the sodium content in the groundwater, the corrosion potential subsequently increases (Kar Sing, et al. 2013).

The concentration of chemical properties such as sodium, potassium, calcium and magnesium, carbonates, bi-carbonates, sulphates, and nitrates also aid in the corrosion potential (Wasim, et al. 2018) and must be carefully considered during the design.

In the Eurocodes, exposure classes have been defined to classify the soil types or the environment into different levels of aggressiveness, and the minimum grout cover required is given in Table 1 below.

Table 1: – Guidance for minimum grout cover according to EN 14199 - Table B1 (DIN EN 14199 2015)

Exposure Class[a]	Chemical aggressiveness	Bearing element with grout cover [mm]		Bearing element with mortar [mm]	
		Compression	Tension	Compression	Tension
X0	with permanent casing	10	10	25	25
X0, XC1 – XC4	Not existing	20[b]	20[b]	35	40
XD1, XD2	Chloride except salt water	20	20	35	40
XS1	Chloride from salt water	20	20	35	40

[a] For other exposure classes in EN 206 minimum cover are given in EN 1992-1-1:2004, Clause 4, and the valid National Annex

[b] For service life of maximum 5 years minimum grout cover may be reduced to 10mm.

These exposure classes are further explained in EN 1992-1 based on EN 206, which details the following - X0 – no risk to corrosion or attack, XC1 – XC4 – Corrosion induced by carbonation, XD1 – XD3 – Corrosion induced by chlorides (not sea water), XS1 – Corrosion induced by chlorides from sea water.

The FHWA, offers guidelines for evaluation of the soil corrosion potential. The ground is considered to be aggressive or to have a strong corrosion potential if any of the following conditions, given in Table 2 below, are met. The norm goes on to explain that the possibility of a change in the aggressivity throughout the service life, should also be considered.

Table 2 – Criteria for Assessing Ground Corrosion Potential - Table 5-5 *(Federal Highway Administration (FHWA) 2012)*

Test	Units	Strong Corrosion Potential/ Aggressive	AASHTO Test Method
pH	-	<5, >10	T289
Resistivity	ohm-cm	<3,000	T288
Sulfates	ppm[(1)]	>200	T290
Chlorides	ppm	>100	T291

Note(1) ppm = parts per million

CORROSION PROTECTION

The codes and norms provide very broad corrosion protection methods but in actuality the applicability of these methods have not been thoroughly analysed specific to Type E micropiles. The methods for corrosion protection include protective coatings, concrete encasement, cathodic protection, plastic sleeves and sacrificial steel.

AASHTO also mentions that the technique of adopting a larger steel section is such that the structural requirements are met even after corrosion (AASHTO 2020). This particular technique works great for steel piles and other elements such as sheet piles, however the effect on elements with coupled connections is often grouped with other steel piles and overlooked. The FHWA Micropile Guideline does not list this method of sacrificial steel as a corrosion protection to the micropile but only applicable to the steel casing that may be considered. The corrosion protection can be provided through physical or chemical protection or a combination of both (Federal Highway Administration (FHWA) 2012). Minimum grout cover should be 1 inch in soil and 0.5 inch in rock, however if galvanization is used, a minimum grout cover of 0.25 inches is acceptable.

In Europe the micropile execution norm EN 14199 (DIN EN 14199 2015), also provides similar guidelines stating that the corrosion protection should consider the environmental conditions, the type of micropile, the steel and required working life. The corrosion protection may consist of sacrificial thickness, grout, mortar or concrete encapsulation and specific precautions such as surface coatings.

The type of connection used is very important when considering the different corrosion protection options presented to steel elements. The way corrosion develops, thickening of the section and allowing the bar to corrode uncontrollably, is rather allowable corrosion and not an effective corrosion protection method. Allowing the bar to corrode can eventually result in corrosion of the threads within the coupled connection during the design life of the project. Once the coupled connection is lost, the bar can no longer meet its designed intent and must be completely replaced.

SACRIFICIAL CORROSION LOSS

The industry has fallen into only accepting the technique of section thickening in order to provide a corrosion protection. In effect, this concept is to allow the section to corrode such that the element that remains will still reach the structural requirement of the design. This concept is particularly effective in sheet piles and buried structures like underground pipelines, however the practicality in a micropile system with a coupled connection is yet to be explored and argued.

This concept although it is widely accepted, has not been widely tested investigated to determine the feasibility of such a practice. However, situation have been encountered where the thickening of the section and allowing the reinforcement to corrode uncontrollably, has resulted in failure of elements.

The main concern lies at the connection. These elements transfer the forces and are extended through coupled connections. Once the grout cracks and corrosion begins, there is no way to ensure that the corrosion does not occur within the coupler. Once the threads are corroded the coupled connection is lost so too the structural capacity of the element. Effectively allowing the bar to corrode is not corrosion protection but uncontrollable corrosion allowance.

GROUT ENCAPSULATION OF TYPE E MICROPILES

Grout encapsulation is a simple method that is often rejected by the industry designers that have only designed steel elements using sacrificial corrosion. The concept however is nothing new but rather adapted from that of the corrosion protection used in reinforced concrete structures. These elements are designed to limit their cracking and thus ensure the durability of the reinforcement within.

According to EN1992-1-1 (EN 1992-1 2004), as part of serviceability limit state, the crack width of the reinforced elements must be verified, such that the cracking is limited. This ensures the durability or capacity of the element is not compromised. In reinforced concrete, an adequate concrete cover is required to reduce the cracking and essentially this same concept can be applied to a micropile.

This method can present as a single layer of corrosion protection. It can also be combined with galvanizing to provide a double corrosion protection in cases of highly aggressive soils.

In describing the practices for concrete structures, AASHTO states that reinforcement detailing is important to control the crack width. Furthermore, many small cracks are more favourable to few very wide cracks.

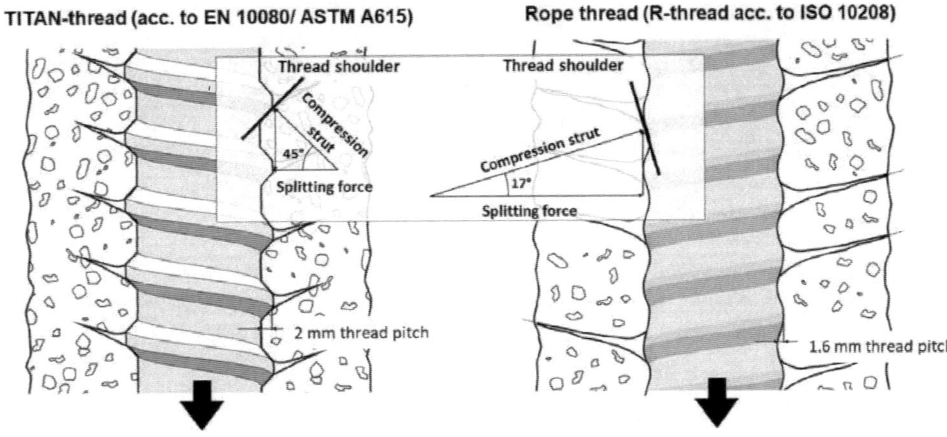

Figure 3: Cracking pattern observed under tension loading for TITAN-thread (left) and Rope thread (right). (Lopez, Nottingham and Gadermann 2020)

Considering the two most common hollow bar sections available – Rope Thread and TITAN Thread - the cracking patterns can be directly analysed based on the thread patterns presented. Under axial loads, the grout is expected to crack, the magnitude of which is directly proportional to the thread pattern. Once the crack width can be limited to less than 0.1mm, the alkalinity of the grout ensures the protection of the steel member and in some cases these small cracks are even considered self-healing.

Figure 3 presents the different cracking patterns subjected to axial forces in tension. Based on the thread patterns alone, the rope threads have a much greater splitting force which creates more frequent and larger cracks that extend to the outer edge of the grouted body.

The shape and extent of the cracking pattern along the TITAN threads have been extensively studied which confirmed that the cracking pattern is much like that observed in reinforced concrete elements, as shown in Figure 4. According to Issue 525 of the German Committee for Reinforced Concrete, the crack widths at the edge of the grout body can be assumed as significantly larger than those at the hollow bar due to the warping of the cracks.

One such test featured the installation of some micropiles and then subsequent exhuming of the grouted micropiles once the grout reached its maximum strength. Tensile tests were then performed where the elements were incrementally loaded, under lab conditions and the cracks were marked and measured at each load step. The tests confirmed that the measured cracks were all less than 0.1mm (0.039 in).

A separate test was also performed in Sweden were TITAN threaded bars were exhumed and the cracking pattern studied. The elements were cut into sample pieces and then impregnated with an epoxy containing fluorescent properties. The epoxy was allowed to penetrate the samples and then viewed under a UV light. A maximum crack width of 0.01mm (0.00039 in) was measured in this experiment.

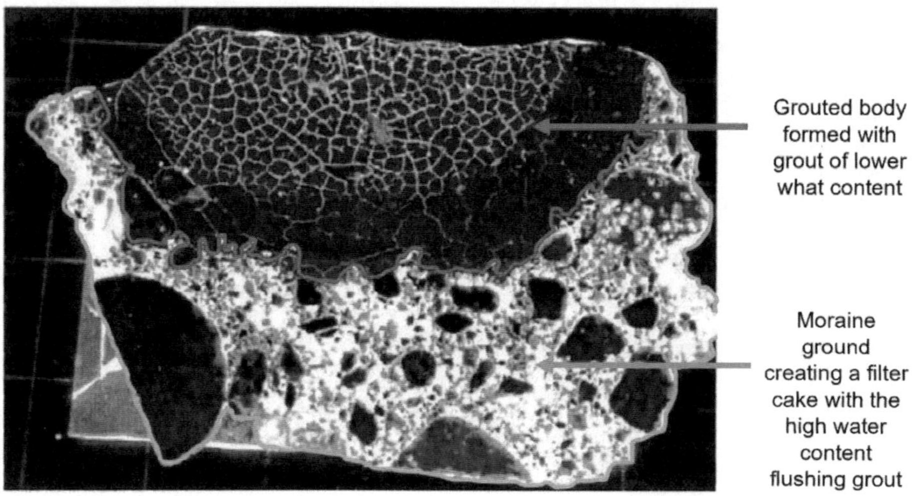

Figure 4 – Sample impregnated with an epoxy and viewed under a UV Light.

The uppermost part of the image represents the grouted body closest to the reinforcement, where many microcracks were observed. As we move outward, away from the steel, the cracking is reduced and only few cracks make it to the edge of the filter cake (highlighted in red). The outermost layer between the red and blue boundaries represents the filter cake formed during drilling where a cement suspension with a high-water content was used as flushing fluid which is very apparent under the UV lighting.

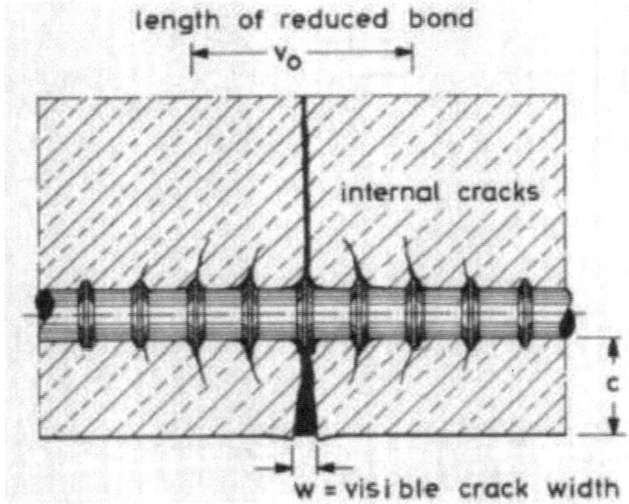

Figure 5: Microcracks forming at every thread crest in reinforced concrete elements and only some cracks extend to the outer edge of the concrete. *(Lopez, Nottingham and Gadermann 2020)*

According to the National Technical Approval document of TITAN, adequate grout cover needs to be provided, based on all the testing carried out to ensure that the corrosion protection is maintained over the serviceability life of the project.

CONCLUSION

The industry has adopted the sacrificial corrosion technique as the main method for corrosion protection when it is in fact not at all feasible where coupled connections are considered for permanent installations. Many professionals assume the same method used for steel piles and sheet piles can also be utilized in micropiles without considering that the applications are quite different.

Although some of the codes and norms mention several methods to provide corrosion protection, it is the responsibility of the designers to determine the most feasible option to ensure that the structural integrity is maintained through the serviceability life of the elements. Allowing the bar to corrode as the method of protection in sheet piles or steel piles also allows the threads to corrode. This can and will result in a loss of the coupled connection and structural capacity of the element.

The two most common thread types for micropiles, namely Rope Threads and TITAN threads are both adopted in practice. The TITAN threads however, adopt the technique of grout encapsulation while the rope threads adopt the method of sacrificial corrosion without considering the grout encapsulation. In many instances a further barrier such as a galvanized coating is proposed as an additional corrosion protection measure.

This paper presents some of the test results for the TITAN threads and also explains the concept behind the thread geometry. This thread type has been designed according to the well-known behaviour of reinforced concrete. Herein the feasibility of using grout encapsulation as an effective corrosion protection is more than adequate once the integrity and size of the grout body is accurately attained. However, very little is available to support the feasibility of sacrificial corrosion in coupled elements as adopted in the case of rope threads.

For the corrosion protection of Type E micropiles, sacrificial corrosion should not be the industry standard as this is not a method of corrosion protection but rather corrosion allowance. Many studies have been performed to show that an adequate grout encapsulation as well as the use of coatings such as galvanizing, and epoxy provide more than adequate corrosion protection in permanent applications. The use of sacrificial corrosion as a form of corrosion protection in permanent elements has yet to be deeply studied in terms of their long-term effects and the likelihood of decoupling after corrosion loss.

References

AASHTO. *LRFD Bridge Specifications.* Washington DC, 2020.

American Galvanizers Association. *American Galvanizers Association.* n.d. https://galvanizeit.org/corrosion/corrosion-process/corrosion-rate.

DIN EN 14199. *Execution of special geotechnical works - Micropiles.* Berlin: European Committee for Standardization, 2015.

EN 1992-1. *Eurocode 2: Design of concrete structures.* Berlin: European Committee for Standardization, 2004.

EN 1997-1. *Eurocode 7: Geotechnical Design.* Berlin: European Committee for Standardization, 2004.

Federal Highway Administration (FHWA). *Micropile Design and Constuction Guidelines.* Bibliogov, 2012.

Hobst, L., und J. Zajic. *Achoring in Rock and Soil.* Elsevier, 1983.

Kar Sing, Lim, Nordin Yahaya, Siti Rabeah Othman, Siti Nor Fariza, and Norhailan Md Nora. "The Relationship between Soil Resistivity and Corrosion Growth in Tropical Region." *The Journal of Corrosion Science Engineering*, 31 July 2013.

Lopez, Freddy, Racquel Nottingham, and Ole Gadermann. "Design of Self Drilling Micropiles for Permanent Applications." Ennepetal, 2020.

Merzah, Amel S., Mohammed H. Hafiz, und Sarah K. Mohammed. „Corrosion Control of Buried Low Carb on Steel Structure by Using Alteration Media Method.“ *Al-Khwarizmi Engineering Jornal* , 15. April 2014: 65-74.

Wasim, Muhammad, Shahrukh Shoaib, N.M. Mubarak, Inamuddin, and Abdullah M. Asiri. "Factors influencing corrosion of metal pipes in soils." *Environemntal Chemistry Letters*, 15 March 2018.

A DISCUSSION OF INCONSISTENCIES BETWEEN LRFD AND ASD METHODS FOR GEOSTRUCTURAL DESIGN APPLICATIONS OF THE AASHTO BRIDGE DESIGN SPECIFICATIONS

Franklin M. Grynkewicz, P.E.[1], Leili Javidannia, P.E.[3], Dimitrios Palantzas, P.E.[2], Collin E. Sweeney, E.I.T.[3]

[1]Senior Principal Geotechnical Engineer, [2]Senior Geostructural Engineer, [3]Staff Geostructural Engineer
GeoEngineers, Inc.
4 California Ave, Suite 204, Framingham, MA, USA 01701

ABSTRACT

In the latest AASHTO design specifications, the Load and Resistance Factor Design (LRFD) method is prescribed for the design of retaining walls. However, inconsistencies arise when LRFD is used for retaining wall design versus the traditional Allowable Stress Design (ASD). Required toe embedment and wall loading demand vary greatly when the two design approaches are compared, all else being equal. This is due to application of linear load factors to properties that vary non-linearly in the design. These inherent inconsistencies are further intensified when introducing factors such as layered soil profiles, different wall types, groundwater elevation, and surcharge requirements. When applied correctly, these two design approaches, in theory, should lead the designer to the same result with little discrepancy, as they do with steel structure design. However, the use of the current AASHTO LRFD method for various geostructural retaining wall designs yields contradictory results that suggest existing retaining walls that have been successfully designed and constructed are unstable. Fundamentally, LRFD should be built on reliability-based calibrations; the successful past performance of existing retaining structures designed using ASD suggests that the LRFD approach requires considerable modification. The authors of this paper recommend that the current LRFD load factors be reviewed and revised based on more comprehensive wall performance data and wall type such that the difference between retaining structures designed using LRFD and those designed using ASD is eliminated.

Keywords: ASD, LRFD, AASHTO, load factors, retaining wall design

DESIGN METHODOLOGIES

Allowable Strength Design (ASD) and Load Resistance Factor Design (LRFD) are common frameworks for analyzing the loading of a structural element. These approaches are both widely used in the field of civil engineering. Both follow the same thesis: the available strength of a structural element must exceed the strength required by a design loading. The difference between the two methods lies in the calculation of these available and required strengths.

Allowable Stress Design

The Allowable Stress Design (ASD) method has been in wide use since the early 1800s (Regan et al., 2013). In the ASD approach, the available strength is known as the allowable strength, which must exceed the required strength for all limit states. The allowable strength is calculated by dividing the nominal strength of a structural member by a factor of safety, often Ω. All uncertainty in loads and material resistance is encompassed by the factor of safety, which is an empirical measure used to reduce the potential for adverse performance. The required strength and factor of safety are both calculated per the applicable building code or specification. The required strength is determined by ASD load combinations in the relevant code; however, the ASD load combinations for retaining wall design apply a factor of 1 to dead, live, earth, and groundwater loads.

Load and Resistance Factor Design

Load and Resistance Factor Design (LRFD) is based on more realistic loading conditions and material properties. In the LRFD approach, the available strength still must exceed the required strength for all limit states, but the approach is calculated differently. The uncertainty in material resistance is represented by a resistance factor that commonly has a value less than 1, and the uncertainty in loads is represented by load factors that commonly have a value greater than 1. In other words, material strength is reduced, and load effects are increased. The LRFD available strength, or design strength, is calculated by multiplying the nominal strength by the resistance factor, often Φ. Again, both LRFD load and resistance factors are per the applicable building code or specification.

The LRFD method has undergone significant improvements since its inception. It has mainly gained traction in structural engineering, most commonly in design of steel and concrete structures. Early use of LRFD for geotechnical applications can be readily tracked back to 1950s (Hansen 1953, 1956) when the industry experienced increased demand for more economical designs.

LRFD in Geotechnical Practice

It is well-known that geotechnical problems are subject to many inherent uncertainties in site characterization, soil behavior, and construction quality. One of the key difficulties in all existing codes is the application of LRFD to geotechnical problems due to this uncertainty. It appears that not all of the associated resistance factors that were developed for geotechnical applications are sourced from reliability-based calibrations. As a result, some current LRFD specifications for geotechnical applications are of limited use; most applicable to this paper is the AASHTO LRFD manual regarding cantilever retaining wall design. As outlined above, the premise of LRFD is based on a rational evaluation of performance reliability. It represents an approach in which applicable failure and serviceability conditions can be evaluated considering the uncertainties associated with loads and material resistance. AASHTO no longer publishes the ASD code, only the LRFD code.

In each approach, load combinations and safety and resistance factors are calibrated such that the two approaches lead the designer to the same result: according to the American Institute of Steel Construction (AISC) Steel Construction Manual (15th ed.), the "two approaches are equally valid for any structure for which the Specification is applicable." This statement highlights the utility and interchangeability of both approaches in steel construction applications. But though both LRFD and ASD approaches lead the designer to the similar result in steel design, the same is not true for some geotechnical and geostructural applications.

Often in geostructural design, the live and dead loads are prescribed by the structural designer. As a result, geotechnical engineers have limited control over design loadings. In geotechnical design, resistance factors depend on the uncertainties associated with the variability and reliability of different elements, including:

- extent of soil exploration and type of sampling and testing used to characterize a site,
- inherent soil variability,
- soil property measurements,
- groundwater conditions,
- procedures or models used for design, and
- measures employed to monitor and test construction processes.

Thus, prescribing a single soil load and resistance factor is not appropriate in geostructural applications. Geotechnical engineers are best equipped to calibrate a design based on their control over the extent and type of sampling and testing used to characterize a site, and the procedures or models used for design.

GEOSTRUCTURAL DESIGN APPLICATIONS

A common geostructural design is the cantilever retaining wall. Per the latest AASHTO design specifications, the LRFD method is prescribed for the design of retaining walls. However, a retaining wall design per LRFD is often inconsistent with traditional force-moment balance design using the ASD methodology. Required toe embedment and wall loading demand may vary significantly when the same wall is analyzed with the two different methods. ASD has proved its reliability and merit as a design method: countless ASD-designed permanent retaining walls have been constructed and served their purpose successfully before the advent of the LRFD methodology. For this reason, the discrepancy between the two design methods for retaining wall applications should be attributed to the prescribed load and resistance factors of LRFD.

As previously discussed, LRFD is based on its own unique load combinations. These load combinations are more variable and specific than their ASD equivalents. In the case of a cantilever retaining wall, load factors for the Service 1 load case (AASHTO, 2020) are 1 for both the active (unfavorable) and passive (favorable) earth pressures that the wall experiences below the bottom of excavation. Since all load factors for this case are 1, it is essentially equivalent to an ASD approach. But for the Strength 1a load case, the load factors are 0.75 for the favorable earth pressures and 1.35 for the unfavorable earth pressures. As expected, this increase in active earth pressure and decrease in passive earth pressure results in larger required toe embedment. However, when this result is compared with an ASD analysis done with a factor of safety (Ω) of 1.5, the LRFD toe embedment is considerably different.

We hypothesize that the resultant difference is twofold: first, the LRFD approach applies linear factors to a non-linear parameter. Passive earth pressure is most accurately modeled as a log-spiral shape, so application of a linear load factor is inappropriate and undervalues passive earth pressures. Second, it appears that LRFD load factors have not been fully calibrated as they do not vary with soil type, wall type, or groundwater conditions. As a result, the published LRFD load factors are not adjustable for soils with differing friction angles, and do not consider a number of variable site conditions, including groundwater levels and surcharge loads.

CANTILEVER WALL ANALYSIS

To illustrate the problem, we analyzed both continuous (sheet pile) and discontinuous (soldier pile and lagging) cantilever retaining walls. The analyses consider a 15-foot-high retaining wall with cohesionless soil of unit weight $\gamma = 125$ pcf. The discontinuous wall considers an 8-foot pile spacing with piles set in a 3-foot-diameter drill hole and passive soil arching equal to 2.66 times the drill hole diameter. Variable friction angles from 26 degrees to 36 degrees are considered to study the effect of AASHTO LRFD load factors on the required wall toe embedment depth. Additionally, various site factors, such as external surcharge loads and a groundwater table, are added to examine their effects on the required embedment. All analyses are limit equilibrium (LEM) (force-moment balance) analyses performed in the DeepEx 2021 software. Each model targets a factor of safety of rotation of 1.5 and 1.0 (unity) for the ASD and LRFD methods, respectively.

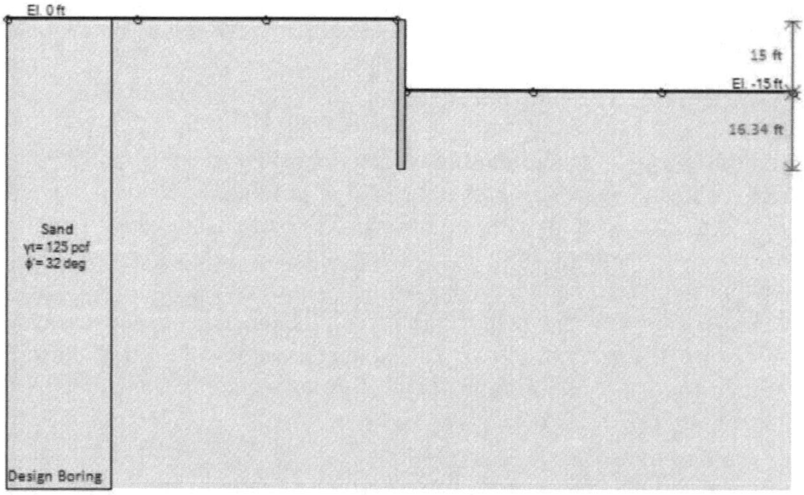

Figure 1 – Base Model in DeepEx

Base Model

The base model considers Rankine active and passive earth pressures to find the minimum required toe embedment to meet the required factors of safety. No wall friction, external surcharge, or groundwater is considered in this design section. The required wall length increased by 10-19% for a continuous wall and 7-10% for a discontinuous wall when considering the AASHTO LRFD load factors compared to the traditional ASD method of analysis.

Wall Friction Model

The wall friction design section considers the same geometry and variables as the base model with the introduction of wall friction on the passive (favorable) side of the retaining wall. This model considers log-spiral earth pressures with a wall friction angle equal to 1/3 of the soil friction angle on the passive side and 0 on the active side. The required wall length increased by 7-15% for a continuous wall and 6-9% for a discontinuous wall when considering the AASHTO LRFD load factors compared to the traditional ASD method of analysis.

Surcharge Model

The surcharge design section considers the same geometry and variables as the base model with the introduction of a 250-psf x 20-ft surface surcharge located 2-ft from the rear face of the wall. Boussinesq lateral surcharge pressure for a flexible wall system is considered in the model to determine lateral earth pressures due to the surcharge. The required wall length increased by 11-19% for a continuous wall and 8-11% for a discontinuous wall when considering the AASHTO LRFD load factors compared to the traditional ASD method of analysis.

Groundwater Model

The groundwater design section considers the same geometry and variables as the base model with the introduction of groundwater to the model at depth of 15-ft below top of wall. Groundwater is considered equal on the active and passive side of the retaining wall (i.e. no net water pressure is considered). The required wall length increased by 13-26% for a continuous wall and 9-13% for a discontinuous wall when considering the AASHTO LRFD load factors compared to the traditional ASD method of analysis.

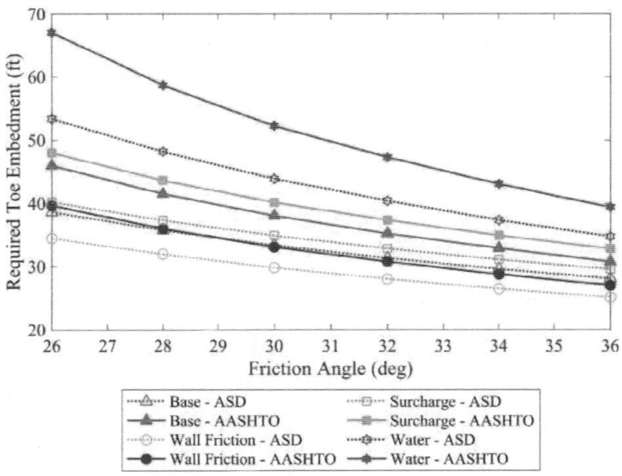

Figure 2a – Friction Angle vs Required Wall Length – Continuous Wall

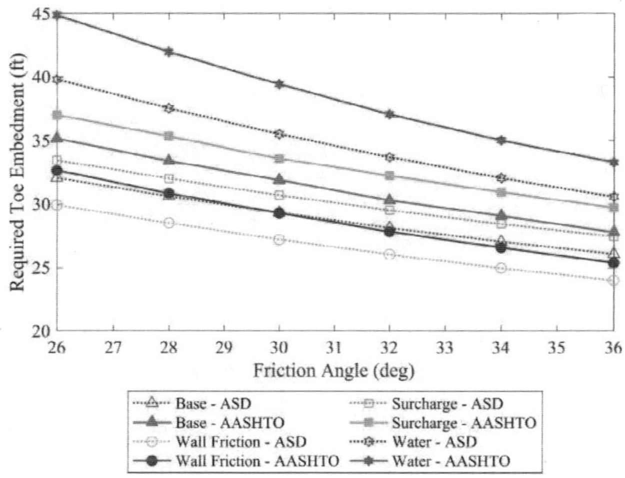

Figure 2b – Friction Angle vs Required Wall Length – Discontinuous Wall

The required wall length is not a linear function of soil friction angle and varies depending on the model type. In general, as the soil friction angle increases, the discrepancy between ASD and LRFD analyses decreases; however, this difference remains significant.

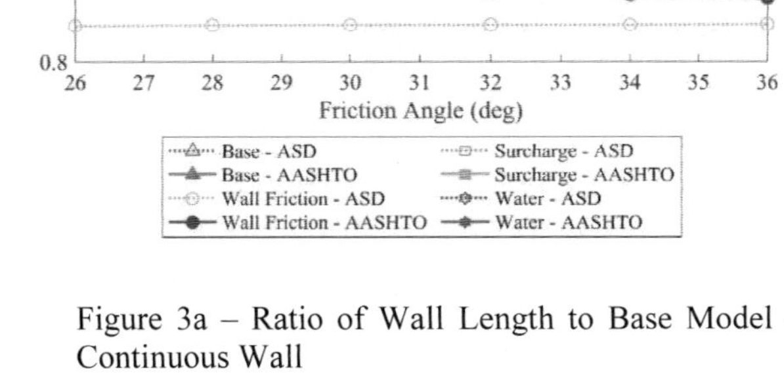

Figure 3a – Ratio of Wall Length to Base Model – Continuous Wall

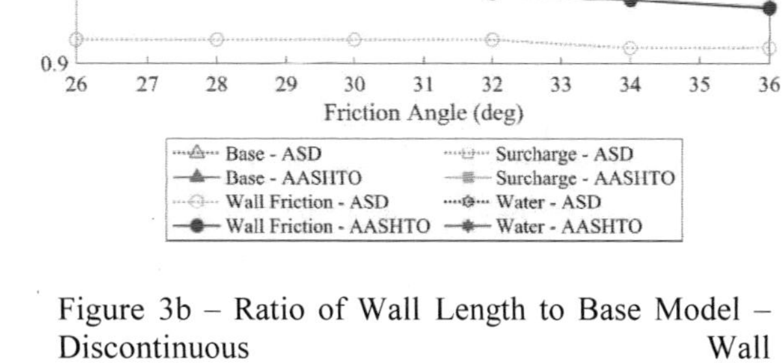

Figure 3b – Ratio of Wall Length to Base Model – Discontinuous Wall

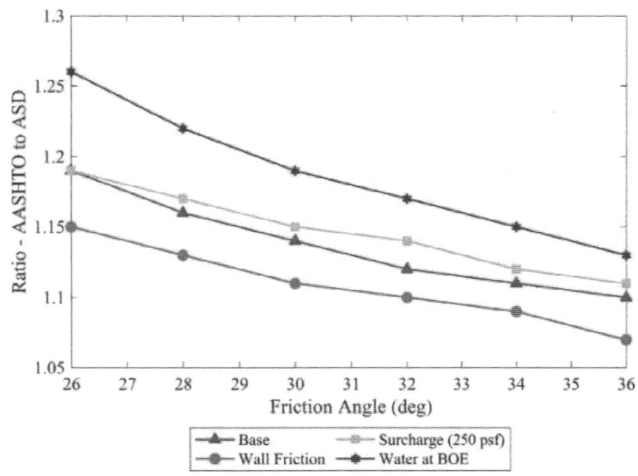

Figure 4a – Ratio of Wall Length AASHTO to ASD – Continuous Wall

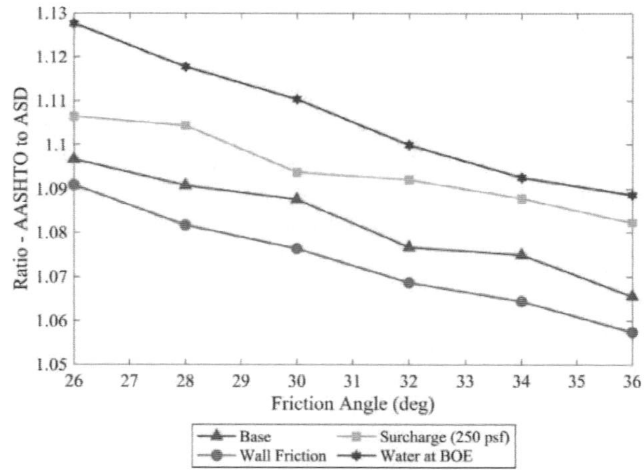

Figure 4b – Ratio of Wall Length AASHTO to ASD – Discontinuous Wall

As shown in Figures 4a and 4b, use of AASHTO LRFD load factors yields longer walls by 7-26% and 6-13% for continuous and discontinuous walls, respectively. It should be noted that the relationship between friction angle and wall length is not linear; at lower friction angles, the difference between the AASHTO LRFD and ASD approaches is amplified. The influence of wall friction and surcharge is nearly linear. However, the influence of groundwater is not linear and varies significantly with friction angle.

Regardless, it is clear that there is a significant difference between AASHTO LRFD-designed walls and ASD-designed walls. A 7% increase in toe embedment may not seem significant, but when carried over a wall that is hundreds of feet long, the direct increase in project costs is substantial. The results indicate that a single soil load and resistance factor is not appropriate for variable soil types and that the existing reliability-based AASHTO LRFD load factors require modification.

LATERAL EARTH PRESSURES

The above analyses are necessary to determine the effects of various individual parameters on cantilever wall design. In practice, there are inherently more in-situ variables and conditions that will amplify the above referenced discrepancies. Considering multiple variables in the same analysis compounds the differences between ASD and AASHTO LRFD. We have developed two models that reflect practical designs in which multiple variables are considered. These models consider a 15-ft high cantilever wall with 32-degree soil friction angle, groundwater at a depth of 15-ft, a 250-psf x 20-ft surcharge located 2-ft behind the wall, and wall friction on the passive side equal to 1/3 of the friction angle.

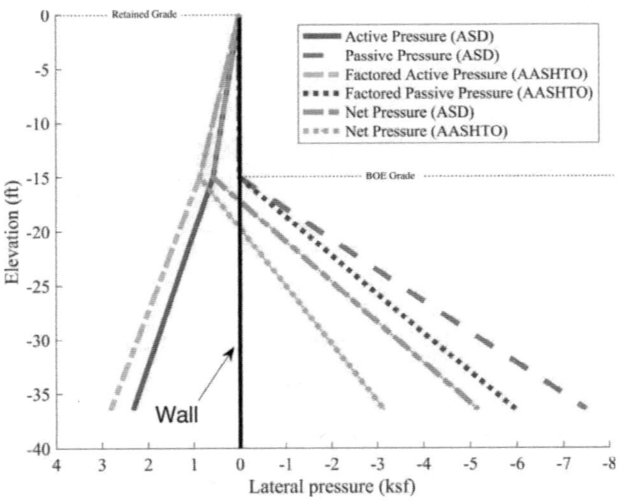

Figure 5a – Lateral Pressure vs Wall Elevation – Continuous Wall

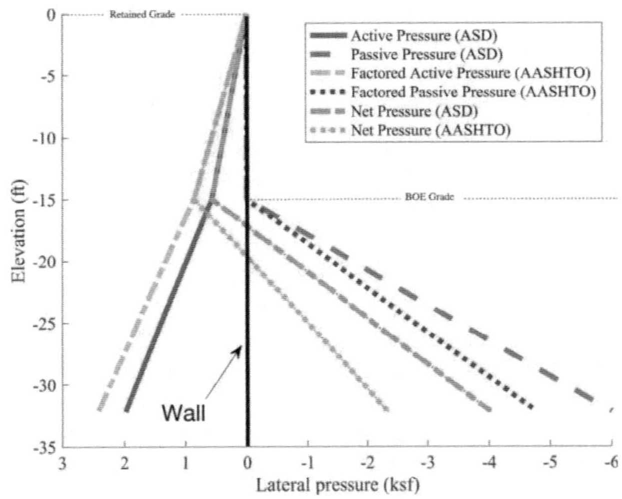

Figure 5b – Lateral Pressure vs Wall Elevation – Discontinuous Wall

The resultant pressures indicate that the ratio of the AASHTO Strength 1a to ASD design cases for required toe embedment is 1.14 and 1.12 for continuous and discontinuous walls, respectively. In both cases, the point of inflection of the net pressure is pushed lower when using AASHTO load factors, which directly drives larger toe embedment. This deepening of the net pressure inflection point can be attributed to an increase in driving pressures and corresponding decrease in the available passive resistance, as shown in Figures 5a and 5b. The ratio of the factored driving pressure to unfactored driving pressure resisting pressure is about 1.2. The ratio of the factored resisting pressure to unfactored resisting pressure is about 0.8. This means that the total driving pressures are increased by 20% and the total resisting pressures are decreased by about 20%, resulting in a 14% and 12% increase in required toe embedment for a continuous and discontinuous wall, respectively.

FURTHER STUDY

The above analysis does not consider differences in structural design of wall elements. In practice, cantilever walls are often controlled by the service condition to control deflections, in which case the structural designs will be the same. Additionally, this paper only considers cantilever walls; the use of LRFD load factors also effects braced walls, but to a lesser extent. Finally, the analysis does not consider cohesive soils in an undrained condition; further evaluation is necessary to study the effect of AASHTO LRFD load factors on these soils. We suspect several factors will affect the wall design in undrained conditions including the height of cut relative to the soil shear strength. These trends may be difficult to observe.

DESIGN PRACTICALITY

The simplification of using constant load and resistance factors is not appropriate for geostructural applications. As demonstrated, they drive a direct increase in both material and schedule project costs. In our experience, the AASHTO methodology for retaining walls has caused a great deal of unnecessary construction. Based on the above analysis, the load and resistance factors should be more flexible or customizable to account for site-specific variables, such as soil type, wall type, groundwater conditions, and surcharge type, such that consistent reliability-based results are attained. We recommend that AASHTO load and resistance factors be recalibrated based on these additional variables. Further, we recommend that use of the AASHTO LRFD approach be suspended in favor of ASD for retaining wall design until further research is performed, calibrations are made, and discrepancies are rectified.

REFERENCES

AASHTO. 2017. *LRFD Bridge Design Specifications*. American Association of State Highway and Transportation Officials, Washington, DC.

AASHTO. 2020. *LRFD Bridge Design Specifications*. American Association of State Highway and Transportation Officials, Washington, DC.

AISC, 1994. Load and Resistance Factor Design, *Manual of Steel Construction*, American Institute of Steel Construction, Chicago, IL.

ASCE, 1993. Minimum Design Loads for Buildings and Other Structures. *ASCE 7-93 (formerly ANSI A58.1).*

Hansen, B. 1953. *Earth Pressure Calculation*. Danish Technical Press, Copenhagen, Denmark.

Hansen, B. 1956. *Limit Design and Safety Factors in Soil Mechanics*. Bulletin No. 1, Danish Geotechnical Institute, Copenhagen, Denmark.

Regan, J.E., Hamblin, S.H., and Palantzas, D. 2013. *Comparison of LRFD Resistance Factors to Tension Pile Load Tests.* The Deep Foundations Institute Conference Proceedings.

DeepEx [Computer software]. (2021). Retrieved from https://www.deepex.com

LOW OVERHEAD SECANT PILES DRILLED FROM WITHIN A SHAFT

Stanley L. Worst, Schnabel Geostructural Design & Construction, Cary, IL, USA, 847-639-8900, stanley.worst@schnabel.com
Michael Grimm, Schnabel Geostructural Design & Construction, Cary, IL, USA, 847-639-8900, michael.grimm@schnabel.com
Parthiv Brahmbhatt, Schnabel Geostructural Design & Construction, Cary, IL, USA, 847-639-8900, parthiv.brahmbhatt@schnabel.com
Allen B. Stanton, Schnabel Geostructural Design & Construction, Cary, IL, USA, 847-639-8900, allen.stanton@schnabel.com

ABSTRACT

Constructing a circular shaft consisting of secant piles has its own complications which make it more difficult than typical drilled shaft construction. Add to that performing under low overhead powerlines and sitting inside of a 40-foot deep, 50-foot diameter shaft and it becomes more about logistics than the actual drilling. The complexities associated with the Romeo Arm Interceptor Segment 5 Lining project are the focus of this paper, delving into the details that made it work successfully.

Keywords: secant piles, low-overhead, constructability

PROJECT BACKGROUND

Major cities in the Great Lakes area of the Midwest have a wide-spread deep sewer system. Large diameter concrete sewers over thirty feet deep are common throughout the metropolitan areas. Suburbs of Detroit, Michigan have had multiple rehabilitation projects in recent years to upgrade and repair sewer systems that were built decades ago.

Unfortunately, Sterling Heights, Michigan, a suburb of Detroit, had a massive sinkhole occur in 2016 from a leak in one of those deep sewers. Fines migrated into cracks in the sewer and eventually led to ground collapse above it, reaching the ground surface and causing property damage. This event not only required an emergency repair of a collapsed sewer but future preventative maintenance. Five years after the disaster, one of those preventative measures taken to mitigate another sinkhole from forming was a re-lining project. This would provide the necessary seal to keep it active for years to come.

Fig. 1. Original Sinkhole from Collapsed Sewer

A UNIQUE APPROACH

Different from typical sewer repairs, access proved to be the most difficult part of the re-lining project. The sewer was originally constructed in 1973 of unreinforced concrete, approximately 18 inches in thickness and 11 feet in diameter. Located 50 feet below the ground surface, normal means of construction were not possible. Minimizing disruption to local traffic along 15-Mile Road, access was provided in a utility corridor off the side of the road. This corridor contained multiple high voltage power lines.

Ground conditions consisting of soft clays and sands below the water table were such that a water-cutoff wall was required. Under normal circumstances, driven sheet pile, slurry walls or secant piles could be relatively easily constructed to facilitate the excavation to access the sewer at depth. However, given the overhead power lines and the constraints of normal construction equipment, this was not possible. Not only was the necessary material itself taller than the overhead clearance, but the heavy equipment used in these operations typically have tall masts which would not physically fit in these circumstances.

Common practice for shoring in the upper 40 feet of soils in the Detroit area is to use soldier beams and wood lagging as there is mainly cohesive material. Below that depth is where a water cutoff system is needed in more granular material below the water table. The design team on this project developed a composite tiered system to combine both shoring types. A circular shaft of steel ribs and wood lagging would be used for the first 42 feet, and below that would be a water-tight circular secant pile shaft. The secant piles for the lower cutoff wall could be installed from within the upper shaft of ribs and lagging. Doing so would allow for an additional 42 feet of equipment mast to fit below the power lines. The upper shaft was designed to be 45 feet in diameter to allow for mobilization of a large diameter rotary drill rig inside of it. The lower secant pile shaft would be drilled from within and have a diameter of 34.5 feet.

The logistics of lifting a 200,000-pound drill rig into this shaft, placing concrete and steel reinforcement from the ground surface, and removing drill spoils and ground water from within the shaft would prove to be more difficult than drilling the secant piles themselves.

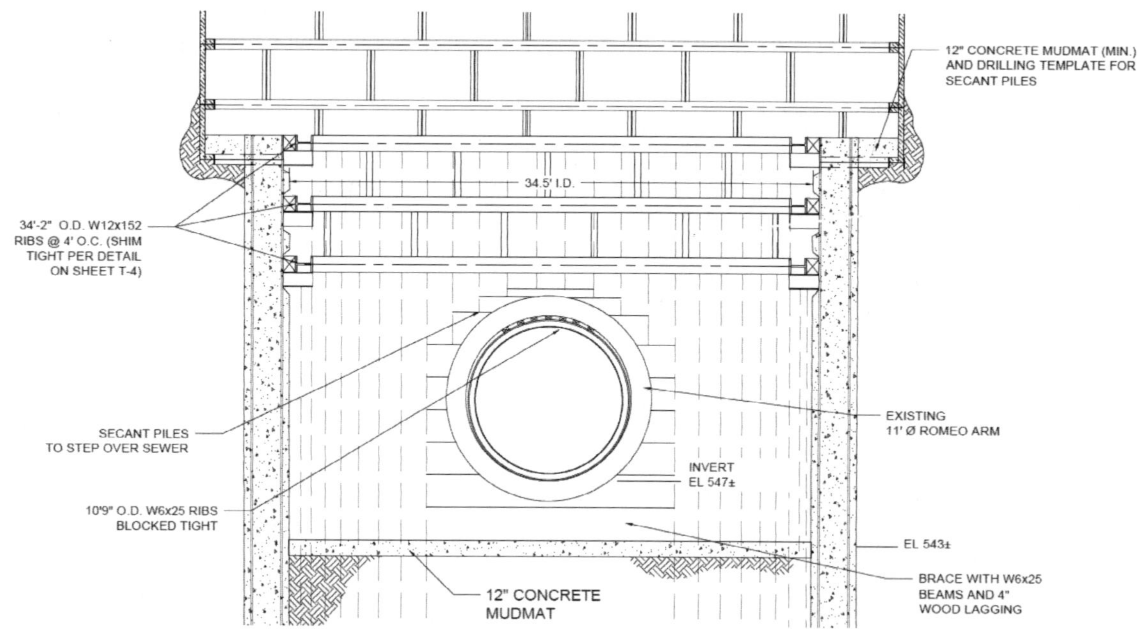

Fig. 2. Elevation View of Upper Shaft and Secant Piles around the Sewer in the Lower Shaft

SECANT PILE TECHNIQUE AND DESIGN

Secant piles are a construction technique used when a cutoff wall is needed, and ground conditions are difficult and need to be drilled as opposed to driven. Although ground conditions were not necessarily difficult at this site, using the geometry of a circular access shaft lent itself to secant piles as they could be used as the compression ring beam themselves without an additional ring-wale.

The concept of a circular shaft consisting of secant piles is that horizontal hoop stresses are resolved with equal and opposite forces around the shaft circumference. No internal braces or external tiebacks are needed in this scenario. This leaves the excavation unobstructed, eliminating any internal bracing and easing the excavation process itself. Not having external tiebacks also speeds up the excavation process, avoiding the need to stop for potentially weeks at a time to install lateral support elements at varying elevations.

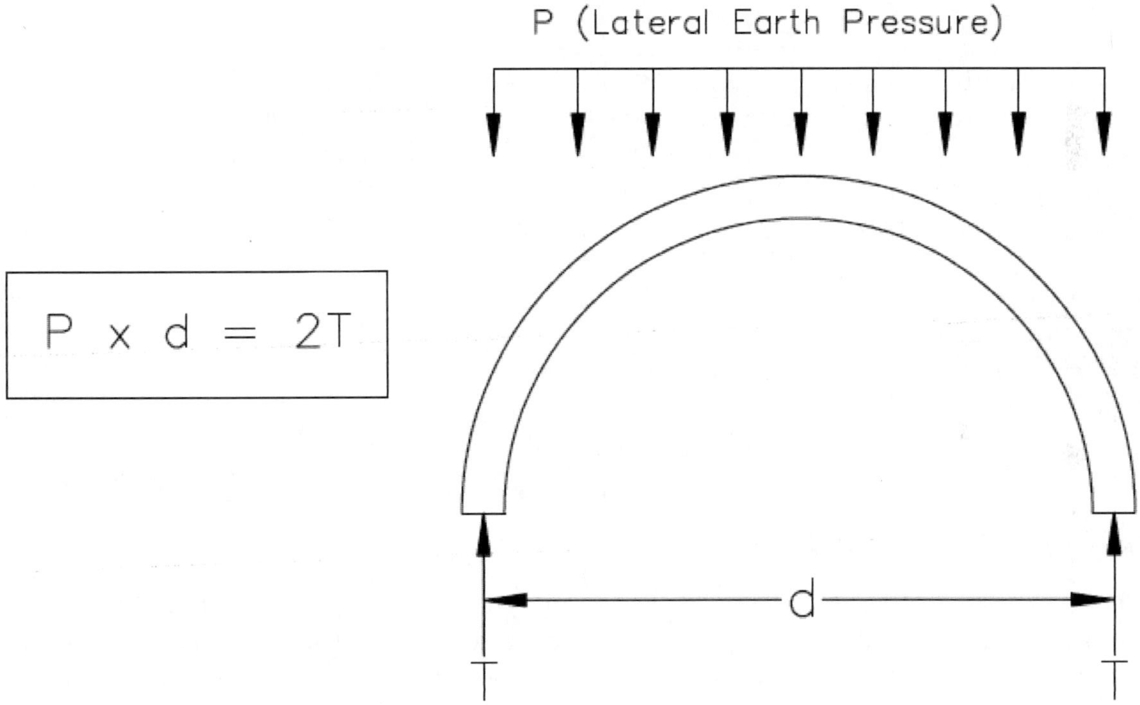

Fig. 3. Hoop Stresses in a Circular Shaft

The secant pile shaft was designed to have a 34.5 foot inside diameter and to be excavated to a depth of 28 feet. Sixty overlapping secant piles ranged in depth from 8 feet to 51 feet. The shallower piles were to be drilled directly on top of the sewer where it crossed through the excavation. The deepest piles were reinforced with steel beams, W21x182 in size. The unreinforced secondary piles were drilled to a depth of 38 feet. Concrete for the secant piles was specified to be 4,000 psi in strength.

For a circular secant pile wall, a guidewall is first constructed. The guidewall serves the purpose of initially aligning the piles so that adequate overlap is achieved per the design. It also aides in maintaining verticality of the piles during drilling, pinning the casing in place at the ground surface. The design of the secant piles is predicated on what is referred to as effective thickness. This thickness dictates the dimensions of a compression ring beam which resolves the lateral earth, water and surcharge pressures acting on the overall shaft. Schnabel used circular Styrofoam cutouts set on a mud mat as the template. The guidewall was poured with concrete surrounding the Styrofoam. A one-foot-thick concrete working pad was also poured inside the guidewall, which allowed for the drill rig to situate itself in the center of the shaft without tearing up the ground surface and potentially disturbing the guidewall.

The required effective thickness for this design was 20 inches, which means that is the minimum thickness needed at the secant pile overlap at the bottom of excavation. The designed number of piles was 60, which meant the starting overlap was 25.5 inches. To meet the requirement, a verticality of 0.7% was needed to limit the deviation to 2.4 inches of two adjacent piles drifting in opposite directions. To confirm that these tight requirements were met, Schnabel measured the verticality of each individual secant pile. This was performed using the Jean Lutz PRAD system, which measures the deviation of the shaft from top to bottom in both the X and Y axis. The driller can see this real-time when sending the sensor into the shaft. Plotting the as-builts of adjacent piles allowed for an assessment of the effective thickness at different depths.

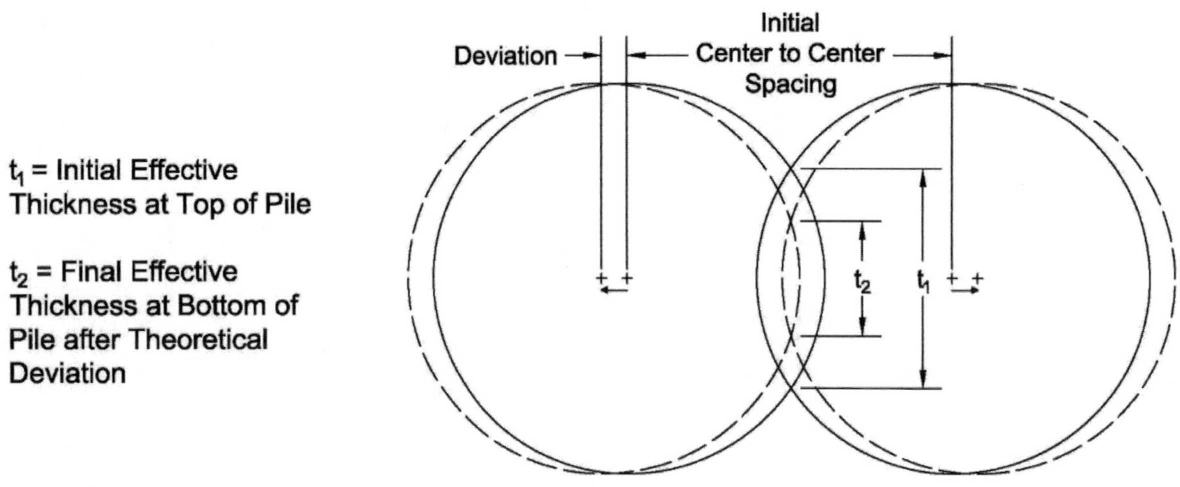

Fig. 4. Effective Thickness of Adjacent Secant Piles

The stiffness of the circular secant pile system is very high due to the low compressibility of concrete and the main component of load is working in compression, as opposed to tension or shear. This relative stiffness usually dictates the use of at-rest earth pressure since the lack of deflection does not allow for active pressures to develop. Using Rankine theory, earth pressure increases with depth. However, given the construction technique of secant piles, the effective thickness has the potential to decrease with depth. The result is a potential for the highest load to be where there is the smallest value of thickness. This can easily be designed for but is an ironic challenge of geotechnical design and construction.

GETTING INTO THE HOLE

A major piece of equipment to be used was a BG-26 Bauer hydraulic rotary drill rig with an approximate total weight of 200,000 pounds. This rig had a mast which was 70 feet tall. Lifting the rig into the upper shaft of ribs and lagging required a detailed lift plan with specific constraints. First, the crane could not sit closer than 15 feet from the edge of the shaft to restrict excessive surcharge pressures per the design. This would have resulted in horizontal pressures which were too unsymmetric around the perimeter of the ribs and lagging shaft. Second, a maximum crane boom height was enforced due to limitations set forth by the International Transmission Company (ITC), the owner of the power lines. The power lines were approximately 63 feet above the ground surface, although this fluctuated with both the ambient temperature and the power draw on the lines themselves. A proximity limit of 20 feet was imposed, which meant any work located beneath the power lines could not be more than 43 feet off the ground surface. Based on the lifting weight, horizontal distance, and height restrictions, it was determined that a 400-ton crane would be used for the operation of hoisting the drill rig. The critical lift would be the base of the Bauer BG-26 drill rig, which weighed 82,000 lbs. Maintaining a low boom angle with a 50-foot swing radius allowed for staying under 75% of the crane's capacity.

After the base was set, the next step was to connect the lower mast to the base of the drill rig. Under normal circumstances this would require one hundred feet of space where the mast would be lifted in a horizontal position on the ground and connected to the rig base prior to being tilted vertical. The inside diameter dimension of the upper shaft prevented this from happening. Complicating things even more, the lower mast was not designed to be lifted in the vertical position. The only possible assembly scenario remaining was a series of two lifts for the lower mast: the first lowering it into the upper shaft horizontally; the second, lifting it on a seventy-degree angle and connecting it to the base. This required adjustable chain slings to be used and took multiple shifts to assemble. Schnabel performed test lifts at their equipment yard to ensure that the rigging made just for this one lift was sufficient, minimizing potential downtime on site. After pinning the lower mast to the base, the rest of the components of the drill rig including counterweights, upper mast, rotary, etc... were relatively easy to assemble, yet still took more time than usual due to the site constraints.

Fig. 5. Lowering Base of Rig into Shaft

DRILLING AND PLACING MATERIAL

Three main soil layers were encountered during drilling. The upper 25 feet consisted of a silty clay, which was dry and stayed on the auger flights. The shallow secant piles above the sewer stayed in this layer and were drilled relatively easily. From 25 to 45 feet, sand below the water table was encountered. To effectively excavate this material, the drill hole was kept flooded during drilling. This, in addition to advancing the casing ahead of the drill tooling, kept the sand from heaving up into the casing and losing ground. A muck bucket was used to drill in this layer to remove the sand, as granular material does not stay on auger flights when submerged in water. Water alone had enough weight to prevent the sand from heaving at the bottom of the drill hole; therefore, a drilling slurry such as polymer was not needed. At 45 feet below grade, a stiff, impermeable clay layer was encountered. The method Schnabel chose to drill the secant piles consisted of flush-joint casing extending full depth. Due to extending the casing full depth and the watertightness of the casing joints, once the lower clay layer was encountered, flooding the hole was no longer required. The casing was able to seal off the water and the bottom of the hole could be drilled conventionally with augers.

The flush-joint, double-wall segmental casing was also used for maintaining verticality of the piles. The high durability and section modulus of the casing itself minimized deviation. One technique to drilling which maintains verticality and basal stability within the drill hole is to lead with the casing. The drill casing can itself be viewed as a relatively small circular excavation with side support. If the support walls only extend to the bottom of excavation, there is a risk of ground and/or water flow below the casing and up into the excavation. If, however, casing is extended below the bottom of excavation, it acts as a cutoff wall for soil and water movement. On a larger scale, cutoff walls extend below the excavation as necessary to either cutoff water flow or minimize the factor of safety against basal heave. Drilling does this on a smaller scale every time the auger removes soil from within the casing.

Fig. 6. Drilling Secant Piles inside the Upper Shaft

A self-dumping box to spin spoils into and then lift out of the upper shaft with a service crane was utilized. There was inherently some drill spoil which did not get into the spoil box by mechanical means. This was addressed with ground workers who had to shovel any extra spoil by hand. Due to safety concerns, only two people were allowed on the ground inside the shaft at a time. No other normal support equipment, such as a skid steer, loader, forklift or manlift were allowed inside the shaft as there physically was not enough room.

Fig. 7. Taking Verticality Reading in the Cab

After drilling an individual secant pile, an inclination measurement was taken using the PRAD system by Jean Lutz. This measures any deviation of the shaft from top to bottom in both the X and Y axis and can be seen in real time by the driller in the cab of the rig. Based on the specifications and a minimum effective wall thickness to be maintained, a tolerance of 0.7% deviation for the shaft length was imposed and achieved. Once the pile was confirmed to comply, the shaft was concreted. The concreting operation was setup to be pumped from the top of the shaft down to the bottom using a trailer pump. To guide the concrete, a hardline pipe was

Fig. 8. Setting a Soldier Beam below Power Lines

Fig. 9. Completed Access to Existing Sewer

secured to the inside of the upper shaft, which transitioned to a rubber line at the bottom. During concreting of the 38-foot-deep shafts, which would have to stay flooded, a sectional tremie pipe was used to fill the secant pile. On the 51-foot-deep shafts which were sealed into the lower clay layer, ground water was pumped prior to pouring the shaft.

With the limitations of the overhead power lines, setting the steel soldier beams were also very challenging. The crane was set up in the only location possible, which was between the two sets of power lines that ran overhead. With this configuration, there was a 6-foot-wide window in which to lift the beams vertical and lower them into the shaft. Once the beam was lowered into the upper shaft, the crane would boom horizontally while staying under the ceiling height imposed by the power lines and place the beam in the secant pile where it was needed.

CONCLUSION

Unique challenges faced this project, making it more complicated than usual. Excavating below the water table and the circular access shaft geometry lent itself to secant pile technology for the lower shaft. Overhead power lines allowed for only one way in and out of the upper shaft, one method of concrete and soldier beam placement, and decreased production as compared to drilling from the ground surface. The secant pile cutoff wall successfully sealed off ground water from within the excavation and provided a very stiff system which limited ground movement of the surrounding soil mass.

REFERENCE LIST

FK Engineering. 3-13-2020. Shop Drawings for CS-12 Temporary Earth Retention System, Romeo Arm Interceptor Segment 5 Lining.

FK Engineering. 4-21-2020. Geotechnical Investigation Data Report, Proposed Control Structure No. 12, Sterling Heights, MI.

DEVELOPMENTS IN THERMAL INTEGRITY TESTING PRACTICE AND INTERPRETATION

A. Fisher, Cementation Skanska Ltd, Doncaster, South Yorkshire, UK, +44 (0) 7753 830 049, anthony.fisher@skanska.co.uk

ABSTRACT

The use of thermal integrity testing to verify the quality of deep foundations has increased significantly in the UK and other world markets over the last 5 years. The introduction of distributed fibre optic sensing to this field has broadened the range of applications in which it can be deployed and the types of foundations which can be tested. This paper presents how fibre optic sensing technology can be successfully deployed in the construction environment the range of foundation techniques to which it can be applied. Alongside developments in the practical aspects of thermal testing, progress has been made in establishing assessment criteria which can be used to evaluate the results. This paper presents recent examples of comparative studies on foundations which include both thermal integrity testing, cross hole sonic logging and intrusive coring. The use of distributed strain measurement during pile load testing presents an alternative means to verify the assessment criteria used for integrity testing. This new approach can achieve comparable results to foundation exhumation or intrusive investigation and at a fraction of the cost. This paper presents the findings of several such studies, highlighting the benefits and opportunities for future improvement of the method.

Keywords: thermal integrity testing, fibre optics

INTRODUCTION

Pouring freshly mixed concrete into deep holes in the ground is a technical challenge and a significant feat of engineering. Whilst the enormous pours associated with nuclear power stations described by Moore (2021) and the incredible heights to which concrete is pumped for the construction of tall buildings described by Aldred (2015) will grab the headlines, placing concrete below ground demands no less attention. The concrete is required to flow or be pumped through pipes, maintain sufficient viscosity to displace any fluid within the bore, scour debris from the base whilst also flowing through dense reinforcement. It must remain workable throughout the duration of concreting and, once poured, it must not bleed excessively or segregate as it cures. It must be self-compacting as no further attention can be given to it once it has been placed. Should the concrete not perform any one of these duties, significant defects may occur in the completed foundation, creating the potential for differential settlement under load and, in the worst of cases, significant structural damage.

Thankfully, this occurrence is relatively rare because of good material science, workmanship and planning coupled with quality controls and testing. However, in light of the potential hazard, even if the likelihood is remote a sensible consideration of the risks will often conclude that there is a need for some in-situ testing of the concrete to demonstrate that the completed foundations meet the required standards. The decision on which type of tests to use, the frequency of testing and, theoretically at least, the acceptance criteria for a test should be based on an assessment of the risks. More testing can be specified for foundations where either the risk during construction is greater (such as foundations constructed using support fluid or in unstable ground) or if there is less redundancy in the foundation system (single pile caps or increased building sensitivity to differential settlement). A fuller consideration of the factors influence decisions on testing strategy is provided in French and Turner (2012).

Unlike above ground concrete pours where the finished product can be inspected once the formwork is removed, sub-surface concrete pours and, in particular, those associated with deep foundations remain obscured once constructed and so indirect methods of testing are required.

A number of different methods have been developed with assessment criteria predominantly being established through case history. There are many publications dedicated to the selection and use of these methods such as French and Turner (2012) and Turner (1997). This paper focuses on thermal integrity testing, which is a method of testing the full length of a foundation element. In this respect, it is an alternative to cross hole sonic logging (CSL), which is also discussed. These methods are more frequently specified for higher risk scenarios where more convenient options such as low strain integrity testing are deemed insufficient. Other alternative methods such as Gamma-Gamma logging are not discussed in this paper on account of their infrequent use in the UK.

CSL has, according to French and Turner (2012) been carried out since the 1960s and consequently the number of case histories available is relatively high. Thermal integrity testing is a more recent development, being first carried out in the early 21st century as described by Mullins (2010). As a result of this difference in maturity, whilst thermal integrity testing is certainly not new, it is often less well understood than CSL and sometimes considered inferior.

THE DEVELOPMENT OF THERMAL INTEGRITY TESTING EQUIPMENT

Thermal integrity testing uses measurements of the heat produced by the hydration of cement as an indicator of the integrity of a concrete foundation. These measurements can be made using a variety of sensing systems. Initial development of the technique was carried out using temperature sensitive probes that were lowered into the foundations through reservations tubes cast into the foundation as described by Mullins (2010). These measurements were made at regular intervals during the concrete hydration period and from these, an assessment of pile integrity was made. More recently, the development of Thermal Wire described by Becker et al. (2015) has allowed measurements to be made using thermistors embedded within the concrete. These are installed on the reinforcement of the foundation and make much more frequent measurements which are recorded and processed by an automatic datalogger at the pile head.

The most recent development in thermal integrity testing technology is the adoption of distributed fibre optic sensing (DFOS) and in particular, distributed temperature sensing (DTS), for this purpose. This method is discussed in (Fisher & Bell, 2017). In this approach fibre optic cable is used as the sensor which is attached to the foundation's reinforcement. Upon completion of the reinforcement installation the fibre optic cable is joined together into a single length which is then connected to a measuring instrument. This instrument makes regular measurements of temperature along the length of the fibre throughout the concrete hydration period. Measurement duration is dependent on the length of fibre to be measured and the measurement parameters, however, a typical foundation integrity test is carried out with less than 2km of fibre, for which a measurement may take in the order of 5 minutes.

There are a number of benefits associated with the used of DTS technology which make it particularly well suited to thermal integrity testing. Despite the perception that glass fibres are fragile, they are actually very durable and therefore well suited to the construction environment. For conventional DTS systems, standard multimode 50/125, loose tube cable can be used. Due to its applications in the telecommunications industry, it is readily available and consists of an outer protective sheath (4 – 6mm in diameter) which contains a nylon braid and provides considerable tensile strength. The fibres themselves are contained within a plastic tube that is filled with a silicone gel, isolating them from the effects of mechanical strain. The net effect is that the fibres are very well protected from the construction environment. Even if broken, measurements can be made either side of a broken section and data spliced together at the processing stage.

Cable strength and durability are especially important in the case of a pile with limited cover. In these situations, the small diameter and smooth sides of the fibres give them the best chance of avoiding damage during the installation process. Fibre robustness is important for constructing (and testing) low cut off piles including those containing plunge columns. In these scenarios, fibres can be lowered into the bore with the reinforcement, acting like a rope and so there is no need to extend reinforcement to the ground surface. Fibre optic cables contain no moving parts or electronics and so are not affected by mechanical shock. This is important when installing reinforcement using a cage vibrator. The low profile of fibre optic cables also minimises the impact of the testing infrastructure on the process of plunging reinforcement into the fresh concrete.

One potential drawback of the fibre optic technique is that a cable needs to be run from each pile tested to the location of the measuring instrument. However, recent advances in equipment manufacture have resulted in rugged instruments which are less affected by ambient temperature fluctuations and which can be powered by batteries or solar power. This allows the instrument to be moved around on site and positioned to best suit the foundations to be tested.

THE DEVELOPMENT OF THERMAL INTEGRITY TESTING ASSESMENT CRITERIA

As has been described, the requirement to test the integrity of a foundation arises from a very real and potentially significant hazard. Detailed work is therefore required to ensure that the methods of analysis and interpretation of testing data are appropriately robust and can be substantiated. This is especially important for indirect tests where considerable weight is given to the interpretation of the data. In order to build a framework with which to develop these assessment criteria, it is helpful to consider some general points which relate to both CSL and thermal integrity testing.

Firstly, these methods rely on the hypothesis that the majority of foundations are satisfactory in their integrity and so the purpose of the test is to detect unexpected variations or anomalies from the "normal" result. Assessment criteria are therefore based on deviation of the measured value from the established normal for the element undertest rather than an absolute value of a measured parameter.

In order to develop a general criteria it is important to determine the threshold at which variations from the established normal become significant. The starting point for this assessment is to consider the magnitude of variation which is to be expected as a result of the measurement accuracy itself along with any external influences on the measured value. This approach is summarized by Turner (1997) for the evaluation criteria for CSL. He concludes that first arrival time (FAT) may be increased by 15 – 20% as a result of variation in the measurement. The implicit conclusion is that variations of greater magnitude of this should be subject to further investigation. This is consistent with current practice as described by Linkins et al. (2007), where FAT increases of less than 10% are not typically subjected to further analysis.

For thermal integrity testing, ASTM D7949-14 (2016) requires that instruments used for measuring temperature be accurate to within 1 degC and have a precision of 0.1 degC. A commonly used product, Thermal Wire produced by Pile Dynamics (2019) states an accuracy of +/- 0.5degC and whilst this author has not verified the statement, over the relevant temperature range (0 to +60 degC) this accuracy should be readily achievable with modern electronic components.

For fibre optic monitoring systems, the accuracy achieved will be a combination of the quality of instrument used, the measurement parameters selected, the precision of calibration and the quality of any fibre optic connections. Manufacture specifications for instruments confirm that accuracies of +/-1 degC are achievable (SilixaLtd, 2021). For the data presented within the paper, site calibration exercises were carried out with a section of each fibre optic cable run through a water bath in which the temperature was gradually varied from 15 to 40 degC. A calibrated temperature probe was also used to measure the temperature of the water and the data collected used to calibrate the frequency response of the fibre to the changes in temperature. This was carried out using a Brillouin optical time domain analysis (BOTDA) analyser and recored an accuracy of +/-0.45 degC. Although more sophisticated laboratory calibration techniques can be used as are described in (Kechavarzi, et al., 2017), it is recommended that a simple calibration method is also used on site to ensure that measurement accuracy does not degrade with time.

Another potential source of error in thermal integrity testing is variation in the position of the sensor in relation to the foundation geometery. Sensors are typically placed in the cover zone where temperature gradients are steepest. Mullins et al. (2021) calculate that these variations can be in the order of 0.9 – 1.1 deg /cm. Any slack in the sensor cable or non-symmetrical cage shape can easily cause variations of a few cm and consequently a 1 – 2 degC variation could be expected. Consideration of both measurement accuracy and sensor position would suggest that temperature variations of +/- 2 degrees Celsius can readily be expected and considered normal in thermal integrity test results. However, to base acceptance criteria on the potential accuracy of the system used is only valid if these errors are below the threshold at which significant defects can be identified. The following example presents the results of a pile tested by both CSL and thermal integrity testing in which a serious defect was present.

The test pile presented in Fig. 2, PTP01, was 1180mm in diameter and constructed using rotary drilling techniques with segmental casing. The drill casing was advanced through 26m of made ground and clay overlying interbedded sandstone and mudstone with a 1050mm digging tool forming a socket in the rock from which the majority of the pile's capacity was derived. The bore was filled with groundwater and so a full length of 4 inch, steel tremie pipe was used to place the concrete. Whilst the second load of concrete was being placed, there was a blockage in the tremie pipe and so it was removed and cleared. Once it had been lowered again into the bore it was not able to penetrate the first load of concrete and was instead discharged directly on top of it. Without sufficient embedment, a layer of debris and weak concrete was encapsulated in the pile. Following assessment of the integrity of PTP1, an identical pile, PTP1a was formed immediately adjacent to it but with no such defect.

The results of the thermal integrity testing clearly show that at +78mAD, there is a difference between PTP01 and PTP01a in excess of 15 degrees Celsius. This represents a 50% reduction in temperature at this location. The drop off in temperature above 98mAD and below 72mAD is related to increased heat dissipation through the base and cap of the pile. The reduction in temperature below 78mAD seen in both PTP01 and PTP01a is a feature of the increased heat dissipation in the rock and reduced cross sectional area outside of the drill casing. The results of the CSL of PTP01 show a significant increase in FAT from 0.15ms to 0.35ms, and a reduction in relative energy of up to 54db at 21m below the pile cap. This corresponds with the anomaly identified in the thermal integrity testing and the concreting records which indicated this to be the location of the defect. In this case, the anomaly was clearly identifiable and the variations observed were significantly in excess of the levels to be expected from measurement inaccuracies.

METHODS OF FURTHER INVESTIGATION

Having established that an anomaly is not caused by inaccuracies in the measurement system or by external influences on the measured value, the second, and perhaps most important part of any acceptance criteria is to determine whether the anomaly could have a significance impact on the foundations performance. In the case of PTP01, the evidence of the construction records, combined with the integrity testing results was enough to confirm that the pile was defective and needed to be replaced. As should always be the case, the decision was not based on the test results alone, but also from consideration of the construction records and the design requirements for the foundation. For PTP01, the defective pile would not be sufficient to prove the geotechnical performance of the ground below the defect and so could not function as a test pile. However, in many cases and even after a thorough inspection of the construction records, the evidence may not be as clear cut, making a decision about the foundations acceptability more complicated. If more information on the nature of the anomaly is required, additional testing can be deployed.

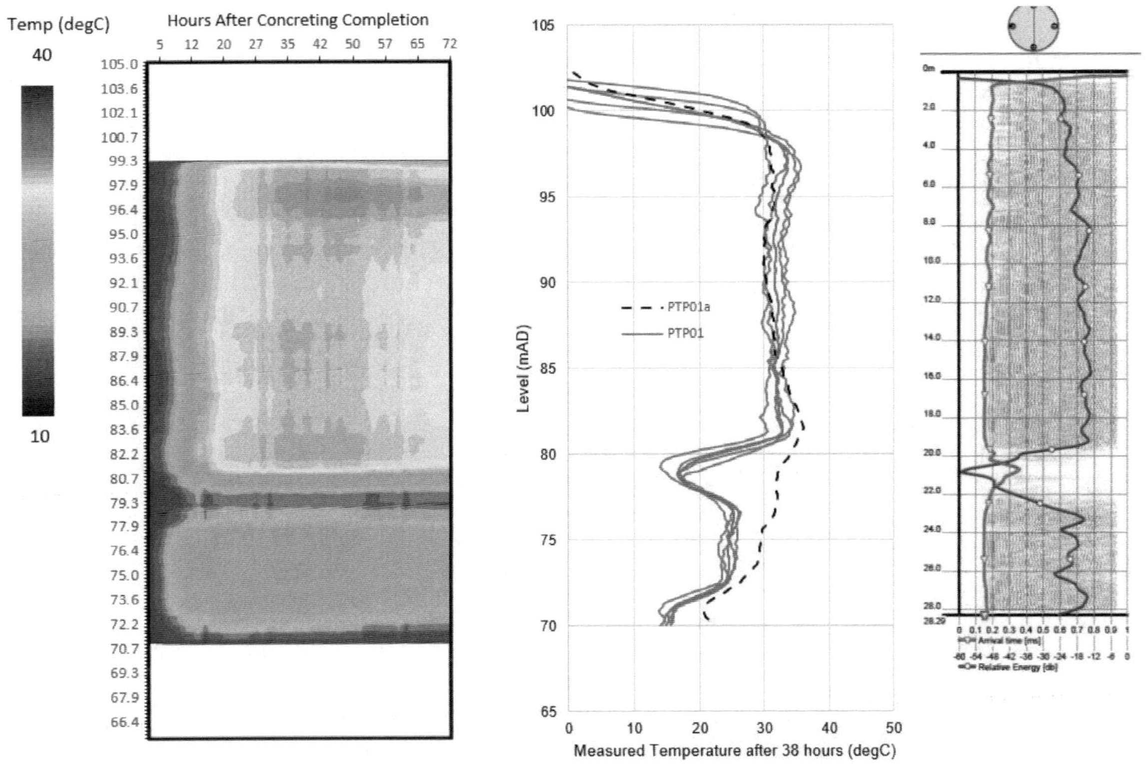

Fig. 1. Left) Temperature vs time and depth for PTP01. (Centre) Measured temperatures in PTP01 and PTP01a 38 hours after concreting. (Right) CSL results from PTP01

If it is possible, repeating the original test will further verify if the anomaly was due to variability in the testing method or if it is a permanent feature of the foundation. This conclusion will be further supported if there can be some small changes in the method employed, such as different equipment or a different testing technician. In some cases, repeating the testing at a later date allows the concrete to gain more strength and may produce a less anomalous result.

For cross hole sonic logging, different methodologies for retesting can be adopted. for example 3D tomography can be deployed to visualize the extent of an anomaly. The details of this method are described in (Linkins, et al., 2007). Figure 2 presents the results of 3D tomography carried out on the anomaly identified in PTP01.

For thermal integrity testing, repeating the test is not possible because once the concrete hydration has finished, no more heat is produced by the cement. However, carrying out additional testing using an alternative test method can be useful for verifying results. For example, if anomalies are relatively near to the ground surface, a low strain integrity test can prove a useful means to corroborate any anomalous results observed. Equally, a dynamic load testing can provide information about the integrity of the pile to support the conclusions of other integrity testing methods.

A more intrusive means of collecting more information is to drill and recover a core through the foundation concrete. The recovered core can be subjected to further analysis and testing whilst the core hole can be used for camera inspection and, if necessary, remedial solutions. However, although the core holes can indicate the condition of the concrete and be a means of remediation, they can rarely reveal the full extent and impact of a defect as only a small section of the pile is sampled and it is difficult to be certain of the position of a core within a pile.

Thermal integrity testing offers a new approach for the detailed analysis of results. With knowledge of the volume of concrete placed and the pile length, it is possible to derive an average diameter for the constructed foundation. By approximating the temperature variation in the cover zone as linear and applying a derived temperature to radius conversion factor to the measured data a plot of inferred pile radius can be produced. This method is described in detail by Piscsalko et al. (2016) and the resultant plot is described as an equivalent pile radius. At the location of an anomaly, a reduction in radius can be calculated and this information used to determine the loss in structural capacity and reinforcement cover. This calculation has been caried out for PTP01 and the results are presented in Fig. 3. In this case, the correction for expected temperature reduction at the cap and base of the pile has not been carried out due to insufficient information being available.

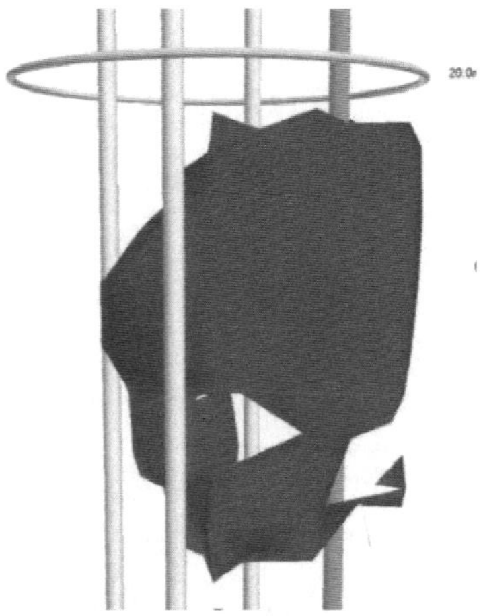

Fig. 2. 3D tomography of PTP01

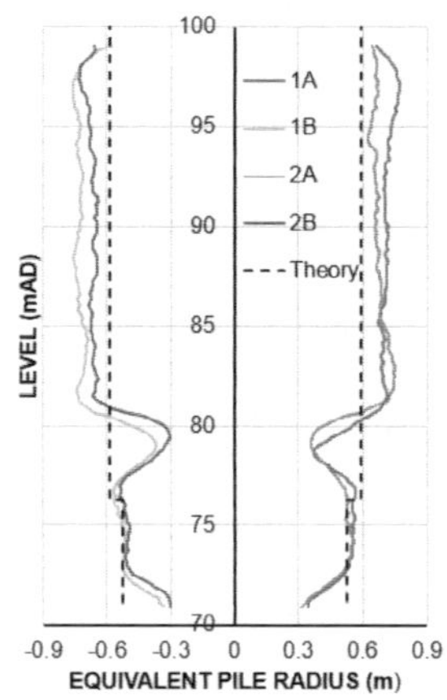

Fig. 3. Equivalent pile radius for PTP01

One immediate observation from this method of analysis is that it assumes that the poor quality concrete is present from the cover zone inwards. By comparison, the results presented in Fig. 2 which are from 3d tomography suggest that the anomalous area extends to within the core of the pile. One conclusion from this example is that whilst both methods of analysis are able to provide some quantitative information on the extent of an anomaly, neither can be considered as an exact representation and care should be taken not to extrapolate the results beyond reasonable limits.

Work is currently on going by authors such as Sun et al (2020) to develop additional ways of modelling the observed temperature changes in a pile using 2D and 3D finite element analysis to allow better decisions about the significance to be made. Although it has yet to be thoroughly tested, this approach has the potential to produce more refined models of anomalous areas as well as extend the approach to non-circular foundations such as diaphragm walls.

A NEW TECHNIQUE VALIDATION OF ASSESMENT CRITERIA

Despite the availability of case history data and the growing levels of expertise, interpreting integrity testing data still presents a significant challenge to the foundations industry. One new technique which can be used to improve the understating of integrity testing results and calibrate the criteria for determining the significance of anomalies is the use of DFOS for strain monitoring. The benefits of a distributed measurement of strain are considerable in geotechnical engineering because, unlike conventional point sensors, foreknowledge of the locations of interest is not required. This advantages allows DFOS data to capture the influence of foundation defects in pile load tests. More details on DFOS strain measurement can be found in publications by Kechavarzi (2017). As a load is applied and the stress in the foundation increases, any significant defects will cause a change in the structural properties of the foundation at that level, typically a reduction in stiffness. These changes can be measured and quantified.

Figure 4 presents strain measurements from a compression load test carried out on a 500mm diameter CFA pile constructed in clay. These results demonstrate the potential for DFOS to capture the influence of defect at an unexpected location. As the load was increased to 4,200kN, the pile failed structurally at +93mAD.. This can be seen from the increased strain measured at this level. At the point of failure, the measured values of strain show a localised increase of almost 100%, pointing to the significance of this feature. As the load was increased, the foundation failed, to the extent where it severed the fibre optic cables in the pile at this location.

Where load tests are carried out on foundations which have had thermal integrity testing or CSL carried out, any anomalies identified can be correlated with the results of DFOS measurements. An example of this is presented in Figs 5 and 6.

The pile in question was a 900mm diameter rotary pile, 47m deep and constructed using a polymer support fluid and 18m of segmental casing. After installation, both thermal integrity testing and CSL were carried out. 22 days after the pile was concreted, an extended proof load test was carried out in accordance with Institution of Civile Engineers specification (2017). During the load test DFOS of strain was employed and these results are presented in Fig. 6.

At -34mAD, a flaw was identified by the CSL test on the basis that there was a reduction in relative energy of more than 9dB. This feature was notable on 2 out of the 3 CSL traces carried out but did not appear to have an effect on FAT. A localised reduction in temperature change at this level was identified in fibre 1A of the thermal integrity test which was approximately 3 degrees Celsius. In both tests, although the variation was identified as an anomaly, it was not deemed to be significant due to only being present in a limited number of the results and relatively small in magnitude.

Within the thermal data other features can also be identified. At -5mAD, there is a notable increase in temperature in the thermal integrity test measurements. This was attributed to localised overbreak in the soft overburden.

During the load test, variations in strain can be observed along the length of the pile. Small variations in strain over the length of the pile are typical of the results from DFOS monitoring. In Fig. 6, a significant reduction in strain can be seen, centered at -4mAd. This is coincident with the increase in temperature measured by the thermal integrity test and is indicative of an increase in section size. At -34mAD, a localised increase in strain can be seen and this is at the same location as the anomaly identified in the integrity testing. Further examination of this result shows that the increase was present in only 1 of the 4 fibres monitoring strain within the pile. This is consistent with the results of the thermal integrity test in which fibre 1A appears to show more deviation than the other fibres.

The magnitude of the increase in strain is relatively small but, proportionally, it is significant and does suggest that there has been some change in the section properties at this location. Based on the location of this anomaly, the actual load experienced by the pile at this level was small and so the test was continued without any structural failure at this location. After the load test was completed, a full length core of the pile concrete was carried out. At the location of this feature there was a 300mm zone of discoloration of the concrete and the drillers notes described the material as "slightly crumbly and sandy". However, other, similar variations in material were found at other levels in the pile with no identifiable integrity testing or strain monitoring anomalies

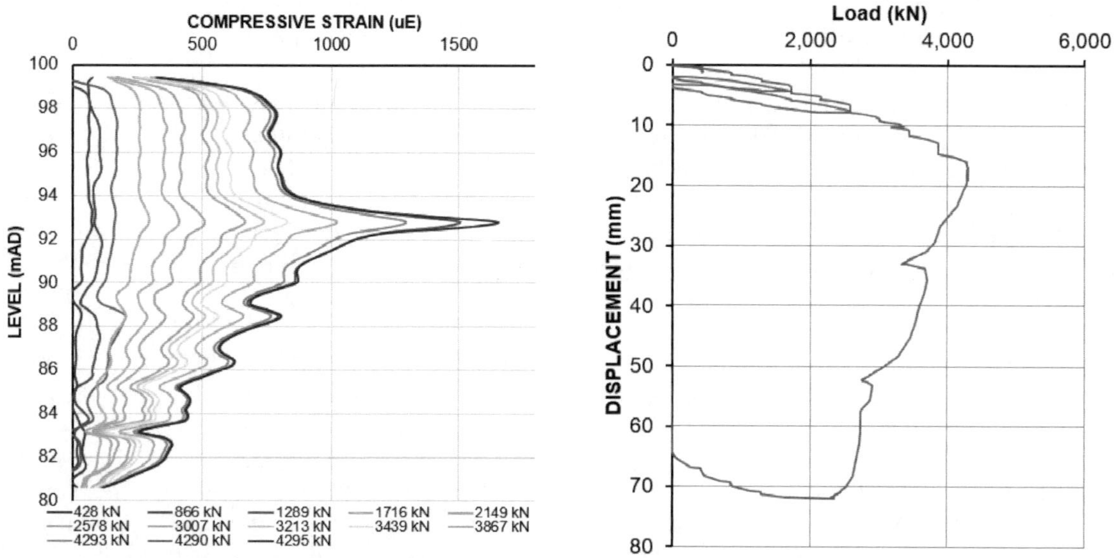

Fig. 4. Left) DFOS results showing strain vs level for a 500mm diameter CFA test pile. Right) Load vs Displacement for the same pile

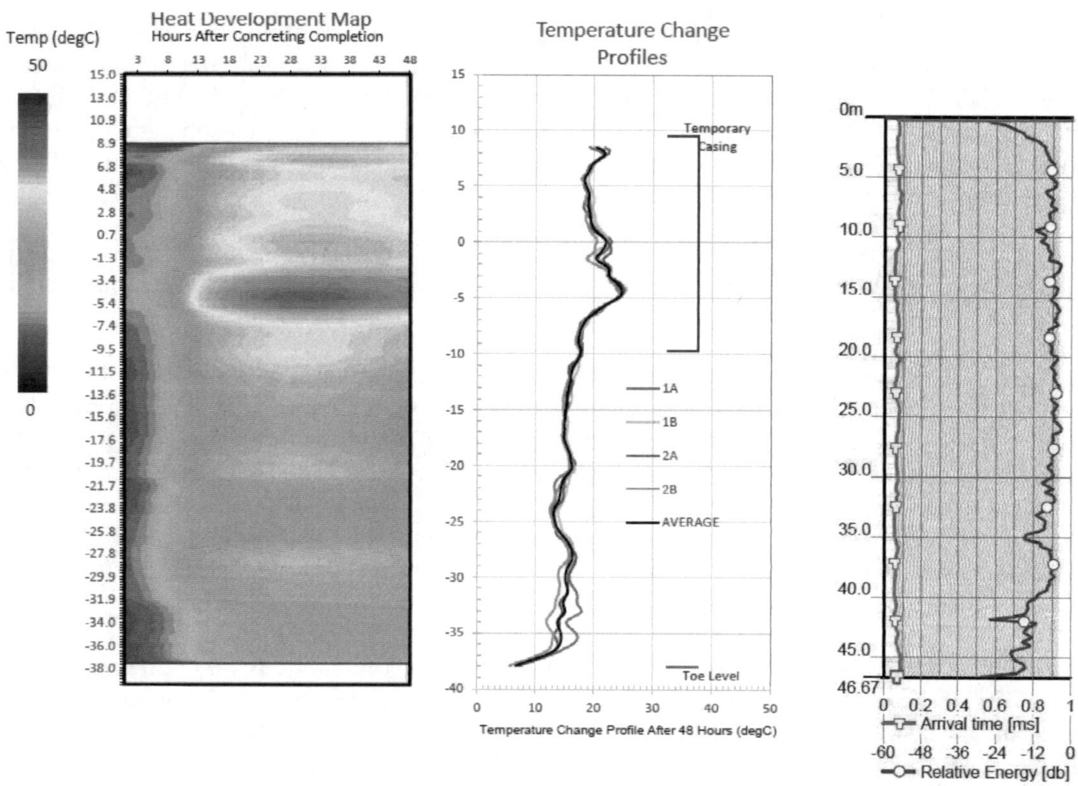

Fig. 5 Integrity testing records for a preliminary test pile. Left) Measured temperature vs time and level, Centre) Measured temperature change at peak temperature. Right) CSL result

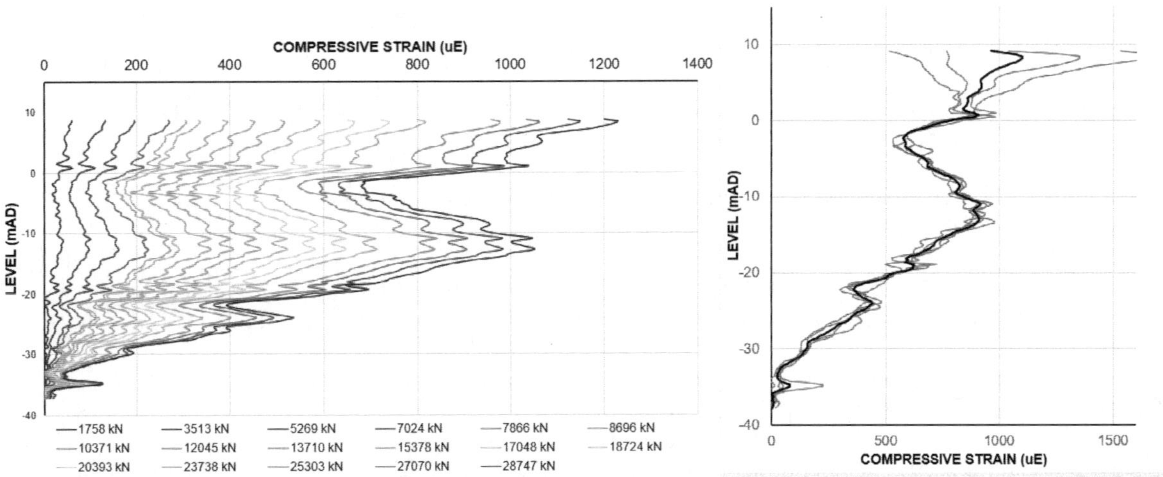

Fig. 6 DFOS measurements of strain of a preliminary test pile. (Left) Average strain across the section and each increment of load. (Right) Average strain (black) and individual fibre strain (grey)

The conclusion of the test piling regime was that the pile did not contain any defects which would have a significant impact upon its performance. It was also noted that correlation between coring, CSL and thermal integrity testing was not always possible, however, in this case, both CSL and thermal integrity testing provided a satisfactory means of verifying the integrity of the pile.

This example does not contain sufficient information to allow a detailed assessment of the validity of acceptance criteria for integrity testing methods. However, as part of a growing data set it does demonstrate a future direction of research which, combined with improvements in numerical modelling has potential to significantly improve the methods used to determine acceptance criteria.

CONCLUSION

The process of thermal integrity testing can be carried out using a number of different techniques and this includes the use of DFOS which has significant practical advantages in some situations. There is a growing body of work describing and providing examples of the appropriate assessment criteria and methods and so, despite the relatively short history of its use, thermal integrity testing is well supported by technical and practical resources.

Consideration of the inherent variability of the thermal testing methods identifies that a similar inherent variability in the measurement exists for thermal testing as it does for CSL. However, in the examples presented, this value is significantly below the threshold at which significant defects are identified.

Further work can be carried out to verify the assessment criteria used for integrity testing through the use of DFOS for strain measurement in test piles. There is now a growing data set of such tests which will provides a valuable tool with which to verify the appropriateness of the assessment criteria used. Examples included in this paper suggest that there is a reasonable correlation, even between relatively minor anomalies but that further work is required to incorporate these conclusions into rigorously defined criteria.

REFERENCES

Aldred, J., 2015. Burj Khalifa - a new high for high-performance concrete. *Civil Engineering,* 163(2), pp. 66-73.

ASTM, 2016. *Standard Test Methods for Thermmal INtegrity Profiling of Concrete Deep Foundations,* Online: ASTM International.

Becker, M., Coleman, T. & Belardo, D., 2015. *Thermal Integrity Profiling of ACIP Piles.* San Antonio, Proceedings of the international foundations congress and equipment Expo.

Fisher, A. & Bell, A., 2017. *The Application of Fibre Optic Sensor Technology in the Integrity Testing of Deep Foundations.* Rome, Deep Foundations Institute.

French, S. & Turner, M., 2012. Pile integrity testing. In: J. Burland, T. Chapman, H. Skinner & M. Brown, eds. *ICE Manual of Geotechnical Engineering: Volume 2.* London: ICE Publishing, pp. 1419 - 1450.

Institution of Civil Engineers, 2017. *ICE Specification for Piling and Embedded Retaining Walls.* 3rd Edition ed. London: ICE Publishing.

Kechavarzi, C. et al., 2017. *Distributed Fibre Optic Strain Sensing for Monitoring Civil Infrastructure: A Practical Guide.* London: ICE Publishing.

Linkins, F., Rausche, F. & Webster, K., 2007. *Defect Analysis for CSL Testing.* Denver, Geotechnical Special Publication No 158.

Moore, C., 2021. Hinkley Point C concrete pour dubbed 'feat of engineering'. *New Civil Engineer*, 13 April.

Mullins, G., 2010. Thermal Integrity Profiling of Drilled Shafts. *The Journal of the Deep Foundations Institute,* 4(2), pp. 54 - 64.

Mullins, G. et al., 2021. Selection of Thermal Integrity Data Regression Parameters. *Geotechnical Testing Journal,* 44(3), pp. 741 - 755.

Pile Dynamics, I., 2019. *Thermal Integrity Profiler (TIP) Specifications.* [Online] Available at: https://www.pile.com/wp-content/uploads/2021/11/TIP_Technical-Spec-Sheet.pdf [Accessed 26 April 2022].

Piscsalko, G., Likins, G. & Mullins, G., 2016. *Drilled Shaft Acceptance Criteria Based Upon Thermal Integrity.* New York, DFI 41st Annual Conference on Deep Foundations.

SilixaLtd, 2021. *Datasheet: XT-DTS.* [Online] Available at: https://silixa.com/wp-content/uploads/XT-DTS-Datasheet-2021-1.pdf [Accessed 26 April 2022].

Sun, Q. et al., 2020. Thermal integrity testing of cast in situ piles: An alternative interpretation approach. *Structural Health Monitoring.*

Turner, M., 1997. *Integrity testing in piling practice.* London: Construction industry research and information association.

HIGH CAPACITY DRILLED DISPLACEMENT PILES AS AN ALTERNATIVE TO CONVENTIONAL DRIVEN METHODS

Patrick Granitzki, Dynamic Earth LLC, Chester, New Jersey, USA, 908-879-7095, pgranitzki@dynamic-earth.com
Alicia Plinio, Dynamic Earth LLC, Chester, New Jersey, USA, 908-879-7095, aplinio@dynamic-earth.com
Janitha Batagoda, Dynamic Earth LLC, Chester, New Jersey, USA, 908-879-7095, jbatagoda@dynamic-earth.com
Pete Howell, Dynamic Earth LLC, Chester, New Jersey, USA, 908-879-7095, phowell@dynamic-earth.com

ABSTRACT

The New York City Housing Authority (NYCHA) Red Hook Houses was subjected to damage during Superstorm Sandy. As a result, the NYCHA has an ongoing project known as Restoration Associated with Sandy Recovery Program to improve the resistance against future storms. The specific project included 28 residential buildings and several service buildings within the Red Hook Development. The subsurface conditions within the eastern portion of the site included existing fill material underlain by varying stratums of very loose, loose and medium dense silty sands followed by very soft to medium stiff clay or dense silty sands. The western portion of the site included existing fill material, very loose to medium dense silty sands and organic clayey silts underlain by very soft to medium stiff sandy silts. Based on the geotechnical report, a conventional driven steel pile system was specified containing two pile depths. As part of the project bidding, an alternative pile type was recommended and analyzed to provide quicker installation, reduce vibrations, reduce noise, and provide relatively large cost savings. Due to relatively high lateral loading, an innovative double steel casing drilled displacement piles was proposed as the alternative pile option. Field testing proved the piles capable of the required project capacities and the use of drilled displacement piles as the best alternative for the project and accepted by the NYCHA.

Keywords: driven piles, drilled displacement piles, high capacity

INTRODUCTION

The Red Hook Houses development was designed by architect Alfred Easton Poor in 1938, which came to realization in 1939. Upon completion, the Red Hook Houses represented the latest trends in modern living at the time. However, over time the area of the Red Hook Houses was reclassified as a flood zone. Hurricane Sandy brought heavy flooding of up to six feet causing major damage to Red Hook Houses residential buildings, including support structures and utilities. Major components of electrical and mechanical systems in boiler rooms were under water causing massive damage requiring the New York City Housing Authority (NYCHA) to use temporary backup systems to keep the housing in livable condition.

The NYCHA was offered Federal Emergency Management Agency (FEMA) funding to improve and rehabilitate damaged infrastructure due to Hurricane Sandy. NYCHA launched the Restoration Associated with Sandy Recovery Program as a response. The rehabilitation program focused on improving the long-term sustainability of the Red Hook Houses and increase the resistance and resilience of the buildings against future storms. The proposed restoration included construction of two new high risers namely East Building and West Building, support structures, new electrical and mechanical rooms above design flood elevations, new boiler systems with standby generators, flood proofing, and permanent flood barriers.

As part of the construction bidding, an alternative deep foundation system was proposed which consisted of drilled displacement piles. The innovative system was designed to carry high axial loads and lateral loads. The drilled displacement pile was designed using both micropile design and helical (screw pile) design methodologies. This paper details the design and testing of the piles.

SUBSURFACE CONDITIONS

The rehabilitation of the Red Hook Housing area was divided to two sections: east building and west building. The subsurface condition investigations reported existing fill material to depths ranging between approximately 10 feet and 25 feet below the ground surface, corresponding to elevations of four feet and negative three feet. Beneath the existing fill material, natural organic deposits were encountered at one boring location to a depth of approximately 25 feet below the ground surface, corresponding to an elevation of 11 feet below the referenced datum. Beneath the existing fill material and/or organic deposits, relatively loose natural sandy soils were encountered to depths ranging between approximately 25 feet and 40 feet below the ground surface, corresponding to elevations of 11 feet below the referenced datum and 26 feet below the referenced datum. Beneath the natural loose sandy soils, variable strata of relatively dense to medium dense sands and stiff clays were encountered to termination depths ranging between 92 and 122 feet, corresponding to an elevation ranging between -78 and -108 feet. Groundwater was encountered at depths ranging between approximately eight feet and 10 feet below the ground surface; corresponding to elevations ranging between four and six feet. Figure 1 represents the summarized subsurface profile in the east building area.

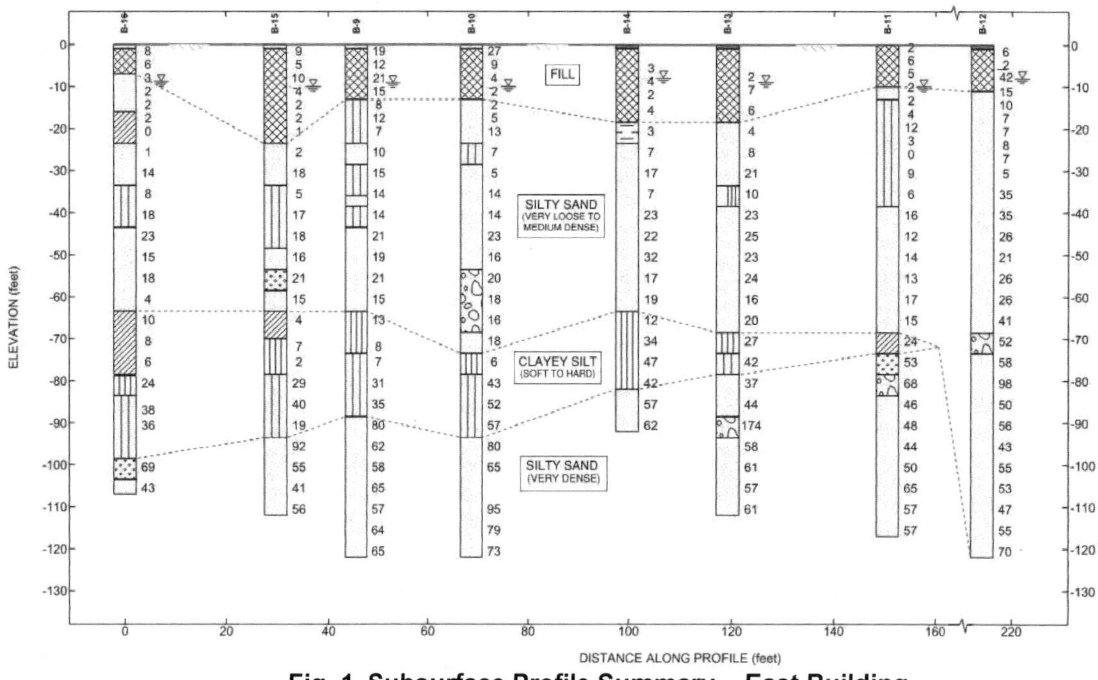

Fig. 1. Subsurface Profile Summary – East Building

The west building subsurface conditions encountered reported to generally include existing fill material to depths ranging between approximately seven feet and 15 feet below the ground surface, corresponding to elevations ranging between two and eight feet. Beneath the existing fill material, organic materials were encountered to depths ranging between approximately 15 feet and 27 feet below the ground surface, corresponding to elevations ranging between 5 feet and 16 feet. Beneath the existing fill material and/or organic deposits relatively loose sands were encountered to depths ranging approximately 22 feet and 45 f9eet below the ground surface, corresponding to elevations ranging between 12 and 38 feet. Beneath the natural loose sandy soils, variable strata of relatively dense to medium dense sands and stiff clays were encountered to termination depths. Groundwater was encountered at depths ranging between approximately seven feet and 10 feet below the ground surface, corresponding to elevations ranging between two and negative 1 feet. Figure 2 represents the summarized subsurface profile in the west building area.

9

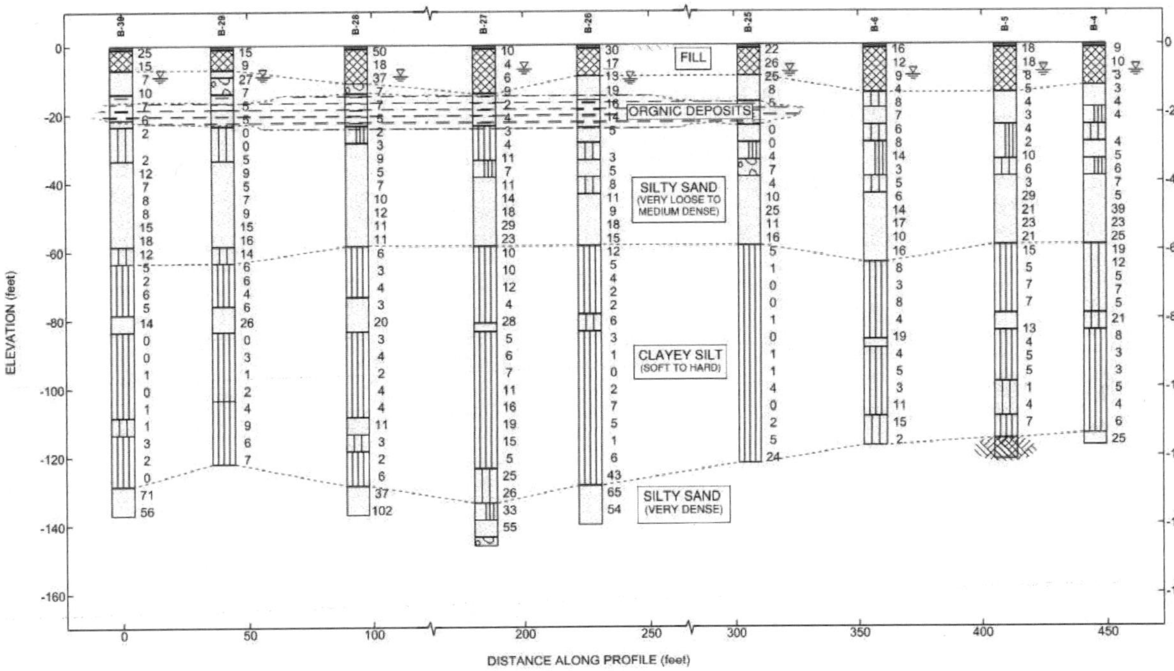

Fig. 2. Subsurface Profile Summary – West Building

DEEP FOUNDATION SYSTEM

The subsurface conditions indicated unfavorable conditions for the proposed development due to deep loose fill layers and loose and soft natural soil deposits. Hence, the initial development consideration included use of driven pile elements to support the proposed buildings and supporting structures. The east building required design capacities of 210 kips compression load, 60 kips uplift and 20 kips lateral loading. The west building required design compression capacities of 64 kips and lateral capacity of 1.5 kips. The initial proposed driven piles for the east building consisted of open-ended 18-inch diameter piles with a wall thickness of 0.5 inches driven to a minimum depth of 110 feet. Due to the relatively low loads on the west building, a 16-inch diameter, 0.5-inch-thick wall close-ended driven pile was suggested.

As part of the construction bidding, an alternative deep foundation element was proposed. The proposed alternative included a pressure-grouted drilled displacement pile with an innovative double steel casing to combat the high lateral loads. The installation of this pile generally includes pumping grout through a hydraulic driving head. The grout is pumped through the center of the pile shaft as it is advanced to the calculated bearing depth. Grout exists the pile shaft through one or more grout ports at the leading and/or trailing edge of the bottom helix/flight of the pile. As the pressurized grout exits the grout ports the grout displaces the soil away from the pipe shaft, creating an outer grouted annulus. The resulting pile is a composite steel/grout/soil-mix pile element.

s applied to the area of the individual helices proposed for the piles (Trofimenkov and Maruipolshii 1965, Adams and Klym 1972). The uplift capacity of the pile was estimated only using micropile design method. Due to the variable nature of the existing fill and compressible nature of organic deposits encountered (as reported in the subsurface investigation), skin frictional capacity values within the existing fill and organic deposits were neglected for axial compressive pile capacity estimates. Figure 3 presents the typical cross section of the drilled displacement pile.

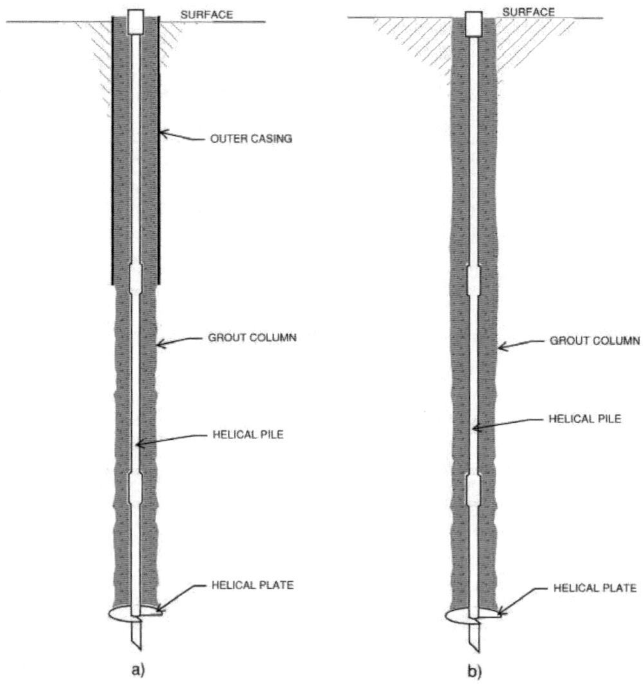

Fig. 3. Typical Section of Drilled Displacement Pile a) East Building b) West Building

The Lateral analysis was conducted using a p-y deformation approach utilizing Lpile (2019). As requested by the structural engineer, a free-head (pin head) condition was assumed as part of the analysis. Figure 4 represents installation of the test piles.

Fig. 4. Test Pile Installation

The analysis estimated the possible downdrag on each proposed drilled displacement pile'

. The material encountered and reported in the geotechnical investigation indicated minimum impact on the downdrag loads. Hence, the downdrag loads on the piles were neglected. The settlement analysis of individual piles was estimated to be less than one inch.

IN-SITU TESTING

The pile theoretical design was followed up by a field testing program prior to installation of all the deep foundation elements. Seven test piles identified as E1, E3, EP TP-1, S1, S2, W1 and W2 were installed in both east and west building locations. These test locations were selected based on the critical load and ground conditions observed within the proposed development. The two new high rise building locations, east building and west building, were proposed as test pile locations. Test pile installations are summarized in Table 1.

Table 1. Summary of Test Piles Installed

Test Pile ID	Test Pile Location	Pile Installation Depth (feet)	Compression Load Test Performed	Lateral Load Test Performed
E1	East Building	42.7	Yes	No
E3	East Building	38.3	Yes	No
EP TP-1	East Building	42.5	No	Yes
S1	East Building	52.6	Yes	Yes
S2	East Building	52.8	Yes	No
W1	West Building	38.9	Yes	Yes
W2	West Building	40.5	Yes	No

The helical pile installation torque, grout volume and grout pressure were monitored during the test pile installation. These parameters were observed to ensure proper installation of the piles closer to the design. The piles were terminated when the piles reached an approximate installation torque of 19,000 to 23,000 pound-feet. The grout strength was tested at seven, 14 and 28 days of the casting of each pile. The strength tests indicated grout strength greater than 5,000 psi at day seven of testing. The drilled displacement pile testing was performed a minimum of seven days after the pile installation.

AXIAL COMPRESSION LOAD TEST

Static compressive load tests were performed on three selected piles in general accordance with *Standard Test Method for Deep Foundations Under Static Axial Compressive Load (ASTM D1143/D1143M-07, (Reapproved 2013) – Section 8.1.2 (ASTM D1143/ D1143M-07, Section 8.1.3 Procedure A: Quick Test and Procedure B: Maintained Test)*. The test criteria was developed according to the New York City Building Code (2014). The ultimate pile load was determined based on Davisson criteria (1972) and the elastic shortening of the piles. Figure 5 represents the axial load testing setup used in the field.

Fig. 5. Axial Compression Load Testing Setup

Table 2 presents the summary of the axial compression test results obtained. The allowable axial pile capacity is derived from load test results based on methods accepted by the New York City Building Code (2014). The deflection of the pile related to the applied load is shown in Figure 6.

Table 2. Axial Compression Load Test Summary

Pile ID	Pile Embedment Depth (ft)	Ultimate Capacity (tons)	Approximate Deformation at Ultimate Capacity (inch)
E1	42.7	128	0.52
E3	38.3	128	0.40
S1	52.6	420	0.48
S2	52.8	420	0.71
W1	38.9	128	0.36
W2	40.5	128	0.40

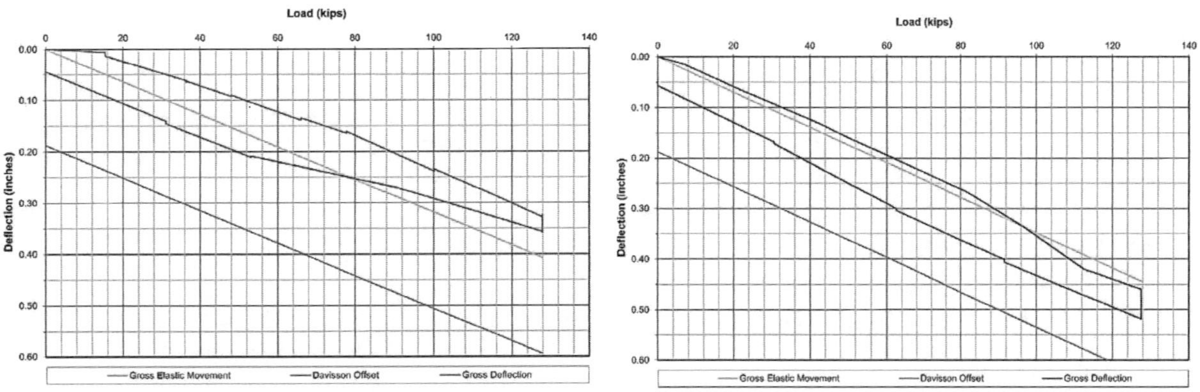

a. Test Pile – W1 **a. Test Pile – E1**

Fig. 6. Axial Compression vs Deflection graph (Allowable Capacity 64 kips)

The results indicate that the ultimate capacity of the piles is greater than the ultimate capacity observed during the load testing.

As part of the axial compression test for higher capacity piles, four strain gauges were installed on the drilled displacement pile identified as S2 to obtain the load transfer from the pile to the surrounding soil. The strain gauges were installed at three feet (gauge 4), 16 feet (gauge 3), 28 feet (gauge 2) and 38 feet (gauge 1) from the pile tip. The load transfer curve observed during the test is presented in in Figure 7.

The load transfer observed during the test indicated the applied load is mainly transferred to the surrounding stratums at gauges one two and three. The load transfer closer to the ground surface indicates relatively smaller transfer amounts.

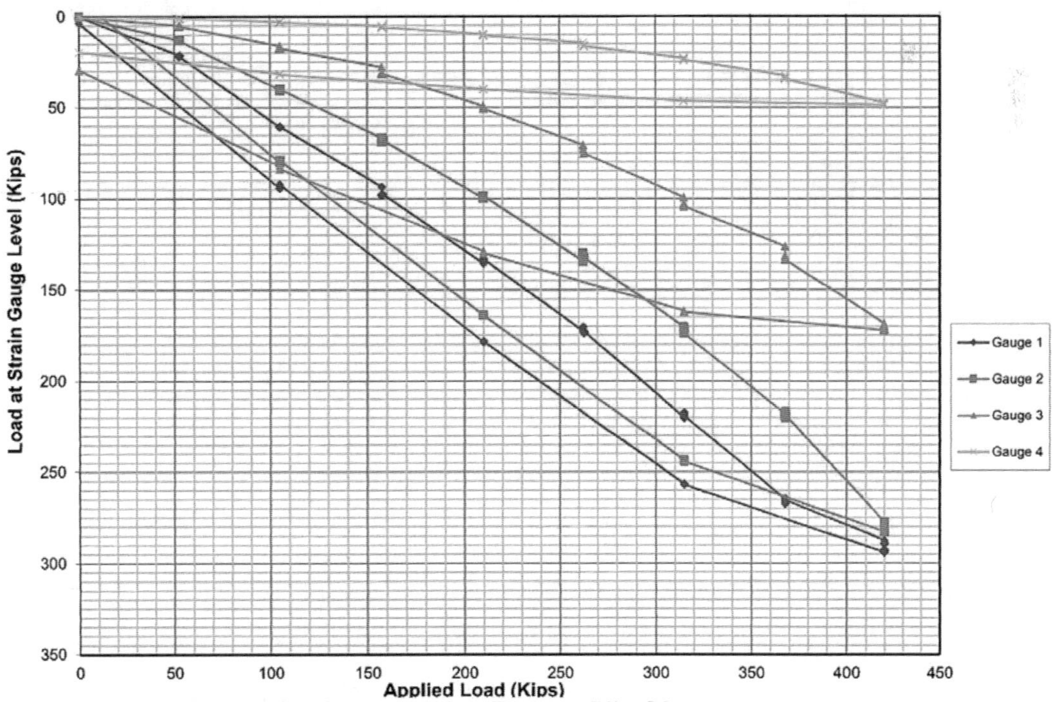

Fig. 7. Load Transfer Curves for Compression Test on Pile S2

LATERAL LOAD TEST

Lateral load testing was performed on one test pile. The lateral load test was performed in general accordance with *Standard Test Methods for Deep Foundations Under Lateral Load (ASTM D3966/D3966M-07 (Reapproved 2013) – Section 8.1.2 Procedure A: Standard Loading Schedule)*. The test results are presented in Table 3.

Table 3. Laterally Loaded Test Pile Installation and Test Results

Pile ID	Pile Embedment Depth (ft)	Ultimate lateral Capacity (kips)	Approximate Deflections at Ultimate Capacity (inch)
W-1	38.9	3.6	0.48
S1	52.6	20	0.47
EP-T1	42.5	20	0.44

Figure 8 presents the observed deflection of the pile due to applied lateral load on piles W1 and EP-T1.

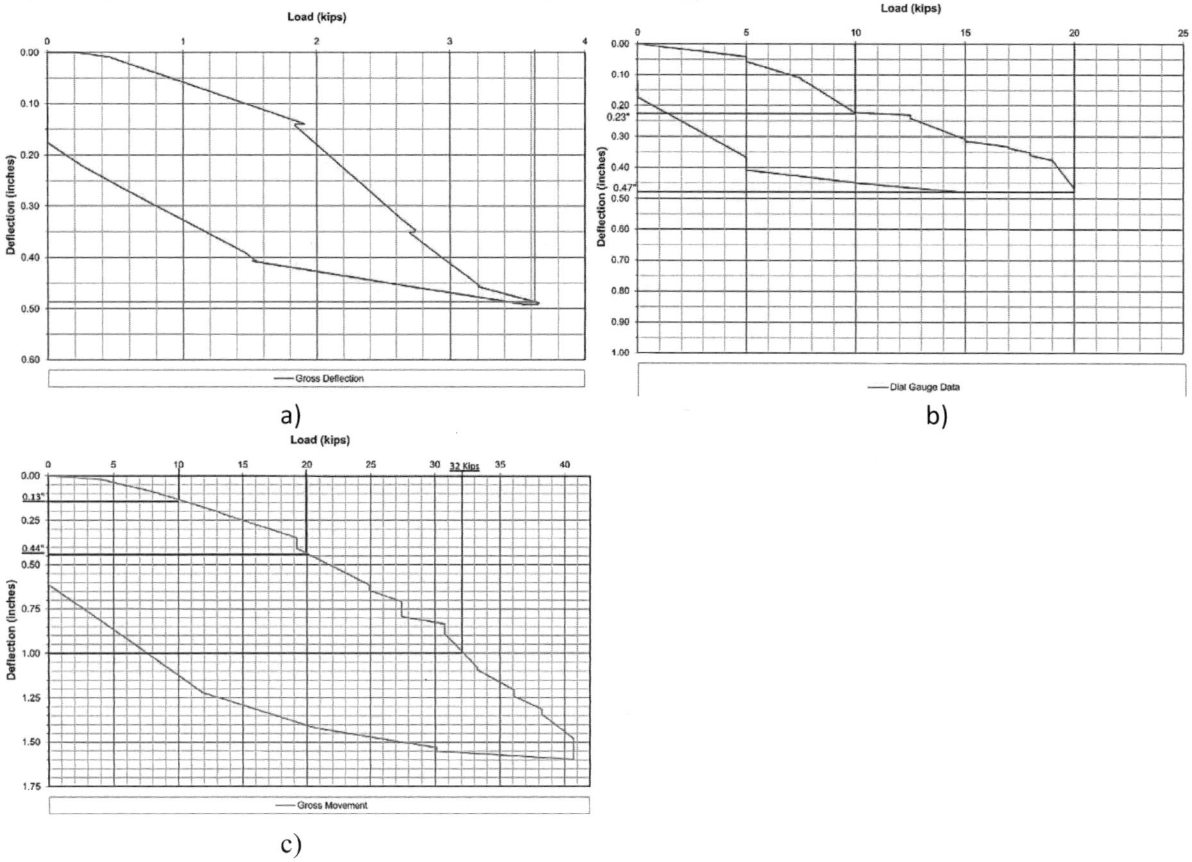

Fig. 8 Test Pile a) W1, b) S1 and c) EP-T1 Lateral Load vs Deflection Graph

The lateral load test indicated deflection under 0.5 inch at the ground surface for the ultimate lateral loads to which the piles were subjected. The theoretical analysis indicated similar deflection in the drilled displacement piles.

SUMMARY AND CONCLUSIONS

Based on the data obtained during the design and testing of the drilled displacement pies, the drilled displacement piles offer multiple advantages. From a design perspective, the piles were capable of carrying larger loads compared to conventional driven piles with relatively shallow embedment depths. Similar observations were reported by Tavenas, F.A (1971, 1977). The use of drilled displacement piles has determined to be an economical solution for this project which included moderately high loading within relatively loose/soft subsurface conditions. Conventional driven piles require larger steel sections with deeper embedment to achieve similar capacities. Drilled displacement piles can be a reasonable and sustainable alternative to a conventional driven pile. As seen in this project, it is possible to reduce construction costs and increase the production speed of the piles. The total cost savings is approximated to be about one million dollars with foundation element installation completed before project estimated timelines. One of the key elements we observed during the construction was the fast phase construction of these piles and minimal amount of spoils generated. There are other added advantages, such as less vibration generated during installations of the drilled displacement piles. Generally, driven piles create excessive noise and vibrations which will require added preventative measures, often leading to community complaints and construction time restrictions in urbanite areas. Using drilled displacement piles will reduce the added costs of vibration monitoring required. Due to smaller pile elements, these systems can bring added advantage in space-restricted construction sites.

However, suggesting an alternative pile system to a conventional driven pile system can cause doubts within design teams who have not experienced these systems. Hence, extra testing or additional design documentation will be part of the final design and testing submissions. The testing of these piles indicated using drilled displacement piles performing well within the requirements. The innovative use of hybrid pile design methodologies can provide better economical and swift construction solutions.

REFERENCES

Adam J.I. and Klym T.W., 1972. A study of anchors for transmission tower foundations. Canadian Geotechnical Journal, 9(1):89-104

ASTM D1143 /D1143M-07 2007. Standard test methods for deep foundations under static axial compressive load, ASTM International, West Conshohocken, PA, www.astm.org

ASTM D3689 / D3689M-07 2013. Standard test methods for deep foundations under static axial tensile load, ASTM International, West Conshohocken, PA, www.astm.org

ASTM D3966 / D3966M-07 2013. Standard test methods for deep foundations under lateral load, ASTM International, West Conshohocken, PA, 2013, www.astm.org

Davisson, M. T., 1972. High Capacity Piles, Proceedings of the Lecture Series on Innovations in Foundation Construction, ASCE, Illinois Section, Chicago, March 22, pp 81-112.

Federal Highway Administration (FHWA) (2005) Micropile design and construction guidelines. US Department of Transportation, Washington, DC

FHWA 2016. Design and construction of driven piles foundations – Volume I & II. Pub. No. FHWA NHI-16-009

New York City Building Code 2014

Tavenas, F.A., Audibert, J.M.E., 1977. Application of the wave equation analysis to friction piles in sand. Canadian Geotechnical Journal, 14(4) 34-51.

Tavenas, F.A., 1971. Load testing results on friction piles in sand. Canadian Geotechnical Journal, Vol. 4, pp 7-22.

Trofimenkov, J.G. and Maruipolshii, L.G., 1965. Screw piles used for mast and tower foundations. In proceedings of the 6th International Conference on Soil Mechanics and Foundation Engineering Vol. 2, pp. 328-332

ADJACENT TOP-LOADED BI-DIRECTIONAL TEST AND CONVENTIONAL BI-DIRECTIONAL LOAD TEST RESULTS – SITE II

Patrick J. Hannigan, GRL Engineers, Inc. phannigan@grlengineers.com
Cleveland, OH, USA, 216-831-6131

Seth Robertson, GRL Engineers, Inc. srobertson@grlengineers.com
Cleveland, OH, USA, 216-831-6131

Rozbeh B. Moghaddam, RBM Consulting Group, Inc. rbm@rbmcgroup.com
San Antonio, TX 726-777-4558

ABSTRACT

The Top-Loaded Bi-Directional Test ("TLBT") and the conventional Bi-Directional Load Test ("BDLT") are full-scale static load test methods. In both methods, the test loads are applied bi-directionally to the deep foundation. The advantages of the conventional BDLT have been well established for cost efficient static load testing of deep foundations, particularly for large diameter, high-capacity drilled foundations. A conventional BDLT applies the bi-directional test loads using an embedded jack and bearing plates. The TLBT method also applies the bi-directional test load to embedded bearing plates but does so using reusable high-strength threaded bars attached to an above-grade hydraulic jack and load transfer assembly. The new TLBT method is well suited for load testing shafts of a moderate diameter, length, and nominal resistance or ultimate load. TLBT advantages include economic considerations resulting from the method's reusable load application components and constructability considerations as it readily accommodates a center concrete tremie pipe or pump line location.

This paper presents comparison results from full-scale load tests performed using both bi-directional load testing methods. Adjacent 3-foot diameter by 30-foot long BDLT and TLBT shafts were tested. Details regarding subsurface conditions and test shaft construction are included for the comparison of the TLBT and BDLT results. Corresponding test data and analyses for both test shafts are presented and discussed in detail. The BDLT shaft had a maximum uni-directional load of 240 kips at upper and lower bearing plate displacements of 0.5 and -3.8 inches, respectively. The TLBT had a maximum uni-directional load of 300 kips at upper and lower bearing plate displacements of 1.0 and -5.9 inches, respectively. These results as well as the load-transfer behavior and equivalent top loading curve results from the adjacent BDLT and TLBT drilled shafts are discussed. The paper also discusses the strain gage instrumentation on the TLBT load application bars and on the embedded bearing plates.

Keywords: bi-directional load test, top-loaded bi-directional test

Introduction

The conventional bi-directional load test ("BDLT") has been well documented by Osterberg (1998), Schmertmann et. al., (1998), and numerous others. The test method requirements are further delineated in ASTM D8169/D8169M-18. Details on the top-loaded bi-directional test ("TLBT") method have been presented in Moghaddam, et. al., (2021a, 2021b). Similar to a BDLT, the TLBT method also applies the bi-directional test load to embedded bearing plates but does so using reusable high-strength threaded bars attached to an above-grade hydraulic jack and load transfer assembly. An initial comparison of conventional bi-directional load test and top-loaded bi-directional load test results from Site I was presented in Hannigan, et. al., (2021). At Site I in Converse TX, 4-foot diameter drilled shafts were installed 20 feet apart center-to-center. The current paper presents results from Site II in Solon OH where a conventional bi-directional load test and a top-loaded bi-directional load test were performed on 3-foot diameter test shafts located 30 feet apart center-to-center. The conventional bi-directional load test shaft at Site II had been installed in 2017. The top-loaded bi-directional load test shaft was installed in 2021 for an additional result comparison. For research and development purposes, the TLBT shaft at Site II had strain gage instrumentation added to the TLBT load application bars and to the embedded bearing plates.

A soil boring near the test shafts indicated the subsurface conditions consist of approximately 18 feet of stiff to very stiff lean clay overlying a 5-foot thick layer of loose sand. The sand layer was in turn underlain by 10 feet of stiff clay. Glacial till deposits were encountered below 33 feet with a 10-foot layer of stiff sandy clay followed by an 18.5-foot thick layer of very stiff to hard clay that extended to the end of the boring at 61.5 feet. Two cone penetration tests were also performed in the vicinity of the test shafts. Neither CPT test encountered the loose sand layer noted in the boring. Unfortunately, neither CPT could be advanced to the test shaft base level 30 feet below grade with one CPT encountering a cobble at 14.5 feet and the other refusing on a cobble at 22.0 feet.

Test Shaft Construction Details

At the Solon OH test site, two 3-foot diameter, 30-foot-long test shafts were installed 30 feet apart center-to-center. Both shafts were drilled in the dry without casing. The BDLT shaft containing a conventional 600-kip bi-directional cell with a 6-inch stroke was installed in September 2017. A second test shaft, containing the top-loaded bi-directional assembly, was installed in May 2021. In both test shafts, the top of the lower bearing plate was located at the same elevation approximately 7 feet above the shaft base. After excavation, both shafts were filled with concrete having a 28-day compressive strength of 4000 psi.

The TLBT shaft was constructed to accommodate six 2.5-inch O.D., 150 ksi threaded steel bars; three base mobilizer bars and three shaft mobilizer bars. Each shaft mobilizer bar was encased in a 3.0-inch I.D. schedule 80 PVC pipe to isolate it from the shaft concrete. Similarly, each base mobilizer bar was isolated from the shaft concrete with a 3.5-inch I.D. schedule 40 steel pipe. A larger diameter pipe around the base mobilizer bars was used to accommodate data transmission cables from instrumentation added to the mobilizer bars.

Fig 1. Drill rig and 300-ton bi-directional cell in cage for 2017 conventional BDLT

Fig 2. Drill rig and bearing plate assembly in cage for 2021 TLBT

Comparison of Conventional and Top-Loaded Bi-Directional Test Results

The conventional bi-directional load test and the top loaded bi-directional load test were performed 20 and 32 days after shaft installation, respectively. The tests followed the procedures outlined in ASTM D8169/D8169M-18, even though the conventional bi-directional load test was performed prior the standard being officially adopted. A 400-kip jack with a load cell and spherical bearing plate was used to apply the TLBT loads. Plots of the upper and lower bearing plate movements versus applied load are presented in Figure 3 for the conventional BDLT shaft and Figure 4 for the TLBT shaft. The conventional bi-directional load test was terminated at an applied load of 240 kips when the combined shaft and base resistances below the lower bearing plate failed geotechnically. In the TLBT, similar load-movement responses occurred up to 240 kips of applied load. However, the shaft base materials appeared to be stiffer at the TLBT shaft location. This, along with the jack having a 13-inch stroke, allowed the TLBT to be taken to greater upper and lower bearing plate movements and thereby mobilized a greater soil resistance.

Figure 5 presents a photograph of the top-loaded bi-directional test in progress including the loading frame, hydraulic jack, load cell, shaft and base mobilizer bars, and associated instrumentation. This design accommodates multiple shaft diameters (generally 3 to 6 feet) and rebar cage sizes as well as an adjustable loading frame height based on the instrumentation extending out from the shaft top. The illustrated top-loading system is capable of applying uni-directional loads up to 1,800 kips (3,600-kip bi-directional load).

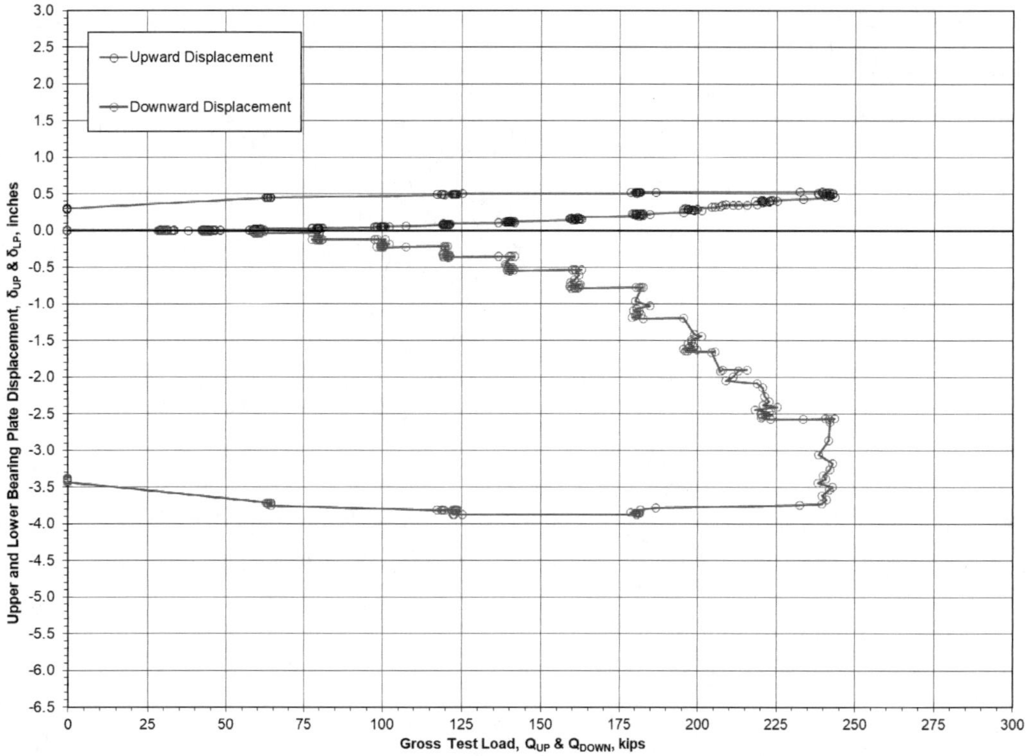

Fig 3. 2017 Conventional BDLT results - bearing plate displacements vs gross test load

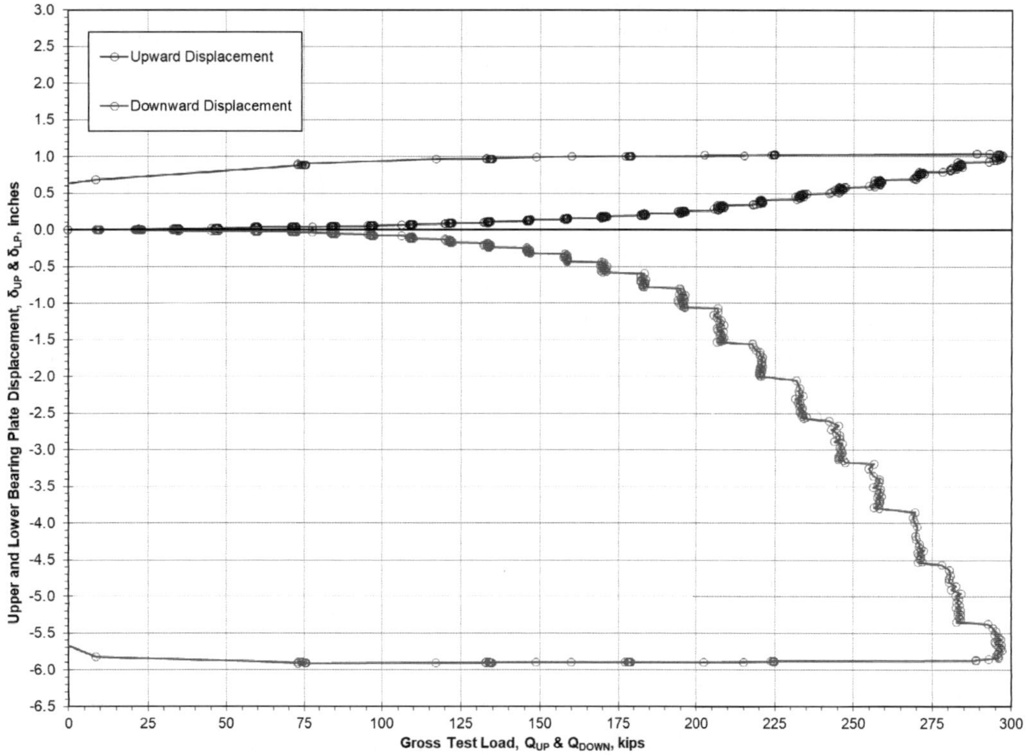

Fig 4. 2021 TLBT results - bearing plate displacements vs gross test load

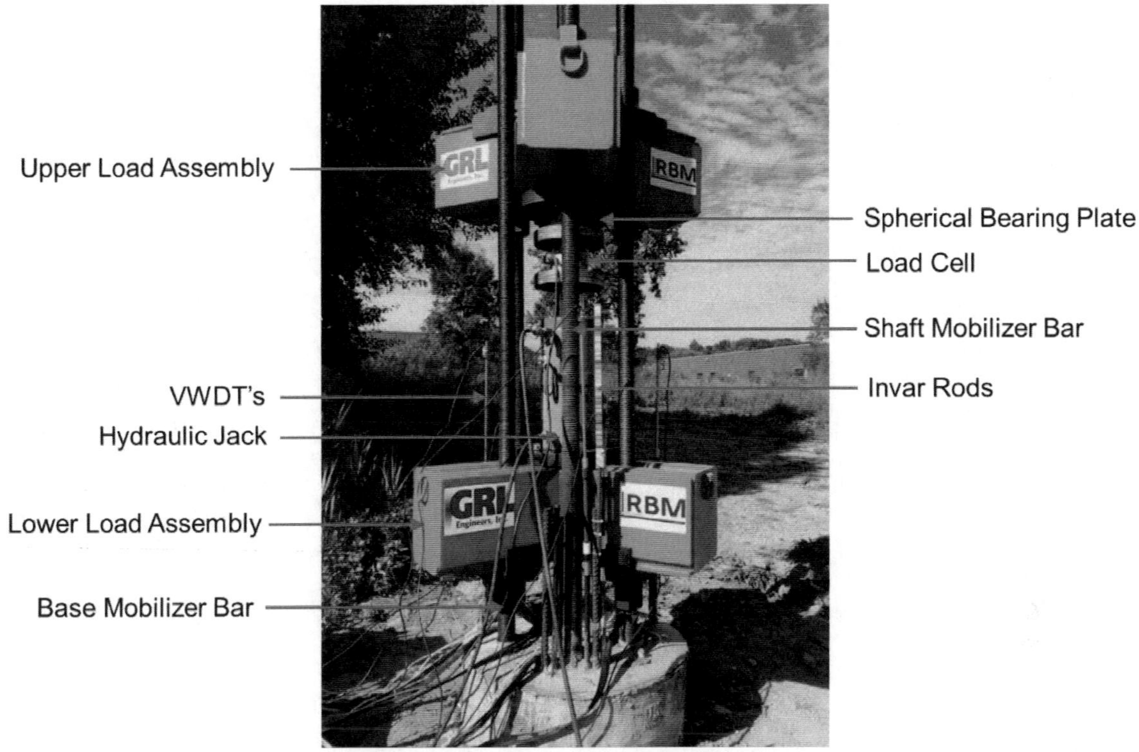

Fig 5. TLBT - load assemblies, hydraulic jack, load cell, and other components

Load Transfer Profiles

In both tests, the measured strains were acquired from at least two sister-bar mounted vibrating-wire strain gages attached at diametrically opposite locations on the reinforcing cage. Identical strain gage elevations or levels were used in both tests. In the TLBT, an additional strain gage level was added at Elevation 103.5, a distance of one diameter above the bearing plate. This additional level, as well as the strain gage levels in the TLBT at Elevation 96.5 (one diameter below the bearing plate) and at Elevation 108.5 (two diameters above the bearing plate) each had four sister-bar mounted vibrating-wire strain gages to better delineate load transfer. The average measured strain at each level was converted to calculated internal forces using the incremental rigidity method, Komurka and Moghaddam (2020), Komurka and Robertson (2020). The resulting load transfer profiles versus elevation (site datum) are presented in Figure 6a for the conventional BDLT and in Figure 6b for the TLBT test shaft.

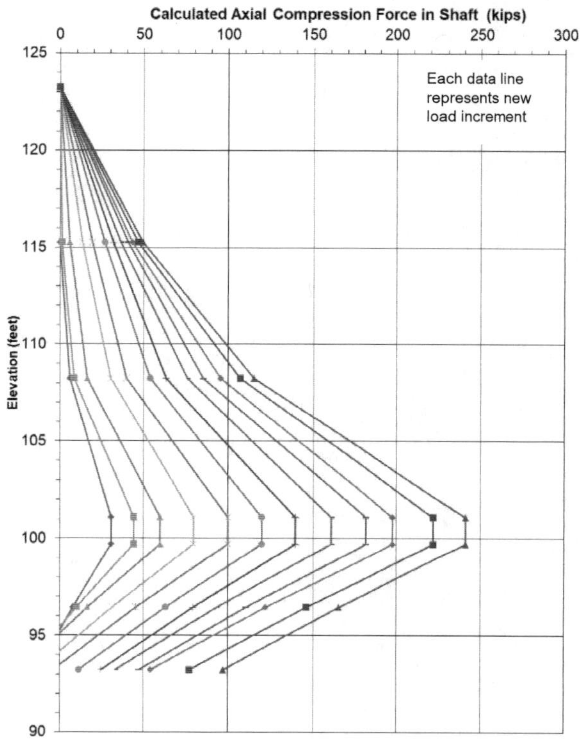

Fig 6a. Internal force profiles versus elevation for conventional BDLT.

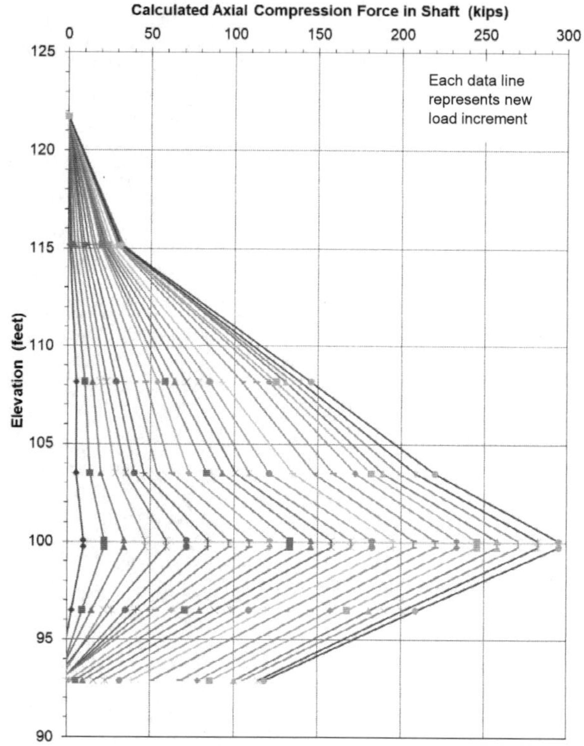

Fig 6b. Internal force profiles versus elevation for TLBT.

Equivalent Top Loading Curves

Figure 7 presents the equivalent top loading (ETL) curves for the conventional BDLT shaft, as well as for the TLBT shaft. The ETL curves were generated using the modified ETL method described by Seo, et al (2016). In the conventional BDLT, the maximum equivalent top load was 375 kips at a shaft top displacement of 0.55 inches. In the TLBT, the maximum equivalent top load was 490 kips at a shaft top displacement of 1.05 inches.

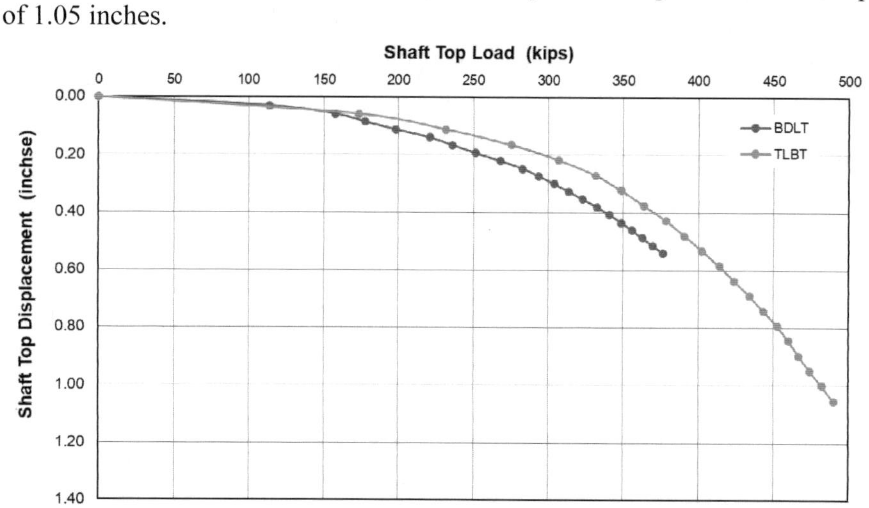

Fig 7. Equivalent Top Loading Curve Comparison for BDLT and TLBT

Measured Strains on Base Mobilizer Bars and Bearing Plate in TLBT

The strain near the bottom of each base mobilizer bar was measured at the center of a two-foot-long unthreaded section using four active gages in uniaxial stress field; two aligned with maximum principal strain and two "Poisson's" gages. The strain was then converted to force using the smooth steel bar cross-sectional area of 5.31 in^2 and the high strength bar manufacturer's recommended elastic modulus of 29,700 ksi. Figure 8 shows the instrumented base mobilizer bars with an epoxy coating over the instrumented section each with the flat cable used for signal transmission to the SLT data logger at the surface. Figure 9 presents a plot the individual force determined from each bar versus elapsed time. Note that two of the three bars have the same load throughout the test but that the NW bar, Bar 3, carries a reduced load until such time that the signal is lost from Bar 3 during the final load interval.

Fig 8. Strain gages attached to 2-foot-long unthreaded portion of TLBT base mobilizer bars

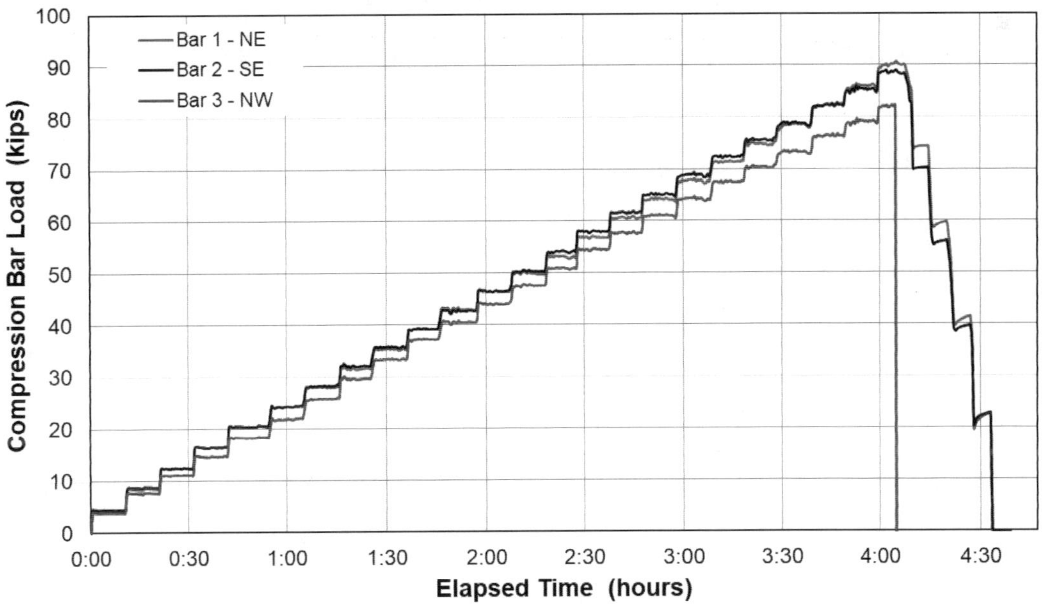

Fig 9. Loads on individual TLBT base mobilizer bars

Figure 10 contains a plot of the "measured" load vs elapsed time as determined from the calibrated jack, calibrated load cell, and the load summed from the three compression bars. Even though Bar 3 carried less load throughout the test, these "measured" load readings were reasonably similar over the first half of the test (up to 150 kips). After that point, the base mobilizer load summed from the three compression bars deviated from the calibrated hydraulic jack and load cell. As noted above, the reading from Bar 3 became unusable as the wire running from the instrumented section up along the bar was crushed. The 3.5-inch ID of the pipe surrounding the 2.5-inch OD compression bars was essential to pass the data cable to acquire the base mobilizer bar readings. Unfortunately, this was also detrimental in allowing greater than normal annular space for lateral movement. This annular space likely contributed to the difference in bar versus jack readings.

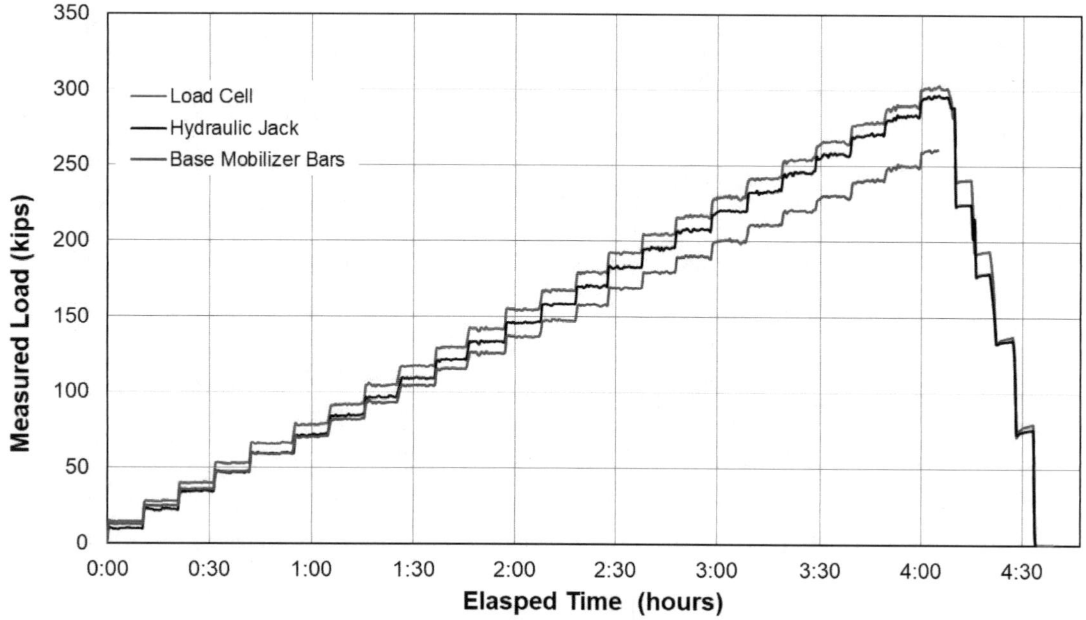

Fig 10. Comparison of hydraulic jack, load cell, and average base mobilizer bar loads in TLBT

Since the conventional BDLT incorporates one or more embedded hydraulic jacks, the applied jack loads are usually assumed to be uniformly distributed to the deep foundation. However, experience has shown that this may not always be true. Therefore, additional instrumentation was incorporated on the TLBT lower bearing plate for assessing anticipated plate deformations under a series of point loads and the influence of potential non-uniform stresses at some distance away from the load application. Stiffener plates were added below the base mobilizer bar locations solely for this purpose to characterize bearing plate response to the TLBT applied load. Only one stiffener plate was instrumented with multiple strain measurement locations. The instrumented stiffener plate was located below Bar 1, the NE bar.

Strain location S3 was located at the vertical center of the stiffener plate immediately below the load application point of a base mobilizer bar. Strain location S4 was located at the vertical center of the stiffener plate near the opposite, inside edge of the plate. Strain locations S1 and S2 were located at the horizontal center of the stiffener plate with S1 on the side face of the stiffener plate near the top edge and S2 on the bottom edge of the stiffener plate. These strain measurement locations are pictured in Figure 11 along with an orientation diagram. Strain locations S1 and S2 were oriented perpendicular to the load application direction and strain locations S3 and S4 were oriented parallel to the load application direction. A plot of these strain measurements versus elapsed time is presented in Figure 12.

Strain measurements were also made on three locations near the center of the lower bearing plate, oriented 120 degrees apart. Strain locations S5, S6, and S7 were located on the bearing plate between the load application points of the base mobilizer bars. These strain locations were not directly in-line with the load application points but were slightly offset toward the shaft center to accommodate the plate cutout locations for the shaft mobilizer bar hex nuts. Figure 13 presents the measured microstrain versus elapsed time for strain locations S5, S6, and S7. The change in microstrain around the 4:10 timestamp corresponds to the beginning of the unloading cycle. The measured microstrain for S5 and S6 terminated early due to transmission cable or gage loss. As anticipated by the strain measurement locations and orientations, the lower bearing plate was in compression throughout the entire test. The relative magnitude of the measured strains on the stiffener and bearing plate characterized the expected plate deformation under the applied loading condition. However, the influence of these observed deformations on shaft and base resistance mobilization cannot be solely assessed from these measurements.

Fig 11. Strain gages attached to bearing plate and stiffener plate in TLBT

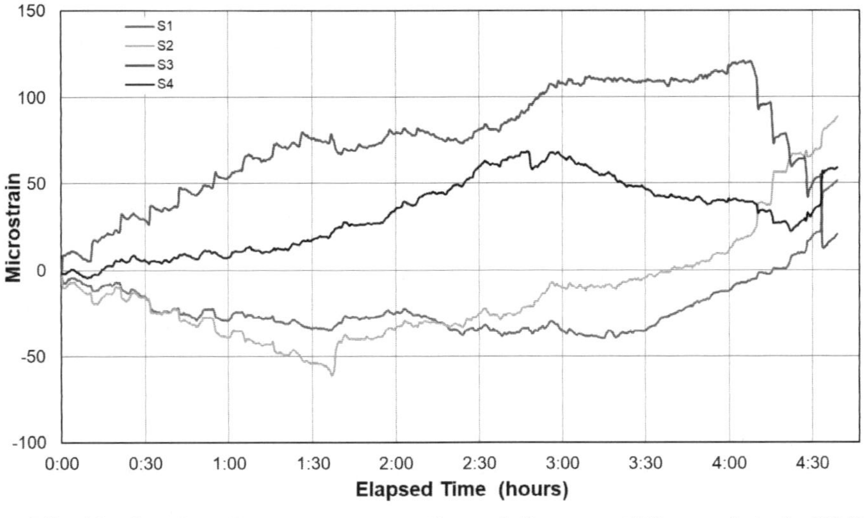

Fig 12. Strain measurements vs elapsed time on stiffener plate in TLBT

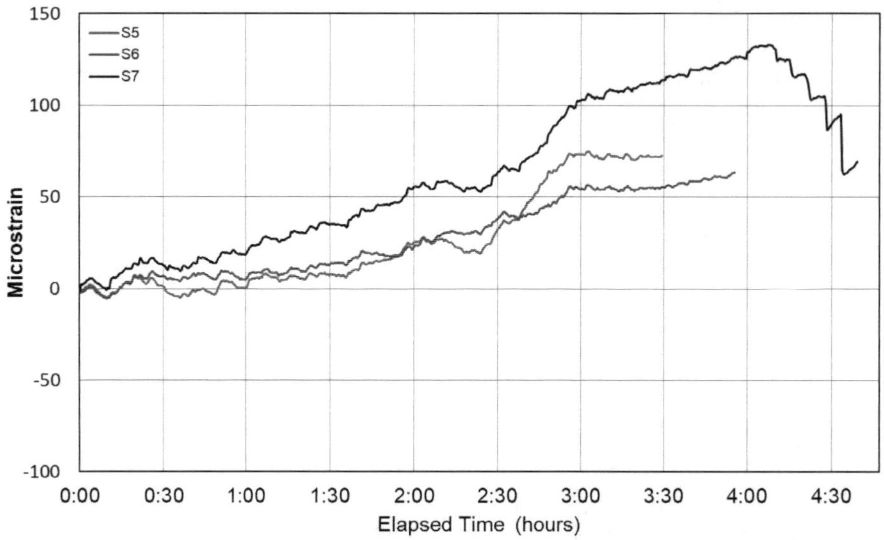

Fig 13. Strain measurements vs elapsed time on lower bearing plate in TLBT

As visible in Figure 11, two additional sister-bar mounted vibrating-wire strain gages were installed on offset rebar approximately nine inches inside the reinforcing cage towards the shaft center. These offset strain gages were located approximately 40 inches below the lower bearing plate. The purpose of these offset strain measurements was to compare the offset strain values with the average strain measured from four sister-bar mounted vibrating wire strain gages on the reinforcing cage. Figure 14 presents the cage and offset strain measurements vs time. Even though plate deformations were indicated by the strain measurements on the lower bearing plate, the strain distribution below base mobilizer bar point loads through the bearing plate and into the concrete was relatively uniform, indicating reasonable strain compatibility across the shaft cross-section under the TLBT applied load.

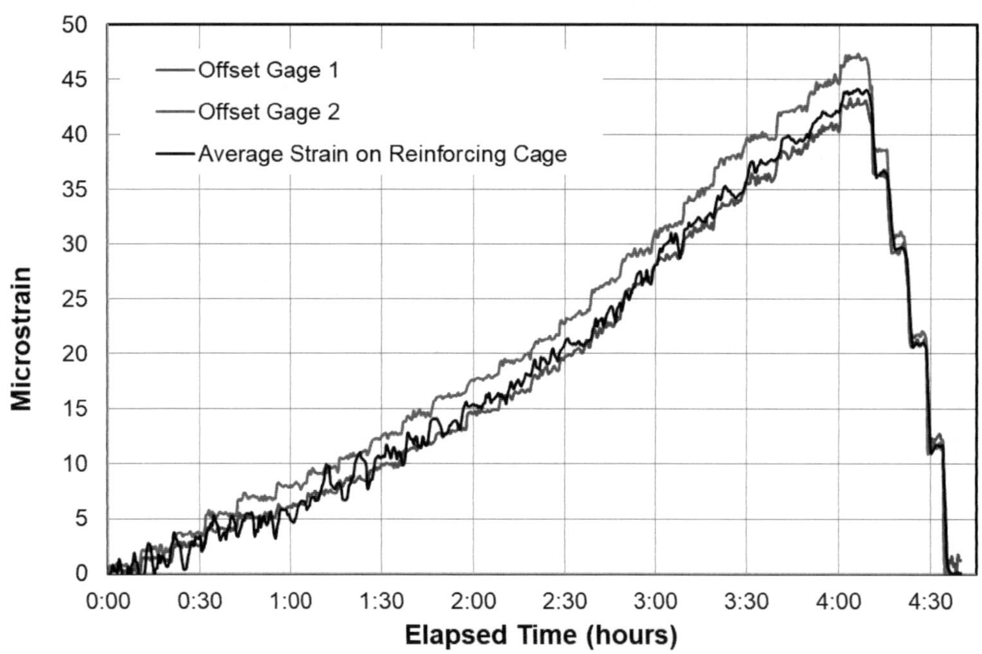

Fig 14. Comparison of average strain measurements on the reinforcing cage vs strain measurements offset near the shaft center in TLBT

Conclusions:

Bi-directional load tests were performed on two 3-foot diameter test shafts at a test site in Ohio. One shaft was load tested using a conventional BDLT and the other test shaft was load tested using a TLBT. In both test shafts, the top of the lower bearing plate delineating the break plane was located at the same elevation approximately 7 feet above the shaft base in the 30-foot-long shafts. The conventional BDLT shaft had a maximum uni-directional load of 240 kips at upper and lower bearing plate displacements of 0.5 and -3.8 inches, respectively. The TLBT had a maximum uni-directional load of 300 kips at upper and lower bearing plate displacements of 1.0 and -5.9 inches, respectively.

The BDLT and TLBT equivalent top loading curve responses at the maximum BDLT applied load were very similar when the encountered variation in base material resistance and typical construction variations are considered. At the maximum equivalent top load of 377 kips available from both ETL curves, the estimated shaft top displacement was 0.54 inches in the BDLT shaft and 0.43 inches in the TLBT shaft.

For research and development purposes, additional instrumentation was added to the TLBT base mobilizer bars and lower bearing plate. These measurements confirmed reasonably uniform load application behavior occurred in the TLBT.

At the maximum applied load, the load on two of the three base mobilizer bars was within 0.2% of one another whereas the load on the third base mobilizer bar, Bar 3, was 8.3% less than the average of the other two bars. This difference in bar load may be the result of horizontal alignment issues in the above grade top loading assembly, in the embedded bearing plates, or some other unknown cause. Even with this variation in bar load, the difference in jack load and the average bar load was only 12.0% at the maximum applied load

The TLBT method offers potential cost savings and reduced construction risk due to the elimination of the embedded hydraulic jack(s) and associated hydraulic lines. Embedded instrumentation requirements are essentially the same for both the TLBT and BDLT methods.

References

American Society of Testing and Materials (2018). ASTM D8169/D8169M-18, Standard Test Methods for Deep Foundations Under Bi-Directional Static Axial Compressive Load.

Hannigan, P., Moghaddam, R., Rausche, F., and Robertson, S. (2021). Top-Loaded Bi-Directional Test and the Conventional Bi-Directional Load Test, A Direct Comparison. 46th Deep Foundations Institute Annual Conference, Las Vegas, NV.

Komurka, V.E. and Moghaddam, R.B. (2020). The Incremental Rigidity method – More-Direct Conversion of Strain to Internal Force in Instrumented Static loading Tests. ASCE GeoCongress 2020, Minneapolis, MN.

Komurka, V.E., and Robertson, S. (2020). Results and Lessons Learned from Converting Strain to Internal Force in Instrumented Static Loading Tests Using the Incremental Rigidity Method. ASCE GeoCongress 2020, Minneapolis, MN.

Moghaddam, R., Rausche, F., and Hannigan, P. (2021a). Top Loaded Bi-Directional Testing System and Method of Using the Same. United States Patent Application Publication No. US2021/0102863.

Moghaddam, R., Hannigan, P., and Rausche, F. (2021b). Top-Loaded Bi-Direcitonal Test, a New Approach to Deep Foundation Load Testing. 46th Deep Foundations Institute Annual Conference, Las Vegas, NV.

Osterberg. J.O. (1998). The Osterberg load test method for drilled shaft and driven piles—The first ten years. Deep Foundation Institute, 7th International Conference and. Exhibition on Piling and Deep Foundations, Vienna, Austria, 17 p.

Schmertmann J.H., Hayes, J.A., Molnit, T., and Osterberg, J.O. (1998). O-Cell Testing Case Histories Demonstrate the Importance of Bored Pile (Drilled Shaft) Construction Technique. Proceedings: Fourth International Conference on Case Histories in Geotechnical Engineering, St. Louis, Missouri.

Seo, H., Moghaddam, R.B., and Lawson, W.D. (2016). Assessment of Methods for Construction on an Equivalent Top Loading Curve from O-Cell Test Data. The Japanese Geotechnical Society, Soils and Foundations Journal, Volume 56, Issue 5, pp 889-903.

ASSESSMENT, REMEDIATION AND ACCEPTANCE OF DRILLED SHAFTS THROUGH THE USE OF CSL TESTS

Bernardo J. Jiménez Vega, H Solís, Costa Rica, bjimenez@hsolis.com
Edy L. T. Montalvan, PhD, edytemon@gmail.com

ABSTRACT

The experience gained during more than 900 Crosshole Sonic Logging (CSL) tests on drilled shafts shows the importance of establishing proper project acceptance and repair protocols. Drilled shaft evaluation is a multidisciplinary task that should not be left solely to the outcome of the CSL test. This test allows to characterize and delimit areas of different qualities of concrete, and it is important that all the interested parties understand the difference between flaws and defects. The information presented here is intended to give a guideline to contractors, inspectors, engineers, owners and clients on how to deal with acceptance and remediation of defective drilled shafts, detected using CSL tests. A database of flaws is presented depending on their location, intensity and remediation method. Procedures that allow the flaws to be analyzed are presented in order to dismiss or classify them as defects. In many cases, the CSL test may indicate an anomaly, but by coring the concrete and analyzing the structural demands and conditions, the flaw can be classified as non-severe. When flaws are classified as defects, their repair will depend on the area where they are located in the drilled shaft pile. In this document two procedures are described for repairing defects depending on their location. For defects in the drilled shaft head area, the procedure takes into account the interaction between the new and old concrete. For repairing the middle zone of the drilled shaft by grouting, the major concerns are the difficulty of accessing through the drilled shaft at deep depths and the communication between boreholes. For defects at the drilled shaft tip, two proposed remediation methods are presented. Finally, tomography tests are presented to compare the effectiveness of remediation by grouting.

Keywords: **CSL test, drilled shaft flaws, drilled shaft evaluation, drilled shaft remediation, grouting**

INTRODUCTION

Most public infrastructure projects where drilled shafts are used require some control to ensure that the finished product meets established quality criteria specifications. In a drilled shaft, unlike, for example, a precast concrete element, it is not possible to visually verify if the integrity of the concrete is adequate, i.e., if there are no voids, soil, inclusions or disaggregation of the concrete in the drilled shafts. Due to this circumstance, it is necessary to use methods to verify the integrity and quality of the concrete. Commonly used Non-Destructive Tests (NDT) allow for identifying, imaging, and characterizing areas of lower quality that must be analyzed to determine if this difference in quality affects the expected performance of the drilled shaft.

The Cross-hole Sonic Logging (CSL) test is the most widely used NDT method for quality assurance of drilled shaft concrete. Basically, the integrity of the concrete is evaluated on the basis of signal travel time and signal strength between an acoustic transmitter and a receiver probe that travels the drilled shaft through access tubes. This arrival time in the CSL test is known as the First Arrival Time (FAT) and can be correlated to the concrete strength (Jalinoos et al., 2005). Additionally, a tomograph of the shaft can be created by the interpolation of the CSL test, giving a graphic localization of any affected area, and can be used to determine where coring might be conducted or to aid in remediation efforts. For CSL terminology the reader is advised to refer to the document of Sellountou et al.(2019).

In this paper, a database of more than 900 CSL tests from 3 road infrastructure projects from Costa Rica is used to summarize ways of approaching the analysis, repair and acceptance of drilled shafts. The intent of this work is to provide tools and guidelines for the technical discussion that may occur in the evaluation of drilled shafts where CSL tests encounter a flaw.

ACCEPTANCE OF DRILLED SHAFTS BASED ON CSL TESTING

For any infrastructure project where drilled shafts are used, the evaluation and acceptance criteria for the drilled shaft must be established before the beginning of the project. One of the methods to evaluate the integrity of the concrete in the drilled shafts is through CSL tests. The evaluation criteria of this test vary between agencies and projects, and its selection must be made by experienced personnel. For those involved in CSL testing, and with little experience with this test, it is recommended to review the criteria presented by Brown et al. (2018) and DFI (Sellountou et al., 2019). All evaluation criteria should specify the parameters used, ranges and number of CSL profiles affected for each concrete rating. For the analyzed cases in this document, the used criterion is shown in Table 1.

Table 1 – CSL evaluation criterion.

Concrete rating	Arrival time reduction (%)	Commentary (m)
Good	0-20	-
Questionable	20-30	A flaw is indicated if this range of velocities is present on at least 50% of the profiles at the same depth.
Poor/defect	>30	A flaw is indicated if this range of velocities is present on at least two profiles at the same depth

Any discussion about a CSL result, which classifies an anomaly into a flaw, should be analyzed using Fig 1. This figure modified from Brown et al. (2018), shows the scope of NDT tests for the drilled shaft assessment analysis. These tests should be used only for the location, characterization and extent of any flaw in the drilled shaft. Once the presence of the flaw has been determined, its effect on the geotechnical and structural resistance must be determined in order to decide whenever it is necessary to repair the fault.

The assessment phase begins after a flaw is detected in any drilled shaft. The acceptance or rejection of any flaw within the drilled shaft corresponds to a multidisciplinary analysis between the geotechnical engineer and the structural engineer and also between contractors and designers, where the consequences of the flaw in the performance of the drilled shaft are verified.

In many cases, during the assessment/evaluation phase, a delay in the project is created because it is not possible to continue with the superstructure until it is verified that if the performance of the drilled shaft has not been detrimentally affected due to the presence of the flaw. These evaluations must be carried out quickly and the key to this is having the information available, which can include: updated structural models, drilled shaft execution reports, tremie concrete test reports and internal coring reports. These data will allow the team to perform a forensic analysis to best describe the current condition of the drilled shaft. It is highlighted that due to the type of project, design-build projects may present faster evaluation times than normal bid projects.

All parties involved in the project must be aware that there may be flaws that do not affect the performance of the drilled shaft, that is, if there are no major geotechnical or structural consequences and the drilled shaft should be accepted in this condition. Also, any interested party must be able to verify possible false positives of the test. These two cases will be exemplified in this paper.

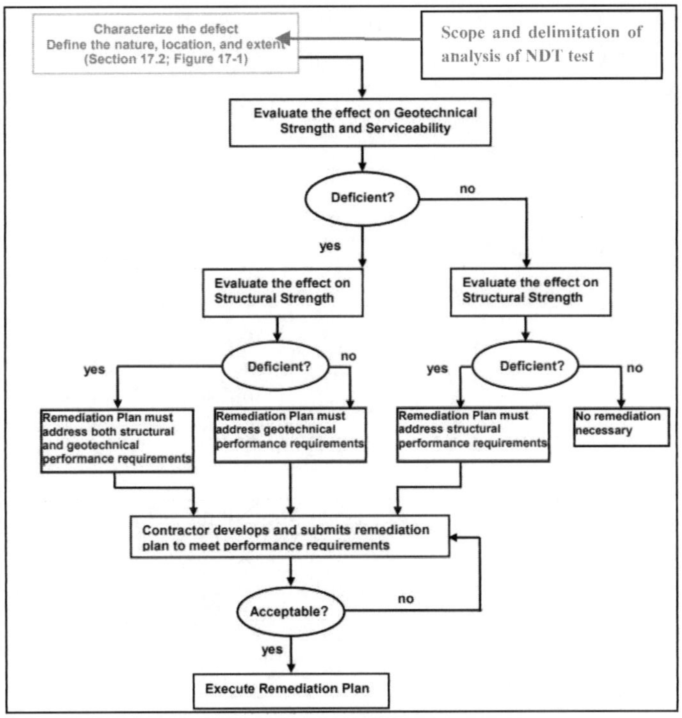

Fig 1. Flow chart of the evaluation and remediation process, adapted from GEC-10 (Brown et. al, 2018).

Summary of Flaws

The information in this document is based on more than 900 CSL tests on three major road infrastructure projects. Drilled shaft diameters varied between 1.0 m (3.3 ft) and 1.2 m (3.9 ft), and depths varied between 7.0 m (23 ft) and 42 m (138 ft). The geological conditions during the construction of drilled shafts were variable, predominantly silty clays, besides sandy soils and volcanic rock conditions were encountered. The database includes 38 drilled shafts (4.2%) that were constructed using a polymer to stabilize the excavation and 26 drilled shafts (2.9%) that utilized temporary casing.

Table 2 presents the obtained flaws divided into three categories: flaw relative position, FAT delay, and remediation method. It is advised that the number presented in the table be used as a guide and not as a trend since projects can vary greatly between contractor practices and geological conditions. A total of 44 tests resulted in the detection of a flaw inside the drilled shaft; this corresponds to 4.9% of the CSL tests. Flaws located at the tip represented 57% of the cases; these cases were due to one of the most frequent errors in the construction of drilled shafts, which is the inadequate cleaning of the bottom of the excavation. The use of proper tools and methods (clean-out buckets and correct drilling fluid manipulation) are necessary to ensure that no significant amounts of debris or sediment will be left at the base of the shaft. Body flaws (upper 4,0 m) were the second most common cause found in the CSL test with 36% of the cases. The presence of these flaws was mostly attributed to problems during the pouring of the concrete which, in several cases, were due to problems in the work platform and delayed the placement of concrete between one truck and another, causing a cold joint. In any project, special care should be given to the issue of working platforms, especially during rainy seasons, because flaws due to cold joints are rarely repairable due to their severity. For further information about working platforms the reader should review EFFC/DFI (2019). Only 7% of the cases indicated a flaw at the shaft's head caused by inadequate flushing of concrete at the top of the shaft, resulting in contaminated concrete with low P-wave velocity (Vp) and increased FAT.

Table 2 – Data base flaw summary.

	Flaw position	
Drilled shaft position	Percentage (Number of cases)	Most common causes
Head	7% (3)	Lack of concrete overflow or lack of tremie pipe movement after casing removal
Body	36% (16)	Problems related to working platforms, delays between concrete trucks or problems in tremie pipes
Base	57% (25)	Inadequate bottom cleanliness or long delay between cleaning and concrete pouring
	FAT delay	
Range of FAT delay (deviation from mean arrival time)	Percentage (Number of cases)	
>100%	41% (18)	
100%-50%	27% (12)	
50%-30%	27% (12)	
30%-20%	5% (2)	
	Type	
Remediation	Percentage (Number of cases)	Commentary
E: assessment analysis showed no negative effect on shaft performance	27% (12)	90% of these cases corresponded to drilled shafts with considerable lateral load but low axial loads, commonly the flaws were encountered at the shaft's base
G: remediation by grouting procedure	50% (22)	In 60% of these cases, it was necessary to perform for at least two grouting attempts
R: replacement of the defective shaft	16% (7)	80% of these cases corresponded to shafts with problems encountered during concrete pouring
D: demolition and reconstruction of shaft's head	7% (3)	100% of these cases corresponded to drilled shafts constructed using polymer and the flaw was encountered at the shaft's head

The FAT delays were separated into 4 categories: FAT delays of over 100% of the mean arrival time are the common detected flaws. A FAT delay>100% means a Vp of concrete less than 1800 m/s (5905 ft/s) in which the expected concrete resistance is lower than 2.5 MPa (360 psi). With this concrete resistance, it is probable that the structural resistance of the drilled shaft is more critical than the geotechnical resistance. The remediation of this type of case involves extensive analysis between different parties and it is the most common type of damage repaired through the construction of additional shafts. Flaws that present lower FAT delays (<75%) may be suitable to be repaired by means of grouting. For the data presented in this document, more than 50% of the detected flaws were repaired by one or two stages of grouting. It is recommended that, when the grouting process is used, a longer time should be allowed to observe important changes in the CSL test. It is recommended to wait at least 7 days to get better results. The remediation method must consider the execution time if the drilled shaft to be repaired is on the critical path of the project; the addition of more elements can bring a better cost-effective solution rather than a grouting process that takes longer.

All cases with head flaws corresponded to drilled shafts that required polymer and casing. When the casing is removed, the concrete level drops, decreasing the stabilization pressure inside the drilled shaft and increasing the possibility of soil or contamination inside the drilled shaft. This effect is shown in Fig 2, where a mixture between concrete and mud (soil with polymer) is observed in the vicinity of the reinforcement. To avoid this, more concrete must be poured after the casing is removed, so that the tremie pipe remains within the concrete column. In this way, a positive concrete head is created and allows these external materials to be flushed and expelled from the drilled shaft. Another solution for head flaws is to use a concrete vibrator in the upper zone.

Fig 2. Mud contamination detected in the drilled shaft during demolition.

EVALUATION OF DETECTED FLAWS

To evaluate the flaws found in a drilled shaft and determine if they are classified as defects, it is necessary to evaluate the effect of the flaw on the geotechnical and structural performances. The person responsible for this analysis must have a good understanding of the project and the loads to which the drilled shaft is going to be subjected (axial, lateral or both).

The analysis must include the scope of the test (in this case CSL) where information on the quality of the concrete outside the reinforcement cage will not be available. A CSL test may indicate a drilled shaft with good integrity, but the concrete may not have flowed to the walls of the excavation due to the high density of the reinforcement cage. Therefore, given the lack of information in this area, the project must require the best practices both in the design of reinforcement cages and in the pouring of tremie concrete. On this subject, it is recommended to consult EFFC/DFI (2018). The flaw evaluation philosophy dictates that it should not be based solely on an NDT test, but various quality aspects such as strength of concrete, cleaning of the excavation bottom, density of the reinforcement cage, use of spacers and a correct extraction maneuver for temporary casing should be also considered.

Assessment Through Coring

NDTs are used in project quality verification because they are an indirect but fast data source to obtain a quality result of the drilled shaft. The use of coring to verify the quality of the concrete is a direct measure of quality, however it is expensive and time-consuming, so it is generally used only to corroborate NDT results, as can be seen in Fig 3 a.

There is the possibility that an NDT will give a false positive result, like the one shown in Fig 3 b: two or three coring samples on the area with the supposed failure should be enough to determine the real quality of the pile. Because drilling is a direct measurement, if the coring result is redundant, it should be considered above the NDT result. Coring analysis must include a visual inspection of the concrete samples to be classified as good concrete and cylinders must come without soil or foreign materials, no segregation, recovery percentages close to 100%, and must be a solid sound element. If the concrete cylinders meet these conditions, simple compression tests are carried out on the samples. A very important aspect is the care of the samples during transportation to the laboratory to avoid alterations in resistance. It is also recommended to consider scale effects according to the diameter of the extracted cylinder (Arioz et al., 2007). If the test results indicate strengths of at least the value stipulated by design, then the integrity of the concrete in the drilled shaft must be classified as adequate. If the cylinders, although they appear to be intact visually, show their resistance is less than the designated value, the analysis must proceed to a structural analysis phase considering the resistance found. Coring can also be used to analyze the effectiveness of the grouting remediation process as seen in Fig 3 c.

a) b) c)

Fig 3. Coring observations a) Concrete flaw b) False positive c) Interaction between grout and concrete

Structural Analysis of Drilled Shaft Condition

The structural analysis consists in verifying the performance of the drilled shaft with the encountered resistance conditions. If the structural performance is not affected, nor is the geotechnical one, the integrity of the drilled shaft can be considered adequate.

There are some cases, such as cold joints, that, due to the length of the flaw, are not worth analyzing. Fig 4 shows the results of CSL tests on two drilled shafts with cold joint problems where the non-continuity of the concrete makes it impossible for the drilled shaft to transmit loads. In this case, the remediation should focus on the structural part of the foundation cap and, at least in the cases analyzed, it was always necessary to build other elements.

A drilled shaft interaction graph can be made using the stipulated reduction factors of the project (e.g., ACI or AASHTO). The obtained real concrete resistance in the drilled shaft will be placed in this graph. With these conditions, the shaft's resistance is compared against the axial forces and moments coming from the structural model corresponding to the detected flaw's area. If the demand is less than the strength, the drilled shaft has adequate integrity. Fig 5 exemplifies two cases of structural analysis; in the first case (Fig 5 a) a retaining wall drilled shaft has enough strength to withstand the demand (low axial load) and was classified as adequate. In the second case (Fig 5b), a drilled shaft where the computed strength is insufficient to withstand the demand (high axial load and moments), and for this case it was necessary to build additional elements. In every case, the deformations due to the reduction in the concrete resistance should be analyzed for the service load combination.

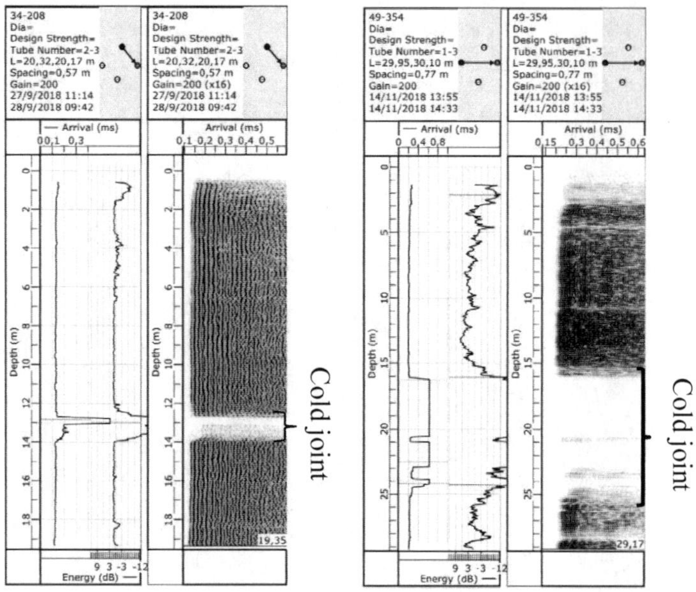

Fig 4. Examples CSL test of cold joints in drilled shafts.

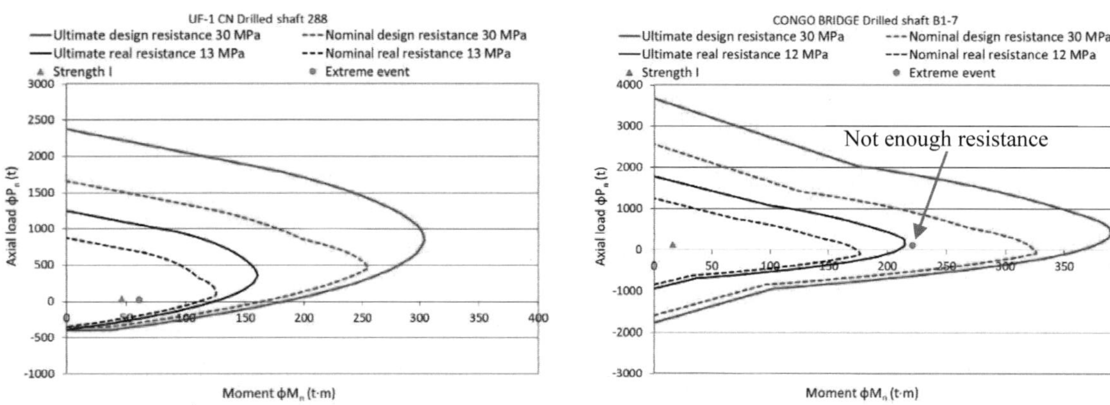

a) b)

Fig 5. Interaction graph for drilled shaft structural analysis a) Concrete resistance classified as adequate b) Concrete resistance classified as defect

REMEDIATION PROTOCOL

The flaws that have been defined as defects must be repaired to guarantee the expected performance of the drilled shaft. The remediation protocol must be previously defined before the beginning of the project and it is recommended to adapt it to the technologies available for the project. This section will detail the repairing procedure by concrete replacement for flaws in the drilled shaft head and the repairing procedure by injection of cement grout, which was the most commonly used method to repair tip and body flaws. For further remediation methods, it is recommended to review section 17.5 of GEC-10 (Brown et al., 2018).

Flaws at the Head of the Drilled Shaft

The defects in the head are characterized by being the upper areas of the drilled shaft where it is possible to excavate the ground to access the defective area. For the analyzed drilled shafts, it was possible to repair head defects up to 3.5 m (11.5 ft) deep.

The repair procedure begins with the definition of the depth of the defect. Subsequently, the excavation of the material around the drilled shaft must be carried out, once the exposed drilled shaft area begins the demolition process by using manual hydraulic or pneumatic concrete demolishers; this process is illustrated in the Fig 6 a. It is not recommended to use mechanical breakers from excavators, as their impact force can deform and damage the reinforcement. After the demolition process, it is necessary to carry out a visual analysis of the conditions of the reinforcement and, sometimes, it is necessary to replace segments of reinforcement (especially hoops or spirals). It is recommended to brush the reinforcement with a steel brush to maintain adhesion conditions. After this step, a casing is placed around the drilled shaft that guarantees the minimum concrete coverage of the reinforcement. As shown in Fig 6 b, a concrete tube with the same cross-sectional area of drilled shaft is often used and it is advisable to provide lateral brace support to the casing.

For the concrete pouring of the head, consideration must be given to the interaction between the old and new concrete; therefore, it is necessary to place an epoxy on the horizontal surface of the concrete that joins the old concrete with the new concrete. The pouring of the concrete can be done with tremie concrete or with conventional concrete using vibrators to guarantee the fluidity of the concrete in the pile. Within the analysis prior to this repair, it is recommended to not account for the geotechnical resistance of this repaired section.

Flaws at the Body or Bottom of the Drilled Shaft

Flaws in the body or bottom of the pile were the most frequent type in the analyzed piles. Most of them were repaired by injecting cement grout. In order to grout this material, it is necessary to borehole the drilled shaft up to the affected area. The first method is by drilling through the CSL access tubes; this is used for tip defects only. The second way is by coring inside the drilled shaft to the defected section; this method can be used for body or tip defects and can be difficult to execute for long piles, due to the difficulty of drilling long sections of the drilled shaft without colliding with the reinforcement. It is recommended to make at least 3 boreholes near the affected area and it is very important that the boreholes communicate, that is, that there is flow between them.

a) b)

Fig 6. Demolition and casing of concrete in the head of drilled shaft.

Before starting with the grouting, the drilled shaft must be thoroughly washed with clean water to verify that the water coming out is free of debris/soil. Once the boreholes have been made, the packers are lowered to guarantee the pressure in the grout. During the injection, the aim is to push any defective materials that may be in the drilled shaft from one borehole to another, promoting the expulsion of the material, so it's necessary to always leave at least on borehole free without the packer to guarantee the displacement of the soil outside the pile. The grouting should go through the perforations in one direction of flow. It is important to record the volumes and maximum pressures reached, which is an indication of the effectiveness of the treatment. For the analyzed cases, the maximum pressure was always close to 1.4 MPa (203 psi) and the injection volumes varied between 0.15 and 0.9 m^3 (5.3-31.8 ft^3).

CSL Test Comparison after Grouting

The effectiveness of a grouting treatment can be evaluated by using CSL tests. For a single drilled shaft depth, the conditions before and after grouting are analyzed. In this analysis it is important to focus only on the affected section and be clear about the initial condition because, due to the drilling process, the adherence between the tubes and the concrete can be affected, creating false positives. The concrete increases its strength over time, so there is no physical sense to find new defects in more recent periods.

Fig 7 a shows a grouting remediation which was adequate to almost completely eliminate the defects, to be adjusted within the quality criteria of the project. On the other hand, Fig 7 b shows a drilled shaft that does not exhibit a greater improvement in the quality of the concrete after three grouting processes. For this case, the injection process had to be abandoned and another drilled shaft was built.

CONCLUSIONS

In this document, the results of analysis and evaluation of 900 CSL tests on drilled shafts were presented. Of the 900 tests carried out, 4.9% indicated some flaw in the drilled shaft. The most common flaws were found at the tip of the drilled shaft, produced by issues of inadequate cleaning of the bottom. The defects in the middle of the drilled shaft were associated especially with problems during the pouring of the concrete. Problems at the drilled shaft head were the least frequent, but there was a relationship between these flaws and the use of polymer and/or sleeves. After the detection of the flaw, drilled shaft evaluation and remediation techniques were demonstrated.

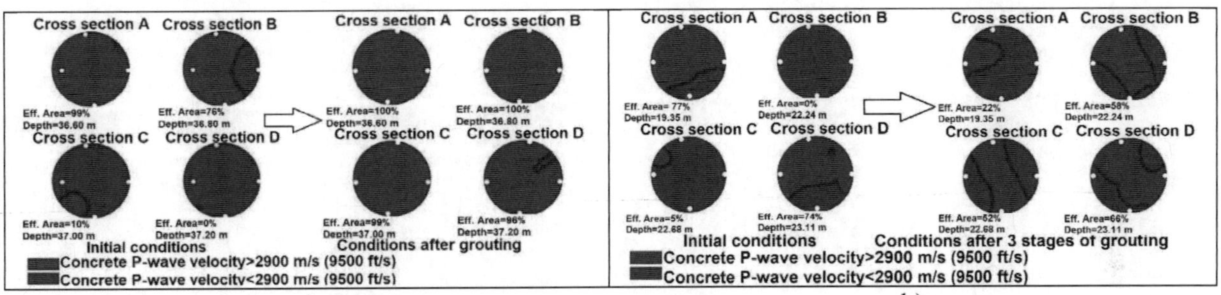

a) b)

Fig 7. Tomograph comparison of remediation grouting process.

In the evaluation of flaws, two tools were presented to confirm or discard the flaw. The first corresponded to coring, which is a direct measure of the quality of the concrete and can show false positives of the CSL test. The second tool is the structural analysis of the concrete conditions found in the drilled shaft. Two drilled shaft repair procedures were presented, the first used for head defects by demolition and concrete replacement. The second used for defects in the middle or bottom of the drilled shaft by grouting. Finally, tomographies of CSL tests were presented to demonstrate the effectiveness of the grouting treatment in the repair of drilled shafts.

REFERENCES

Arioz, O., Ramyar, K., Tuncan, M., Tuncan, A., & Cil, I. (2007). Some factors influencing effect of core diameter on measured concrete compressive strength. *ACI Materials Journal*, *104*(3), 291.

Brown, D.A., Turner, J.P., Castelli, R.J, and Loehr, E.J. (2018). *Drilled Shafts: Construction Procedures and Design Methods*, Geotechnical Engineering Circular No. 10 (FHWA GEC 10), Federal Highway Administration, Washington, D.C., 754p.

EFFC/DFI: *Guide to Tremie Concrete for Deep Foundations*. European Federation of Foundation Contractors (EFFC) and Deep Foundations Institute (DFI) (2018)

EFFC/DFI: *Guide to working platforms*. European Federation of Foundation Contractors (EFFC) and Deep Foundations Institute (DFI) (2019)

Jalinoos, F., Mekic, N., Grimm, R. E., & Hanna, K. (2005). *Defects in Drilled Shaft Foundations: Identification, Imaging, and Characterization* (No. FHWA-CFL/TD-05-003). United States. Federal Highway Administration. Central Federal Lands Highway Division.

Sellountou, A. E., Amir, J., Canivan, G., Chernauskas, L., Hertlein, B., Kandaris, P., ... & Likins, G. (2019). Terminology and Evaluation Criteria of Crosshole Sonic Logging (CSL) as applied to Deep Foundations. *Deep Foundations Institute—PIMS Committee.*

DYNAMIC LOAD TESTING OF NON-UNIFORM PILES

Thai Nguyen, Ph.D., P.E., Author, H2R Corp. Pinellas Park, Florida, USA. 727.270.1297.
tnguyen@h2rcorp.com

ABSTRACT

Dynamic Load Test (DLT) on driven piles has been performed routinely all over the world. In several recent projects, even 100% testing has become a norm where every single pile is dynamically load tested. In these cases, most piles are of uniform sections, making the field analysis (CASE method) straight forward. The CASE method, which has been used for the last 50 years, however, is only applicable to uniform piles to monitor pile driving stresses as well as pile capacities. Signal matching analysis can handle different configurations. However, the analysis used to require user inputs (i.e., non-automatic). Manual signal matching analysis, which can solve non-uniform solutions, traditionally takes hour(s) to complete for one blow. As such, the real-time pile driving stresses and resistances had traditionally been relying upon the CASE method, using typical range of CASE damping (J_C). Until 2021, due to the lack of a rigorous instant signal matching analysis for non-uniform piles, no tool was available to help engineers evaluate non-uniform piles in real-time. An inexperienced field engineer may overlook this shortcoming when evaluating peak driving stresses. Furthermore, ultimate bearing resistances for non-uniform piles may require further evaluation to account for the non-uniformity of piles if relying on the CASE method. This paper highlights recent advances in instant Signal Matching Analysis on non-uniform piles, allowing field engineers to monitor stresses and resistances of non-uniform piles during driving.

Keywords: dynamic load test, signal match, non-uniform, driving stresses, pile capacity, real-time

1 INTRODUCTION

Dynamic Load Test (DLT), also known as high strain dynamic test, is performed following ASTM D4945 method [1, 2, 3, 4]. Based on the DLT signal, pile capacity can be evaluated by the following methods:

1) CASE method [5] – only applicable to uniform piles. The CASE method is fast, which can be computed in a fraction of a millisecond and is traditionally used to monitor pile driving (i.e., instantly, in real-time). CASE method results are only for preliminary use.
2) Signal matching method, which is divided into two sub-groups:
   a. Manual signal matching method, such as CAPWAP™ [6] or N_GAPA™ [7]. This method can be used for any pile configuration (uniform or non-uniform). The traditional CAPWAP requires users to enter the pile configuration after data acquisition, while the N_GAPA (which is a part of GPC Suite) requires users to enter that prior to data acquisition. Per [5, section 10.6], "The nominal resistance determined by signal matching is a more accurate assessment of the nominal resistance than Case Method results" (1). Various authors [8,9,10] reported that the ratio of CAPWAP results and Static load test results are close to 1.0. The Florida Department of Transportation (FDOT) did a blind study (unpublished, with guideline published in [12]) and found that the ratio of CAPWAP and N_GAPA results are also close to 1.0.
      The processing time for this method can take up to an hour, and as such not suitable for real-time monitoring.
   b. Instant signal matching method, which can be used for real-time monitoring, such as:
      * iCAP [6, 11] is "currently only applicable to uniform piles, and while not as powerful and accurate as CAPWAP, iCAP is still better than the CASE method, and is expected to gradually replace the CASE Method", per [6](1), and "when applicable, the iCAP solution is more reliable than the Case Method capacity result", per [11].
      * or iN_GAPA [7], which uses the exact same model and algorithm as N_GAPA, therefore, can be used on any pile configuration. As such, N_GAPA and iN_GAPA are similar. iN_GAPA (also

---

[1] Thus, the CASE method is considered to be outdated.

a part of GPC Suite) was initially developed in 2010.

2 WAVE PROPAGATION AND WC DETERMINATION FOR UNIFORM PILES

The overall wavespeed (WC) is the rate that the force wave travels from the top to the toe of the pile. The pile top wavespeed (WS) is the local speed of the force wave that travels through sensor locations, which are typically mounted near the top of the pile. When the pile is homogeneous and hence uniform, WC is equal to WS.

When a force wave propagates from the hammer impact down the pile (also known as Wave-Down, which is a characteristic of the hammer system), there will be a force wave returning up the pile (also known as Wave-Up, which is a characteristic of the pile-soil system). Wave-Up, or WU, plotted as a red curve in Fig. 1, is computed as follow:

$$WU = (F - ZV) / 2 \quad (1)$$

where, F – Measured force (calculated from measured strain)

V – Measured pile velocity (integrated from measured acceleration)

Z – Impedance = Modulus (E) * Cross Sectional Area (A) / Wave Speed (WS). A constant Z means that the pile is uniform and homogeneous.

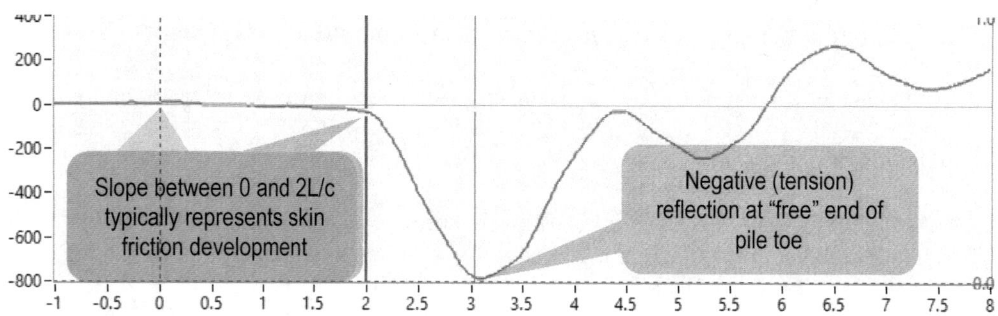

Fig. 1. Homogeneous Pile Responses using Two Sets of Sensors (Pile 6-2, Blow 63)

In a homogeneous pile, when the sensors are placed far enough below the pile top (to avoid local stresses imparted from the hammer to the pile top), then two sets of sensors (on two opposite sides of the pile) are needed. When the distance from the sensors to the pile top is smaller than 2 x pile diameter, such as the case of drilled shafts (bored piles), local stresses will affect the portion of the pile that receive the direct impact from the hammer. As such, four sets of sensors may allow for averaging out the local stress effect.

Soil skin friction reflection will cause F and ZV to separate from each other [5, Figure 10-11], i.e., WU = (F-ZV)/2 to increase, from start of impact (time 0 in Fig. 1) to time 2L/c (plotted as 2 on X-axis in Fig. 1). As such, the following WU scenarios may occur:

- going down (piles having minimal or no skin friction nor end bearing, driven by diesel hammer with high pre-ignition gas pressure). (F-ZV)/2 is negative in this case because of the pile toe responses due to the pre-ignition gas pressure.
- going flat (piles having minimal skin friction), Fig. 1,
- going up (piles having noticeable skin friction),
- going up, then very gradually (i.e., no sudden change from blow to blow) going down smoothly to 2L/c (very long piles, having large skin friction, with upper portion of the pile in unloading skin friction during dynamic testing event – evident by negative velocity record).

At time 2L/c, the pile impedance changes suddenly from full pile impedance to zero (end of pile), which causes WU to dip. As such, the overall wavespeed (WC) can be properly determined in the beginning of the pile driving by evaluating the distance from 0 to the WU dip. As the pile driving progresses, change in the location of the toe response can be used to evaluate the pile integrity (i.e., damages if it occurs).

3 WC DETERMINATION FOR NON-UNIFORM PILES

3.1 WC Determination for Composite Piles

For composite piles (such as concrete with an H-stinger piles), the WU dip will correspond to the 1st major change of impedance (i.e., change from concrete part to the stinger part). This WU dip can be used to evaluate the wave speed of the concrete portion of the pile (WS_1).
Then, the overall wave speed (WC) is calculated as follows [6, 7]:

$$WC = \frac{L}{\Sigma \frac{L_i}{WS_i}} \quad (2)$$

L = Total effective pile length
L_i = Length of each pile section in the composite pile
WS_i = Material wave speed of each pile section in the composite pile

3.2 Overall Wave Speed (WC) Determination for Heterogeneous Piles

In a heterogeneous pile, the force propagation is not proportional. On each side of the pile, the force is propagating downward along different directions depending on the orientations of the non-uniform pile material grains. Two examples of heterogeneous pile types are:

- Timber piles,
- Spiral welded pipe piles (which has uniform section, but heterogeneous property).

While many testers have successfully used two sets of sensors on these piles, with proportional results. In some cases, if the results turn out to be non-proportional, it may be beneficial to utilize four strain gauges to have a more representative average force, hence a better proportional result.

When the data is proportional, WC for heterogeneous piles will then be evaluated using the same principle as homogenous piles (Section 2 above).

4 PILE DRIVING MONITOR OF UNIFORM PILES

To monitor pile driving in real time, CASE method (and iCAP recently) has traditionally been utilized on uniform piles. As noted in Section 1.2.a above, the manual signal matching method, such as CAPWAP, uses segments of non-uniform piles while calculating pile capacity and stresses. However, due to slow processing time, CAPWAP has not been utilized to monitor pile driving in real time. In the CASE method, the pile driving stresses and static resistances are evaluated as follows [1, 5]:

$$CSX = FMX / A \quad (3)$$

CSX = Pile top average maximum compression stress
FMX = Maximum average force measured by the strain gauge ($F = AE\varepsilon$)
A = Pile cross sectional area
E = Pile modulus
ε = measured strain

$$F_B = RX0 - c\ SFT \quad (4)$$

F_B = Estimated pile toe maximum compression force

RXO = Estimated total resistance (static + damping)
c SFT = Estimated adjusted total side resistance (static + damping)
c = Adjustment factor

$$CSB = F_B / A \tag{5}$$

CSB = Computed pile toe maximum compression stress

$$TSX = F_{TS} / A \tag{6}$$

TSX = Pile minimum tension stress (tension stress is negative, thus a minimum negative value is the largest absolute tension stress)
F_{TS} = net tension force wave

$$RMX = (1 - J_C)\, W_{D1} + (1 + J_C)\, W_{U2} \tag{7}$$

RMX = CASE method capacity
W_{D1} = Wave-down force at time t_1
W_{U2} = Wave-up force at time $t_2 = t_1 + 2L/W_C$
J_C = Lumped CASE damping. It is called "lumped" and is dimensionless, because it is a single value for the whole pile. This is not to be confused with soil segment Smith damping value, which has unit of s/ft. Soil segment Smith damping can vary from other segments along the pile, and is utilized in the Signal Matching method. It is also noted that CASE damping J_C is not only a function of soil type, but also pile configurations and hammer energy. For example, at a same job site, same soil conditions, uniform concrete piles (18-in to 24-in size) usually yield J_C of 0.4 to 0.6, while uniform steel H or pipe piles (also 18-in to 24-in size) usually yield J_C of 0.8 to 1.0.

5 PILE DRIVING MONITOR OF NON-UNIFORM PILES

There are many examples of non-uniform piles: Timber piles, stinger piles, voided (hollow) concrete piles with solid ends, or pipe piles with a constrictor plate in the middle of the pile (constrictor-plate pipe piles). For the constrictor-plate pipe piles, each pile will have two toes: one toe at the end of the pile, and the other toe at the mid-plate. This type of piles is used when the dense soils are too shallow, so that pile penetration is longer (using the open section) to provide the required lateral stability, while achieving a significant end bearing at the upper toe level.

For non-uniform piles, the CASE method may be unrepresentative as the method assumes a uniform pile. This shortcoming can lead the field engineers to incorrectly monitor the wrong values for driving stresses as well as the ultimate bearing resistances for non-uniform piles. To overcome these difficulties, improvement to the signal matching analysis has been made to analyze the wave propagation in real time. The signal matching process involves a vector valued function *f*, which is a simulation of the pile driving process based on one-dimensional wave mechanics:

$$\underline{f}(\underline{x}, \underline{V}_m, \underline{F}_m) = \begin{Bmatrix} \underline{V}_s \\ \underline{F}_s \end{Bmatrix} \tag{8}$$

Where *f* yields simulated velocity and force time histories (V_s and F_s), and requires a vector of pile-soil parameters *x*, where the dimension of the *x* is (*p* x 1), and the measured velocity and force histories (V_m and F_m). The pile-soil material vector *x* contains parameters such as soil resistance R_i, soil quake q_i, soil damping j_i, etc.

The measured data is interpolated and resampled at the simulation intervals such that simulated data and measured data are synchronized. Writing equation (8) in residual form we have

$$\underline{r}(\underline{x}) = \underline{f}(\underline{x}) - \begin{Bmatrix} \underline{V}_m \\ \\ \underline{F}_m \end{Bmatrix} \tag{9}$$

So, we seek a vector x which will yield a local minimum (i.e., lowest Match Quality MQ number in signal matching program). A solution will be obtained by considering the expansion of r about the regular value $x + \delta x$

$$\underline{r}(\underline{x}+\delta\underline{x}) = \underline{r}(\underline{x}) + \frac{\partial \underline{r}}{\partial \underline{x}}\delta\underline{x} + \ldots \tag{10}$$

Assuming the left-hand side yields a null vector in the residual form above, then truncating the right-hand side so as to include only the linear term, we have:

$$J\,\delta\underline{x} = -\underline{r}(\underline{x}) \tag{11}$$

Where J is the Jacobian of rectangular matrix:

$$J = \left[\frac{\partial r}{\partial x_1}\ \frac{\partial r}{\partial x_2}\ \cdots \frac{\partial r}{\partial x_p}\right] \tag{12}$$

$$\frac{\partial r}{\partial x_i} = \frac{r(x+\Delta x_i) - r(x_i)}{\Delta x_i} \tag{13}$$

Solving equation (8) for vector x is made possible by the minimum-norm solution to a linear least squares problem in Linear Algebra. With advance in computer power, it takes a fraction of a millisecond to solve equation (8) in matrix form. Therefore, instant signal matching analysis involving between 100 and 500 iterations will consume approximately 0.1 to 0.5 seconds of computing time using a regular PC or laptop, which is less than the time interval between the field hammer blows during the pile driving process. The signal matching process has been coded into the instant Nguyen Goble Automatic Pile Analysis (iN_GAPA) program ([2] and [7]), which is part of the Goble Pile Check (GPC) suite. In this process, the lumped CASE damping (J_C) is irrelevant. However, it is often printed out as a preliminary result for comparison. Once the pile-soil parameters vector x is established through multiple iterations to search for the local minimum of the residual vector r(x), then pile stresses (compression and tension) are tabulated based on the 1-D wave propagation through the non-uniform pile.

6 CASE STUDY OF TIMBER PILE

Timber pile is non-uniform by nature: the pile is tapering down from top to bottom. Additionally, due to different orientations of timber grains, data from 2 strain gauges was not sufficient in our case study. As such, 8 channel data (i.e., 4 strain gauges) was utilized for acceptable data processing[2]. In this case study, the pile top diameter is 11 inches, and pile butt diameter is 10 inches. The instant signal matching program is successfully implemented in analyzing the pile capacity in real-time. Fig. 2 presents plot of pile capacities as the pile is being driven. The signal matching analysis (iN_GAPA) result indicates that if the CASE method were to be used, then a J_C of approximately 1.4 would need to be utilized. This J_C value is outside of typical range of 0.4 to 0.9. PDA data was not collected on this pile. As such, iCAP results are not possible. However, if iCAP were to be performed (on PDA data) using a uniform pile model (i.e., same pile diameter from top to toe of pile), it is believed that iCAP would produce similar capacity results because the pile diameter (hence pile impedance) variation (i.e., taper) is small for timber piles.

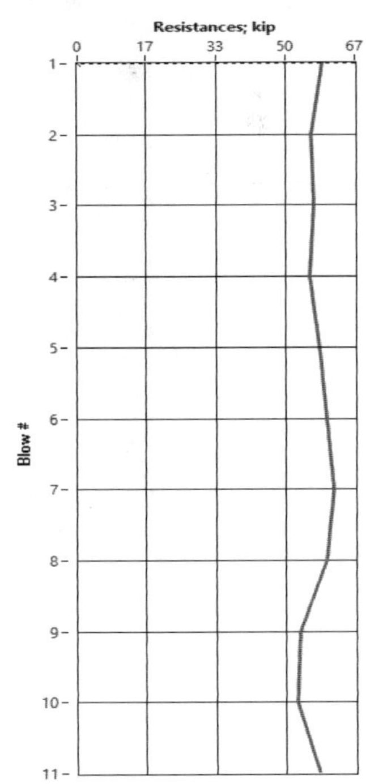

Fig. 2. GPC Timber Pile Monitoring, Pile 47

---

[2] It is noted that many others have successfully test timber piles using 2 strain gauges.

7 CASE STUDY OF VOIDED CONCRETE PILE

Fig. 3 presents the highly non-uniform 155-ft pile: The pile consists of eleven individual sections, starting from a solid 30x30 in^2 concrete section, tapering down to an 18-in void within the 30x30 pile, then tapering up to the other solid end. The purpose of the voided (hollow) sections are to reduce the total weight of concrete pile, so that it is manageable for transport truck and crane to handle the pile prior to and during pile driving. Despite of the void, each pile section still weights 27.5 tons. Due to soft soil condition, the pile length is excessive, such that it was spliced by a steel mechanical splice (KieLock), with 2 sections weighting a total 55 tons.

Using the traditional CASE method, the wave propagation is always assumed to be through uniform section with the exact same properties at the sensor location. As such, the CASE pile integrity is unknown (either by BETA method or by user's visual inspection of the WU) due to the Wave Up reflection encountered while the forces are travelling through the void within the pile (red WU curve, Fig. 4).

With the ever-increasing power of the computer speed and mathematical model, instant Signal Matching Analysis on non-uniform piles is made possible. Based on signal matching simulation (blue WU0 curve, Fig. 4), the pile integrity BTA factor is estimated to be 95% with the reflection being at the mechanical splice.

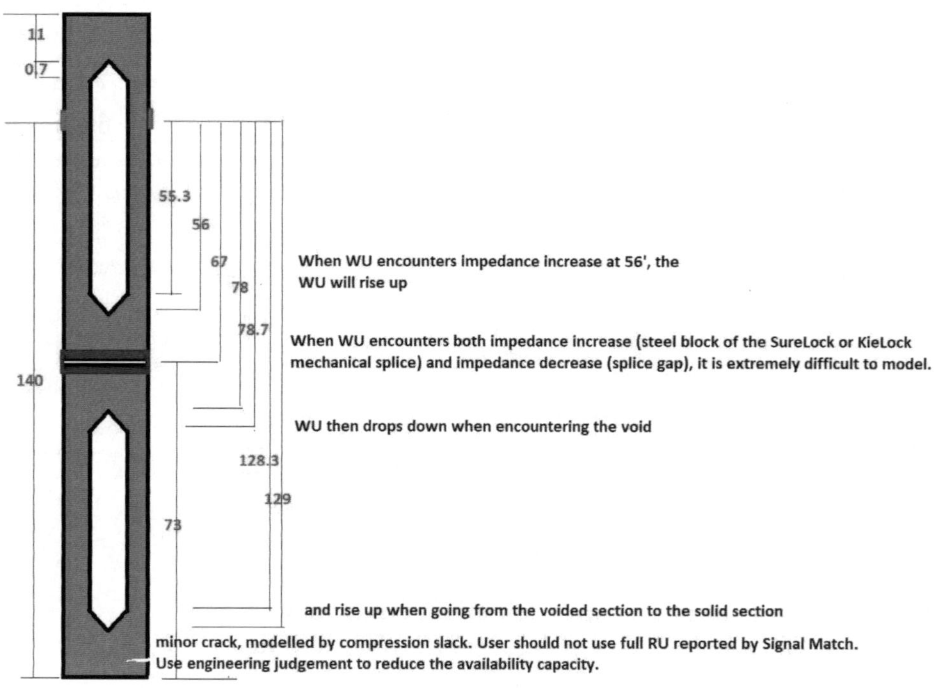

Fig. 3. Hollow Spliced 30-in Square Concrete Pile

Notes: *1) All distance values are in feet.*
2) Sensors (as red rectangles) were attached at a distance of 15-ft from pile top.

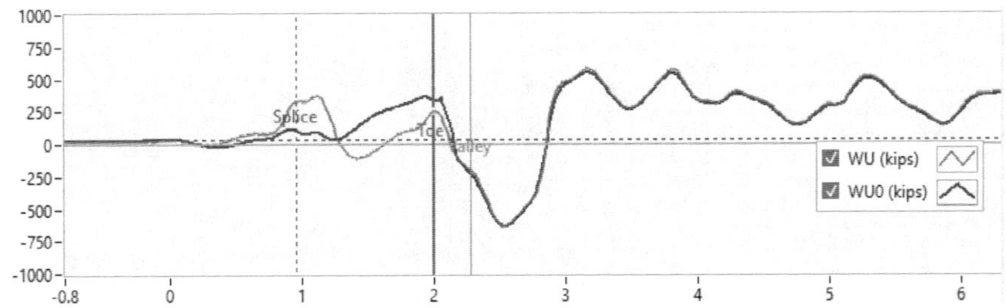

Fig. 4. WU responses – GPC 4 - Blow 1936

CSX	CSB	TSX
3.67 ksi	2.56 ksi	0.73 ksi

Fig. 5. CASE method results – GPC 4 - Blow 1936

Notes: CSX = Measured compression stress at sensor location
CSB = Computed compression stress at pile toe
TSX = Computed maximum tension stress anywhere along the pile depth
It should be noted that the CASE method does not estimate maximum compression stress along the pile depth.

Real-time monitoring of stresses and pile resistances can be reasonably evaluated via instant signal matching analysis for every single blow. Signal Matching Analysis results (Fig. 6) indicate that the pile actually experienced higher pile driving stresses than those of CASE method (Fig. 5):

- CSX is the same as 3.67 ksi (CS NGAPA cyan curve at 0-ft depth Fig. 6.b)
- CS = 4.01 ksi (Fig. 6.b), the maximum compression stress occurs at depth 55-ft, which is right at the end of the 1st hollow portion of the pile (Fig. 3).
- TS = 1.40 ksi, which is about double the CASE method value. Fig. 6.b indicates that the maximum tension stress occurs at depth 79-ft, which is the start of the 2nd hollow portion of the pile (Fig. 3).

Therefore, utilizing the instant signal matching analysis procedure may be useful to identify pile stresses in non-uniform piles to avoid over-stressing the pile. Real-time monitoring of signal matching pile resistance is also available instantly as shown in Fig. 6.c.

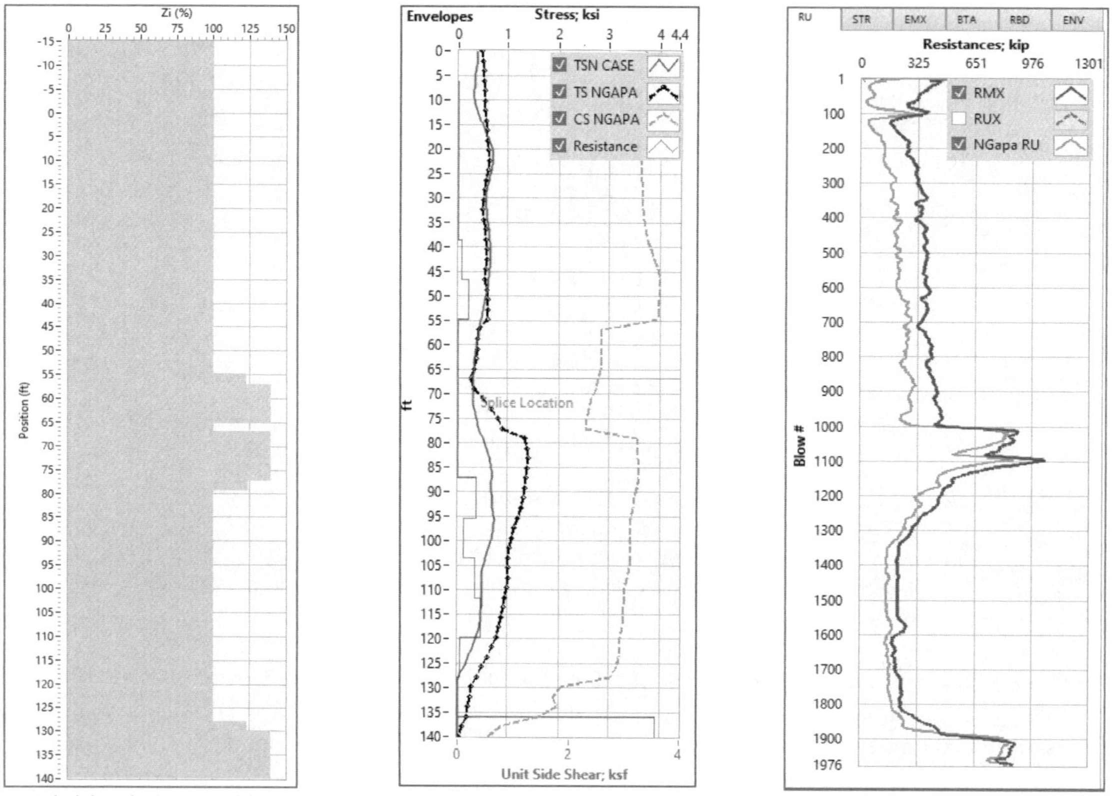

a. Digitized Non-Uniform Pile Model

b. Stress Envelope for Blow 1936

c. Real-time Monitoring of Pile Resistances - All 1976 Blows

Fig. 6. Non-Uniform Pile in iN_GAPA Signal Matching Analysis

Notes: Figure b:

TSN CASE: Tension stress envelope result from CASE method.

TS NGAPA: Tension stress envelope result from signal matching method.

CS NGAPA: Compression stress envelope result from signal matching method.

Resistance: Unit Side Shear (i.e., skin friction) envelope result from signal matching method.

Figure c:

RMX: CASE method capacity results using Lumped CASE damping $J_C = 1.4$*.*

NGapa RU: Signal matching capacity results.

A gap between RMX and NGapa RU indicates that lumped CASE damping J_C *is not a constant. In this example, the upper soil layers may have* $J_C = 2.0$ *to 3.0, while the lower soil layers may have* $J_C = 1.4$*.*

All existing programs plot the CASE results using a single lumped CASE J_C *value.*

8 CONCLUSION

The traditional CASE method is based on 1-D wave propagation, as such it is suitable to evaluate pile resistances as well as pile stresses for each hammer blow for uniform piles. However, it becomes inadequate for non-uniform piles, as the piles may have multiple sections, with different Young's modulus values or pile cross sectional areas, which means different impedance values need to be considered. The piles may also have more than one toe end bearing locations. Because of the complexity of the mathematical wave equation model, signal matching used to take up to an hour to be analyzed. However, with advancement of technology and computer power, instant Signal Matching Analysis is becoming available to properly model the various non-uniform pile properties. Matrix equation is solved via minimum-norm solution to a least squares problem in Linear Algebra. The model has been successfully coded into GPC program to allow real-time monitoring of pile driving stresses and resistances of any complex non-uniform pile configuration.

9 REFERENCES

[1] Pile Dynamics, Inc. (PDI), 2009, "*PDA-W Manual of Operation*", Ohio, USA.

[2] H2R Corp (H2R), 2022, "*GPC Review User Manual*", Florida, USA.

[3] Allnamics, 2016, "*Manual PDR Version 1.1E 2016-06-16",* Amsterdam, Nederland.

[4] Smart-Structures, 2012, "*SmartPile Review User Guide Version 4*", Pennsylvania, USA.

[5] P. J. Hanigan, PE, Frank Rausche, PhD, PE, Gerald E. Likins, PE, Bret R. Robinson, PE, and Matthew L. Becker, EI., 2016, "*Design and Construction of Driven Pile Foundations*", FHWA-NHI-16-009, FHWA GEC 012-Volume II, Washington DC, USA.

[6] Pile Dynamics, Inc. (PDI), 2014, "*CAPWAP Background Report*", Ohio, USA.

[7] H2R Corp (H2R), 2022, "*N_GAPA User Manual*", Florida, USA.

[8] Hussein, M.H., Rausche, F., 1991, "*Bearing Capacity of Deep Foundations from Dynamic Measurements and Static Tests - Ten Correlation Cases*", in *Piletalk International 1991*: K. Lumpur, Malaysia. Pp. 49-53.

[9] Goble, G.G., Rausche, F., and Likins, G., 1980, "*The analysis of pile driving – A state-of-the-art*", in *Proceeding of the 1st International Conference on Application of Stresswave Theory to Piles*. Balkema, Stockholm, Sweden.

[10] Likins, G.E., Rausche, F., 2004, "*Correlation of CAPWAP with Static Load Tests*", in Proceedings of the Seventh International Conference on the Application of Stresswave Theory to Piles 2004: Petaling Jaya, Selangor, Malaysia. Pp. 153-165.

[11] Likins, G. E., Liang, L., Hyatt, T., 2012. "*Development of Automatic Signal Matching Procedure – iCAP*" Proceedings from Testing and Design Methods for Deep Foundations; IS-Kanazawa: Kanazawa, Japan

[12] Florida Department of Transportation, 2022, "*SDB 22-01 Goble Pile Check (GPC) Dynamic Load Testing*", Tallahassee, Florida, USA. https://fdotwww.blob.core.windows.net/sitefinity/docs/default-source/structures/bulletins/2022/sdb-22-01.pdf?sfvrsn=85c0b9d4_2

DATA DRIVES HERBERT HOOVER DIKE (HHD) REHABILITATION EFFORTS

Yoav Rappaport, BAUER Foundations Corp., Odessa, Florida, USA, 813-748-5470, Yoav.rappaport@Bauerfoundations.com
Jignyash Manyam, BAUER Foundations Corp., Odessa, Florida, USA, 669-241-8367, Jignyash.Manyam@Bauerfoundations.com
Michael Arnold, Ph.D., BAUER Foundations Corp., Odessa, Florida, USA, 813-792-3368, Michael.Arnold@Bauerfoundations.com

ABSTRACT

The USACE required construction of over 13 miles of cutoff wall at an average depth of 65 feet for the Herbert Hoover Dike (HHD). To meet project requirements, to provide quality control and make sound operational decisions, automated data capture, retention and visualization is crucial. Accurate and complete data management allows the authors to deliver critical business insights and to improve operational efficiencies. The author's understood the necessity of capturing all their data, and efficiently processing and displaying it for multiple stake holders, for both internal and external purposes.

The authors realized early in the project life cycle that data ownership and management is one of the fundamental features of the construction process. The authors developed their own software and use two in house applications; WEB-BGM online portal and, b-project which store rig production information combined with field quality control. b-project is used to create and deliver the Enterprise Database (EDB) to the USACE. Another biproduct of capturing the production data is the generation of the Geographic Information Systems (GIS) and also business analytics visual interactive display, developed through Microsoft Power BI.

The paper focuses on the business process workflow and various technologies employed and lessons learned from the field operator perspective to achieve higher returns on business software operations.

Keywords: Cutoff Wall, Digitalization, Data Management, Geographic Information System

PROJECT

One major component of the Herbert Hoover Dike (HHD) Rehabilitation project is the installation of a seepage cutoff wall (Fig. 1). The seepage cutoff wall is installed at the southeast, southwest, and partially on the northwest side of Lake Okeechobee, located in south central Florida. As reported elsewhere by Bruce and Sills (2009), Arnold et al. (2011) and Arnold (2015), The contractor was tasked to installed 10 miles (16 km) of cutoff wall at the southeastern shores of the Lake from 2008 to 2013. In 2018, the U.S. Army Corps of Engineers (USACE) issued the Herbert Hoover Dike Rehabilitation Dam Modification Cutoff Wall MATOC contract, a multiple award task order contract (MATOC), in order to finish the remaining 29 miles (47 km) of planned cutoff wall at the west side of the Lake. Over the years 2018 to 2022 the contractor was awarded Task Orders 1, 3 and 5 (TO1, TO3 and TO5) totaling 17.6 miles (28.3 km) of cutoff wall. These three contracts include ten connections to existing concrete structures via jet grouting and six tie-ins to existing cutoff walls at previously installed culverts.

A Project Management Information System (PMIS) was developed to manage and analyze the huge amount of data created by up to ten simultaneously working drill rigs (Fig.2) and the related quality control (QC) activities for more than 45,000 elements. The contractor used the in-house developed b-project data management system. This is a front-end data entry system for the core of the PMIS. It hosts both design and as-built data, has various interfaces for data input/import and output/export and provides comprehensive data analytics via dashboards for the QC team as well as for site and corporate management. Production documents are shared with USACE via a Secure File Transfer Protocol (sFTP) site; using b-project to automatically file these documents on the sFTP, and maintaining a live link between documents and production records.

MAIN DATA MANAGEMENT FOCUS AREAS

Data Collection

Once the design criteria are established and project preconstruction phase is complete, the production phase begins. This phase includes the collection of data from instrumentation on the drill rigs (via the onboard B-Tronic telemetry system), data entered by hand on paper or tablets such as QC observations during drilling and excavations, borehole logging data, etc, and field and laboratory quality control data. Additionally, documents (scanned field notes, photographs and video), design drawings, planned construction elements drawings, and field activity tracking forms are also collected. These are all part of a comprehensive digitalization effort of field activities, which must be properly entered, stored and consistently named. The project included collection of over 45,000 elements, hundreds of thousands of rows of records, which necessitated strict panel/borehole/pile naming convention.

Data Preparation

Data collected directly from the B-Tronic telemetry system is passed to the data analysis stage where data engineers evaluate the raw data from the rig and diligently check for any errors. The purpose of this step is to eliminate the erroneous data (incomplete or incorrect data) and convert the raw encrypted data to structured data. The structured data is then converted into an MS Excel format which provides all the information about depth of the panel, deviation of X and Y, suspension flow, suspension amount and production start and production stop times of every panel.

Data Input

Design data such as panel name, horizontal coordinates, station numbers, type of the element (primary or secondary), technique type, and the platform survey elevation are the primary attributes of the constructed features entered into the PMIS. The clean raw data is then transferred to b-project (Fig. 3), supplemented with additional tables of data generated during exploration and construction. These data include but are not limited to additional as-built information, QC data of binder slurry and soil-cement-bentonite material as well as delivered and consumed materials.

Data Processing

The construction data is processed using algorithms which compare the actual as-built data and the design data. Any discrepancy is brought to the data engineer's attention for further evaluation and/or correction. A set of exceedance rules are preconfigured, and table cells are color coded for exceedance of the tolerance. For instance, if there is a mismatch between the element ID when comparing the as-built as entered by the operator in the field and design data, the operator input can be updated and overwritten with a manual entry to match the design ID.

Data Output/Interpretation

The output/interpretation stage is the stage at which data is finally usable to all the QC Engineers and Managers. A production report is generated which displays all the details related to as-built information of the element such as total drilled depth, deviation of X and Y graphs and much more. The individual element production report (Fig. 4) can be customized based on client's requirements. The b-project application also executes geometric routines to determine degree of verticality and panel overlap to satisfy specifications. Any deviation from the specific thresholds is color coded to alert the data engineer, so that they can investigate the root cause. In some instances, this results in a re-cutting of the panel, while in others this maybe a data entry error or sensor error.

Geographic Information System (GIS) Display

Production data is stored in the HHD GIS geodatabase and is also published as a representational state transfer (REST) endpoint feature services for web mapping purposes. Scripting algorithms have been created by the data management team to automatically convert record from b-project to GIS through an Application Programming Interface (API) and Python language scripting. The GIS is presented as plan view (Fig. 5) and 3D view (Fig. 6). The plan view layer for cutoff wall consists of hyperlinks to photographs, QA/QC data, and laboratory QC results such as unconfined compressive strength (UCS) results, dry densities, and moisture contents. The plan view also includes verification holes and exploratory drill holes locations. The overlap between cutoff wall panels is calculated at 10-ft selectable intervals plus element top, middle, and bottom depths. Using a color scheme for overlaps that do not meet design specifications, the viewer can quickly spatially observe these. The 3D view contains similar information to the plan view but here the individual panels and piles can be seen at depth and overlap, and can be investigated using an interactive slicer tool.

Web-based access to aggregated measurement and production data

Another component of the integrated approach to data management is the web-based, location-independent presentation of evaluations and key performance indicators (KPIs) (Fig. 7). This functionality enables timely access to aggregated performance and quality data even outside the construction office for internal and external project participants. The standardized presentation of the data enables a quick understanding of the current level of performance of projects (Fig. 8). These standardized evaluations form the basis for further and more in-depth evaluations and process optimizations. Using the data entered in b-project the authors perform data analysis and create various visualization dashboards for project management. This provides a summarized view of project status, production productivity and aggregate statistics.

A use case is the presentation of the production data collected on the construction site using these near real-time dashboards which the project team can discover schedule saving opportunities, as well as easily display outliers and trends of the data stored in the EDB (Fig. 9).

Web-based access to activity data

Production reporting through rig telemetry, laboratory, and QC testing information is critical for assessing the project construction quality and consistency of wall material. The collection and display of these parameters provide the client and project team confidence that the cutoff wall has been constructed to specifications. However, this activity accounts for only a portion of the time spent at the site. Activities occurring outside of the instantaneous (ie drilling and wall placement) production hours are tracked by the project team for the purposes of increasing production efficiency and document production delays due to scheduled or unscheduled events (Fig. 10). Events such as harsh weather, equipment breakdown, or material supply issues are documented in the contractor's activity interface. The data entry interface is viewed through a tablet inside the rig and web browser in the back office. This detailed information, once verified by the project quality control manager, is then available through a web dashboard to credentialled team members with aggregate visualizations displayed using charts and user selected filters (Fig. 11).

Data Storage

At the HHD project, all data files are additionally stored in the cloud. The cloud site stores raw data, structured data in excel format, production reports, and also includes drawings and the GIS geodatabase. USACE can access the data files easily through the GIS system. Alternatively, the client has read access to folders and documents in the cloud via a web interface and as specified at an sFTP interface via FTP client software, such as WinSCP.

Disaster Recovery

In case of data loss, the recovery will be performed as a full recovery of the complete system. If necessary, incremental backups can be recovered additionally which guarantees a maximum data lack of 2 hours. This procedure will be executed by the contractor central IT team. Any data added to any of the systems that cannot be restored in this way (e.g., added after the last backup) will be recreated from the original records.

LESSONS LEARNED

During this project we identified that timely communication was one of the main challenges between field QC, subcontractor's and the head office. We resolved this by setting up Scrum meetings every day for 15 minutes to discuss all the issues which occurred the previous day.

Because some of the analytical dashboards were developed by IT, far removed from the HHD project, we found it helpful to review the visualizations with local project leadership and customize to the reports to their specific needs and focus areas. For example, the standard reports lumped all rigs into one productivity report, but the project needed finer resolution displaying production statistics by rig, by shift. Creating easy to use mobile application takes an effort and time observing the field operations greatly assists the developer.

Over the course of the project additional data delivery application developments were necessary to accomplish timely delivery based on USACE specifications. The developments included a mobile form data entry system for batch plant wet slurry testing results and at rig pre-placement wet slurry testing results. As a result of the success of this digital form, the team plans to expand the use of mobile digital forms for other data capture needs in the field.

Over the course of the project adjustments to work distribution were necessary. At the start of the project, the data field engineers were organized by Task Order. However, as the project productivities varied, and the type of work performed varied with time, it was necessary to work more collaboratively and redistribute the work between multiple editors. Weekly work redistribution meetings were held to accomplish this.

CONCLUSIONS

Cutter Soil Mixing (CSM), which had been utilized at HHD before, was used to install another 17.6 miles (28.3 km) of cutoff wall distributed over three task orders. The production of over 45,000 elements required a robust and dependable PMIS. While production reporting, data storage, delivery and display was required as part of the USACE specifications, the contractor also developed an internal QC and production monitoring system to assist project management and leadership to monitor productivity, efficiency, identify quality outliers and track "unproductive" activities. These digitalization tools allowed the team to respond quickly to data entry errors, and to discover methods for increasing team efficiencies and data recording automation.

REFERENCES

Arnold, M., Beckhaus, K., and Wiedenmann, U. 2011. Cutoff wall construction using Cutter Soil Mixing: a case study. Geotechnik 34, pp. 11-21.

Arnold, M. 2015. Seepage cutoff wall installation using Cutter Soil Mixing for Herbert Hoover Dike Rehabilitation. In Proceedings of the Deep Mixing 2015 Conference, June 02-04, 2015, San Francisco, CA, USA, pp 213-222.

Bruce, D., and Sills, G. 2009. Technology Review: Seepage Cutoffs for Levees. United States Society on Dams (USSD), Nashville, TN, April 2009.

US Army Corps of Engineers (USACE) 2018. W912EP-18-R-0019 Herbert Hoover Dike Rehabilitation Dam Modification Cutoff Wall MATOC: Construction Solicitation and Specifications.

FIGURES

Fig. 1. b-project Information Management System plan view showing locations of HHD Task Orders 1, 3 and 5 at the southwestern and northwestern shores of Lake Okeechobee, Florida.

Fig. 2. Kelly predrilling operation including CSM cutoff wall installation (front) and backfilling (back) at HHD TO3 east.

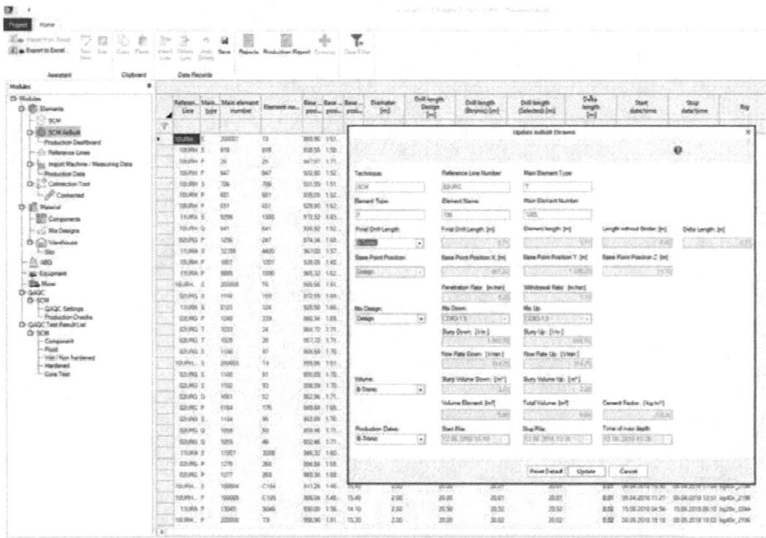

Fig. 3. b-project data entry and data processing interface. Various techniques are available (SCM, SCM, pile and Jet Grouting).

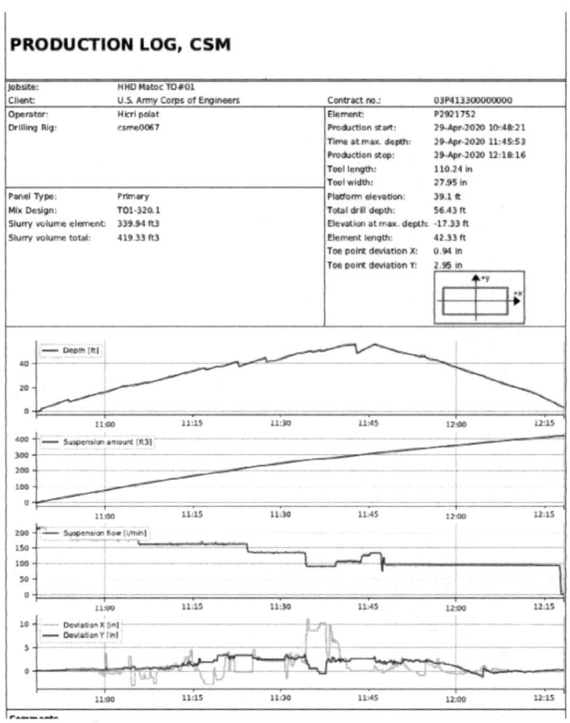

Fig. 4. Production report generated from the b-report interface. Production reports are customized to meet client specifications through Python scripts.

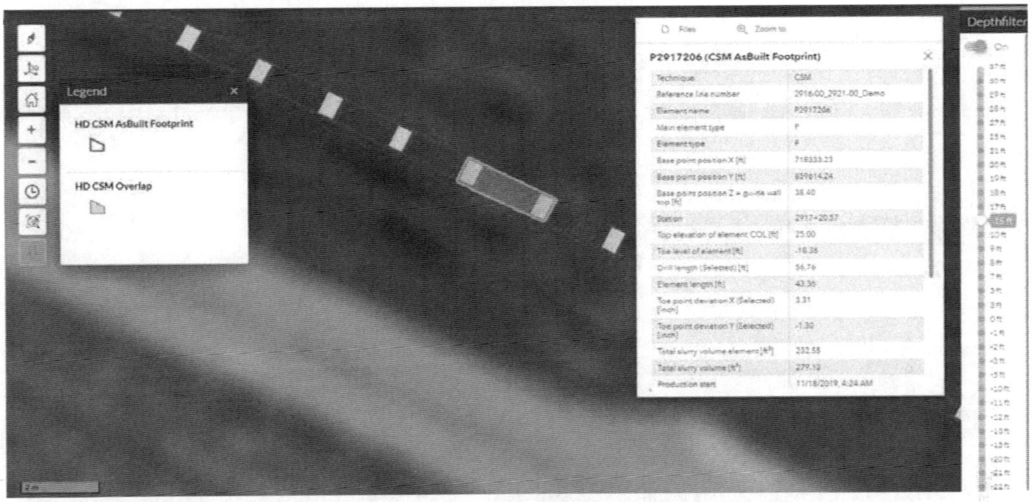

Fig. 5. Geographic Information System web interface Plan View. Individual panels details, hyperlinks and overlaps can be seen against a basemap such as aerial or roads layer. Overlap detail can be selected at various depths using the "depthfinder" tool to the right.

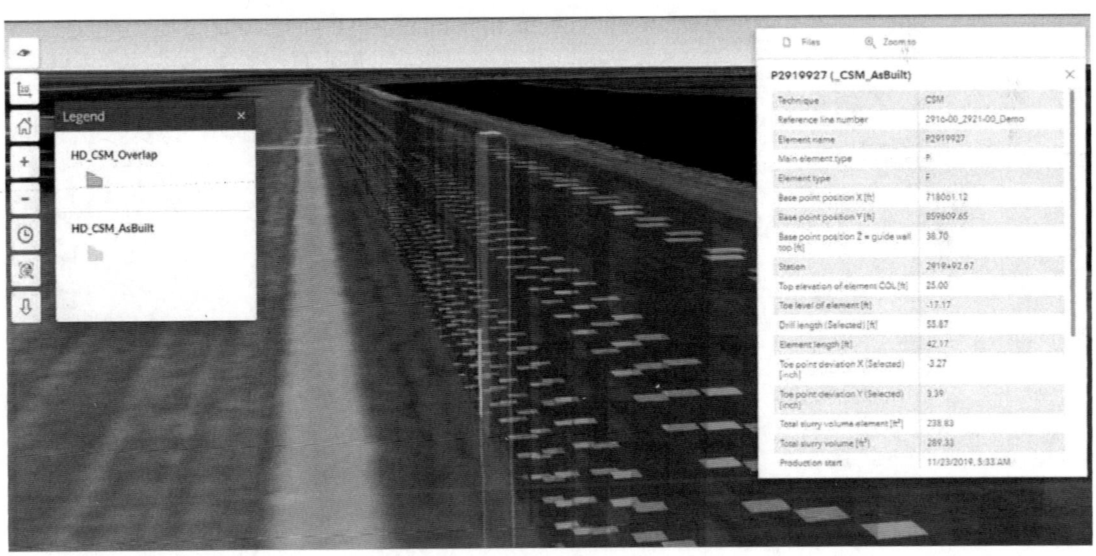

Fig. 6. Geographic Information System web interface 3D view. Individual panels details, hyperlinks and overlaps can be seen. The 3D view can be rotated. In this example, the panels are shown above ground for purpose of above ground feature context, however these will be displayed below ground in future release of the application.

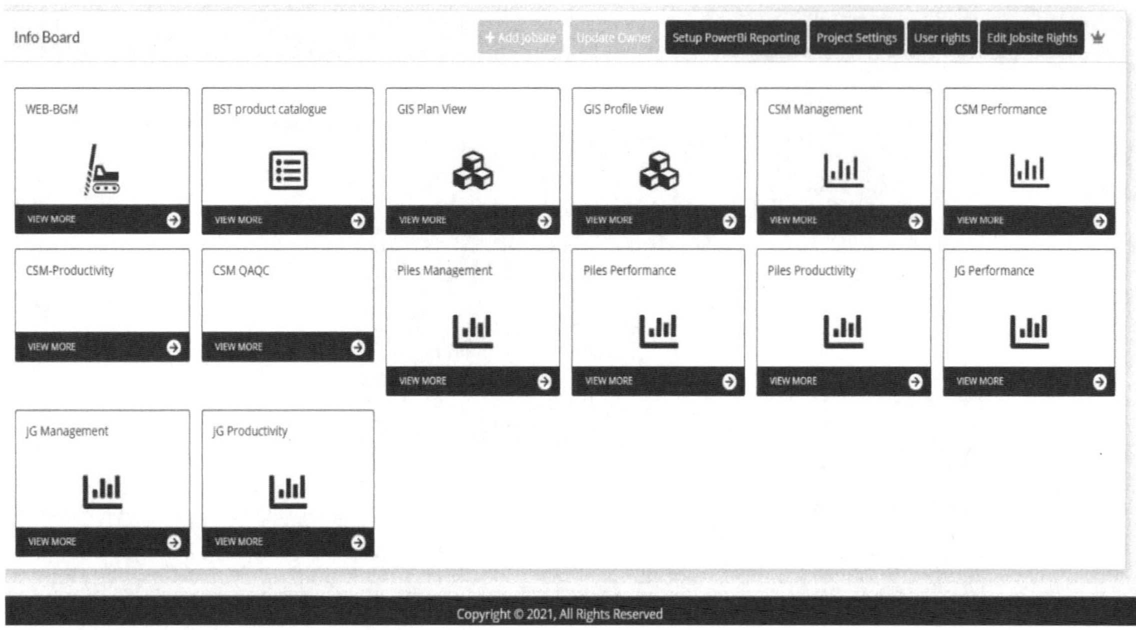

Fig. 7. Web-based dashboard interface. The project has access to several dash board modules for each drilling technique and application through preconfigured tiles.

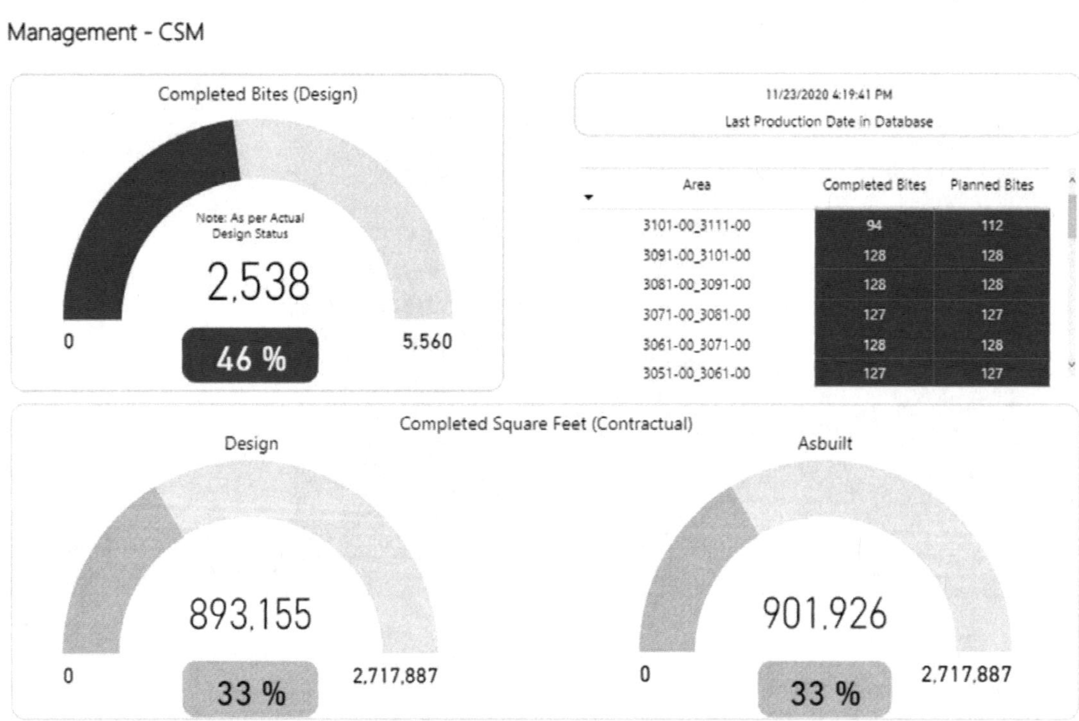

Fig. 8. A standard management dashboard module enables a quick understanding of the current level of performance of projects.

Fig. 9. Use of the project dashboard for the purpose of QC and discovery of outliers. This allows the QC staff to react quickly to either real production issues or data entry errors.

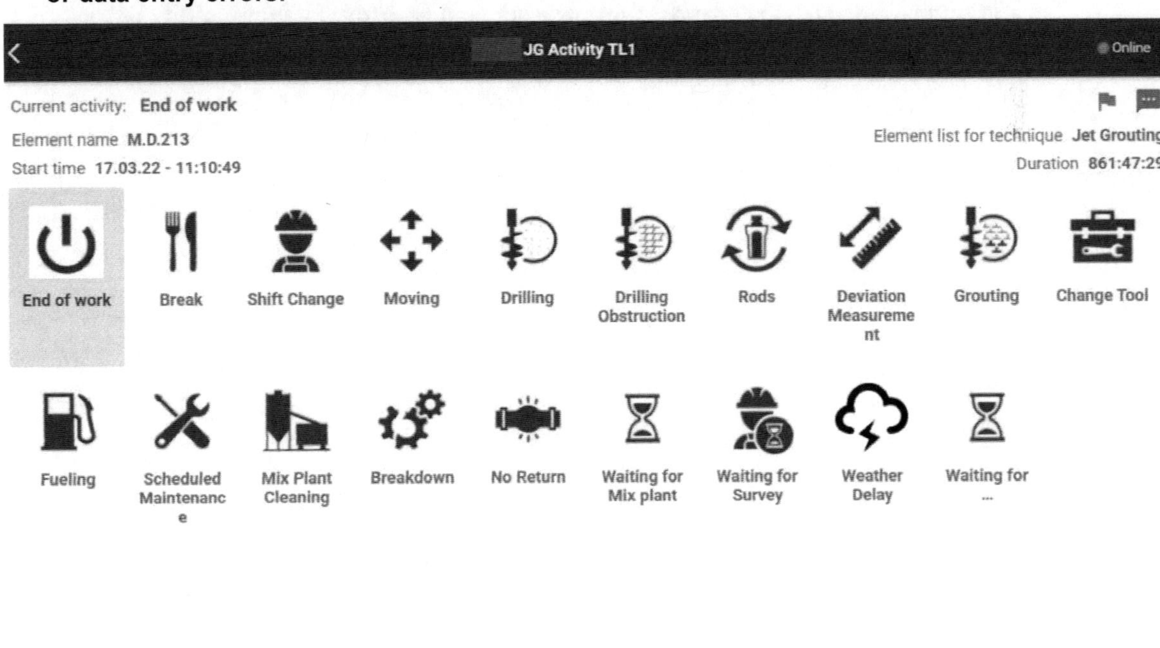

Fig. 10. Mobile application interface for tracking of Jet Grouting Activities. Operators record productive and "unproductive" time spent on site, recorded on a per rig timeline.

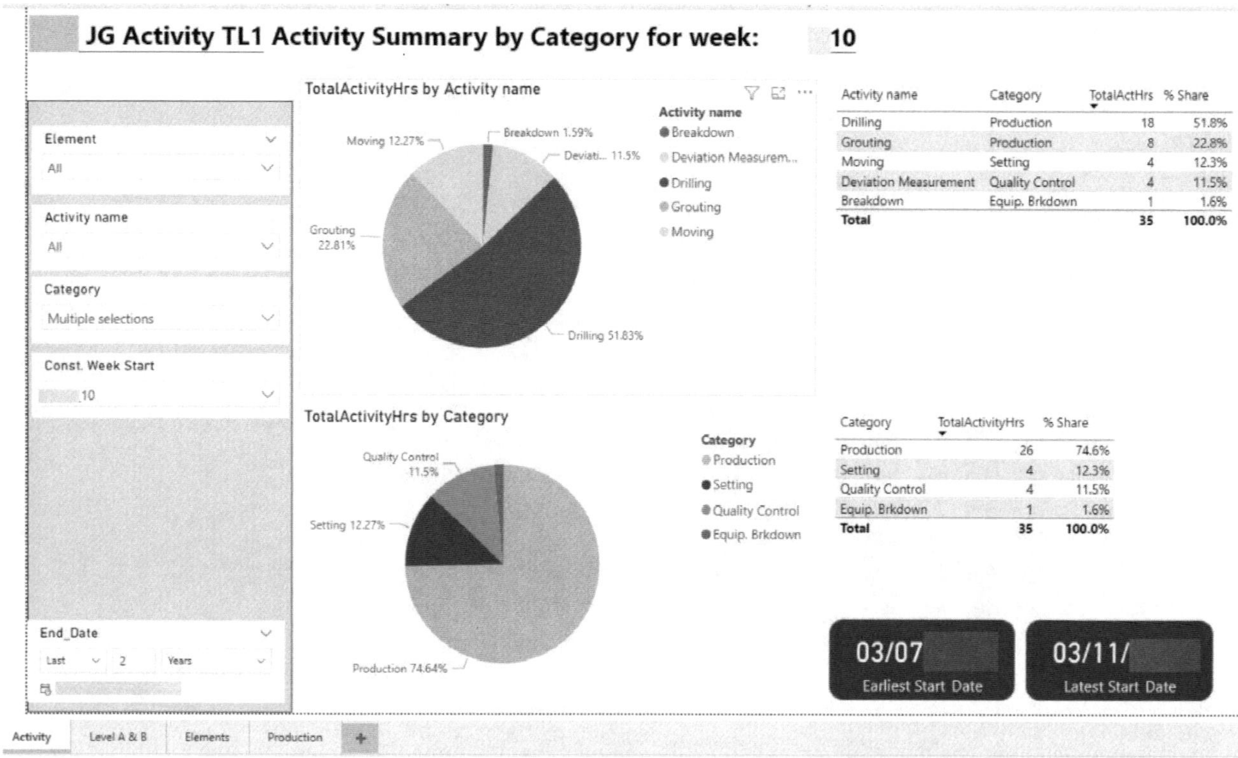

Fig. 11. Activities are summarized and displayed by week or category on a project level dashboard. Information is used to report "unproductive" time activities, and visualize trends over time, as well as provide a time study record for estimation of future projects.

EVALUATION OF EXISTING *P-Y* MODELS FOR CALICHE BASED ON NUMERICAL ANALYSIS OF RAIDERS STADIUM LATERAL LOAD TESTS

Fahim M. Bhuiyan, University of Nevada Reno, Nevada, USA, fbhuiyan@nevada.unr.edu
Raj V. Siddharthan, University of Nevada Reno, Nevada, USA, siddhart@unr.edu
Ramin Motamed, University of Nevada Reno, Nevada, USA, motamed@unr.edu

ABSTRACT

Numerical lateral load analysis based on the beam on Winkler foundation (BNWF) model (i.e., the p-y method) is a simple, and widely used tool to predict lateral responses of drilled shafts from field load tests. The lack of a site-specific p-y model for cemented soil layers such as caliche is one of the major limitations of simulating lateral load tests conducted in Las Vegas Valley. In this study, some of the existing p-y models for rock materials, such as Florida Limestone and Vuggy Limestone have been evaluated to simulate the lateral resistance of caliche. This has been done in the context of four lateral load tests in caliche-dominant sites from the recently constructed Raiders Stadium project in Las Vegas (NV). The uncertainty in characterizing the material properties of caliche has been addressed. The p-y analyses were performed using NVShaft, a MATLAB-based, comprehensive load analysis program. The applicability of the considered p-y models for caliche material was evaluated by quantifying the normalized error margins between measured and predicted lateral responses from the simulations of the mentioned load tests. Correlations between such errors and the depth and thickness of caliche in soil layers affecting numerical predictions were established, for both up to and beyond design load levels.

Keywords: drilled shaft, lateral load test, *p-y* analysis, cemented soil, caliche

INTRODUCTION

The deep foundation system is widely adopted in civil engineering projects to transmit the load from the superstructure into the subsurface. The data obtained from subsurface exploration is one of the key information in the selection of the proper type of deep foundation system. The hard, cemented calcium carbonate material known as Caliche is known for its erratic, intermitted existence across the Las Vegas valley. As explained in Werle and Luke (2007), the fluctuation of the water table in the presence of calcareous deposits causes the formation of discontinuous layers of caliche. Typically classified as an Intermediate Geomaterial (IGM) (Brown et al. 2010; Motamed et al. 2016), the cementation in caliche material varies with the unconfined compressive strength (q_u) values ranging between 30 psi to 20,000 psi (Cibor 1983; Saint-Pierre 2018). Caliche poses major engineering challenges in site exploration techniques, expensive drilling and excavation, and unknown characteristics due to a limited number of studies (Rinne et al. 1996; Werle and Luke 2007; Stanton et al. 2017). Performing standard penetration tests often result in the refusal SPT-N values, and obtaining intact samples of weakly cemented deposits often proves to be problematic (Stanton et al., 2017; Saint-Pierre, 2018). Despite these limitations, the cemented soil in Las Vegas offers high load carrying capacity and strong monolithic interaction with foundation material (Karakouzian et al. 2015; Afsharhasani et al. 2020; Bhuiyan et al. 2021), making the deep foundation system often a viable choice in this region.

Performing lateral load analysis of a deep foundation is a necessary computational step for the design for lateral loading. To simplify complex soil-structure interaction in lateral load analysis, Hetényi and Hetbenyi (1946) introduced a subgrade reaction method where the pile is assumed to behave as an elastic beam, and the soil is simplified by a series of uncoupled elastic springs. Later, McClelland and Focht (1956) incorporated nonlinearity in soil lateral stiffness in the lateral resistance (*p-y*) model, followed by the development of other nonlinear *p-y* models for different soil conditions (e.g., Matlock 1970; Reese et al. 1974). This simplified method to perform lateral load analysis is now widely known as Beam on nonlinear Winkler foundation (BNWF) model, or *p-y* method, and is recommended by design guidelines (e.g., API 2014).

The *p-y* models on specific soil types are generally back-calculated from lateral field load tests data (Matlock 1970; Reese et al. 1974; Reese and Nyman 1978), or finite-element (FE) model simulations (Suryasentana and Lehane 2014; Zhang and Andersen 2017; Byrne et al. 2020). The obtained *p-y* models are validated using case studies from several field load tests conducted on similar soil types (Suryasentana and Lehane 2014; Taghavi et al. 2020). The complex soil-pile interaction and characteristics of key lateral resistance properties (e.g., coefficient of subgrade modulus, k_v, and ultimate soil reaction, p_{ult}) are integrated into the derived *p-y* model. In practice, the API (2014) recommended *p-y* models are widely used in commercial computer programs and by practicing engineers, and as noted by many before, cannot properly capture the lateral resistance mechanisms for site-specific unique soil conditions (Briaud and Wang 2018; Fu et al. 2020). The selection of proper *p-y* models in the numerical lateral analysis is imperative to obtain reliably predicted responses and to reinforce the design process.

The focus of this paper is the evaluation of lateral resistance (i.e., *p-y* model) of caliche in the numerical *p-y* analysis. As of date, there is no site-specific *p-y* model developed for caliche material. The *p-y* model for vuggy limestone (Reese and Nyman 1978) has been used in the past to simulate lateral resistance of caliche for the I-15/US 95 load test program (Rinne et al. 1996; Bhuiyan et al. 2020), but the applicability of the model in this context remains to be investigated. In this study, the *p-y* models for vuggy limestone and Florida limestone (McVay and Niraula 2004) were evaluated for caliche material based on simulations of four lateral load tests from the Raiders Stadium project (Fiorelli et al. 2018a). Apart from the mentioned *p-y* models, the authors also considered the stiff clay *p-y* model (Reese and Welch 1975) for evaluation. The predicted responses from the preliminary analyses using the stiff clay model showed significant deviations from the measured responses and hence the model was excluded from this study. The considered load tests were conducted in cemented soil with the presence of caliche layers, making the load test program a viable choice for evaluating the aforementioned *p-y* models. The authors acknowledge that a limitation of this study is the limited amount of data on the caliche materials at the project location, which has been addressed in the numerical modeling. A MATLAB-based, finite-difference program, NVShaft (Bhuiyan et al. 2020, Bhuiyan et al. 2022) was used to perform the *p-y* analysis of the mentioned load test program. The applicability of the mentioned *p-y* models was further assessed by investigating the error margin in measured and predicted pile head deflections, concerning varying depths to the first caliche layers from the Raiders Stadium load tests. This assessment provided some insight on the reliability of these rock *p-y* models considering the relative distances of first caliche layers and the amplitude of applied lateral load.

RAIDERS STADIUM LOAD TEST PROGRAM

Inaugurated on July 31st, 2020, the new Allegiant stadium in Las Vegas is the home venue to the Las Vegas Raiders football team. Also known as the Raiders Stadium, the facility can accommodate 65,000 spectators, and features a sliding field tray and operable walls towards the views of the Las Vegas strip. Initially, the stadium was planned to be supported by drilled shaft foundations with a 48 inches diameter and 600 kips of axial load capacity, given the previous records of competent cementitious soil conditions in the region (Fiorelli et al. 2018b). Considering the foundation load to be distributed along a large project area, continuous flight auger (CFA) piles were chosen, to shorten the construction schedule and cut down costs (Fiorelli et al. 2018a). Approximately 1600 CFA piles with 24 inches diameters (D), and lengths (L) ranging from 30 to 65 feet were chosen as the final design (Fiorelli et al., 2018b). In this paper, four lateral load tests conducted on CFA piles with 24 inches diameter and around 66.5 ft length were considered. The considered test piles were designated in their corresponding load test reports as follows: 1) test pile LT-EL (LOADTEST 2018d), 2) test pile LT-FL (LOADTEST 2018c), 3) test pile LT-CL (LOADTEST 2018b) and, 4) test pile LT-AL (LOADTEST 2018a). The former two test locations (LT-EL and LT-FL) have caliche layers near the excavated pile head locations, making these load tests ideal to study the effect of caliche in *p-y* analysis. The generalized soil profiles for four lateral load tests from the Raiders Stadium project, along with the recorded SPT-N values, the depths of groundwater tables (GWT), and the locations of applied lateral loads relative to the top of the test piles are shown in Fig. 1. The details of site sub-surface conditions are discussed in the following section.

Site sub-surface conditions

For the test pile LT-EL, from the boring log LB-13, a caliche layer with a thickness of 2.5 ft is reported at the excavated ground level near the pile head location. This is followed by cemented fine to coarse sand, silt, and sandy clay layers up to another caliche layer at the depth of 14.7 ft. A third, 2 ft thick, caliche layer is encountered at a depth of 62.5 ft, overlain by a series of clay, silt, sand, and gravel layers with some cementation.

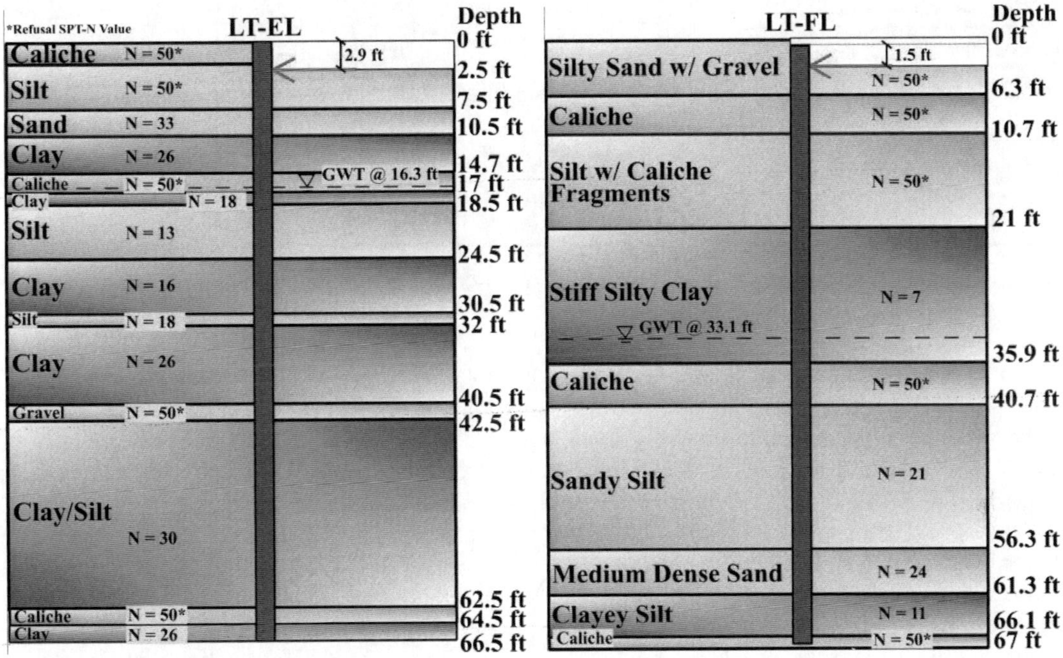

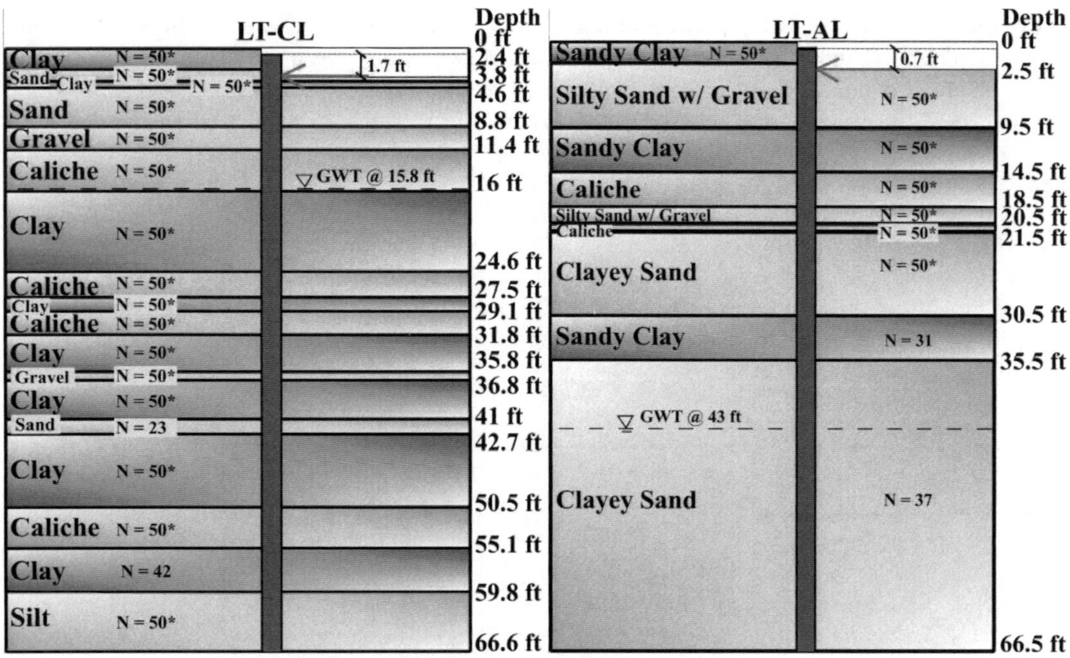

Fig. 1. Schematics of the generalized soil profile for test pile LT-EL (upper left), LT-FL (upper right), LT-CL (lower left) and LT-AL (lower right).

The subsurface condition of the test pile LT-FL indicates the first caliche layer to be located at a depth of 6.3 ft, as reported in the boring log LB-11 from the respective load test report (LOADTEST 2018c). This is overlain by very dense silty sand and coarse gravel layers. The second and third caliche layers are located at depths of 35.9 ft and 66.1 ft, respectively. Slightly cemented clayey silt and silty clay with coarse gravel-sized caliche fragments are reported at several depths.

As reported in the boring log LB-7 in the load test report, the soil layers at the location of test pile LT-CL consisted mostly of sandy clays, silts, and four layers of caliche (LOADTEST 2018b). The site has a series of slightly cemented silty clay, gravelly silty sand, and coarse gravel layers over the first 11.4 ft of depth from the ground level near the pile head location. This is followed by the first caliche layer with a 4.6 ft of thickness. Three more subsequent caliche layers at greater depths with thicknesses of 2.9 ft, 2.7 ft, and 4.6 ft are also present at this particular test site.

The soil profile for the test pile LT-AL can be characterized by the reported boring log B-105 in the load test report (LOADTEST 2018a). The excavated ground level near the pile head is reported to consist of 2.5 ft thick, very stiff, slightly cemented sandy lean clay material. This is underlain by a 7 ft thick, dense silty sand layer with gravel. The first caliche layer is reported to be located at a depth of 14.5 ft with 4 ft thickness, followed by silty sand with a gravel layer and another thin caliche layer.

Details of test piles and lateral load test configuration

All the CFA test piles had nominal diameters of 24 inches and were embedded up to around 66.5 ft of depth into the subsurface. Drillings were done up to a tip elevation of 2093 ft in all cases. The grouting process was immediately followed by the insertion of reinforcing cages into the wet grout. The reaction system consisted of a beam positioned across two similar CFA reaction piles. The lateral load was applied through a hydraulic jack calibrated to a vibrating wire load cell and a set of spherical bearing plates. Both pile head and reaction beam displacement were monitored using attached Linear Vibrating Wire

Displacement Transducers (LVWDT). For each test pile, two inclinometers were installed at the pile head, and oriented in the parallel and transverse loading direction, to monitor tilting.

The lateral load tests were conducted following the standard loading schedule by ASTM D3966 (ASTM 2007). Except for test pile LT-FL, lateral load tests were carried out in two loading cycles. In the first cycle, the load was applied in 10 increments up to 60 kips (200% of design load) of maximum lateral load and then unloaded in four decrements. In the second cycle, a maximum of 90 kips of the lateral load was applied in fourteen increments, followed by unloading in four decrements. For the test pile LT-FL, a third additional cycle of the lateral load was applied, up to 187 kips of maximum load in nineteen increments, and then unloaded in four decrements.

A summary of test pile geometries, location and thickness of caliche layers, and location of the centerline of lateral load below grade level for the Raiders Stadium load test program are shown in Table 1 below.

Table 1. Summary of key parameters for test pile, caliche layers, and location of applied load for the Raiders Stadium lateral load tests

Test Pile	Pile Diameter (ft)	Pile Length (ft)	Depth to Middle of First Caliche Layer (ft)	Thickness of First Caliche Layer (ft)	Centerline of Lateral Load Below Top of Pile (ft)
LT-EL	2	66.5	1.25	2.5	2.9
LT-FL		66.5	8.5	4.4	1.5
LT-CL		66.2	13.7	4.6	1.7
LT-AL		66.54	16.5	4	0.7

ROCK *P-Y* MODELS USED FOR EVALUATION

Lateral resistance (p-y) model for Vuggy limestone

A p-y model for vuggy limestone was developed based on a lateral field load test on a bored pile at Islamorada, Florida (Reese and Nyman 1978). The 4 ft diameter pile was embedded 43.7 ft into the limestone material, with a 14 ft thick overburden fill at the top. The characterization of the vuggy limestone was limited by the challenges in obtaining intact rock samples, and q_u values of only two rock specimens were reported (69.4 ksf and 54.4 ksf). Although the test indicated no failure of rock material, brittle failure is assumed when soil reaction reaches the ultimate value. The derived relationship between the lateral soil reaction (p) and lateral displacement (y) can be expressed as follows,

$$p = 1000 * q_u * y, \; 0 < y \leq 0.0004 * D \quad [1]$$

$$p = 0.4 * q_u * D + 50 * q_u * (y - 0.0004 * D), \; 0.0004 * D < y \leq 0.0024 * D \quad [2]$$

$$p = 0.5 * q_u * D, \; y > 0.0024 * D \quad [3]$$

Lateral resistance (p-y) model for Florida limestone

A total of 12 centrifuge lateral load tests conducted at the University of Florida, on drilled shafts with 6 ft and 8 ft of diameters in prototype, and *L/D* ratios of 2, 3, and 4 were used to formulate a *p-y* model for Florida limestone (McVay and Niraula 2004). Due to the difficulty in coring and fitting the rock material within the centrifuge container, Florida limestone was synthesized with similar properties by mixing ground-up limestone, type I cement, and water. Lateral tests were conducted on material with unconfined compressive strengths of 20 ksf and 40 ksf. To obtain a single trendline, the back-computed p values were normalized relative to $q_u^{0.25}D^{0.9}$, leading to a p-y model given by the following equations,

$$p = 895.41 * \mathrm{D}^{0.90} * \mathrm{q}_u{}^{0.25} * \frac{y}{D},\ 0 < \frac{y}{D} \leq 0.004 \quad [4]$$

$$p = \mathrm{D}^{0.90} * \mathrm{q}_u{}^{0.25}\left(68.57 * \frac{y}{D} + 3.35\right),\ 0.004 < \frac{y}{D} \leq 0.1 \quad [5]$$

$$p = 10.21 * \mathrm{D}^{0.90} * \mathrm{q}_u{}^{0.25},\ \frac{y}{D} > 0.1 \quad [6]$$

where units for p, y, D, and q_u are kips/inch, inch, inch, and ksf, respectively.

NUMERICAL MODELLING OF RAIDERS STADIUM LOAD TESTS IN NVSHAFT

Numerical p-y simulations of four lateral load tests on 2 ft diameter CFA piles from the Raiders Stadium project were done using the finite-difference program, NVShaft. Based on the available subsurface information from load test reports, soil parameters were calculated using empirical formulas and correlations from FHWA (Brown et al. 2010) and Caltrans (2019) manuals. As the maximum measured lateral displacements from the considered load tests is only 0.13 inches, the CFA piles were assumed to have elastic section properties in the numerical model. The lateral loads were applied at their respective locations as summarized in Table 1. For test pile LT-EL, the lateral load were applied slightly below the first caliche layer near the ground level, and for the remaining three test piles loads were applied above the respective first caliche layers (Fig. 1). The mobilization of lateral resistance from caliche can be expected to be small given this test condition, which has been assessed by observing the mobilization of the p-y models at the respective depths (Fig. 4).

One of the major limitations of this study is the very limited information on caliche in the load test reports. In the boring logs, refusal SPT-N values are reported at several caliche depth locations, and no laboratory test data was reported. Stanton et al. (2017) evaluated four different methods used by local engineers and NDOT practitioners to assume caliche material properties. Typically, an SPT-N value of 50 and unit weight of 140 pcf (22 kN/m^3) are assumed in local practice (Stanton et al. 2017). Following the classification of caliche as cohesive intermediate geomaterial (IGM) by Brown et al. (2010), the maximum value of q_u is assumed as 100 ksf (4.79 MPa). In this study, 12 ksf is used as the q_u of caliche at all test locations following a correlation suggested by Abu-Hejleh (2003) based on SPT-N for weak rocks.

The selected p-y models and relevant soil properties such as effective unit weight (γ'), angle of friction (φ), cohesive strength (c_u), unconfined compressive strength (q_u), soil modulus (E_s), and strain at 50% stress level (ε_{50}) for test pile LT-FL is shown in Table 2. The characterized soil profiles for the remaining three load tests are almost similar. Although this study focuses on evaluating existing p-y models for caliche material, a slight degree of cementation reported in the sand and silt layers also presents a unique opportunity to evaluate the p-y model for weakly cemented sand (Juirnarongrit and Ashford 2004). Finally, the lateral resistance of clay was simulated using the p-y model given by the Integrated Clay Method (O'Neill and Gazioglu 1984).

Table 2. Representative soil profile for test pile LT-FL from Raiders Stadium lateral load test

Depth (ft)	Soil Profile	p-y Model Used	γ' (pcf)	φ (Degree)	c_u (psf)	q_u (ksf)	E_s (ksf)	ε_{50}
0-6.3	Silty Sand w/ Gravel	Weakly Cemented Sand	130	43	-	-	-	-
6.3-10.7	Caliche	Vuggy/Florida Limestone	135	-	-	12	-	-
10.7-21	Silt w/ Caliche Fragments	Weakly Cemented Sand	130	43	-	-	-	-
21-35.9	Stiff Silty Clay	Integrated Clay	106	-	1407	-	750	0.007
35.9-40.7	Caliche	Vuggy/Florida Limestone	72.6	-	-	12	-	-
40.7-56.3	Sandy Silt	Weakly Cemented Sand	51.6	35	-	-	-	-
56.3-61.3	Medium Dense Sand	Weakly Cemented Sand	61.6	35	-	-	-	-
61.3-66.1	Clayey Silt	Integrated Clay	56.6	-	2036	-	232	0.007
66.1-67	Caliche	Vuggy/Florida Limestone	72.6	-	-	12	-	-

NVSHAFT PREDICTED RESPONSES AND EVALUATION OF ROCK P-Y MODELS

The comparison between measured data and predicted lateral responses from numerical *p-y* analysis of test piles LT-EL and LT-FL are shown in Fig. 2. For test pile LT-EL, the predicted response obtained after implementing the *p-y* model for vuggy limestone resulted in a reasonably good match with the measured data, particularly up to 200% of design load (60 kips). For this test, the *p-y* model for Florida limestone resulted in a stiffer predicted response compared to measured data at all load levels. The predicted responses for the test pile LT-FL (also shown in Fig. 2) indicate a better match with the measured data, particularly for the Florida limestone *p-y* model. The bar plots for these two tests, showing the error between predicted and measured pile head deflections (y_{head}), normalized by the pile diameter at 100% and 200% of design loads are shown in Fig. 3. The normalized error (e_{head}) was calculated using the following formula,

$$e_{head}(\%) = (y_{measured} - y_{predicted})/D * 100 \qquad [7]$$

where $y_{measured}$ = measured pile head deflection from field load test and $y_{predicted}$ = pile head deflection predicted from NVShaft simulations.

For test pile LT-EL, the *p-y* model for vuggy limestone resulted in 0.006% and 0.01% of normalized error for the design load and two times of design load, respectively. For test pile LT-FL, the overpredicted responses from the vuggy limestone *p-y* model resulted in -0.06% and -0.14% of normalized error at these load levels. As seen in Fig. 3, the Florida limestone *p-y* model yielded relatively smaller normalized errors at design load, which is 0.015% and -0.05% for test piles LT-EL and LT-FL, respectively. At 200% of design load level, using the Florida limestone *p-y* model resulted in 0.08% and -0.13% of normalized error for test piles LT-EL and LT-FL, respectively. These observations indicate that both rock *p-y* models can produce reasonable predicted responses for caliche, at least up to the design load level, in the context of the Raiders Stadium load test.

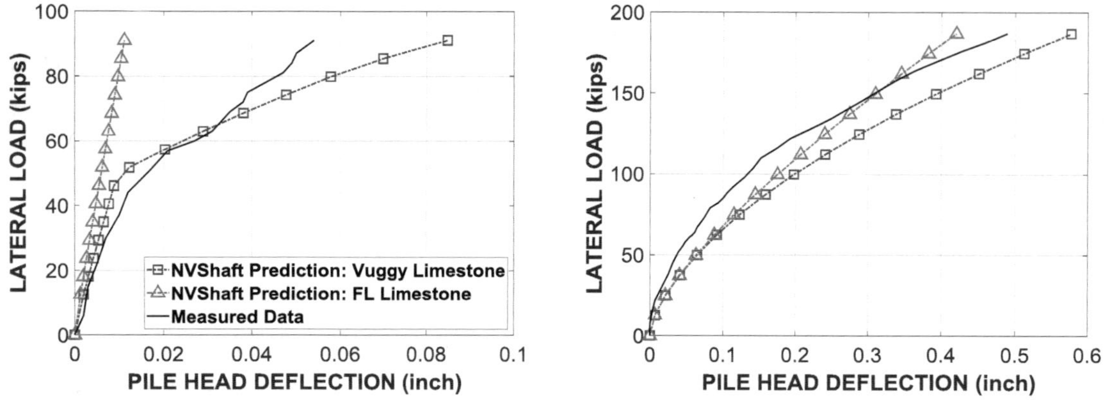

Fig. 2. NVShaft predicted responses of test piles LT-EL (left) and LT-FL (right) from lateral load test simulations with measured data.

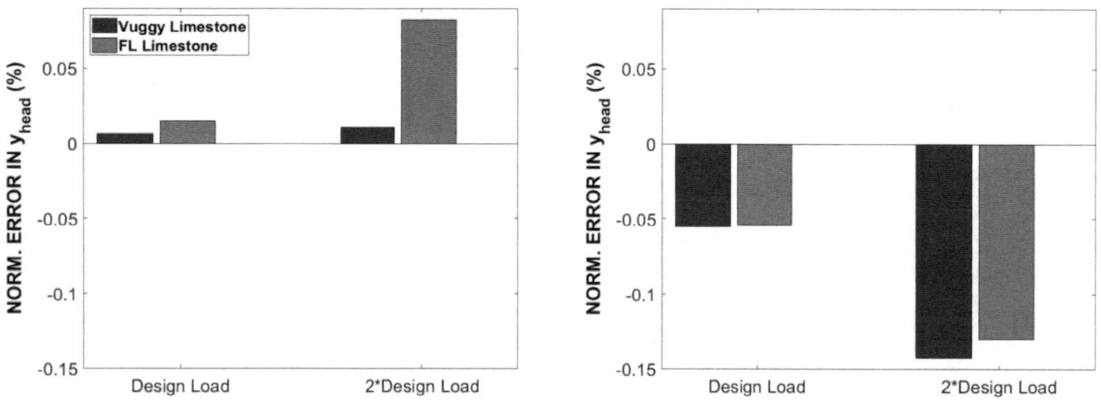

Fig. 3. Bar plots showing normalized error in pile head deflection (y_{head}) at design load and two times design load for test piles LT-EL (left) and LT-FL (right).

For test piles LT-EL and LT-FL, caliche layers were located at shallower depths compared to other lateral field load tests. In practice, to obtain the design length of piles to achieve lateral stability, the initial length of the pile is often taken as 10 to 15 times the pile diameter (ADOT 2010). As observed from past studies, the soil layers at this range of depths (20 ft to 30 ft in this study) for flexible piles exhibit wedge-type soil flow mechanism, affecting a major part of the overall lateral pile response (Hong et al. 2017; Wang et al. 2020). This fact is also reinforced in Fig. 2, which shows that using different *p-y* models for caliche for these two load tests caused major changes in predicted lateral responses from *p-y* analyses. The influence of *p-y* models to represent lateral resistance of caliche on numerical simulations can be further explained in Fig. 4. The NVShaft generated *p-y* plots presented in Fig. 4 show mobilization of lateral soil reactions (*p*) in caliche at lateral displacements (*y*) corresponding to 100% and 200% of design load for the considered lateral load tests. It can be seen that mobilization of lateral soil reaction in caliche is higher for test piles LT-EL and LT-FL, compared to test piles LT-CL and LT-AL, as caliche materials are located at greater depths for the former cases. For test pile LT-EL, a change in lateral stiffness can be seen in the mobilization of the vuggy limestone *p-y* model when the lateral load is increased beyond the design load level. This is also reflected in the predicted response corresponding to the vuggy limestone model, as shown in Fig. 2 (left). Figure 4 also reveals that none of the considered rock *p-y* models achieved full mobilization in the context of Raiders Stadium load test simulations.

The subsurface conditions for test piles LT-AL and LT-CL along the upper 19 ft depth are almost identical, as can be seen from Fig. 1. Performing lateral analyses using different *p-y* models for caliche for these two test piles resulted in almost no variation in predicted responses. The lateral responses for these test piles are mostly affected by the cemented sand, gravel, and clay material overlying the first caliche layer. Also, these two test piles have identical geometric properties and were subjected to the same lateral loads. Considering all these facts, it can be construed that these two lateral tests were almost identical, and the overall responses can be summarized by a composite plot, as shown in Fig. 5. It can be seen that the predicted responses for test pile LT-CL and LT-AL reasonably brackets the measured responses obtained at the field. Particularly for test pile LT-AL, a good match between predicted and measured responses can be observed. Due to the cyclic nature of the loading, the measured response for test pile LT-AL appears to be almost constant around two times of design load level (Fig. 5). The overall load-displacement responses from the numerical simulations of all the mentioned lateral load tests also indicate the reliability of the *p-y* model for weakly cemented sand given by Juirnarongrit and Ashford (2004).

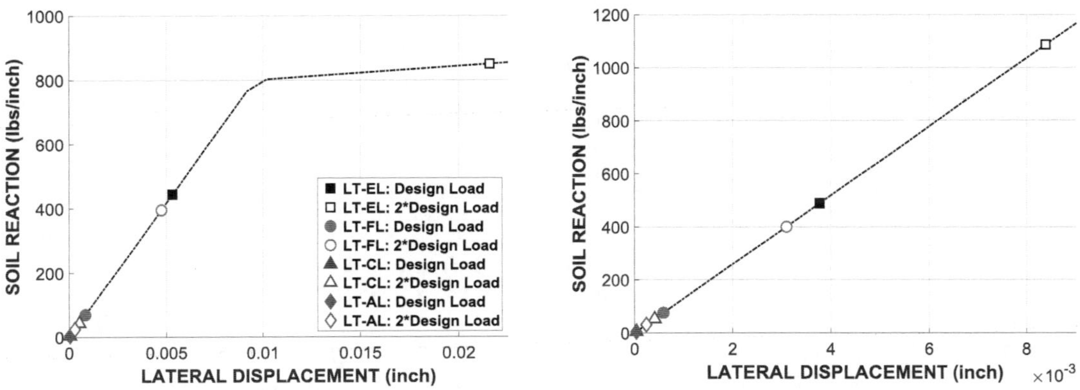

Fig. 4. Mobilization of lateral soil reaction at 100% and 200% of design loads, considering the *p-y* models for vuggy limestone (left) and Florida limestone (right) from the simulations of Raiders Stadium lateral load tests.

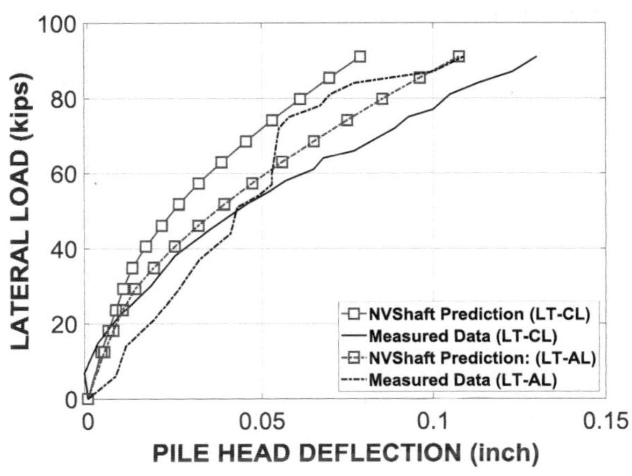

Fig. 5. Composite load-deflection plots for test piles LT-CL and LT-AL.

Relationships between the absolute normalized error in pile head deflection ($e_{abs,head}$) and the ratio of depth to the middle of the first caliche layer to pile diameter, for both design load and twice the design load levels are shown in Fig. 6. For design load level shown on Fig.6(left), it can be seen that $e_{abs,head}$ seems to show a relatively steadily increases with the increase in caliche depth with a maximum approximate value of 0.05%. A similar plot as shown in Fig. 6(right) for twice the design load also presents a smaller $e_{abs,head}$ when the caliche layer is located near the grade level, particularly for the vuggy limestone model. In this case, the $e_{abs,head}$ increases to a maximum approximate value of 0.14% corresponding to the caliche depth to diameter ratio of 4.25 and then decreases for greater depths. These observations indicate the reliability of both considered rock *p-y* models to simulate the lateral resistance of caliche, at least up to the applied design lateral load based on the findings from the Raiders Stadium load test study. In summary, the *p-y* model for vuggy limestone resulted in smaller $e_{abs,head}$, compared to the *p-y* model for Florida limestone in this regard.

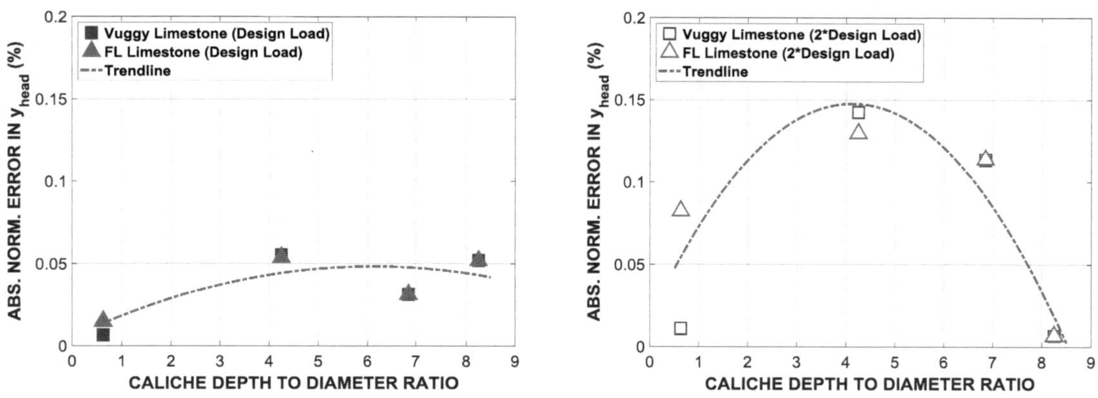

Fig. 6. Variation in absolute normalized error in pile head deflection with increasing caliche depth to diameter ratio for design load (left) and two times design load (right).

SUMMARY AND CONCLUSIONS

In this study, the applicability of two existing lateral resistance (*p-y*) models based on vuggy limestone and Florida limestone to simulate the lateral soil reaction of caliche were evaluated. This was done in the context of four lateral load tests performed in cemented soil conditions, as part of the Raiders Stadium construction project in Las Vegas. Field lateral load tests were carried out on CFA piles with 2 ft of diameter and around 66.5 ft of embedded depth. Unavailability of laboratory test data and limited information on caliche are the major limitations of this study. This limitation was addressed by characterizing the caliche based on some assumptions used in local practice and by using correlations to obtain the strength of the material.

A MATLAB-based comprehensive load analysis program, NVShaft was used to evaluate the mentioned rock *p-y* models. For test piles, LT-EL and LT-FL, the locations of the first caliche layer were relatively closer to the grade level, significantly affecting the lateral responses. For this reason, the simulated lateral responses from these test piles were extensively studied to evaluate the applicability of both rock *p-y* models. The predicted lateral responses of test pile LT-EL and LT-FL from using the considered rock *p-y* models showed reasonable agreements with the measured data, up to the design load level. The error margin between measured and predicted pile head deflection was calculated and then normalized relative to pile diameter. At design load level, the *p-y* model for vuggy limestone resulted in 0.006% and -0.06% of normalized error, for test pile LT-EL and LT-FL, respectively. At the same load level, the *p-y* model of Florida limestone resulted in 0.015% and -0.05% of normalized error for test pile LT-EL and LT-FL, respectively. These observations add to the reliability of the considered *p-y* models to perform *p-y* analysis in caliche-dominant sites.

Even though the lateral responses of test piles LT-CL and LT-AL were not significantly affected by the caliche, good matches between NVShaft predicted response and measured data were observed in these cases. This proves the capability of the *p-y* models used for other soil materials (e.g., weakly cemented sand) to produce reasonable predicted responses in cemented soil conditions. Finally, it has been shown that both *p-y* models are capable of producing reasonable predicted responses for caliche located at shallower depths, at both design load and twice the design load levels. Some discrepancies between measured and predicted responses presented in this paper can be attributed to the lack of site-specific *p-y* model and uncertainty in caliche material properties due to insufficient site characterization. Further research on quantifying lateral resistance of caliche is strongly encouraged in this context.

ACKNOWLEDGMENTS

The authors would like to recognize the Nevada Department of Transportation (NDOT) for providing funding for this project. The load test reports were provided by Peter Faust and Eleazar Sotelo of Malcolm Drilling Company, Inc. which is greatly appreciated. The constructive discussion with Diane Fiorelli (Langan) about the load test program is appreciated as well.

REFERENCES

ADOT., 2010. ADOT policy memorandum: ADOT DS-3. Arizona Department of Transportation. https://azdot.gov/sites/default/files/2019/04/drilled-shafts-lateral-analysis.pdf. Accessed 4-21-22.

API., 2014. API recommended practice 2A-WSD - planning, designing, and constructing fixed offshore platforms – working stress design. 22nd ed. American Petroleum Institute.

ASTM., 2007. Standard Test Methods for Deep Foundations Under Lateral Load. ASTM D3966 - 07

Abu-Hejleh, N., O'Neill, M. W., Hanneman, D., and Atwooll, W. J., 2003. Improvement of the geotechnical axial design methodology for Colorado's drilled shafts socketed in weak rocks. Colorado. Dept. of Transportation.

Afsharhasani, R., Karakouzian, M., and Farhangi, V., 2020. Effect of competent caliche layers on measuring the capacity of axially loaded drilled shafts using the osterberg test. *Applied Sciences*, *10*(18), 6169. https://doi.org/10.3390/app10186169

Bhuiyan, F. M., Siddharthan, R. V., Motamed, R., and Sanders, D. H., 2020. Evaluation of a new *p-y* analysis tool for lateral analysis of drilled shafts using load tests in Nevada. DFI 45th Annual Conference on Deep Foundations, October 13-16 2020, pp. 303–312.

Bhuiyan, F.M., Toth, J., Siddharthan, R.V. and Motamed, R., 2021. Evaluation of existing *t-z* models for caliche based on numerical analysis of bi-directional load tests using NVShaft. DFI 46th Annual Conference on Deep Foundations, Las Vegas, Nevada, October 12-15 2021, pp. 21-31.

Bhuiyan, F.M., Motamed, R., Siddharthan, R.V., and Sanders, D., 2022. Evaluation of a unified p-y method for lateral analysis of large-diameter drilled shafts using NVShaft. *Transportation Geotechnics* (under review).

Briaud, J. L., and Wang, Y. C., 2018. Synthesis of load-deflection characteristics of laterally loaded large diameter drilled shafts: Technical report (No. FHWA/TX-18/0-6956-R1).

Brown, D.A., Turner J.P., and Castelli, R.J., 2010. Drilled shafts: construction procedures and LRFD design methods. US Department of Transportation, Federal Highway Administration.

Byrne, B. W., Houlsby, G. T., Burd, H. J., Gavin, K. G., Igoe, D. J. P., Jardine, R. J., Martin, C. M., McAdam, R. A., Potts, D. M., Taborda, D. M. G., and Zdravković, L., 2020. PISA design model for monopiles for offshore wind turbines: Application to a stiff glacial clay till. *Géotechnique*, *70*(11), 1030–1047. https://doi.org/10.1680/jgeot.18.P.255

California Department of Transportation., 2019. Caltrans geotechnical manual.

Cibor, J. M., 1983. Geotechnical considerations of las vegas valley. *Geological Environmental and Soil*

Properties, 351–373.

Fiorelli, D., Zorne, J., Siegel, T., and Faust, P., 2018a. Value engineering – design risk and reward at major sport arena projects. DFI 43rd Annual Conference on Deep Foundations, Anaheim, CA

Fiorelli, D., Zorne, J., Siegel, T., and Faust, P., 2018b. Value engineering – design risk and reward at major sport arena projects [PowerPoint slides]. DFI 43rd Annual Conference on Deep Foundations, Anaheim, CA

Fu, D., Zhang, Y., Aamodt, K. K., and Yan, Y., 2020. A multi-spring model for monopile analysis in soft clays. *Marine Structures*, *72*, 102768. https://doi.org/10.1016/j.marstruc.2020.102768

Hetényi, M., and Hetbenyi, M. I., 1946. Beams on elastic foundation: Theory with applications in the fields of civil and mechanical engineering (Vol. 16). University of Michigan press Ann Arbor, MI.

Hong, Y., He, B., Wang, L. Z., Wang, Z., Ng, C. W. W., and Mašín, D., 2017. Cyclic lateral response and failure mechanisms of semi-rigid pile in soft clay: Centrifuge tests and numerical modelling. *Canadian Geotechnical Journal*, *54*(6), 806–824. https://doi.org/10.1139/cgj-2016-0356

Juirnarongrit, T., and Ashford, S. A., 2004. Lateral load behavior of cast-in-drilled-hole piles in weakly cemented sand. *Transportation Research Record: Journal of the Transportation Research Board*, *1868*(1), 190–198. https://doi.org/10.3141/1868-20

Karakouzian, M., Afsharhasani, R., and Kluzniak, B., 2015. Elastic analysis of drilled shaft foundations in soil profiles with intermediate caliche layers. *IFCEE 2015*, 922–928. https://doi.org/10.1061/9780784479087.083

LOADTEST, Inc., 2018a. Report on CFA pile lateral load testing, LT-AL - Raiders stadium production piles, Las Vegas, NV. Project No. LT-1839-2, February 14, 2018.

LOADTEST, Inc., 2018b. Report on CFA pile lateral load testing, LT-CL - Raiders stadium production piles, Las Vegas, NV. Project No. LT-1839-6, March 15, 2018.

LOADTEST, Inc., 2018c. Revised report on CFA pile lateral load testing, LT-FL - Raiders stadium production piles, Las Vegas, NV. Project No. LT-1839-4, April 04, 2018.

LOADTEST, Inc., 2018d. Report on CFA pile lateral load testing, LT-EL - Raiders stadium production piles, Las Vegas, NV. Project No. LT-1839-8, April 17, 2018.

Matlock, H., 1970. Correlation for design of laterally loaded piles in soft clay. Offshore Technology Conference. https://doi.org/10.4043/1204-MS

McClelland, B., and Focht, J., 1956. Soil modulus for laterally loaded piles. *Journal of the Soil Mechanics and Foundations Division*, *82*(4), 1081–1.

McVay, M.C., and Niraula, L., 2004. Development of PY curves for large diameter piles/drilled shafts in limestone for FBPIER. No. Final Report.

Motamed, R., Elfass, S., and Stanton, K., 2016. LRFD resistance factor calibration for axially loaded drilled shafts in the Las Vegas valley. Nevada. Dept. of Transportation.

O'Neill, M., and Gazioglu, S., 1984. Integrated formulation of py relationships in clays. *A Report to the American Petroleum Institute, Report PRAC-82-41-2, University of Houston*.

Reese, L. C., Cox, W. R., and Koop, F. D., 1974. Analysis of laterally loaded piles in sand. Offshore Technology Conference. https://doi.org/10.4043/2080-MS

Reese, L. C., and Welch, R. C., 1975. Lateral Loading of Deep Foundations in Stiff Clay. *Journal of the Geotechnical Engineering Division*, *101*(7), 633–649. https://doi.org/10.1061/AJGEB6.0000177

Reese, L., and Nyman, K., 1978. Field load tests of instrumented drilled shafts at Islamorada, Florida.

Report to the Girdler Foundation and Exploration Corporation, Clearwater, FL, February.

Rinne, E., Thompson, J., and Vanderpool, W., 1996. I-15/US 95 load test program, Las Vegas, Nevada. Kleinfelder, Inc., Las Vegas, Nevada.

Saint-Pierre, E. C., 2018. The Development of a material model for engineering behavior characteristics of cemented soils for the Las Vegas valley. M.S. thesis, University of Nevada, Reno.

Stanton, K., Motamed, R., and Elfass, S., 2017. Robust LRFD resistance factor calibration for axially loaded drilled shafts in Las Vegas. *Journal of Geotechnical and Geoenvironmental Engineering*, *143*(6), 06017004. https://doi.org/10.1061/(ASCE)GT.1943-5606.0001664

Suryasentana, S. K., and Lehane, B. M., 2014. Numerical derivation of CPT-based *p–y* curves for piles in sand. *Géotechnique*, *64*(3), 186–194. https://doi.org/10.1680/geot.13.P.026

Taghavi, A., McVay, M., Niraula, L., Davidson, M., and Patil, A., 2020. Axial and lateral resistance coupling in the analysis of large-diameter drilled shafts. *Engineering Structures*, *206*, 110160. https://doi.org/10.1016/j.engstruct.2019.110160

Wang, L., Lai, Y., Hong, Y., and Mašín, D., 2020. A unified lateral soil reaction model for monopiles in soft clay considering various length-to-diameter (L/D) ratios. *Ocean Engineering*, *212*, 107492. https://doi.org/10.1016/j.oceaneng.2020.107492

Werle, J. L., and Luke, B., 2007. Engineering with heavily cemented soils in Las Vegas, Nevada. *Problematic Soils and Rocks and In Situ Characterization*, 1–9. https://doi.org/10.1061/40906(225)8

Zhang, Y., and Andersen, K. H., 2017. Scaling of lateral pile p—Y response in clay from laboratory stress-strain curves. *Marine Structures*, *53*, 124–135. https://doi.org/10.1016/j.marstruc.2017.02.002

NEW INSIGHTS ON DETECTING FLAWS WITH CSL AND TIP

Andrew Z. Boeckmann, Dan Brown & Associates, Columbia, MO, USA, 573-424-0017, aboeckmann@dba.world
David S. Graham, Dan Brown & Associates, Chattanooga, TN, USA, 205-427-0682, dgraham@dba.world
Paul J. Axtell, Dan Brown & Associates, Overland Park, KS, USA, 816-210-1954, paxtell@dba.world

ABSTRACT

In the last fifteen years, several research studies have evaluated the ability to detect various types of deep foundation concrete integrity flaws with crosshole sonic logging (CSL) and thermal integrity profiling (TIP). This paper synthesizes the findings from the research, focusing on two studies that included CSL and TIP testing of drilled shafts with various types of flaws. The first, a 2016 research study of test shafts for the Wisconsin DOT, included inclusions inside and outside the reinforcing cage, intentional tremie withdrawal during concreting, and soft bottom flaws. The second, a 2021 study of production shafts for the Missouri DOT, included one shaft with and another without soft bottom conditions. After synthesizing the ability of each test method to detect each type of flaw, the paper presents recommendations for improving how CSL and TIP results are evaluated in practice. The recommendations highlight DFI's CSL interpretation protocol as well as the importance of evaluating TIP data versus time, not just depth. The paper includes example temperature-versus-time plots for various types of flaws and emphasizes how temperature-versus-time evaluation can be used to distinguish soft bottom conditions from standard bottom-of-shaft temperature roll off.

Keywords: drilled shaft, concrete, integrity test, thermal integrity profiling, crosshole sonic logging

INTRODUCTION

The considerable increase in diameters and depths commonly specified for drilled shafts in recent years has increased the challenges associated with drilled shaft construction, particularly the challenges associated with concrete placement. The increase in drilled shaft size is also accompanied by an increase in per-shaft construction costs. As a result of construction challenges and more valuable shafts, concrete integrity testing has emerged as a high-stakes hold point in many construction projects with drilled shafts. Potential consequences associated with integrity testing include additional investigation of the shaft by concrete coring or other test methods, as-built engineering analysis, and remediation efforts ranging from grouting to installation of supplementary foundation elements. Each of these consequences is associated with direct costs that are significant but commonly dwarfed by indirect costs associated with project delays.

Considering the potential consequences of integrity test results, it is incumbent upon foundation engineers and testing consultants to employ interpretation methods that are consistent, rational, and comprehensive. This paper presents recommendations for improving interpretation of results from crosshole sonic logging (CSL) and thermal integrity profiling (TIP) tests, the two most common integrity test methods. The recommendations are supported by recent research involving integrity tests of drilled shafts with known flaws. The research is summarized and synthesized prior to presentation of recommendations at the end of the paper.

This paper adopts the terminology established in DFI's *Terminology and Evaluation Criteria of Crosshole Sonic Logging (CSL) as applied to Deep Foundations* (2018):

- An anomaly is abnormal data that deviates from expectations and may indicate a flaw.

- A flaw is any imperfection in the planned shape or material of the foundation that may not necessarily affect foundation performance.
- A defect is any flaw that, because of its size, location, and inferred concrete properties, will have a significant adverse effect on the performance of the foundation.

RESEARCH WITH INTEGRITY TESTING OF KNOWN FLAWS

Since TIP was developed in the early 2000s, several research studies have been performed to evaluate the ability to use TIP to detect known flaws. The studies are presented in this section. Primarily, the studies have been funded by state transportation agencies that were considering implementing TIP as an alternative or supplementary integrity test method to CSL, which has a longer history of use for integrity testing. All of the studies involve evaluation of shafts with known defects, with most of the studies including both TIP and CSL in each of the evaluated drilled shafts.

Florida DOT Study

TIP results from a 4-ft diameter, 25-ft long test shaft in saturated sandy soil are presented in a report documenting the original development of TIP methods for the Florida DOT (Mullins, 2007). Sandbags filled with native soil were tied to the outside of the reinforcing cage at two depths. Temperatures were measured using the probe method. Three important findings were evident from the first-of-its-kind study:

- Temperature decreases were observed at each of the sandbag locations (depth and cross-sectional).
- The temperature decreases were limited to 3 or 4° F at the time of temperature measurement.
- The temperature profile with depth and temperature differences across the reinforcing cage both appear to have been strongly influenced by the alignment of the reinforcing cage.

As noted in the remaining research summaries, these findings are supported by subsequent studies.

Iowa DOT Study

Research for the Iowa DOT documented in Ashlock and Fotouhi (2014) that investigated the sensitivity of TIP and CSL to flaws. The research included two 5-ft diameter, 80-ft long shafts installed through approximately 45 ft of sand over interbedded shale, limestone, and sandstone. Each test shaft included two intentional flaws installed on the inside of the reinforcing cage and concentrated on one side of the cage. The flaws were small, at 3 or 4% of shaft cross-sectional area for Test Shaft 1 and 8% of shaft cross-sectional area for Test Shaft 2. For comparison, the flaws by Mullins et al. (2007) represented 10% of cross-sectional area. For Test Shaft 1, the flaws consisted of hardened concrete cylinders with low cement content to achieve compressive strength around 600 psi. The flaws for Test Shaft 2 were cylinders filled with sand, gravel, and water in similar proportion to the first test shaft concrete mix, but without cement.

TIP results did not indicate any effect on temperature from the Test Shaft 1 flaws. A 2 to 3° F decrease in temperatures could possibly be attributed to the flaws in Test Shaft 2, but measured TIP temperatures for the shaft with larger flaws were strongly influenced by cage misalignment, which makes identification of small temperature differences difficult. CSL results for Test Shaft 1 did not indicate any anomalies, while CSL tests for Test Shaft 2 produced an approximately 25% increase in first arrival time (FAT). Decreases in energy were also observed but are difficult to quantify as presented in Ashlock and Fotouhi (2014).

The results could suggest the minimum size of a flaw for detection by TIP and CSL is somewhere between 4 and 8% of shaft cross-sectional area. However, the TIP data were collected using the probe method; it is possible greater temperature differences would have been observed at a different collection time. As discussed in Synthesis of Findings below, the ability to detect flaws with TIP likely depends on several factors, not just the size of the flaw.

South Carolina Study

TIP and CSL results for a 102-ft long by 5-ft diameter drilled shaft with two intentional flaws are documented in Schoen et al. (2018). The flaws consisted of gravel-filled bags attached to the inside of the reinforcing cage; each flaw represented approximately 15% of the shaft cross-sectional area. One flaw was installed 3 ft below the top of the shaft and the other was 21 ft below the top of the shaft. At 14 hours, TIP records indicated a 12 to 14° F decrease in temperatures measured in the wires nearest the gravel bags at the depths of the bags (3 ft and 21 ft). At the time of peak temperatures, 34 hours, the temperature decrease attributed to the gravel bags had decreased to 2 or 3° F for the shallower flaw and 8° F for the deeper flaw. CSL testing successfully identified both flaws, but details are not included in Schoen et al. (2018).

Wisconsin DOT Study

A research study evaluating TIP and CSL for the Wisconsin DOT is documented in Boeckmann and Loehr (2018, 2019). The study involved construction of three 3-ft diameter by approximately 30-ft long drilled shafts, two of which included 4 or 5 intentional flaws (the test shafts) and one of which included one intentional flaw (the control shaft). Conventional TIP testing by both wire and probe methods was performed; CSL testing was also performed. The flaws were intended to replicate common drilled shaft construction issues:

- Soft bottom conditions were mimicked by pouring sand into the bottom of the shaft after placement of the reinforcing cage.
- Soil inclusions within concrete were mimicked by tying sandbags to the reinforcing cage. Some of the sandbags were tied outside the reinforcing cage; others were tied inside.
- A tremie breach was mimicked by removing the tremie pipe during concrete placement and pouring sand in the shaft before re-inserting the tremie.
- Weak concrete was created by mixing additional water into the concrete placed at the top of each of the test shafts.

Results of the study are presented in Table 1. The results indicate the soft bottom conditions in two of the shafts were detectible with CSL methods, but not TIP methods. Soil inclusions outside the reinforcing cage were readily identifiable from TIP results, but not from CSL methods; the opposite was true of soil inclusions inside the reinforcing cage. Effects of the tremie breach were also identifiable in the CSL results but not the TIP results. Weak concrete was clearly identifiable in the TIP results, but not the CSL results.

Several additional observations from the Wisconsin DOT study are noteworthy:

- For some of the flaws that were detectible with CSL, the effect was more evident in the arrival time delay than in the relative energy loss while the opposite was true for others. This observation applies even to two similar flaws: the sandbags within the interior of the reinforcing cage of the control shaft resulted in a more pronounced effect on arrival time, while a similar flaw within Test Shaft 2 produced a considerable energy loss with a relatively modest arrival time delay.
- For several of the flaws, the effect on TIP results was evident at the time of maximum temperature rate of rise, but not at the peak temperature time. This is discussed further below.

- As shown in Table 1, the results included a significant CSL effect for a flaw consisting of sandbags tied outside the circumference of the reinforcing cage. Boeckmann and Loehr cautioned that such a finding is likely not repeatable, although similar results were reported in Iowa by Graham and Axtell (2019).

Table 1. Results of Wisconsin DOT study of CSL and TIP (modified from Boeckmann and Loehr, 2018).

Shaft	Depth, ft	Cross-section Location	% Shaft Area	Description	CSL Results		TIP Temperature Difference, °F	
					Arrival Time Delay, %	Relative Energy Loss, dB	Time of Max Rate of Rise	Peak Time
Control	15	Center of shaft	10%	Inclusion in cage interior	17	7.7	0	0
Test 1	30	Shaft perimeter	50%	Soft bottom	0	0	0	0
Test 1	21	Entire cage perimeter	10%	Similar to necking	26	5.0	*Average of all wires:* 6.7	5.0
Test 1	4.5	Reinforcing cage	N/A	Debonded CSL tubes	13	17	*Not applicable to TIP.*	
Test 1	12	1/3 of cage Perimeter	10%	Defect outside cage	0	0	*From south wire:* 5.2	0
Test 1	0-5	N/A	N/A	5-ft zone of weak concrete	0	0	*Average of all wires:* 12.3	7.4
Test 2	30	Entire shaft bottom	100%	Soft bottom	25	12	0	0
Test 2	21	N/A	N/A	Tremie breach	14	7.2	*From south wire:* 3.4	0
Test 2	12	Center of shaft	15%	Inclusion in cage interior	11	15	*From south wire:* 3.4	0
Test 2	0-5	N/A	N/A	5-ft zone of weak concrete	0	0	*Average of all wires:* 16.7	7.0

A significant technical finding from the Wisconsin DOT study was the utility of TIP time records for identifying drilled shaft flaws. An example temperature versus time plot is shown in Fig. 1, which shows results within a zone of weak concrete in blue and results from a nearby zone of sound concrete in green. Also plotted is the temperature at the peak depth (black dash-dot line) and the temperature difference between the flaw and the unaffected zone (dashed purple line). Comparison of the temperature-time records within the weak zone and the unaffected zone reveals two effects: (1) temperatures are lower in the weak zone and (2) temperature increase in the weak zone lags the increase elsewhere. The second effect indicates the weak concrete started hydrating nearly 9 h after the adjacent zone of sound concrete.

Temperature differences for four of the flaws are plotted versus time in Fig. 2. The plot includes the temperature difference for each of the four flaws versus time (solid lines), with the temperature difference defined between the sensor nearest the intentional flaw and a nearby sensor deemed to be unaffected by the flaw. For reference, a dashed line representing temperature for the average of all sensors at the peak temperature depth is also included in the plot. The temperature differences are generally greatest at a time between one-third and one-half of the time to peak temperatures, which generally coincides with the maximum temperature rise (i.e. the steepest portion of the dashed line in the figure). The temperature difference associated with several types of flaws reduces to near zero by the time of peak temperature. This observation has important implications for interpretation of TIP data and reporting of TIP results, as discussed in Recommendations below.

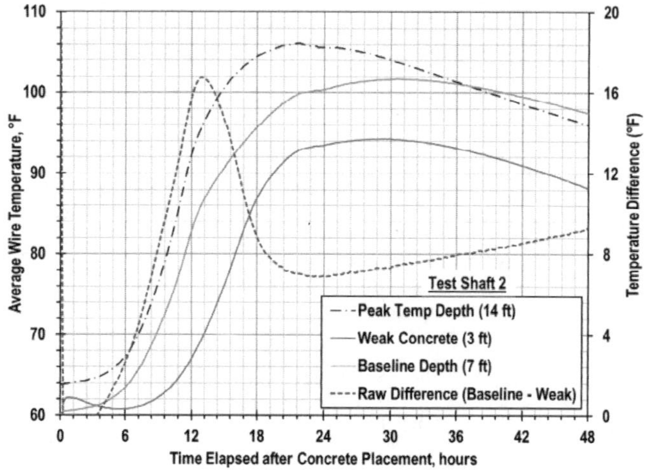

Fig. 1. Example temperature versus time plot for a drilled shaft concrete flaw. Based on Boeckmann and Loehr (2018).

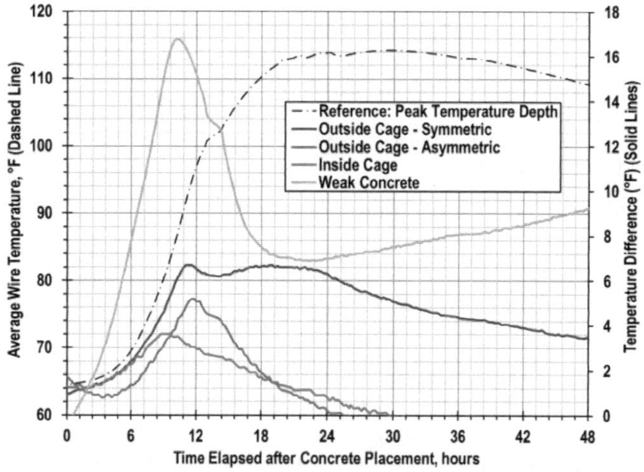

Fig. 2. Temperature difference versus time for different types of drilled shaft flaws. Figure is from Boeckmann et al. (2022), with data from Boeckmann and Loehr (2018).

Missouri DOT Study

The Missouri DOT funded a research study to evaluate potential implementation of TIP methods in 2019. Results of the study are reported in Boeckmann et al. (2022). Unlike the studies documented above, the Missouri DOT research was performed on production shafts rather than research shafts with intentional flaws. Four production shafts were tested, two shafts each at two bridge sites. The first site included 3.5-ft diameter shafts with 3-ft diameter sockets and a total length of approximately 20 ft. The second site included 6-ft diameter shafts with 5.5-ft diameter rock sockets. The total length of the shafts was about 90 ft, with approximately 20-ft long sockets. All shafts were permanently cased down to rock. Pre-pour soundings with a weighted tape indicated one of the production shafts at each site included soft bottom conditions. Field observations of the drilled shaft construction did not indicate any other significant flaws should be expected.

CSL results for the shafts with soft bottom conditions indicated velocity reductions of 15 to 18 percent and relative energy loss of 3 to 6 dB for the smaller soft-bottom shaft (at the first site) and velocity reductions of 12 to 25 percent and relative energy loss of 7 to 15 dB for the larger shaft (at the second site). Application of the DFI white paper criteria results in two 'B' ratings for the smaller shaft and three 'C' ratings for the larger shaft. No anomalies were reported at the base of the other shafts, i.e. those without evidence of soft bottom conditions from the weighted tape soundings.

TIP results for shafts at the second site (with larger shafts) are shown in Fig. 3 (temperature versus depth for half peak time), Fig. 4 (effective radius versus depth for half peak time), and Fig. 5 (temperature versus time at the base of the shaft). For both shafts, the temperature versus depth response is dominated by temperature differences across the shaft. The differences are likely the result of reinforcing cage alignment issues associated with resting the reinforcing cages on the base of the shaft. As is typical of TIP data, temperatures decrease significantly at the base of the shaft, but it is difficult to discern the effect of soft bottom conditions from the typical effect of temperature rolloff. The effective radius profiles of Fig. 4 perhaps show a more significant effect of soft bottom conditions. However, effective radius interpretations at the base of the shaft are sensitive to curve-fitting parameters and volume inputs. The soft bottom conditions are most evident in the temperature versus time records for the bottommost TIP sensors (Fig. 5), with more gradual temperature increases observed for the shaft with soft bottom conditions. The temperature-time plot for the soft bottom shaft shows relatively smooth curves; in contrast, the plot for the shaft without soft bottom conditions shows a relatively steep increase to peak temperature that ends with an abrupt transition to gradual post-peak decreases.

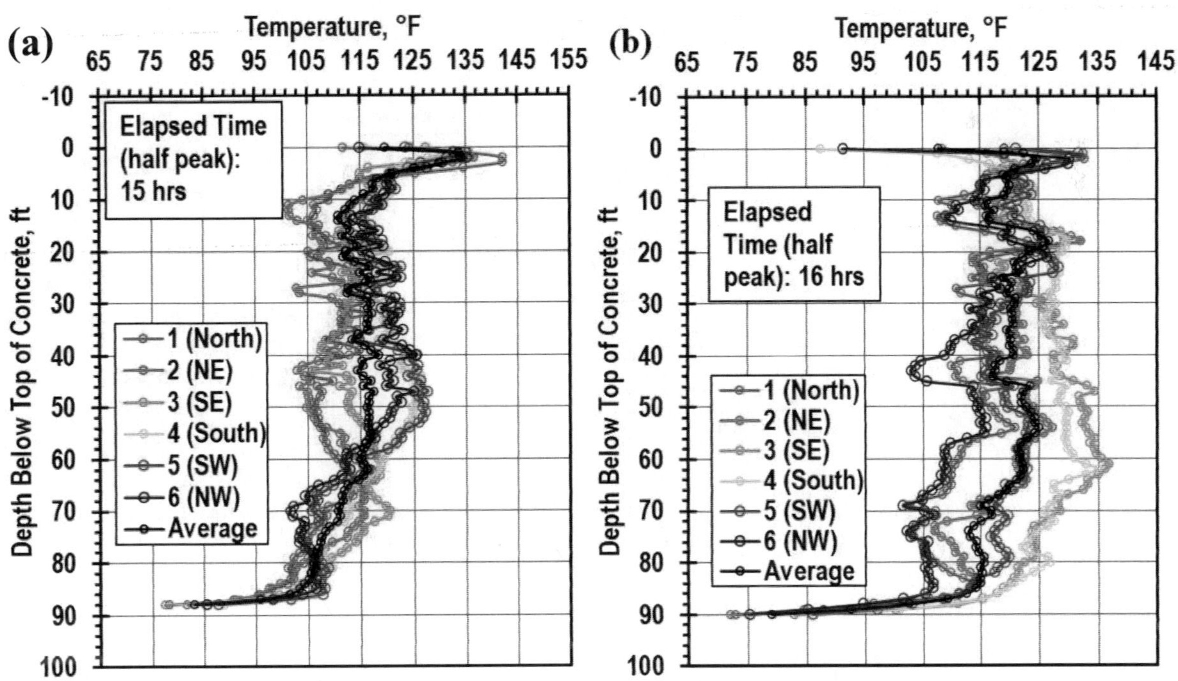

Fig. 3. Temperature versus depth at half peak for Missouri DOT study shafts: (a) shaft without soft bottom conditions and (b) shaft with soft bottom conditions.

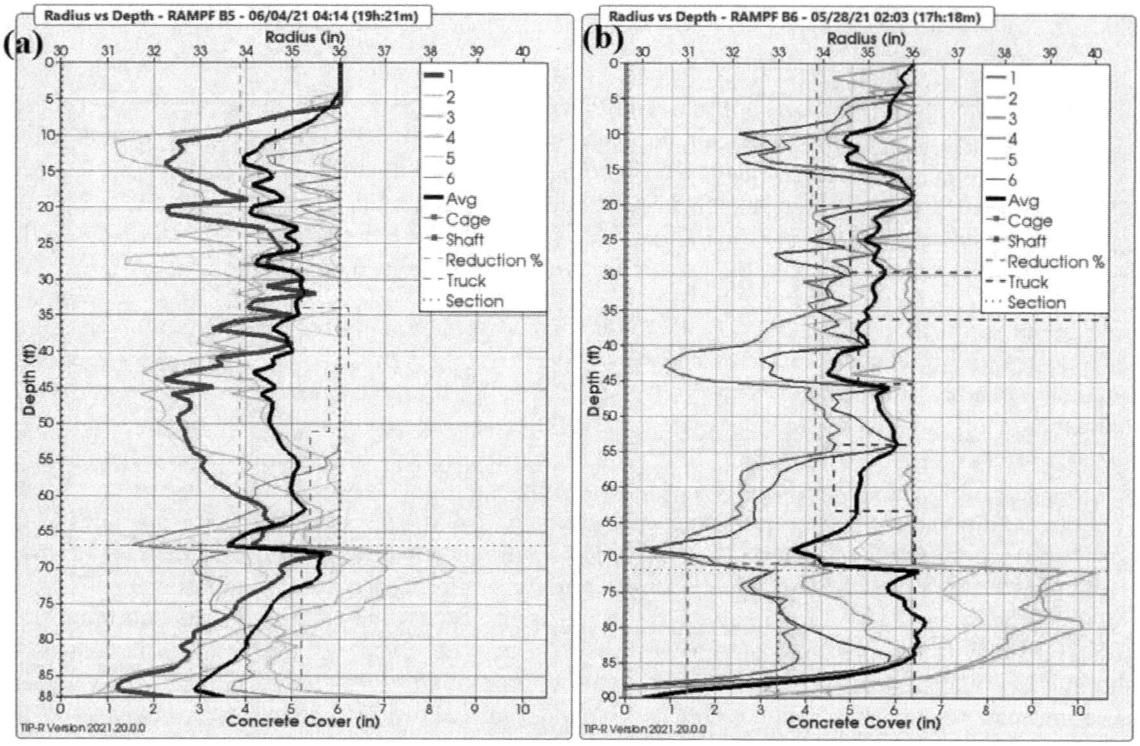

Fig. 4. Effective radius versus depth at half peak for Missouri DOT study shafts: (a) shaft without soft bottom conditions and (b) shaft with soft bottom conditions.

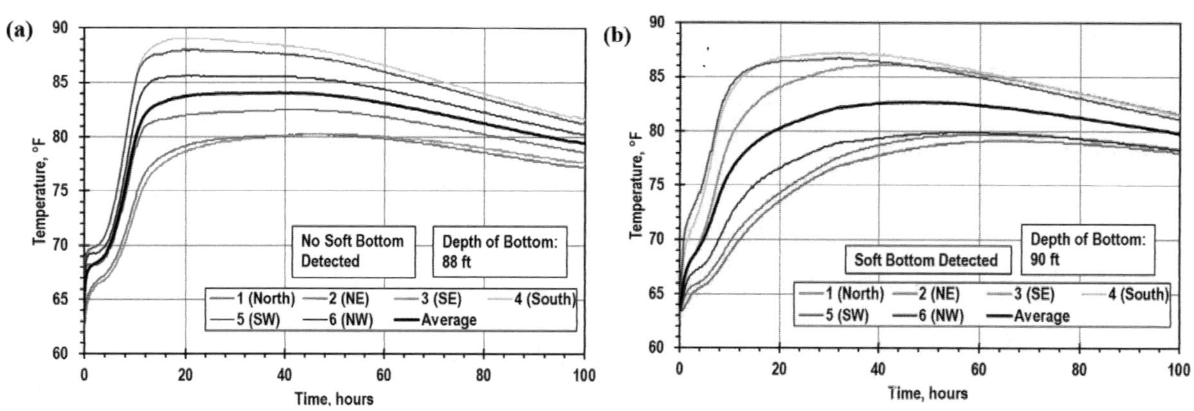

Fig. 5. Temperature versus time at the base of Missouri DOT study shafts: (a) shaft without soft bottom conditions and (b) shaft with soft bottom conditions.

SYNTHESIS OF FINDINGS

Results from the five studies summarized in the previous section support several conclusions with important implications for integrity testing practice. The conclusions are presented in three categories below: ability to detect various types of defects, consideration of energy in evaluation of CSL results, and consideration of time in evaluation of TIP results. Importantly, all of the studies presented in the previous section are based on relatively small diameter shafts, ranging from 3 to 6 ft. Extrapolating to large diameter shafts (i.e. 9 ft diameter or greater) is not supported by research, and evaluation of integrity testing results for large diameter shafts is typically more complicated, particularly for TIP results.

Ability to Detect Various Types of Flaws

An apparent objective of some of the previous research was to identify the minimum size of a flaw in order for it to be detectible with TIP or CSL. However, detectability with any integrity testing method depends on a number of factors, not just the size of the flaw. Detectability with CSL depends on the size of the flaw, its location relative to the CSL access tubes, and likely the stiffness contrast between the flaw and the drilled shaft concrete. For a flaw to be detectible, it must be located along the ray path of a tested access tube pairing and be of sufficient size. Ray paths between access tubes are necessarily within the reinforcing cage, so flaws outside of the cage are generally undetectable with CSL. Previous research has estimated the required size of a flaw for CSL detectability to be on the order of 10 percent of the shaft cross sectional area (e.g. Amir and Amir, 2009; Iskander et al., 2001). For TIP, detectability depends on the location within the shaft cross-section, the nature of the flaw (e.g., a soil inclusion versus trapped laitance versus weak concrete), boundary effects (e.g. flowing water, telescoping casing, roll-off at the top and bottom of shafts), likely the size of the shaft, the relative size of the flaw, and the method of interpretation as discussed subsequently.

Results of the Wisconsin DOT and Missouri DOT studies support the conclusion that CSL and TIP are complementary test methods with respect to their ability to detect different types of flaws. Neither test method provides a perfectly reliable assessment of drilled shaft integrity, but combined, they can be used to detect most conceivable defects. Detectability for various types of flaws is presented in Table 2. The table was originally presented in Boeckmann and Loehr (2018) but was modified after the Missouri DOT study to indicate soft bottom conditions are potentially detectible with TIP. As discussed in Recommendations below, the complementary nature of the two tests has important implications for selecting which test method to perform for a given drilled shaft.

Table 2. Ability to detect various types of flaws with TIP and CSL (Boeckmann et al., 2022).

Type of Flaw	TIP	CSL
Inclusions in cage interior	Difficult to detect	Readily detectible
Flaws outside reinforcing cage	Readily detectible	Not detectible[1]
Soft bottom	Potentially detectible	Readily detectible
Weak concrete	Readily detectible	Difficult to detect
Tremie breach	Potentially detectible	Readily detectible

[1]Conventional wisdom and understanding of the CSL test methodology indicates flaws outside the reinforcing cage are not detectible, but such flaws have been observed in CSL results reported in the Wisconsin DOT study (Boeckmann and Loehr, 2018) and in Graham and Axtell (2019).

Importance of Energy in Consideration of CSL Results

Reporting of CSL data typically includes plots that show arrival times (or velocity) as well as relative energy (or energy). However, interpretations (or rating) of the CSL data within the same reporting is often based only on arrival time delays, without explicit consideration of the relative energy loss. Because CSL data consists of a series of waveforms that vary not only in when the wave begins, but also in the amplitude of the wave, interpretations that neglect energy are essentially leaving information on the table, unconsidered. Although not well understood at this time, it is possible, if not likely, that different types of flaws are characterized by different effects on arrival times and energy. If so, flaws that affect energy more significantly than arrival time would potentially be missed by interpretation criteria that consider only arrival time. In addition, interpretations that consider energy as well as arrival time are less susceptible to interpretation differences resulting from how arrival time is defined (i.e., with threshold values in the algorithms that select arrival times).

Importance of Time in Consideration of TIP Results

Results of the South Carolina, Wisconsin DOT, and Missouri DOT studies all demonstrate the important role of time in interpretation of TIP results. Each of the studies included TIP results that show flaws are considerably more difficult to detect at peak temperature than earlier in the hydration process, for example at half of the peak time. Moreover, the Wisconsin DOT and Missouri DOT studies demonstrated that the temperature effect of flaws is more effectively identified when the TIP data are plotted versus time than plotted only versus depth. The temperature-versus-time plot associated with a flaw typically has a shape that is readily distinguishable from sound concrete: the delay in hydration is more evident in plots versus time than in plots versus depth. Compared to TIP plots versus time, TIP plots versus depth are more susceptible to nuisance effects associated with cage alignment and boundary conditions such as geometry changes, temperature rolloff at the ends of the shaft, and effects from flowing water or groundwater.

RECOMMENDATIONS FOR USE AND INTERPRETATION OF INTEGRITY TEST RESULTS

The findings presented in the previous section lead to several important recommendations for effective use and interpretation of integrity test methods. The recommendations are organized into three categories in the sections that follow: when to use CSL and/or TIP, interpretation of CSL results, and interpretation of TIP results.

One general recommendation regarding the use of integrity test methods is offered first: It is important to recognize and appreciate that drilled shaft integrity testing is not the sole quality assurance measure for drilled shafts. Quality assurance for drilled shafts is also derived from design (e.g., reinforcement spacing restrictions, concrete mix characteristics), construction methods (e.g., methods for maintaining hole stability, slurry control requirements, concrete placement requirements), controls on characteristics of fresh concrete (e.g., trial batching procedures, workability constraints and measurements), and inspection practices (e.g., verification of base cleanliness and monitoring of concrete placement). Integrity testing is therefore but a single tool in the toolbox of measures that assure drilled shaft quality.

Use of CSL and TIP

Both CSL and TIP are useful integrity test methods that can indicate the presence of flaws within deep foundation concrete. Moreover, the complimentary nature of the two tests' abilities to detect certain types of flaws underpins the following recommendations:

- The decision of which test to specify should be made by the Engineer of Record on the basis of which test method will more effectively reduce risks associated with construction defects. The decision should therefore consider project-specific characteristics such as the likelihood of various types of defects, diameter of the drilled shaft, reserve capacity of the design with respect to both lateral and axial loading, and the reliance on base resistance. For example, CSL is probably more appropriate for shafts with design that relies on considerable base resistance or shafts in geologic conditions where base cleanliness has been problematic. In contrast, TIP is probably more appropriate for shafts where defects outside the reinforcing cage are more likely or where lateral reserve capacity is limited.

- In some circumstances, it may be beneficial to perform both CSL and TIP. Such circumstances include method (aka technique or demonstration) shafts, critical drilled shafts (e.g., non-redundant shafts), and drilled shafts where interpretation of integrity test results will likely present challenges. An example of the latter is long, permanently cased shafts constructed using a concrete mix prone to producing significant bleed water. CSL access tube debonding is relatively common for such shafts. The addition of TIP wires to such shafts can be useful in demonstrating

CSL anomalies near the top of the shaft are the result of debonding. However, if both TIP and CSL are performed on production shafts, care must be exercised in the test evaluation. In many cases, it is more appropriate to use one test to demonstrate that anomalies from the other test are false positives, rather than treating all anomalies from both tests as assumed flaws.

- Regardless of which test is used, attentive and knowledgeable observation and documentation of drilled shaft construction is needed. Complete and accurate records of drilling, bottom of shaft conditions, and concrete placement are necessary for evaluation and acceptance of drilled shaft construction, not to mention for agency records. In addition, the construction records are critical for understanding and interpreting integrity test results. The practices recommended in FHWA's Drilled Shaft Manual (Brown et al., 2018) result in proper documentation.

Interpretation of CSL

The approach recommended in the DFI's "Terminology and Evaluation Criteria of Crosshole Sonic Logging (CSL) as applied to Deep Foundations" white paper (2018) should be used to evaluate CSL results. Interpretation of CSL results that considers both arrival time and relative energy likely results in better identification of flaws.

Interpretation of TIP

Several recommendations regarding TIP interpretation follow from the findings presented in this paper. The recommended evaluation practice represents an "all of the above" approach wherein temperature versus depth, effective radius, and temperature versus time are all considered:

1. Records of construction, including drilling records, bottom-of-shaft inspection forms, and accurate concrete volume logs should be reviewed to identify potential zones with flaws (e.g., depths of a tremie breach, potential soft bottom conditions, etc.). If he or she was not on site for construction, the engineer reviewing the TIP results should discuss the installation with on-site inspection personnel.

2. Temperature versus depth, effective radius versus depth, and results of any other concrete integrity tests should be reviewed to identify potential zones with flaws (e.g., depths with abrupt temperature changes). The TIP depth profiles should be reviewed for results from peak time and half-peak time.

    - Effective radius results can be used to evaluate TIP results but should not be used to infer concrete cover or otherwise interpret the precise shape of the shaft.

    - The engineer should attempt to identify causes of suspect TIP results (e.g., abrupt temperature changes) by considering shaft construction records, geologic information, surface water, and groundwater information.

3. Temperature versus time plots should be evaluated for any depths of suspected flaws identified during Steps 1 and 2. The plots should include records from each wire for the depth of the suspected flaw as well as a nearby depth that appears to be uninfluenced by the potential imperfection (i.e., the baseline measurement).

Because of the importance of the temperature versus time response to interpretation of TIP results, TIP testing should only be performed by the wire method. The probe method, which produces temperature information only at the time of data collection by personnel on site, should not be used.

SUMMARY AND CONCLUSIONS

Recent experience and research regarding drilled shaft integrity test methods has produced new insights regarding the capabilities of CSL and TIP as well as more effective techniques for interpreting both types of tests. Several recent research studies involving drilled shafts with intentional flaws and both CSL and TIP testing are summarized in this paper. The results of the studies demonstrate that CSL and TIP are generally complementary methods: each test is well-suited to identifying certain types of flaws but limited or incapable of identifying other types of flaws. Results of the studies also indicate CSL interpretation is more effective when it considers both arrival time and relative energy, and that TIP interpretation is more effective when it is based on temperature versus time response. As a result of the findings, a project-specific approach to selecting the appropriate integrity test method or methods is recommended. In addition, the methodology in DFI's CSL white paper (2018) is recommended for interpretation of CSL results, and evaluation of temperature versus time for all depths of interest is recommended for interpretation of TIP results.

ACKNOWLEDGEMENTS

The authors gratefully acknowledge the financial and administrative support of their agency partners, Missouri DOT and Wisconsin DOT. The authors are also grateful for the technical support and feedback from colleagues and industry friends, especially Erik Loehr, Mark Madgett, and Jim Zammataro.

REFERENCES

Amir, J.M. and Amir, E.I., 2009. Capabilities and Limitations of Cross Hole Ultrasonic Testing of Piles. Proceedings of International Foundation Congress and Equipment Exposition, Orlando, FL.

Ashlock, J.C. and Fotouhi, M.K., 2014. Thermal Integrity Profiling and Crosshole Sonic Logging of Drilled Shafts with Artificial Defects. ASCE Geo-Congress 2014 Technical Papers, GSP 234, 1795-1805.

Boeckmann, A.Z. and J.E. Loehr, 2018. Thermal Integrity Profiling for Detecting Flaws in Drilled Shafts. Final Report, Wisconsin Department of Transportation, WisDOT ID no. 0092-16-07, 113 p.

Boeckmann, A.Z. and J.E. Loehr, 2019. Evaluation of Thermal Integrity Profiling and Crosshole Sonic Logging for Drilled Shafts with Concrete Defects. Transportation Research Record, 2673(8), 86-98.

Boeckmann, A.Z., Yeskoo, R.A., and Axtell, P.J., 2022. Use of Thermal Integrity Profiling (TIP) in Drilled Shaft Evaluation. Final Report, Missouri Department of Transportation, MoDOT ID no. cmr 22-002, 92 p.

Brown, D.A., Turner, J.P., Castelli, R.J., and Loehr, J.E., 2018. Drilled Shafts: Construction Procedures and LRFD Design Methods. Federal Highway Administration GEC No. 10, FHWA-NHI-18-024, 756 pp.

Graham, D.S. and Axtell, P.J., 2019. Case History: Comparison of CSL Results to Physical Observations. Proceedings: Deep Foundations Institute 44th Annual Conference, pp 420-427.

Iskander, M., Roy, D., Ealy, C., and Kelly, S., 2001. Class A Prediction of Construction Defects in Drilled Shafts. Transportation Research Record, 1772(1), 73-83.

Mullins, G., Johnson, K., and Winters, D. 2007. Thermal Integrity Testing of Drilled Shafts. Final Report Prepared for Florida Department of Transportation, 214 p.

Schoen, D.L., Canivan, G.J., Camp, W.M. III, 2018. Evaluation of Thermal Integrity Profiling (TIP) Methods—Probe, Embedded Wire, and Wire Suspended in CSL Tubes. American Society of Civil Engineers Proceedings of IFCEE 2018, GSP 294, 550-560.

HARRY NICE BRIDGE GEOTECHNICAL INSTRUMENTATION CASE STUDY

Kevin Tehansky, P.E. – Author, Foundation Test Group, Inc., Owings Mills, MD, USA, 443-605-8953, ktehansky@foundationtestgroup.com
Jeffrey Goodwin, P.E. – Contributor, Foundation Test Group, Inc., Owings Mills, MD, USA, 410-517-0715, jgoodwin@foundationtestgroup.com
Melih Demirkan, Ph.D., P.E. – Contributor, Rite Geosystems, Pittsburgh, PA, USA, 412-745-8300

ABSTRACT

Just an hour drive down MD 301 from National Harbor, construction to replace the Governor Harry W. Nice Memorial/Senator Thomas 'Mac' Middleton Bridge (HNB) began in spring of 2020. The Maryland Transportation Authority (MDTA) contracted with Skanska-Corman-McLean (SCM) to deliver a new US 301 bridge spanning 1.9 miles from Maryland to Virginia over the Potomac River aiming to improve capacity, safety, and access of the roadway. The project area presented a geotechnical design challenge due to the installation of new bridge foundations in the problematic soils of the Atlantic Coastal Plain alluvial deposits. With the existing bridge remaining in service during construction, and the site's logistical constraints placing new pile driving operations near existing abutments and piers, a complex instrumentation and monitoring plan was developed to ensure public safety throughout the project. This paper discusses the geological conditions of the project area, the collaborative approach of various design team members to develop an appropriate monitoring plan utilizing various instrumentation, an understanding of baseline conditional movements of the existing bridge, and impact response of adjacent construction activity to help deliver effective new bridge foundations while protecting the old ones.

Keywords: geotechnical, instrumentation, monitoring

PROJECT INTRODUCTION

After more than a decade of planning studies with consideration to project costs, schedule, environmental impact, economic impact, access for various modes of transportation, existing bridge condition, and remaining service life, a new US 301 bridge spanning from Maryland to Virginia was planned as a four-lane span with deep foundations that would be constructed less than 150 ft north of the existing alignment of the more than 80-year-old bridge shown in Figure 1. Travel for the new bridge is anticipated to open by early 2023 with a design life of more than 100 years, servicing 12-ft wide lanes with 2-ft shoulders, doubling capacity, enhancing emergency response, upgrading maintenance and inspection activity access, and featuring lane sharing for cyclists.

Fig. 1. Facing MD Shore, existing 2-lane HNB shown on right, construction of new bridge on left

GEOTECHNICAL BACKGROUND

The site geotechnical investigation performed by the project Geotechnical Design Manager, Schnabel Engineering (2020), revealed several construction risk factors that could cause in-situ soil movements along the alignment of the bridge. Beneath the top layer of uncontrolled fill depths up to 34 ft, the Mid-Atlantic Coastal Plain geology of the project site consists of recent terrace and alluvial soil deposits overlying older lower tertiary soils of the Nanjemoy, Marlboro, and Aquia Formations. The terrace and alluvial soils ranged from depths of up to 46 ft at the shores to depths of up to 140 ft below the mudline in the deepest sections of the Potomac and are composed of soft to very soft clays with layers of sand and gravel. The lower tertiary soils encountered below the terrace and alluvial deposits consist of stiff to very stiff or medium dense to very dense clay and sand layering, some of which is cemented.

Due to the existing bridge being kept in service during construction of the new bridge, the composition and layering of the various soil stratums presented two major risk components in relation to deep foundation installations: the effects of pile driving on existing structures, and the effects of embankment construction on existing bridge foundations. Risk analysis for the evaluation of adjacent pile driving lateral displacements and vibrations were performed using cylindrical cavity expansion process by M.K. Chong and pile driving induced vibrations methods developed by the California Department of Transportation. Risk analysis for evaluation of embankment settlement for potential downdrag on existing bridge foundations was performed using soil parameters and methods contained in FHWA GEC-5 Geotechnical Site Characterization (2107) and modeled with 2D finite element analysis.

The risk analysis of site soils determined that even with pile driving operations of 36-inch square pre-stressed concrete piles occurring as little as 35 ft away from the existing bridge and embankment fills of up to 30 ft directly adjacent to the existing bridge, the impact of lateral displacement, vertical heave, soil densification, and vibrations on existing substructure units would fall below structural impact design tolerances. However, to ensure protection of the existing bridge structures and verify the structural impact of adjacent pile driving, geotechnical instrumentation was recommended to monitor pile driving activity.

Historical use and geologic investigations of both the Maryland and Virginia shores determined that each embankment, with fills of up to 30 ft, should anticipate settlement of up to 8 inches due to the very soft, loose, and compressible nature of the underlying soils. In addition, calculations indicated that the weight of embankment fill could impose settlement of up to 0.4 inch along existing bridge foundations and potentially 0.6 inch of lateral squeeze in the soft clay stratum, requiring additional geotechnical instrumentation monitoring for embankment operations.

GEOTECHNICAL INSTRUMENTATION PLAN

As the Geotechnical Instrumentation Specialist, Foundation Test Group's (FTG) role on the project was to provide engineering services to monitor conditions of the existing bridge utilizing the known geotechnical parameters. In the interest of public safety across the length of the bridge and each embankment approach, FTG developed an instrumentation monitoring plan that evaluated the movement of soils and existing bridge structures, including the abutments, bents, piers, roadway deck, walls, and trusses. To understand the impact of adjacent heavy construction on the new bridge and determine pre-construction baseline movements of the existing bridge, a combination of more than 140 instruments of varying types were installed to monitor structural performance.

Automated total stations were installed at three locations to monitor 122 prisms across the bridge structure at various elevations on the deck, piers, and abutments. For stress level monitoring of steel trusses, 8 strain gauges were installed every 500 ft along the main navigation span. A partial view of various instrument type locations around the main span is shown in Figure 2. For rotational monitoring of piers and abutments, 20 tilt sensors were installed at both pier cap and bridge deck elevations. For vibration monitoring of adjacent pile driving, 10 triaxial vibration sensors were installed and rotated as construction progressed. For monitoring of existing deficiencies, 2 automated crack meters were installed on concrete piers. For embankment monitoring, 7 inclinometers, 14 vibrating wire piezometers, and 16 settlement plates were installed between each shore approach.

Fig. 2. Partial Geotechnical Instrumentation locations around existing High-Level navigation span

Along with existing soil parameters, having a thorough background on the structural condition of the existing bridge was critical to the selection of instrumentation monitoring types and locations. To supplement regular structural assessment reports provided by the AECOM design team, FTG performed several pre-construction drone inspections of the bridge. The drone footage provided a continuous perspective of bridge conditions that would not have been reasonably accessible otherwise, and which permitted a targeted approach to instrumentation. After consideration of bridge and geologic conditions, a monitoring plan was implemented within multiple site constraints, both anticipated and unforeseen, that worked for the project design-build team.

Early in the project, budgetary considerations had to be made to accommodate the unique settings of the remote project location with tough environmental conditions. With limited time and resources available to access instrument locations via water vessels or bridge deck aerial rigging, efficient management of determining instrument types and quantities was made in the field to get enough monitoring coverage and successfully mount units to the 80-year-old concrete and steel structure that was difficult to penetrate. Accommodations were engineered to provide protection from bird nesting and droppings. Initially, project plans were to have all instrument units communicate through a site Wi-Fi network that ran the length of the bridge alignment. As it was found, the cost of running the Wi-Fi network was not reasonable and instrumentation was pivoted back into using cellular network connections. After months of programming multiple SIM card types, trying different network providers, exploring other cell tower connections, and overcoming unexplained signal interferences, each instrument type was able to find its 'sweet spot' of communication in time to provide value to the project management team prior to pile driving and embankment operations.

To help overcome a major project challenge of lack of reasonable access to the desired instrument locations, selected automated instrumentation equipment supplied by Rite Geosystems and Sigicom allowed for one-time installation and remote setup for most instruments through a 24/7 web-based platform. This two-way remote communication to instruments, such as the robotic total stations and vibration sensors, provided real-time adjustments to help FTG evaluate and update monitoring levels remotely through the transition from baseline to construction phases while allowing the contractor to maintain schedule.

The line-of-sight conditions for the structural monitoring points were heavily restricted by several site factors, including property limits, proximity of adjacent construction, haze disturbance over the brackish water, temporary installments of contractor as-built prisms overlapping monitoring prisms, and inability to practically install dedicated reference structures in the waterway. To avoid the progression of new construction inhibiting prism sight lines, the robotic total stations were installed between the new and old bridge alignments, creating difficult line of sight angles. Along the length of the 1.9-mile construction limits, project budgets allowed for three total station setups: one installed at each of the MD and VA shoreline and one installed near midway on the existing bridge fender structure at the navigation span on Pier 14.

The available monitoring locations also limited the amount and locations of reference prisms able to be installed on site. With a lack of backsight prism referential control, validity and consistency in engineering analysis of the data had to be created in other ways. One way to reestablish data redundancy and understand the bridge behavior was to overlap the quantity and type of instrument in a single critical location to provide multiple methods of structure evaluation, as shown in Figure 3.

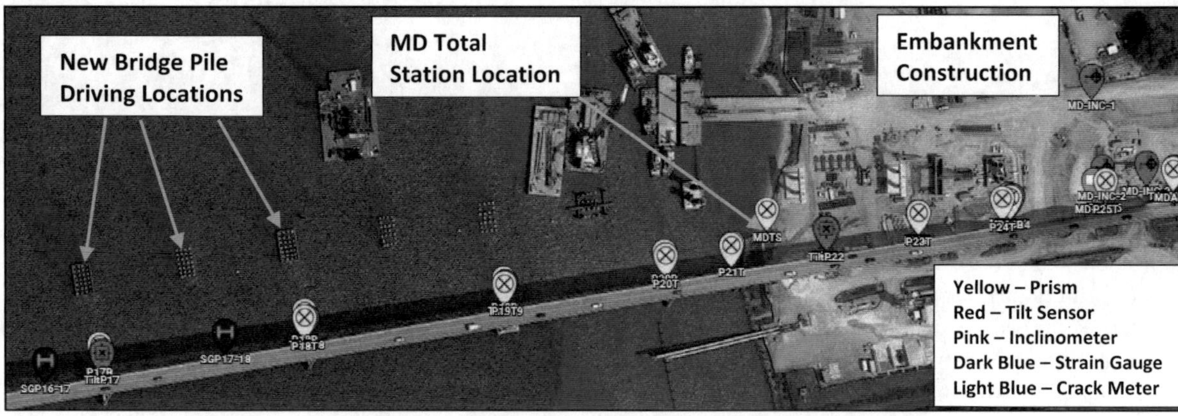

Fig. 3. Maryland Shore and High-Level Approach Instrumentation Location Plan

At one of the deepest bridge locations, navigation span Pier 14 was equipped with:

- Two tilt sensors – one at the bridge deck elevation and one at the pier water level elevation.
- Three structural monitoring prisms – one at the bridge deck elevation and two at the pier water level elevation, each facing a different total station.
- One strain gauge on an existing steel truss member.
- One vibration sensor measuring vibrations on the existing pier cap.

A key role of the Geotechnical Instrumentation Engineer, as in all monitoring projects, was to review the body of data and alerting that came through on instruments and be able to make engineering judgements on what impact the adjacent construction had on the existing bridge.

Through FTG's team of engineers, continuous monitoring of the data trends was performed, sometimes through evenings and weekends during critical construction stages, to evaluate triggered alerts for validity and consideration of the types of causes. Common alert causes observed during the baseline monitoring period included thermal expansion and contraction of the structural material on which the instruments were installed due to temperature cycles, and impact from traffic loads on bridge deck elements that experienced frequent vibrations. With supply chain shortages starting to develop in the summer of 2020 and contractor schedule ramping up, the baseline monitoring phase quickly transitioned into the construction monitoring phase with little opportunity to increase the instrument coverage. Having a strong geotechnical analysis of the instruments in place during the beginning stages of construction was integral to project success.

Throughout the project the instrument data was routinely analyzed for performance and adjustments were made to tighten or loosen design alert threshold levels accordingly. Temperature-induced effects were observed on both a daily and seasonal cycle, with greatest impacts during the winter period as the freezing point was routinely crossed. Another constant factor in data analysis that required early design changes was the effect of dynamic traffic loads on the existing bridge deck designed for anticipated flexural movement. Tilt sensors installed on the bridge deck parapet wall typically experienced tilt fluctuation magnitudes of +/-0.3 degrees greater than accompanying tilt sensors installed at the water level elevation on existing pier caps, as shown in Figure 4.

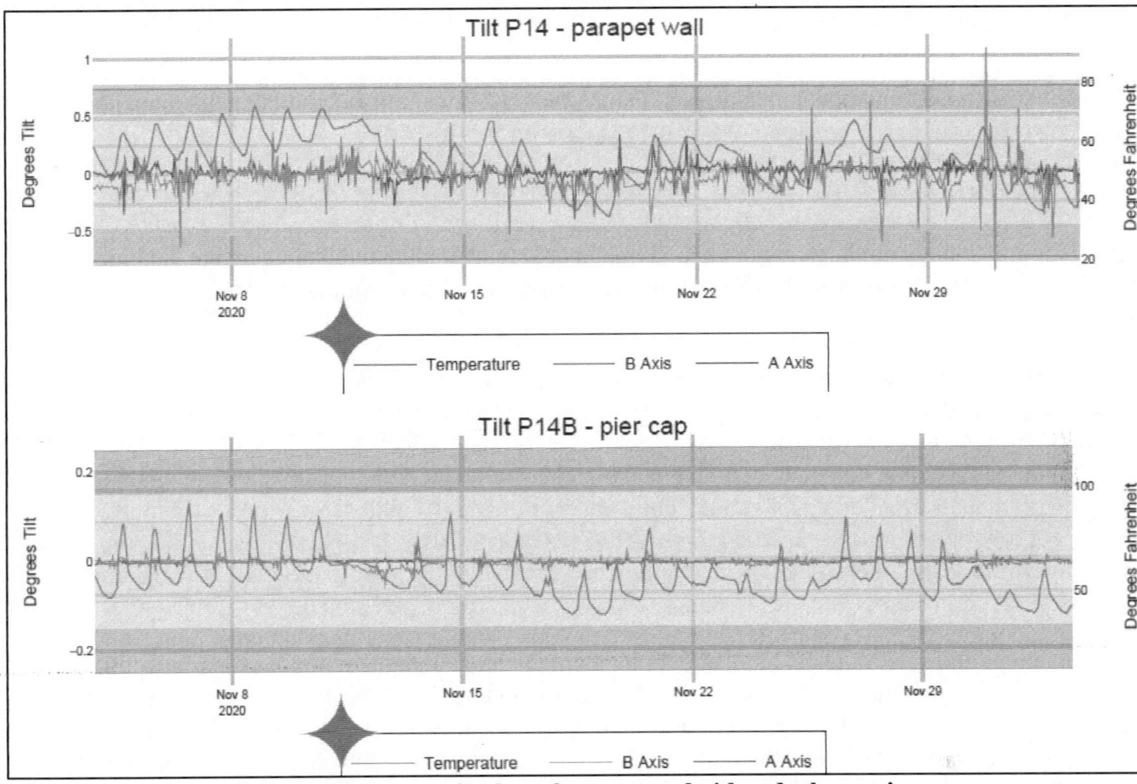

Fig. 4. Comparison of tilt reading magnitudes of sensor on bridge deck vs. pier cap

RESULTS OF GEOTECHNICAL INSTRUMENTATION AND MONITORING

Beginning in May 2020, vibration sensors were installed on the existing bridge from the Maryland Abutment to the Virginia Abutment for structure vibration monitoring. Vibration sensors were regularly relocated to various Bents and Piers to coincide with adjacent pile driving operations as they advanced. An evaluation of each vibration sensor dataset over at least a 20-month long monitoring period, in conjunction with data observed from tilt sensors and prisms at same locations, showed no long-term structural effects on the existing bridge due to construction-induced vibrations. Although construction activity did not trigger alerts, some alerts were generated between several locations throughout the monitoring phase due to impacts from unknown vessels, waves, weather, or debris. In each of those scenarios, the cause of the alert was investigated on site by members of the project team, readings immediately returned to baseline levels, and any necessary repairs to instrumentation units were made.

Over the course of the project, pile driving activity at several locations produced elevated vibration levels in existing structures above baseline levels. However, vibration readings were never observed above design alert threshold level of 0.5 in/s established by the MDTA's Performance Specifications (2018), example shown in Figure 5.

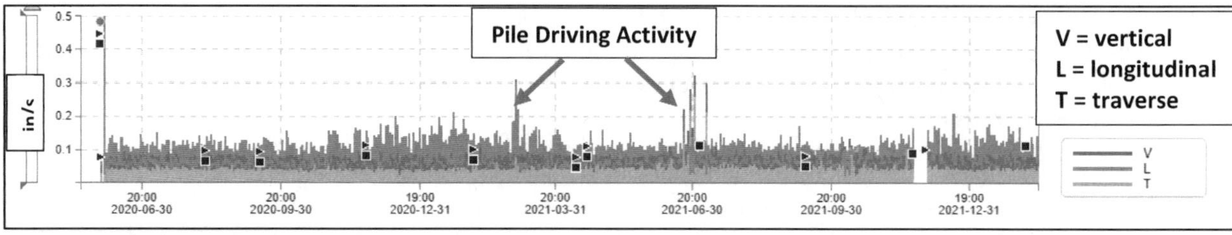

Fig. 5. Vibration readings at VA Abutment from May 2020 to January 2022, with acute measured effects of nearby pile driving during March 2021 and June/July 2021

The precast concrete piles were designed to develop their side resistance in stiff Nanjemoy and Marlboro soils as well as dense to very dense Aquia sands. New pile driving impacts that were expected to occur at the adjacent bridge substructure units, which consisted of a combination of steel, concrete, and timber piling support, had calculated ranges up to 0.7 in of lateral ground displacement and 1.1 in/s of anticipated peak particle velocity.

On July 6, 2021, increased lateral ground movement of up to 2 in was observed at VA-INC-4, at a depth of 55 ft below grade, during and after pile driving activity at the adjacent Virginia Abutment A, shown in Figure 7. The proximity between new abutment pile driving, the inclinometer, and the existing bridge abutment can be seen in Figures 6 and 8. The 36-inch square pre-stressed concrete pile had a required driving design resistance of 954 kips, minimum plan depth of 93 ft, with blows per foot ranging from 4 to 7 at the depths where soil displacement was greatest. Reevaluation of the original geotechnical report analysis determined that the soil displacements were caused by a combination of shoving and heaving between various soil strata with varying consistencies as the nearby pile was driven.

Due to the magnitude of the lateral soil movement profile, data from the inclinometer and other nearby instrumentation on the bridge were analyzed by FTG and Schnabel to determine the pile driving impact on existing bridge foundations. Lateral ground movement at the edge of the existing abutment due to pile driving was predicted to be approximately 0.4 inch from original Geotechnical analysis. Utilizing the measured inclinometer movements and proximity factors, additional finite analysis determined that soil movement at the closest existing pile was approximately 0.6 inch, on average. With expected soil movements above original design tolerance, the instrumentation data at the abutment was analyzed in relation to structure tolerance of +/-0.5 inch total settlement or heave, as set forth in the MDTA's Performance Specifications (2018). After pile driving, the data from the inclinometer indicated no additional movement, the abutment structural monitoring prism indicated no settlement or heave, and nearby tilt sensor data showed no rotational movement, allowing construction and bridge service to safely continue.

Fig. 6. Virginia Abutment Instrumentation Location Plan

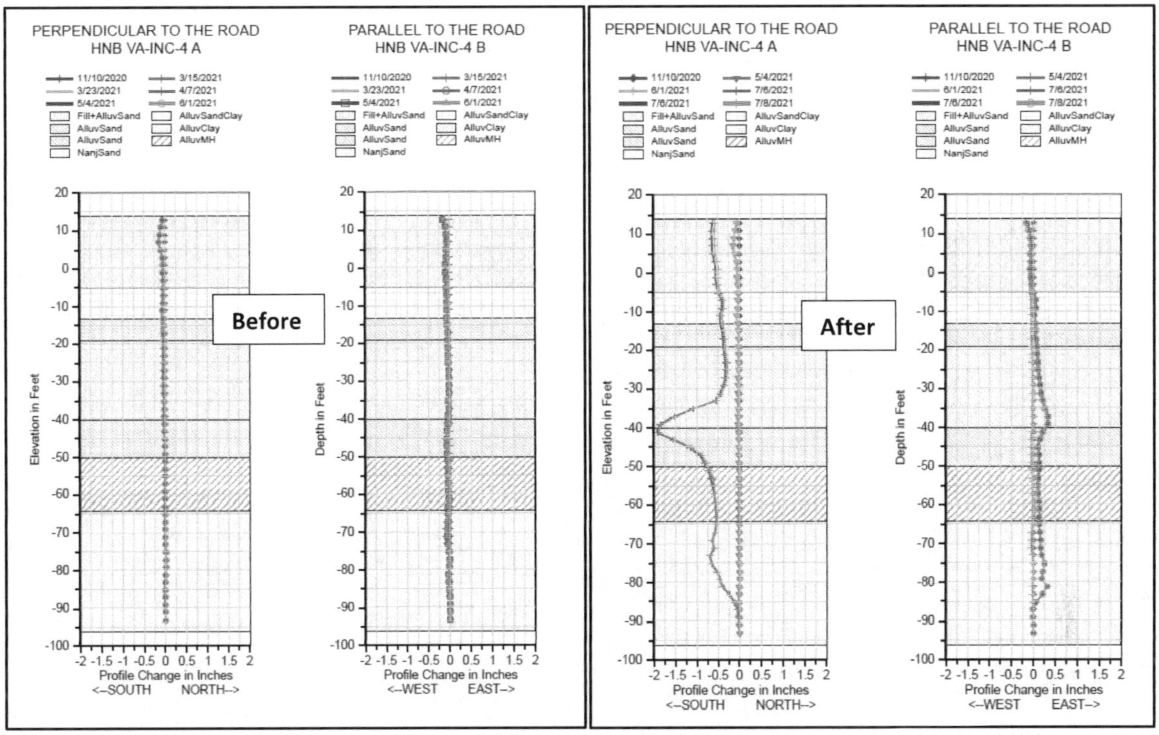

Fig. 7. Inclinometer readings of VA-INC-4 before and after nearby pile driving activities

Fig. 8. Reading of VA-INC-4 between new and existing bridge structures

As the project evolved into phases of simultaneous superstructure and substructure installations on the critical path, an unanticipated construction risk factor developed. Adjacent to existing bridge Pier 14, steel fender pile driving operations and new pier column concrete pour were scheduled to occur within 50 ft of each other on the same day, shown in Figure 9. The concern was that pile driving at that proximity would induce excessive vibration of freshly placed concrete. In a real-time collaboration with the project team, FTG coordinated a vibration monitoring plan for the new column to measure the impact of the pile driving. A vibration sensor was installed on the formwork of the new column to be poured to record data continuously throughout the concrete pour and during the initial 24-hr curing period. Peak particle velocity design tolerance of 0.25 in/s was established for fresh concrete (Freyne, S.F. and Watkins, M.L., 2013) with considerations to drive energies and wave velocity paths through new and existing piles within the fender-to-pier cap layout. Vibration readings were monitored during and after the pour for verification of vibration impact within tolerance, see Figure 10. Collaboration of all project team members determined that pile driving could continue safely adjacent to the newly placed column without the threat of damage and that the column was not subject to adverse amplitudes.

Fig. 9. Pier 14 Fender Pile Driving and new column pour operations

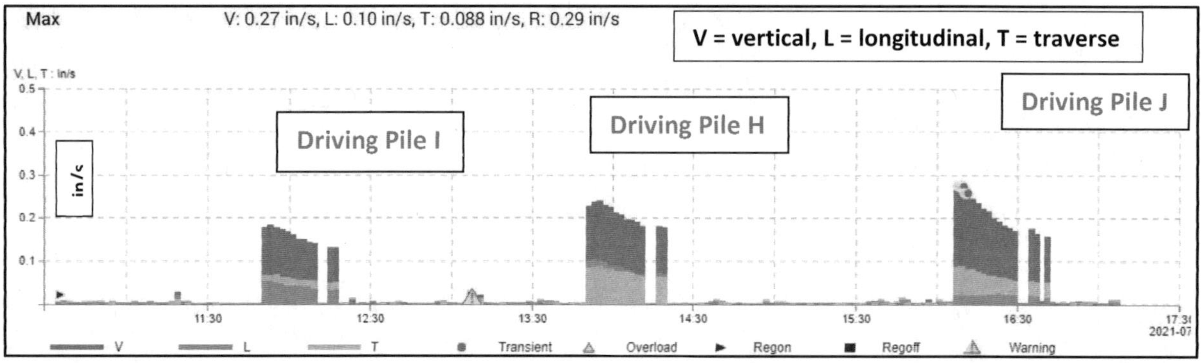

Fig. 10. Vibration data during pile driving activity as measured at the site of freshly placed concrete

Settlement plates for monitoring of the Virginia embankment and surcharge operations from November 2020 through June 2021, and settlement plates for monitoring of the Maryland embankment operation from November 2020 through September produced soil settlement values at or below expected design estimates based on the compressible soil layers encountered in the geotechnical investigation. Total embankment settlement on the Maryland side reached as high as 6.5 in, as seen in Figure 11, while total settlement on the Virginia side reached as high as 2.4 in, as seen in Figure 12.

Maryland and Virginia settlement data over time was reviewed by FTG and confirmed by Schnabel when settlement had substantially dissipated enough at each embankment to warrant removal of surcharge, continuance of fill operations, or approval of roadway construction. Lateral squeeze in the soft clay layers and overall embankment stability were also monitored using deep inclinometers to verify that no drag or lateral loads had any effect on existing bridge abutment and pier piles as close as 35 ft away.

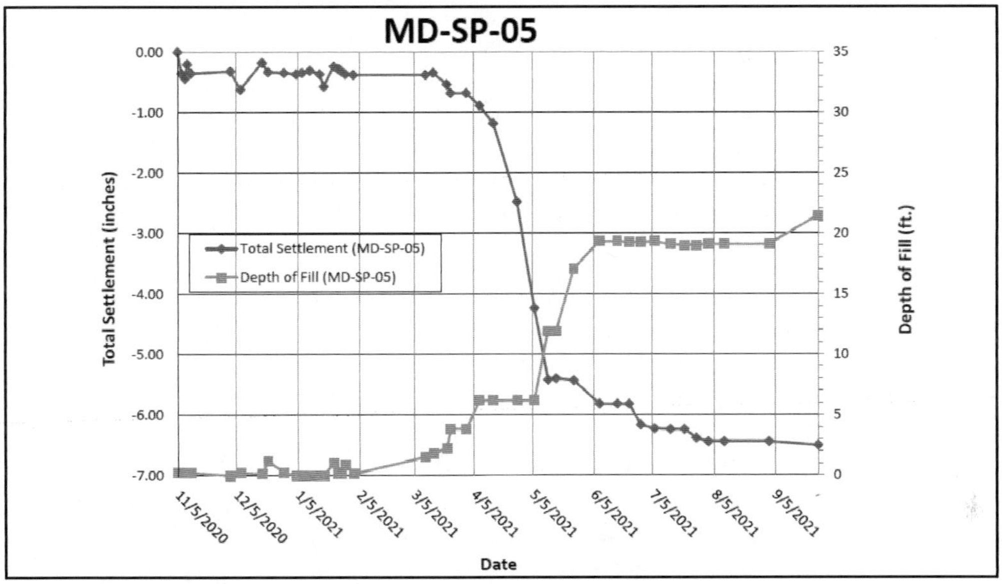

Fig. 11. Readings from Maryland settlement plate MD-SP-05

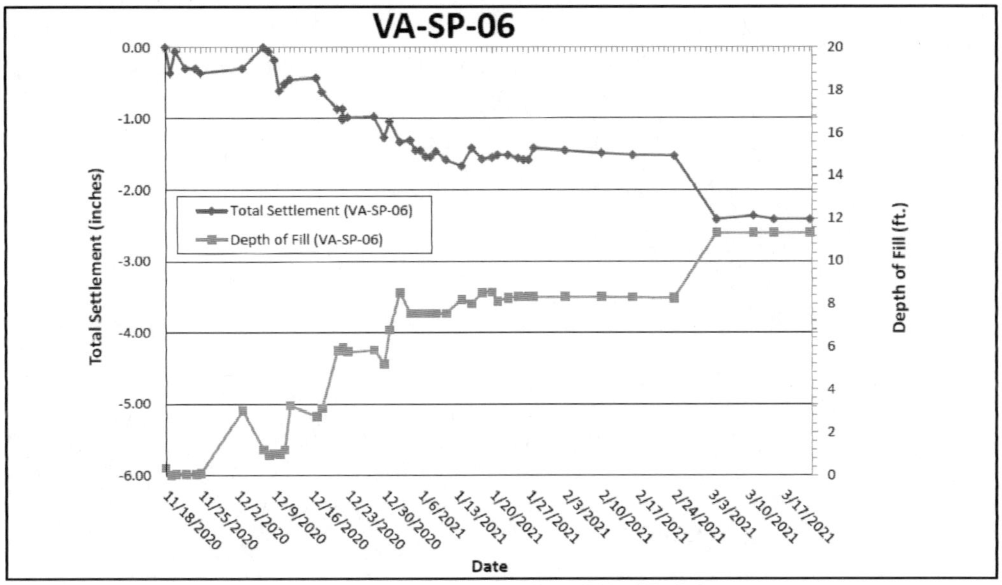

Fig. 12. Readings from Virginia settlement plate VA-SP-06

In conjunction with the settlement plates on the VA and MD embankments, piezometers were used for pore water pressure monitoring during advancement of construction fill operations. Due to the granular nature of most soil strata encountered at piezometer installation elevations, pore water pressures showed minimal increases with each phase of fill loading and typically dissipated to original levels within a few days allowing construction activity to advance at a reasonable pace.

CONCLUSION

The data acquired through geotechnical instrumentation monitoring demonstrated that the two high-risk construction activities on the project, pile driving and embankment fill operations, did not have any long-term impact on the existing bridge over a 2.5-year monitoring period. As demonstrated by the discrete events highlighted above, minor effects of construction were seen, but did not trigger alerts set in place for the protection of the existing bridge and its components. Instead, it was found that weather, traffic, and debris were the primary causes of alerts generated by the instrumentation, with seasonal and daily temperature fluctuations having a measured effect as well.

Successful monitoring of the health and performance of a bridge beyond its intended service life was achieved through the careful planning and utilization of selected instruments. Deploying each instrument for a specific purpose avoided overcollection of data and maintained a manageable level of daily evaluation. The use of various overlapping instrument types in similar areas allowed FTG to continually assess trends and limit risk to the public safety.

The value of geotechnical instrumentation monitoring is demonstrated by what does not occur, and is unique to each type of structure, the work being performed, and conditions surrounding it. For the Harry Nice Bridge, owners and contractors were able to lean on an experienced team of Geotechnical Engineers to manage bridge foundation integrity while the project remained on schedule. With only months of superstructure construction left and monitoring of various instruments still in place, the project team is geared into position for a successful topping off in early 2023. Bridge construction progress from July 2022 is shown below in Figure 13.

Fig. 13. Harry Nice Bridge new construction progress July 14, 2022

REFERENCES

Schnabel Engineering, LLC, 2020. Harry W. Nice/Thomas 'Mac' Middleton Bridge Replacement. Geotechnical Report – Channel Spans NB-0543-DSUB-040.008, Geotechnical Report – High Level Approach Spans NB-0543-DSUB-042.005, Geotechnical Report for MD Approach Roadway, Walls, and Embankments NB-0543-DSUB-049.004, Geotechnical Report – Low Level Approach Spans NB-0543-DSUB-058.004.

Maryland Department of Transportation and Maryland Transportation Authority, 2018. Harry W. Nice/Thomas 'Mac' Middleton Bridge Replacement Design-Build Project Specifications, Contract No. NB 0543-0000, Part II – Performance Specifications, 6.5.8 Geotechnical and Structural Instrumentation Monitoring.

Freyne, S.F. and Watkins, M.L., 2013. Acceptable Vibrations on Green Concrete, Final Report, Mississippi Department of Transportation, State Study 252.

ACTIVE SOIL NAIL WALL UNDERPINNING SOLUTION FOR THE BRUCE MUSEUM PROJECT

Jonathan L. Ernst, P.E., Geosciences Testing and Research, Inc., North Chelmsford, MA, USA
Les R. Chernauskas, P.E., Geosciences Testing and Research, Inc., North Chelmsford, MA, USA
Martin B. Hudson, Ph.D., P.E., Turner Construction Company, Los Angeles, CA, USA
Rolff B. Knobel, Turner Construction Company, Shelton, CT, USA
Andrew K. Cho, Ph.D., P.E., P.G., Turner Construction Company, New York, NY, USA

ABSTRACT

The Bruce Museum, located in Greenwich, Connecticut, is home to an extensive display of art and natural history exhibitions. A large addition was designed to be constructed adjacent to the museum requiring new spread footings to be installed up to 16.5 feet below the existing foundations. The soil conditions at the site consist of highly variable sand and gravel overlying weathered rock and sound bedrock.

The Contract Documents called for traditional foundation underpinning. Traditional underpinning techniques, however, presented two major challenges on this Project; (1) deep underpinning pits would be required involving multiple levels and staging with tieback support and (2) the highly variable soil and bedrock composition and elevations presented uncertainty in the design and construction of the pits. In lieu of traditional underpinning, Geosciences Testing and Research, Inc. (GTR) proposed an "active" soil nail underpinning system. A major reason that the soil nail underpinning option was feasible, related to the fact that the soil nails and facing could be installed slightly in front of the existing footings rather than directly underneath the existing footings. The soil nail wall design involved one to four rows of soil nails with nail positions and wall depth able to be adjusted "on the fly" as excavation progressed depending on the field conditions encountered. To create the "active system," the soil nails were pre-loaded to mitigate wall deformations that would be associated with a standard passive soil nail wall design.

To evaluate the performance of the active soil nail wall and confirm design assumptions, an extensive automated monitoring, on-site testing and inspection program was proposed and executed by GTR to monitor deformations of the wall and the museum in real-time, evaluate and confirm bedrock and soil bond stresses and ensure that each soil nail was pre-loaded and locked off. Careful analysis, planning, and use of the observational method illustrates the importance of collaboration between design and construction personnel and applying engineering judgement throughout construction. The use of the observational method includes the potential for modifying design based on conditions observed. This paper presents a unique case study of the "active" soil nail underpinning system to support and protect sensitive structures, close collaboration between design and construction facilitated by real-time deformation monitoring, and its successful application as a safe and cost-effective alternative to traditional underpinning.

SITE/BUILDING HISTORY AND PROJECT BACKGROUND

The Bruce Museum is an art science and natural history museum that features natural science located in Greenwich, Connecticut. The museum was originally built as a private residence in 1853, and has undergone several renovations, including a 1992 renovation being the most recent prior to the current construction. The current renovation is part of a $60 million capital expansion plan for the "New Bruce". The renovation includes the installation of a 43,000-square-foot, three-story addition that will more than double the size of the museum, adding space for new exhibitions, education and community spaces. The proposed foundations for the renovation sit up to ±16 feet below the Bruce Museum's existing shallow foundations.

PROJECT/GEOTECHNICAL CHALLENGES

To excavate and install the proposed foundations up to 16 feet below the existing foundations, traditional underpinning was proposed by the Geotechnical Engineer of Record. Eastern Excavation, Inc. of Elmsford, New York, was retained by the Owner to perform the required earthwork and excavation support. Geoscience Testing and Research, Inc. (GTR) of North Chelmsford, MA was retained by Eastern Excavation, Inc., to design the temporary underpinning system. Based on the underpinning height, GTR considered the feasibility of traditional underpinning and considered it unnecessarily hazardous and too complex to construct at the site. Several options were discussed with the Eastern Excavation and the Project Engineers and Architects for an alternative to traditional underpinning. Based on the "Soils and Foundation Investigation Report, 2015" (Beattie, 2015) and results from drill probes performed, the depth to competent bedrock was variable across the site, ranging from approximately ±3 feet (1m) to ±16 feet (5m) below existing grade. In addition to the bedrock variability, sloping bedrock was anticipated beneath the traditional underpinning, posing issues with sliding. Excavation would require removal of bedrock beneath the upper underpinning tier and the existing foundations, whereby excessive vibrations due to excavation activities could cause movement and instability.

Based on the required cut heights, the traditional underpinning approach would create the need for an anchor system. Underpinning and supporting ±16 feet of soil with surcharge load from the existing building foundations would require two (2) levels of anchored support, depending on the location of the traditional underpinning in addition to multiple levels of staging to install the underpinning.

Existing foundation elements also posed several challenges. The existing shallow foundations ranged from 3.5 to 7 foot square foundations at column locations, connected by 2.5 foot wide strip footings around the existing building exterior. The bearing pressures beneath the existing footings ranged from 3.4 to 3.7 kips per square foot (ksf). Typically, these bearing pressures would be transferred directly to the underpinning elements. With no immediate access to the backside of the footings, the underpinning contractor would have to: (a) expand underpinning pits deep into the slope to reach the back of the footings, (b) excavate below the basement level on the inside of the building to reach the back of the existing footings, or (c) account for these bearing pressures as additional lateral surcharges on the underpinning. If these foundation pressures acted upon traditional underpinning as surcharges, these loads would create the need for significant additional lateral bracing and underpinning elements. See Fig. 1 for location of required underpinning and lateral earth support layout with respect to the 1992 renovation and existing stone building structure.

Excavation heights of up to approximately 16 feet in additional to the technical challenges described above also poses significant risk to the safety of the workers. They are working directly beneath existing concrete or stone that could potentially fall or soil that could potentially cave in if the underpinning pit was not prepared properly.

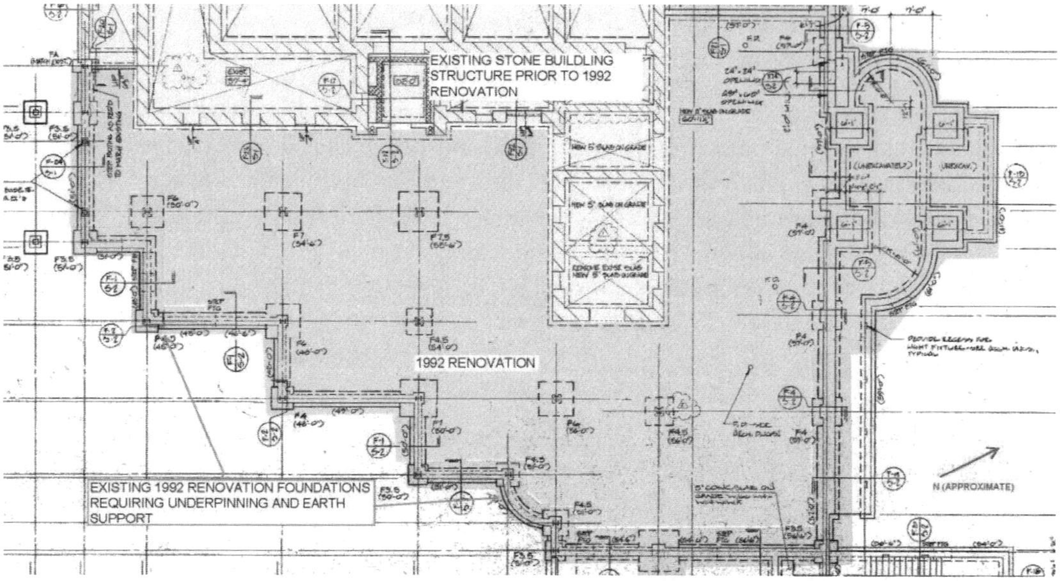

Figure 1: Required underpinning locations (blue) and existing structures (green and yellow).

GEOTECHNICAL BACKGROUND AND RECOMMENDATION

The "Soils and Foundation Investigation Report, 2015" (Beattie 2015) outlined the existing soils as loose to medium dense fill material consisting of sands and silty sands with varying amounts of gravel overlying medium dense to very dense silty sand with varying amounts of gravel overlying fractured gneiss/gneiss bedrock with RQD values ranging from 30 to 50 percent. Additionally, based on historical photos of the 1992 renovation, the base course used beneath the existing foundations was visually similar to roadway base course.

There was no mention of underpinning of existing foundations in the geotechnical report. This report may have been prepared before the proposed foundation elevations were discussed to be below the existing foundations. Initially, the report called for earth support to resist the lateral earth and surcharge pressures and called for the proposed walls to be designed for the lateral earth and building surcharge pressures. Based on discussions with the Project Structural Engineer, the proposed foundations and walls were only designed for lateral earth pressure, and therefore the proposed underpinning was required to be permanent to support the existing building and prevent the existing building from surcharging the proposed renovation foundation walls. Initially, due to the challenges traditional underpinning typically faces and the added permanent design requirement, GTR proposed an active soil nail wall design to be executed to support the lateral earth and existing foundation surcharge pressures. For simplicity in the design that eliminated the need to consider variable soil conditions, a soil profile consisting of only sand was conservatively assumed in the design (no rock used in design).

SOIL NAIL WALL APPLICATION AND FEASIBILITY

Based on a review of the subsurface conditions and the results of additional test pits next to the adjacent existing foundations, soil nailing posed to be a feasible option for the support of the earth and surcharge pressures. Based on the test pits performed, the material beneath the existing footings was confirmed to be a dense well-graded granular subgrade, similar to roadway base course. Below the dense fill material was dense to very dense silty sands, which had varying amounts of gravel. This soil was considered to have some apparent cohesion. These soils were a factor in the proposal of a soil nail wall and were well suited for the application. Conservatively, no cohesion was used in the design of the active soil nail system.

In addition to the favorable soils, several other advantages were considered with the use of an active soil nail wall. An active soil nail wall improves the soil stability. By transitioning the passive nature of a soil nail wall to an active state prior to excavation, the deflections typically seen with a soil nail wall due to reinforcement elongation and activating loads within the nails were mitigated. This was more significant where the location of the proposed active soil nail wall was in close proximity to the existing footings. Other advantages also include safety and minimal soil disturbance. Considering the variable nature of the soil within the building's footprint and the irregularity of the existing footings, there would have been increased risk while excavating beneath the existing footings.

Due to the nature of the construction, the soil nails would be drilled in the shotcrete facing which was placed in shallow sequenced excavated faces in front of the existing footings. Drilling would create low vibrations on the existing structure. During sequencing, the soil nail concrete facing was placed and soil nails were drilled though using casing, which further mitigated soil disturbance and soil loss resulting in limiting settlement of the existing foundations.

The depth to rock was variable across the site, and in turn, variable depths of soil nail wall were needed below the existing foundations to reach the top of rock. A soil nail wall could effectively be terminated at the top of rock, while the nail spacing was maintained (see Figure 1). Although the soil nail wall would be terminated at the top of the rock, the excavation was to continue to around ±16 feet below the bottom of existing footings. A rock shelf in front of the soil nail facing was proposed to support the soil nail facing, while the excavation continued below to the proposed footing depths. See photo in Figure 2 for an example of the soil nail excavated face at the required excavation depth. To facilitate the excavation of rock, rock below grade was line drilled by the contractor in several areas to the bottom of footing elevation just outside the proposed footing footprint. This allowed the rock to be excavated with limited disturbance to the rock beneath the existing footings and soil nail wall.

Figure 2: Photo of soil nail wall facing above excavated rock face.

ACTIVE SOIL NAIL WALL DESIGN

The need to minimize vertical and horizontal movement of the existing foundations was the primary consideration for utilization of an active soil nail wall design. Soil nail wall designs are typically passive systems that require wall deflection to activate and generate resistance along the soil nails. Although there were no performance specifications on the Project, there was a strong concern from the Project Team about the movement of the existing foundations and how it could adversely affect the existing building structure. Therefore, a conservative design approach was taken.

To evaluate the surcharges, the historical building plans from the 1992 renovations were provided by the Project Owner. These plans provided bearing pressures of each footing location along the area to be underpinned. Lateral surcharge pressures were calculated using the program SNAIL using Boussinesq's method, modeling the foundations as strip surcharges at distances behind the soil nail wall. After considering various options for the facing, an increased section was selected with additional reinforcement to reduce flexural deflections. Introducing an active condition was performed by prestressing each soil nail. To properly apply a prestresss load, each nail would require a free length, within the soil failure plane wedge, which in-turn ensured that the bond zone was being activated beyond the failure plane. To accommodate the free stressing length and prestresss loads, the facing was designed for a higher nail head loading, using the maximum stress found in the soil nails during analysis.

Nail spacing was also considered in the design due to the proximity of the existing building and foundations to the soil nail wall. Soil nails were spaced on a 4-foot-by-4-foot grid. Therefore, shallower lifts could be made during construction, reducing the probability of a lift sloughing.

Another aspect of design was the requirement for the wall to be permanent, which eliminated the surcharge loads from the existing building foundations on the proposed addition foundation walls. Considering the wall was to be buried, the contractor elected to use a single shotcrete facing, and install a steel end cap on each nail head with corrosion inhibiting grease. All soil nails installed were manufactured with the Post Tensioning Institutes (PTI) Class 1 Double Corrosion Protection (DCP) with all bar hardware, bearing plates and steel end baps being galvanized to resist potential corrosion. The soil nail wall was also designed for seismic conditions.

Based on the aforementioned design requirements, 4.5-inch diameter soil nails ranging from 22 to 27 feet in length utilizing a #10, Grade 75 threaded bar for reinforcement were selected for the permanent design. The soil nail design loads ranged from 36 kips to 40.5 kips depending on their location and up to four rows were considered for the cut heights anticipated. A 6-inch thick shotcrete facing (minimum f'_c = 4,000 psi) was designed using two layers of welded wire mesh for the support of the soil around the nails.

SOIL NAIL CONSTRUCTION

Typical construction of a soil nail wall consists of excavating typically in 5 foot lifts along the whole length of the proposed wall, installing soil nails, and installing the reinforcing and shotcrete facing after all nails are installed for each lift. After the lift is stabilized with the soil nails, the entire subsequent lift is excavated for the installation of each subsequent row of soil nails and the process is repeated for each level.

Considering the proximity of the existing foundation and building structure, construction sequencing was crucial to mitigate soil loss and disturbance. Shallow lifts, limited to 4 feet, were maintained during excavation with sequencing the excavations of short and precise spans of soil nail rows to limit time for soil to be open cut. The soil nail installation sequencing was more important in areas of concern where the soil nail wall beneath existing footings was in close proximity. See Figure 3 for an example of the soil nail sequencing used for each lift at one portion of the site. Additionally, in soil, the contractor used rotary duplex drilling methods (casing and drill string) to reduce ground loss. In rock, the casing was seated into rock and the rock was open hole drilled to the required depth.

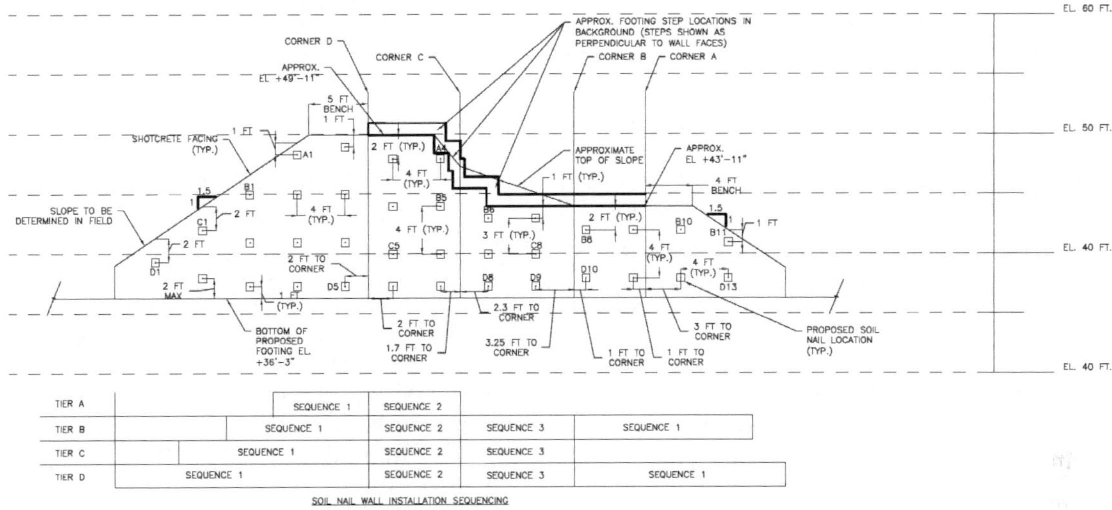

Figure 3: Soil nail lift sequencing drawing with soil nail elevation view.

Facing rebar and shotcrete was assembled and installed on the same day excavations were performed to reduce soil loss before the soil nails were installed for each lift and sequence section. After soil nails were installed, each soil nail was locked off at 50% of the design load to create an active condition. The lock off sequence also was considered a form of modified proof testing of each soil nail. See Fig 4 for facing construction (wire mesh installation and shotcrete placement).

Figure 4: Shotcrete facing – rebar installation and shotcrete placement

During construction, daily on-site inspection was performed by GTR to confirm the soil nails and shotcrete facing were installed in accordance with the design. This allowed the use of the observational method (Peck, 1969) for making field modifications to the design on an as-needed basis in real-time through communication with a representative of the design engineer present on site. The layout of the foundation and excavation heights were modified during construction due to varying rock elevations and existing site conditions not indicated on the plans. GTR observed these changes and the wall design was analyzed on a case by case basis resulting in a modified layout consisting of installing additional nails or eliminating nails where they were unnecessary.

GTR observed and documented all soil nail testing, including 4 verification tests and 7 proof tests, to confirm the soil and rock bond stresses were consistent with our design methodology. To confirm there was limited movement of the wall and the existing structure, GTR's Automated Deformation Monitoring program was implemented using two automated robotic total stations (AMTS), accurate to 1 second (vertical and horizontal) with an accuracy of 0.05 inches at 1000 feet (using optical survey prisms and total station) – see Figure 5. Deformation monitoring points installed on the existing structure showed minimal northing, easting and elevation fluctuations throughout construction and after the completion of the soil nail wall, confirming the soil nail wall was performing as intended. Figure 5 shows the elevation changes of the existing structure above the underpinning elevations throughout the monitoring program of the Project.

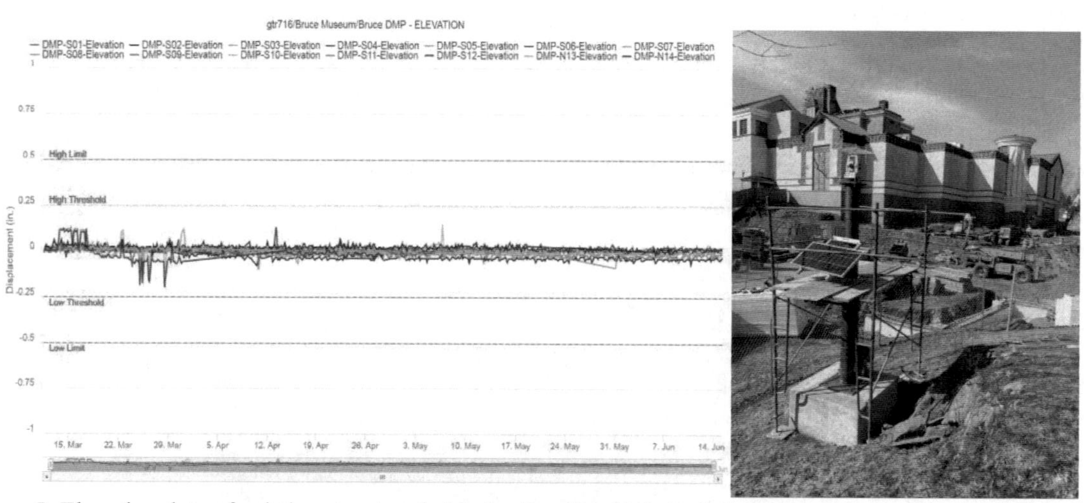

Figure 5: Elevation data of existing structure behind soil nail wall through the monitoring program and one of the AMTS setups at the project.

CONCLUSIONS

An efficient alternative underpinning solution that was executed through careful analysis and planning. The design was coordinated using effective communication between the Architect, Engineer, Contractor and Earth Support Designer. Through automated near real-time deformation monitoring, this active soil nail wall design proved to be an effective and safe solution given the existing and proposed foundation geometry and complex ground conditions. With GTR's on-site inspection, the use of the observational method illustrated the importance of exercising engineering judgment throughout the construction phase. The observational approach allowed for a more organized and efficient design and acknowledged potential for modifying the design during the construction process based on in-situ conditions as well as promoting a safer alternative than traditional underpinning. Given its safe application and capability to tolerate large surcharges with little deformation, an active soil nail wall is considered a viable option when underpinning existing foundations.

ACKNOWLEDGEMENTS

Eastern Excavation, Inc,. the underpinning contractor on the project and gave a practical view during the design process of what was achievable from a construction standpoint. Turner Construction served as the Projects General Contractor with Eskew Dumez Ripple as the Project Architect and Guy Nordenson and Associates serving as the Projects Structural Engineer.

REFERENCES

Beattie, J. H., P.E., & Horowitz, T. E., PE (2015). *Soil and Foundations Investigation*. Greenwich, CT: Bruce Museum.

Lazarte, C. A., PhD, PE, GE, Robinson, H., PE, Gomez, J. E., PhD, PE, Baxter, A., PE, PG, Cadden, A., PE, & Berg, R., PE. (2015). *Geotechnical Engineering Circular No. 7 Soil Nail Walls - Reference Manual* (Rep. No. FHWA-NHI-14-007). Washington, DC: National Highway Institute, US Department of Transportation, Federal Highway Administration.

Peck, R.B. (1969). Advantages and Limitations of the Observational Method in Applied Soil Mechanics. *Geotechnique, 19*, 171-187.

Tavrow, B. (n.d.). Retrieved April 15, 2022, from https://www.newbruce.org/

INTERDISCIPLINARY AND INVENTIVE APPROACHES TO DRILLED SHAFT REMEDIATION

Seth H. Hamblin, P.E., Geosciences Testing and Research, Inc., 55 Middlesex St # 225 No. Chelmsford, MA 01863, 978-301-3236, shamblin@gtrinc.net
Les R. Chernauskas, P.E., Geosciences Testing and Research, Inc., 55 Middlesex St # 225 No. Chelmsford, MA 01863, 978-301-3231, les@gtrinc.net
William D. Oder, Geosciences Testing and Research, Inc., 55 Middlesex St # 225 No. Chelmsford, MA 01863, 978-301-3242, bill@gtrinc.net
Jonathan L. Ernst, P.E., Geosciences Testing and Research, Inc., 55 Middlesex St # 225 No. Chelmsford, MA, 01863, 978-301-3243, jernst@gtrinc.net

ABSTRACT

Large diameter drilled deep foundations are a widely utilized solution throughout North America and the world. Numerous design and construction techniques developed over the past 100 years have made this type of foundation a robust, reliable, and efficient foundation option to support a wide variety of structures. From bridges, to marine structures, to high-rise buildings, drilled shafts are a time-tested approach for all types of infrastructure projects.

Yet, what remains ambiguous in the current state of engineering practice is how to respond when a shaft is constructed that, for potentially different reasons, may or may not meet the design requirements? In this case, what are the best practices for Contractors and Engineers when a structural, geotechnical, or material integrity issue arises during construction? In our experience the answers to these questions are not widely known, but are profoundly important so that engineers can have tools available to confidently and efficiently address concerns, and provide a path forward for the project.

This paper seeks to answer the above questions by presenting details of different remediation approaches using the results of several case studies where structural and geotechnical engineering methods, combined with non-destructive integrity testing (NDT), were employed to evaluate the acceptance of "questionable" drilled shafts. Through multidisciplinary techniques like Cross-hole Sonic Logging and 3-D Tomography which enhance the resolution of anomalous zones and defects in concrete, the development of concrete strength estimates from non-destructive testing data, and executing innovative structural and geotechnical analyses to evaluate the as-built conditions, drilled shafts that might have historically been rejected were ultimately accepted by the project team. This paper presents a series of unique case studies that together depict a roadmap for a comprehensive approach to reliably evaluate and guide the industry to address these types of drilled shaft construction issues.

KEYWORDS

Remediation, Non-destructive Testing, Drilled Deep Foundation Repair,

DRILLED DEEP FOUNDATION INTEGRITY

Confirming the integrity of drilled deep foundations has proven to be invaluable in today's engineering practice. Testing methods such as Cross-hole Sonic Logging (CSL), Cross-hole Tomography (CT), Low Strain Integrity Testing, also known as Sonic Echo (SE) and Thermal Integrity Profiling (TIP) have all been widely implemented with success to determine the as-built integrity of drilled deep foundations. Determining the integrity, however, is only the first step when an anomaly is observed in the NDT data.

Anomalies do not always result in a deep foundation being rejected. CSL and TIP both have established acceptance criterion. Acceptance criterion for CSL testing are well established but can vary widely. In general, this paper refers to the criteria developed in "Terminology and Evaluation Criteria for CSL of Deep Foundations" for CSL testing (Sellountou et al., 2019). As TIP is a relatively young testing method, the acceptance criteria is not as well established, however, a criterion was developed based on the shaft radius reduction and local cover reduction (Piscsalko et al., 2016). These criterion do not tell the user what to do when a drilled deep foundation has anomalies.

After the initial observation of an anomaly in the NDT data, various courses of action may be employed dependent on the location and size of the anomaly as well as the design parameters and as-built conditions of the drilled deep foundation. The path forward depends on the severity of the anomaly encountered. Each situation is different and may require its own unique solution. Our goal is to provide a general understanding of what options are available and the best situations for those options to be implemented. We will discuss four case studies from projects we have worked on recently to provide practical application of these options and practices.

CONCRETE STRENGTH EVALUATION

In a concrete strength evaluation, we estimated the compressive strength of the overall shaft concrete by back-calculating from the overall shaft wave speed in the CSL results. This is achieved by first plotting the break strengths from the concrete compressive break strength logs. The plot is used to scale off the estimated strength of the concrete on the day of CSL testing. The estimated strength is then used as the target to match the wave speed obtained by the CSL testing. The concrete elastic modulus (E_c) is related to the concrete strength (f'c) through the use of the following empirical formula:

$$E_c=57\sqrt{f'c}. \qquad [1]$$

Where E = Elastic Modulus and f'c = concrete strength.

The elastic modulus is related to the wave speed (C) using elastic wave theory based on elastic parameters of unit weight (γ), unit weight, Poisson's ratio (ν), and the ratio between elastic to dynamic modulus (E_c/E_d). The dynamic elastic modulus is calculated with equation 2 and the Static Elastic Modulus (E_c) is calculated with equation 3.

$$E_d = C*\gamma*(1+\nu)*(1-2*\nu)/(1-\nu) \qquad [2]$$

Where E_d = dynamic elastic modulus, C = wave speed, γ = unit weight and ν = Poisson's ratio.

$$E_c = E_d*E_c/E_d \qquad [3]$$

Where E_c = static elastic modulus, and E_c/E_d = the static to dynamic modulus ratio.

The values of Poisson's ratio, unit weight and E_c/E_d were varied to obtain a back-calculated or estimated f'c value equal to the estimated compressive strength on the day of the CSL testing using equation 1. The above process of "calibrating" the elastic parameters is then used for estimating the strength within the anomalous zone based on the reduced CSL wave speeds (Neville 1996).

CASE STUDY 1 – ROUTE 18 OVER THE MBTA RAILROAD

The route 18 bridge over the MBTA Railroad in Weymouth Massachusetts was replaced and the abutments and piers were designed to rest on 4 foot and 6.5 foot diameter drilled shafts. CSL testing was performed by GTR on all 16 Drilled Shafts. The shafts ranged between 40 and 50 feet in length. A few of these shafts required further investigation. The process for evaluating and repairing/ redesigning shafts DS-10 and DS-14 are explored below.

In shaft DS-10, two anomalies were observed in the CSL results. The calculated average apparent wave speed of the drilled shaft was 12,900 ft/s, based on the tube distances and the measured FATs in good concrete. We performed CT testing in order to better define the anomalous zone. It was determined that the majority of the cross-sectional area was compromised from 35.5 to 37 feet below the top of tubes. From 43 to 45 feet below the top of tubes the reduced wave speeds were isolated near tubes 1 and 6. The CSL and CT results for DS-10 are summarized in Table 1. Based on the CT data it was also estimated that the lower anomaly was only about 5% of the cross-sectional area.

Anomaly Location, Depth below top of tubes (ft)	CSL Results		CT Results
	FAT increase (%)	Energy Reductions (dB)	Wave speed Reduction
35.5 to 37	15 to 50	30 to 36	9,000 to 11,000
43 to 45	15 to 35	9 to 16	9,500 to 11,000

In this situation, a CSE is not practical or cost effective due to the size and significance of the FAT increases (wave speed reductions) in the upper anomaly. For this reason, we suggested coring and performing hydro flushing remediation procedure on the anomalous zone. After the defect was found with the core and the anomalous material was completely removed with the hydro flushing method, the void was verified with a down-hole camera. The shaft was CT tested again to show the results of the cleaned zone (filled with water). High strength grout was then placed and the shaft was CT tested a third time to verify that grout had filled the zone.

These additional tests are necessary in this process for a few reasons. The Pre-grout CT test verifies that the zone has a wave speed approaching that of water. In the post-grout CT test, we are looking to see that the wave speed in the anomalous zone increases from the pre-grout CT test. This proves that the grout has filled the zone. There is another reason for all of these tests. Wave speeds in concrete are typically between 11,800 and 14,400 ft/s (ASTM D6760). Wave speeds of grout are typically lower, in the range of 8,000 to 10,000 ft/s. If CT testing is only performed after grouting is completed a zone of reduced wave speeds will still be present where the anomaly was and at first glance, it would appear that the defect was not repaired. Testing before and after grouting allows the engineer to verify what the results show before grouting so that an improvement can be seen in the post-grout CT testing. Additionally the apparent wave speed observed in the post-grout CT testing should be around the typical wave speeds for grout (8,000 to 10,000 ft/s). See Figure x for representative slices from each CT test performed on DS-10.

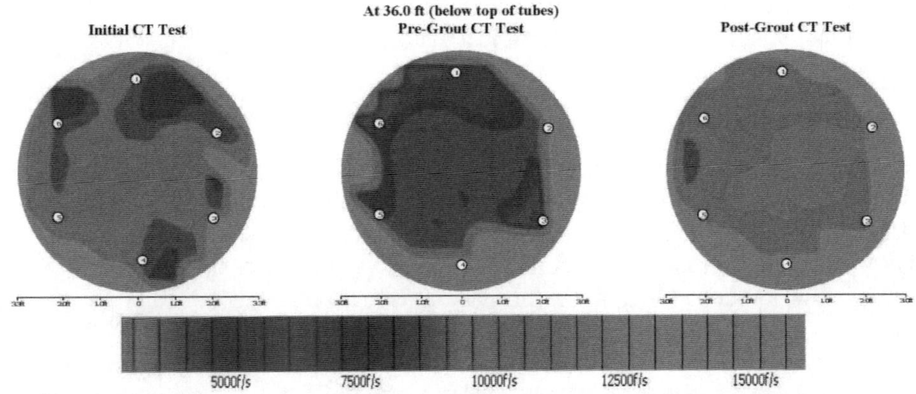

Figure 1: CT Slices from Initial, Pre-Grout and Post-Grout CT Testing of DS-10

Another shaft in which anomalies were observed in the CSL results was DS-14, where FAT increases of 25 to 40% were observed in the bottom few inches to 2.5 feet in all profiles. Energy reductions of 24 to 45 dB were associated with these FAT increases. This zone was further evaluated with CT testing. The results of the CT testing indicated wave speeds in the bottom 1 to 2 feet of tubes were between 5,500 and 8,500 ft/s (around 25% to 45% reductions) across most of the shaft cross sectional area.

Due to the location of the anomaly and the time constraints on the project schedule an alternate approach to repair and remediation was taken. Rather than coring and flushing the shaft and performing multiple tests to confirm the repair was proceeding as planned, we proposed installing an 8-inch diameter drilled micropile through the center of the drilled shaft into the bedrock below to mitigate the anomaly at the bottom of the shaft. In the process, the anomalous zone was flushed. After reviewing the design loads and calculations, we determined that the as-built shaft was capable of developing a factored axial design load of 346.4 kips and DS-14 required a factored axial design load of 659 kips. Therefore, the micropile had to be designed to a minimum factored axial design load of 313 kips. Using a micropile design already approved for the project allowed us to further expedite the remediation. A bond length of 16 feet was required to develop 351.9 kips. The decision to use a micropile to remediate the shaft was a quick and efficient way to keep the project moving forward.

CASE STUDY 2 – MOHEGAN BRIDGE CENTER PIER SHAFTS

The replacement of the Mohegan Bridge, Bridge number 673, in Burrillville, Rhode Island involved the installation of a new center pier using staged construction. The new center pier was to be supported by four (4) drilled shaft foundations that would be constructed through the existing footing (shallow foundation). Two (2) drilled shafts, designated as DS-1 and DS-2, were installed during Phase 1. The drilled shafts were installed with a 20 foot long, 36”x0.75” wall (Grade 50 steel) permanent casing to the top of rock. Below the casing, a 34.5-inch diameter rock socket was drilled ranging from 6.15 to 6.3 feet. A rebar cage was installed inside the drilled shaft consisting of 22 - #11 bars with #4 spiral reinforcement. The Contract Documents required CSL testing to be performed on the drilled shafts to confirm the integrity of the in-situ shaft concrete.

The results of the initial CSL testing, performed by others, indicated anomalies in both installed drilled shafts. GTR was retained to evaluate the integrity of the concrete in the two (2) shafts and determine what measures were required for repair and/or remediation. Additional CSL retesting was recommended with CT testing to better understand the extent and location of the anomalies. A Concrete Strength

Evaluation, CSE was also performed on both shafts to estimate the strength of the concrete in the anomalous zones. Finally, structural and geotechnical evaluations of the as-built shafts were completed.

Shaft DS-1

The CSL retest results of DS-1 indicated FAT increases of 85% to greater than 100% in the concrete over the bottom 1 foot of CSL tubes in all profiles with energy reductions associated with these FAT increases between 30 dB and 40 dB. CT testing results indicated that the wave speeds in the anomalous zone ranged from 4,000 ft/s to 6,500 ft/s. The overall concrete wave speed was approximately 8,100 ft/s. This is, in general, very low for concrete but the primary reason is that the sound wave has to propagate through a larger percentage of water (in the tubes) as compared to the distance of the concrete in the body of the shaft between the tubes. This is much more pronounced when the tube spacing approaches 12 inches. The low overall wave speed, however, is representative of 5,500 psi concrete based on the 21-day concrete strength testing performed.

A CSE was performed to estimate the concrete strength in the anomalous zone at the bottom of DS-1. Due to the significant FAT increases, the concrete in around the bottom 1 foot of the CSL tubes in DS-1 had an estimated 28-day compression strength of 474 psi. In our experience, this indicates that the material is non-cementitious and has little to no strength. Due to this zone having no strength, a geotechnical evaluation of the shaft was required to understand the reduced axial capacity of the drilled shaft.

Shaft DS-2

The CSL retest results of DS-2 indicated FAT increases in two locations. FAT increases of approximately 30% to 40% were observed between 13 feet and 15 feet below the top of tubes (upper anomalous zone) in two (2) profiles with energy reductions associated with these FAT increases of 9 dB or lower. FAT increases of approximately 25% were observed between 19.5 and 21 feet below the top of tubes (lower anomalous zone) in one profile with energy reductions associated with these FAT increases of less than 9 dB. CT testing results indicated that the wave speeds in the upper and lower anomalous zone ranged from 5,500 ft/s to 7,000 ft/s. The overall concrete wave speed was approximately 8,400 ft/s. This is, in general, very low for concrete but the primary reason is that the sound wave has to propagate through a larger percentage of water (in the tubes) as compared to the distance of the concrete in the body of the shaft between the tubes. This is much more pronounced when the tube spacing approaches 12 inches. The low overall wave speed, however, is representative of 5,500 psi concrete based on the 21-day concrete strength testing performed.

Another CSE was performed to estimate the concrete strength in the upper and lower anomalous zones in DS-2. Due to the FAT increases the concrete in the upper and lower anomalous zones had an estimated 28-day compression strength of 1,962 psi. Based on this lower concrete strength in the anomalous zones, a structural evaluation of the drilled shafts cased and uncased sections was required.

Geotechnical Evaluation – Shaft DS-1

Due to the anomalous zone over the bottom 1 foot of DS-1, the rock socket length was reduced to 5.3 feet and no end bearing was considered. The skin friction of the remaining rock socket and tip resistance of the casing on rock was used in the geotechnical resistance evaluation. No structural evaluation was required as the shear and moment at the bottom of the socket were expected to be minimal. A nominal unit side resistance of 43.5 ksf was calculated in accordance with AASHTO procedures based on the actual uniaxial compressive strength of rock limited by the 28 day strength of concrete (6.2 ksi) of the competent shaft. The factored side resistance is estimated to be 1145 kips, which was greater than the required factored axial load of 1084 kips and satisfied the Project Requirements.

Structural Evaluation – Shaft DS-2

During CSL and Tomo testing, two (2) anomalous zones were indicated in DS-2, one in the cased section, and one in the rock socket, just below the casing. Due to the reduced concrete strength calculated for the anomalous zones in Shaft 2, a structural evaluation was performed to ensure the load carrying capacity of the drilled shafts. Due to the location and concrete strength, the geotechnical resistance was assumed to be minimally affected.

The loadings based on the Contract Design Calculations for the drilled shafts were used to confirm the structural performance of the cased section with a reduced concrete strength of approximately 2 ksi. See Table 2 for the Loadings provided and location. A combined stress and shear analysis was performed in accordance with AASHTO for the cased section. The analysis indicated that the cased section with a reduced concrete strength was structurally adequate for the required loading.

The casing was originally expected to extend to elevation 1 foot into rock. During construction, the casing was extended approximately two (2) additional feet into rock. To account for the additional cased length into rock and the 1.5-foot long reduced strength anomalous zone below the cased section, the socket was modeled and analyzed using the program LPile to evaluate the moment in the reduced concrete zone. The loading provided was input into LPile at the top of the cased section.

Table 2. Contract Provided Loads

Shaft and Location:	Full Axial Load (kip)	Resultant Moment(kip*ft)	Shear (kip)
DS-1 - At Tip	1083.67	Not Used	Not Used
DS-2 - at upper anomalous zone	1155	1569.2	76.1
DS-2 - at lower anomalous zone	1155	1643.3	76.4

Based on the results of the LPile analysis, a moment of 850 kip-feet was determined at the top of the anomalous zone, below the cased section. A concrete interaction diagram was developed using SPcolumn based on ACI 318 code with the reinforcement cage and the reduced concrete strength of the anomalous zone (approximately 2ksi) were incorporated into the analysis. The required factored moment from LPile and required factored axial loading were plotted on the interaction diagram. See Figure 2 for the Interaction Diagram of DS-2. A shear analysis was performed in accordance with AASHTO for the uncased socket section using the shear load provided. The analysis indicated that the uncased section with a reduced concrete strength was structurally adequate for the required loading.

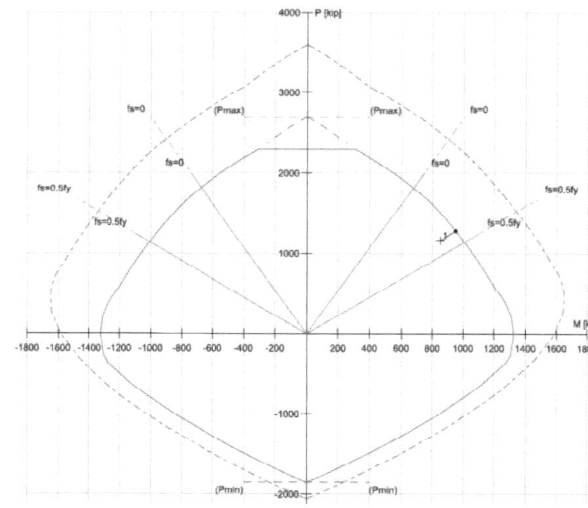

Figure 2: DS-1 Concrete Interaction Diagram

Conclusion

Using careful evaluation and applicable analyses and design standards, the shafts were approved by the Rhode Island DOT and their Engineer. Based on our discussions with the General Contractor, the remediation prevented 6 to 8 months of down time that would have incurred due to environmental restrictions and project logistics.

CASE STUDY 3 – NJTA CONTRACT T600.481A

The NJTA contract T600.481A project consisted of constructing 28 sign structures at various locations in Northern New Jersey. These sign structures consisted of 4 to 6 drilled shafts per structure and 2 to 3 drilled shafts per foundation cap. The shaft were either 30-inches, 36-inches or 42-inches in diameter and were between 20 and 60 feet in length. During CSL testing at Structure 17 and Structure 19 anomalies were observed. Structure 17 Right Side, North Shaft was observed to have FAT increases of 20% to 25% in the bottom 1 foot of tubes in profiles 14 and 34 with energy reduction of 22 dB to 27 dB observed in all profiles. Structure 19 consisted of three (3) drilled shafts with the following anomalies observed in Table 3.

Table 3. Structure 19 CSL Anomalies

Shaft:	Location	FAT Increases (%)	Energy Reductions (dB)
Structure 19 North	Bottom 2 ft	20 to 35	21 to 27
Structure 19 Middle	Bottom 2.5 to 3 ft	35 to 40	18 to 21
Structure 19 South	Bottom 2 to 2.5 ft	35 to 45	9 to 18

Structure 17 Right Side, North Shaft

After the initial CSL results CT testing was performed and analysis on this shaft. The nominal average apparent wave speed for the shaft overall was around 13,385 ft/s. We found that the anomaly consisted of a zone of reduced velocities in the lower 6 to 12 inches of tubes mostly between tubes 3 and 4 with a small portion near tube 1. The wave speeds associated with these zones were around 8,500 ft/s (blue areas in Figure 3) and 11,000 ft/s (green areas in Figure 3). Based on the CSL results the anomaly encompassed around 20 to 40% of the cross-sectional area over the bottom 6 to 12 inches (6% to 7% blue area and 17% to 33% green area).

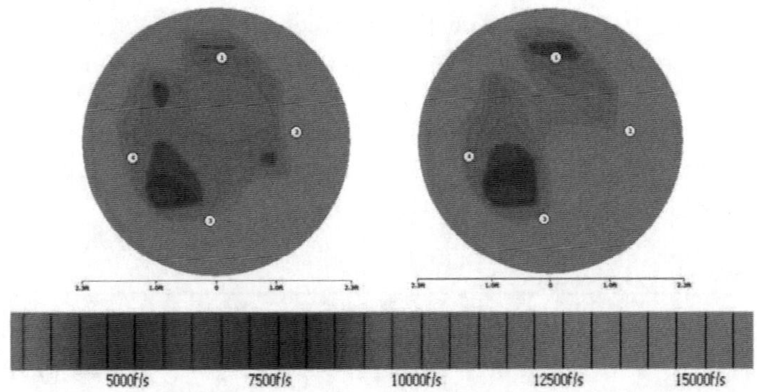

Fig. 3. Structure 17 Right Side, North Shaft CT Slices at Bottom of Tubes

Following the CT analysis we performed a CSE. The CT testing was completed 58 days after the concrete pour. Based on the Concrete Compressive Break Strength Logs provided by Key Tech we estimated that the strength of the overall shaft was approximately 7,132 psi 58 days after the concrete was poured (See Figure 4). Using the CSE equations described previously, we estimated that the concrete strength in the blue zones (wave speeds of around 8500 ft/s) was approximately 1,160 psi and the concrete strength in the green zones (wave speeds of around 11,000 ft/s) was approximately 3,250 psi. With this analysis the design engineer was able to accept the shaft in its as-built condition and avoid additional and more costly options for remediation or repair.

Structure 19 Right Side

Structure 19 Right Side consisted of three 42-inch drilled shafts. As stated above, all three of the drilled shafts had anomalies in the foot 2 to 3 feet of tubes. In lieu of drilling new shafts, a 90 inch diameter 6 foot long reinforced "donut" was installed around the top of the drilled shafts. The skin friction and end bearing on the donut carried the loads neglected at the drilled shaft tip. The donut was be filled with concrete and reinforcing steel in accordance with the proposed repair procedure and the contract documents. To fulfill this plan we performed CT testing, a CSE, geotechnical calculations, reinforcing size calculations and adhesion calculations.

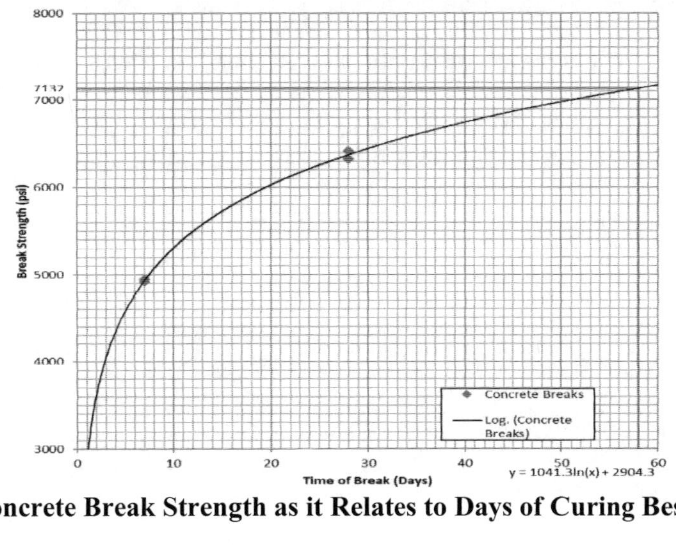

Fig. 4. Concrete Break Strength as it Relates to Days of Curing Best-Fit Curve

CT testing and CSE

CT testing was performed on the North and the South Shafts. Based on the design, the anomaly in the bottom of the Middle Shaft did not have a significant effect on the structures performance so a CT test was not required. The CT results for the North Shaft indicated a zone of reduced velocities in approximately the lower 12 to 18 inches of tubes affecting a portion of the cross-sectional area primarily near the west portion of the shaft (near tube 4). The CT analysis of the North Shaft was enough for the shaft to be accept. See Table 4 for the results of the CT testing for Structure 19 Right Side North and South Shafts and Figure 5 for the CT horizontal slices.

Table 4. Structure 19 CT results

Shaft:	Anomaly Location	Average reduction in wave speed (ft/s)	Portion of Cross Sectional Area (%)
Structure 19 Right Side North	Bottom 2 ft	7,500 (Blue)	16 to 32
		11,000 (Green)	21 to 36
Structure 19 Right Side South	Bottom 2 to 2.5 ft	6,000 (Red)	4
		7,500 (Blue)	28 to 47
		11,500 (Green)	13 to 30

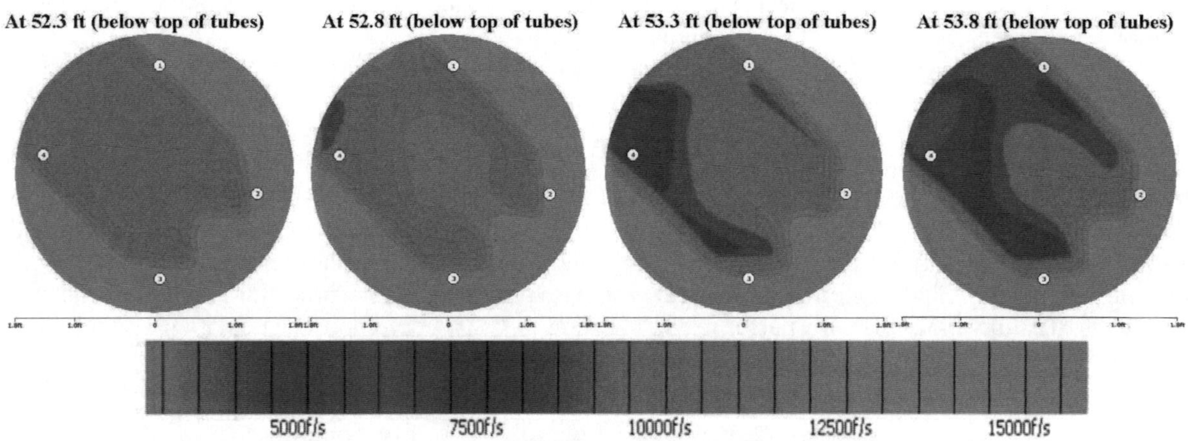

Fig. 5. CT Slices For Structure 19 Right Side, North Shaft

A CSE was performed on the South Shaft to estimate the concrete strength in the anomalous zone. The green areas were estimated to be 3,400 psi and the blue areas were estimated to be 750 psi (Figure 6).

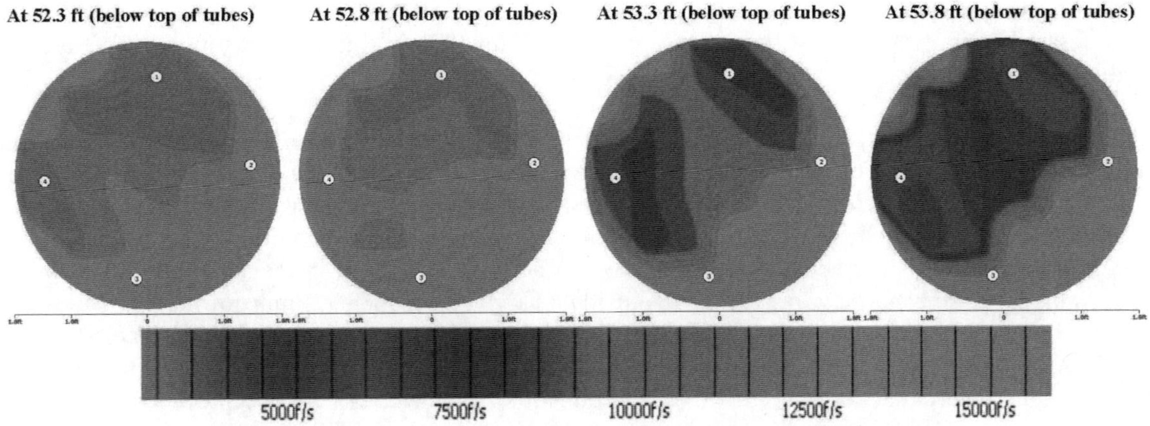

Fig. 6. CT Slices For Structure 19 Right Side, South Shaft

Calculations and Redesign of Structure 19 Right Side, South Shaft

Based on the design calculations, the ultimate tip resistance is 38.1 kips. Based on the tomography and concrete strength evaluation, approximately 32 to 51% of the drilled shaft tip consists of low/minimal strength concrete. Therefore, 50% of the drilled shaft tip could still adequately carry load. The Geotechnical resistance of the donut was calculated based on back-calculating the ultimate unit resistances presented in the Engineers SHAFT output. Based on 65% of the drilled shaft tip carrying load, the ultimate capacity of the drilled shaft tip is 24.7 kips. Based on the unit friction resistances in the SHAFT output, the 90 inch diameter donut can carry 14.2 kips in friction. Based on the unit end bearing resistances in the SHAFT output, the 90 inch diameter donut can carry 57.4 kips in end bearing. The three ultimate capacities total to 96.3 kips, greater than the 38.1 kips required. By redesigning and adding the Donut on the South Shaft we were able to prevent the client form having to drill 3 new shafts which would have been more time consuming and more expensive.

The Donut was designed to carry the geotechnical resistance and therefore, the larger section at the Donut has a greater structural capacity than the 42 inch diameter drilled shaft. A reinforcement cage was sized for the donut, to satisfy the minimum requirements per AASHTO LRFD 7th edition. The reinforcement cage will consist of 26 #11 vertical bars with #5 stirrups at a 6 inch pitch. The load will be transferred from the 42 inch diameter drilled shaft to the 90 inch diameter donut through the adhesion resistance between the concrete surface.

SUMMARY

There are many options to explore when determining the appropriate remediation, repair and/or mitigation for drilled shaft after defects are found and confirmed. The question is, where do you start? We have review three case studies, which were approach differently due to the specific and unique conditions and situations. In general, starting with the simplest option can prove useful and, at least, may provide additional information.

When an anomaly is identified in a drilled shaft and is potential categorized as a defect, we typically perform some form of additional NDT to confirm the anomaly/defect. Typically, this is the simplest first step. The method used in the original testing program will help determine what test to perform and what tests can be performed to confirm the anomaly/defect. For example, if Thermal Integrity Profile (TIP) was used for initial testing and there are no access tubes, then CSL testing is not possible but Low Strain

Integrity Testing may be a great method to use instead. Some options for further testing include but are not necessarily limited to, CSL testing or retesting, low strain integrity testing and Cross-hole Tomography testing.

If the anomaly is relatively small based on the respective data and criterion established for a specific testing method, a retest or alternative test may prove useful to determine how significant or insignificant the anomaly actually is. If retesting or secondary testing shows improvement in the testing results, then the drilled foundation element may be accepted in its as-built condition without the need for more destructive and expensive remediation measures or further time and effort in reevaluation of calculations and designs. When the anomaly proves to be outside of a given acceptable limit, further evaluation may still be beneficial before coring or replacement is required. CT testing can provide a more accurate determination of the size of the affected zone. In some cases, the size or location of the zone is enough information to determine that the drilled deep foundation element can be accepted. If the zone is determined to be too large or the location is in a concerning zone then a CSE can provide an estimate of the strength of the anomalous zone. If concrete break strengths from sample are used to "calibrate" the wave speed to the as-built concrete strength. This calibration can then be used to estimate the strength in the anomalous zone to be used in as-built calculations. This process is extremely valuable for anomalies that are isolated, have smaller reductions in wave speed but are present over a larger area or have NDT results that are beyond the acceptable limit by a small margin. It is important to note that, in general, this process is not suited for anomalies that encompass the entire cross-sectional area of a drilled shaft or have values far exceeding the acceptable limit. In those cases, a CSE will likely be irrelevant and time-consuming when the client needs progress. Skipping to perform as-built design evaluations or repair will be more efficient and likely will be required regardless as the CSE will likely show the zone has little or no estimated strength.

CONCLUSIONS

The goal is to provide the most efficient and effective plan to evaluate a drilled deep foundation element with anomalous NDT results for each project that requires it. Unfortunately, each project and specific situation is different and one process or plan will not meet the need of every project. In addition, many solutions may work but one solution may be much more costly or time consuming, inefficient, than another. It is important to keep the project and clients schedule in mind when making a remediation or repair plan. Avoid unnecessary evaluations when the anomalies are more significant but also avoid significant repair efforts when less expensive evaluations may provide understanding on the drilled deep foundations as-built condition and ability to perform as intended.

REFERENCES

Sellountou et al., 2019, Terminology and evaluation criteria for CSL of deep foundations, www.DFI.org

Piscsalko G. et al., 2016, Drilled shaft acceptance criteria based upon thermal integrity profiling, DFI 41st Annual Conference on Deep Foundations, New York, NY

Neville, A. M. 1996, "Properties of Concrete" 4th Edition, John Wiley and Sons, Inc., New York, NY

American Association of State Highway and Transportation Officials, 7th Edition, 2014

DESIGN AND FIELD COMPARISON OF GROUND IMPROVEMENT AND BORED PILES FOR LARGE DIAMETER STEEL WATER TANKS

Vijaya Bhushan Rao, Author, BAUER Spezialtiefbau GmbH (Dubai Branch), United Arab Emirates, +971 (56) 545 0821, vijay.bhushan@baueruae.ae
Ali Semercioglu, Technical Engineer, BAUER Spezialtiefbau GmbH (Dubai Branch), United Arab Emirates, +971 (56) 507 0815, ali.semercioglu@baueruae.ae
Jason Redgers, Contributor, BAUER Spezialtiefbau GmbH (Dubai Branch), United Arab Emirates, +971 (56) 996 0403, Jason.Redgers@baueruae.ae

ABSTRACT

Large diameter steel water tanks with high storage capacities impart heavy loads deep into the ground, generally requiring deep foundations to adequately support them. The use of ground improvement as a sustainable and cost effective alternative is tempting but the design of the foundations for such heavily loaded, thin-shelled structures have to meet several performance requirements such as bearing capacity, limiting the specified total settlements, typically limiting the radial differential settlement to 13mm over 10m along the circumference and dishing settlement from edge to center of 1/120 as per standard API 650 Appendix B as well as the mitigation of liquefaction induced settlement risk. Any inadequacy in the soil investigation, geotechnical design, or misrepresentation of stresses of the foundations may lead to serviceability compromise or potentially catastrophic failures. Therefore, the selection of the foundation is dependent on the availability of suitable ground report and the limitations imposed by operational requirements. Two such case studies are evaluated and compared in this paper, with tank diameters ranging from 41 to 106m in the Kingdom of Saudi Arabia, (KSA) which were founded on two types of soils and supported on deep foundations i.e., stone columns and bored cast-in-situ piles.

Keywords: Steel water tanks, High storage, Ground Improvement, Settlements, Field Tests

INTRODUCTION

Storage tanks are not always built on the best sites and the sub-soil may not be able to resist heavy loads. Most commonly, mixed bearing soils and fresh fills are encountered in the project sites. Such stratum below tank foundations can cause differential settlements, offers maintenance challenges, and can lead to catastrophic consequences. Accordingly, tanks are rested either on shallow improved ground with grade beam, or ring beam or on deep foundations with piles connected to the slab. Generally, pile foundations are adopted as the most suitable foundation type to support these tanks in difficult and sensitive grounds. However, this may significantly impact the project's cost and timeline depending on the performance criteria and sub-soil conditions. While piling is the premiere and well-established solution for supporting tanks in poor ground conditions, ground improvement can also be considered as a reliable alternative on case-by-case basis. When the in-situ soil beneath the tank is a relatively homogeneous material, a uniform settlement may occur, and the tanks may not undergo significant problems. However, it may cause damage or distress at piping connections. If the sub-soil settles differentially, there will be sagging, circumferential and tilting settlement. Sagging causes uneven settlement of tank bottom plate and may result in buckling of tank bottom plate. Tilting settlement causes an increase in liquid level on one side and an increase in hoop stress on the shell. Circumferential differential settlement also allows tank shell to settle in a non-planar configuration inducing additional stresses in the tank shell. All such deformations effect the serviceability condition of tanks. Tank Loads are fixed based on the fluid carried by the tank, appurtenances, and its self-weight. Whereas many specifications provide limits on tank settlement, the most common specifications such as American Petroleum Institute (API 650) provide guidelines and minimum requirements for tank foundation deformations, which may differ according to the tank builder.

The limits vary based on the type of tank roof: fixed, floating or mixed of tanks. This paper covers 2 case histories of tanks supported on both shallow and deep foundations and their performance verified by suitable field tests and interpretation of test results

CASE STUDY 1 – TANKS AT TERMINAL AND PUMPING STATIONS

The project site located in Makkah province in Kingdom of Saudi Arabia, KSA, where the site is split into two distinct areas of a pumping station and a terminal station built as part of the proposed water transmission system project. The two stations are situated about 3.2 km apart and therefore founded on 2 different soil conditions. Whilst the Pumping station development consisted of building a steel storage water tank of 41m diameter with a capacity of 25,000m3, the Terminal station consisted of a 50m diameter tank with a capacity of 50,000 m3 along with associated structures that required ground improvement to support the structures.

Table 1. Settlement design criteria for steel storage tanks at Terminal and pumping stations

Area	Structure	Size [Dimensions	Loads	Allowable Settlements
Terminal Station	Tank	50 m Dia. & 26.25m Ht ($50{,}000\ m^3$ capacity)	253 kPa & 318 kPa on Ring beam	250mm at center of Tank
Pumping Station	Tank	41 m Dia. & 20.75m Ht ($25{,}000\ m^3$ capacity)	225 kPa & 245 kPa on Ring beam	250mm at center of Tank

SOIL CONDITIONS

Following are the representative borehole and pre-improvement Cone Penetration Tests, PreCPTs profile at the location of 50m Diameter tank at the Terminal station location.

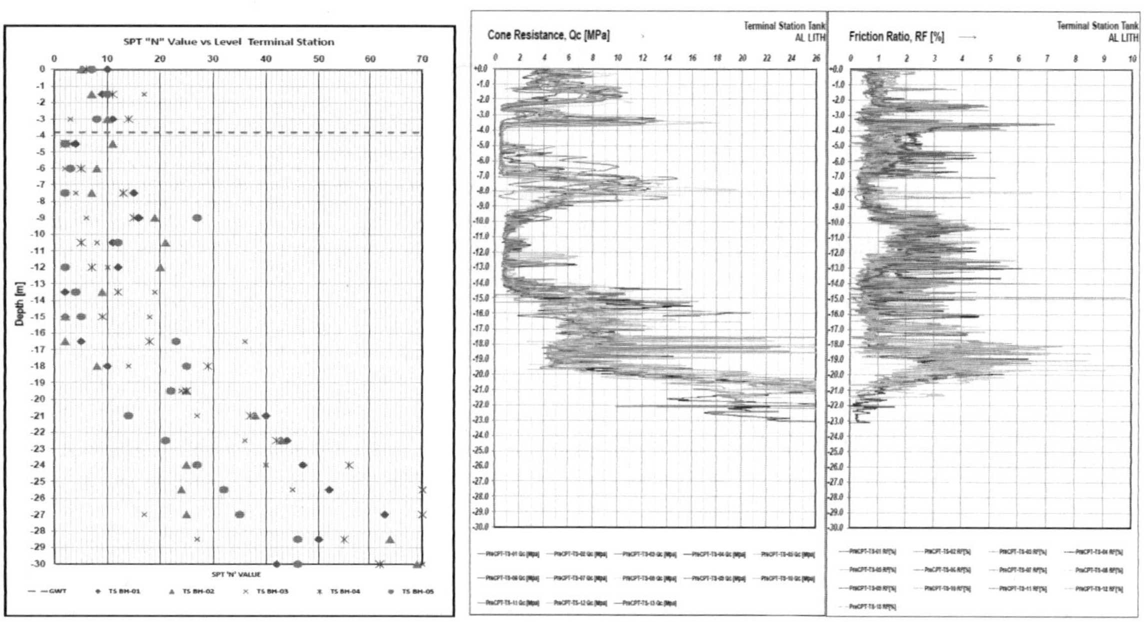

Figure 1. Sub soil profile from Borehole and CPTs at Terminal station

In addition to the available boreholes from the soil investigation report undertaken by the client, additional pre-improvement Cone Penetration Tests with pore pressure measurement, PreCPTus were taken at the Terminal station to make a detailed assessment of the soil conditions and to correlate the boreholes and CPT test results. An analysis of the soils using Ic, the Soil Behavior Type (after Robertson 2010), showed that the soil conditions at the Terminal station consisted of relatively weak and soft layers extending to 15m depth, with alternating layers of loose to medium dense sands and soft clays [in-situ compactness of very loose to medium dense condition in general with SPT N in the order of 2~29 in sands and Cone resistance, qc [MPa] measuring between 0.5 to 1.0 MPa in Clays).This layer is underlain by medium dense to dense sands up to 20m depth before encountering refusal at 22/23m depth from Ground Level.

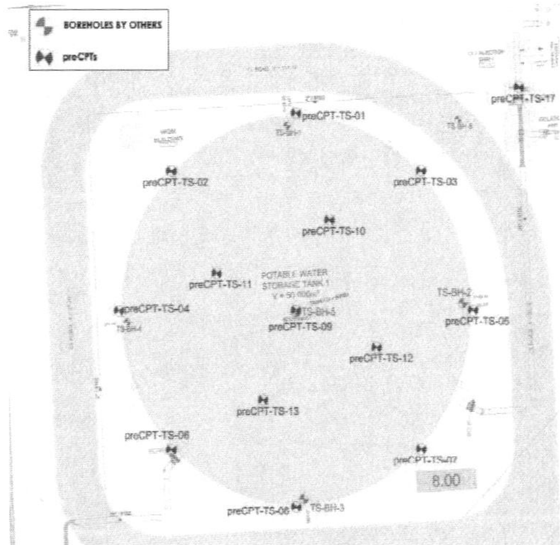

Figure 2. Borehole and PreCPT layout at Terminal station

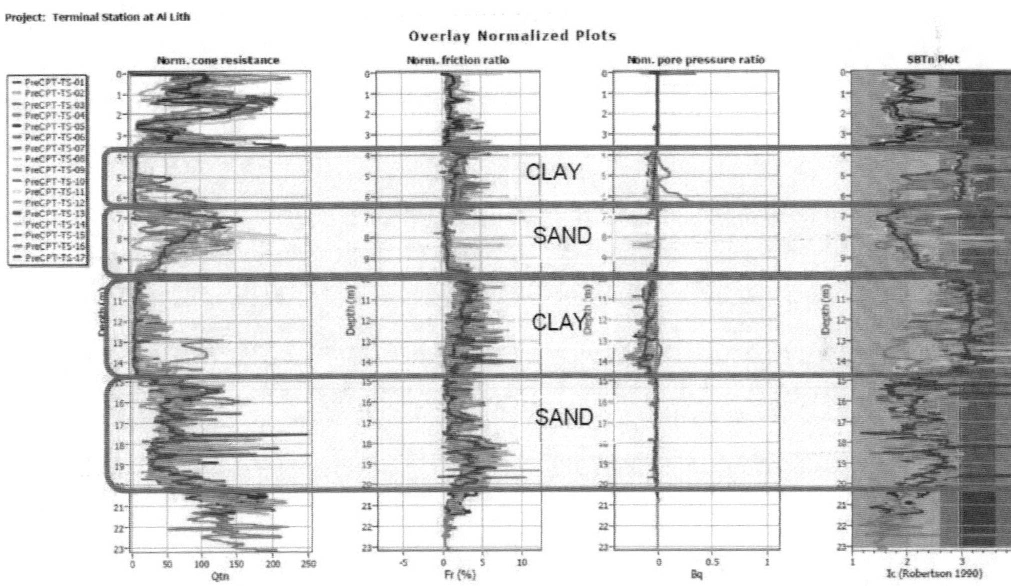

Figure 3. Robertson Soil Behavior Type Index, SBT Ic Plots at Terminal station

Similarly, in addition to the available boreholes from the soil report, additional preCPTU tests were taken in the 40m diameter tank at the Pumping station.

Using the same assessment method on both stations (i.e., the SBT Ic plots) showed weak soils extending up to 30m depth with the top 10m depth consisting of alternating layers of sands and very soft clays underlain by a mixture of silts, sandy silts, and clayey silts as shown in the figure 5 and 6 below.

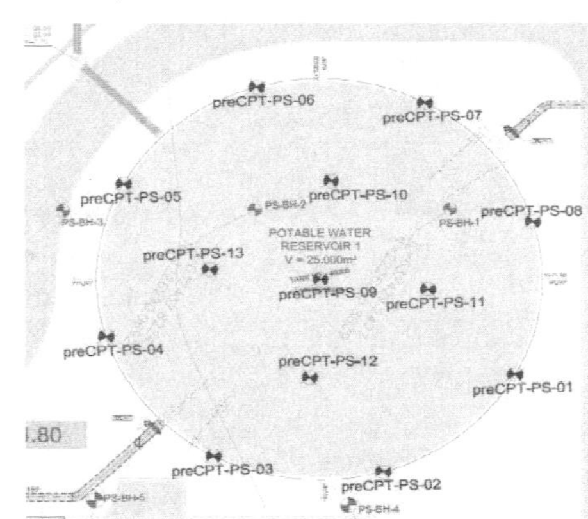

Figure 4. Borehole and PreCPT layout at Pumping Station

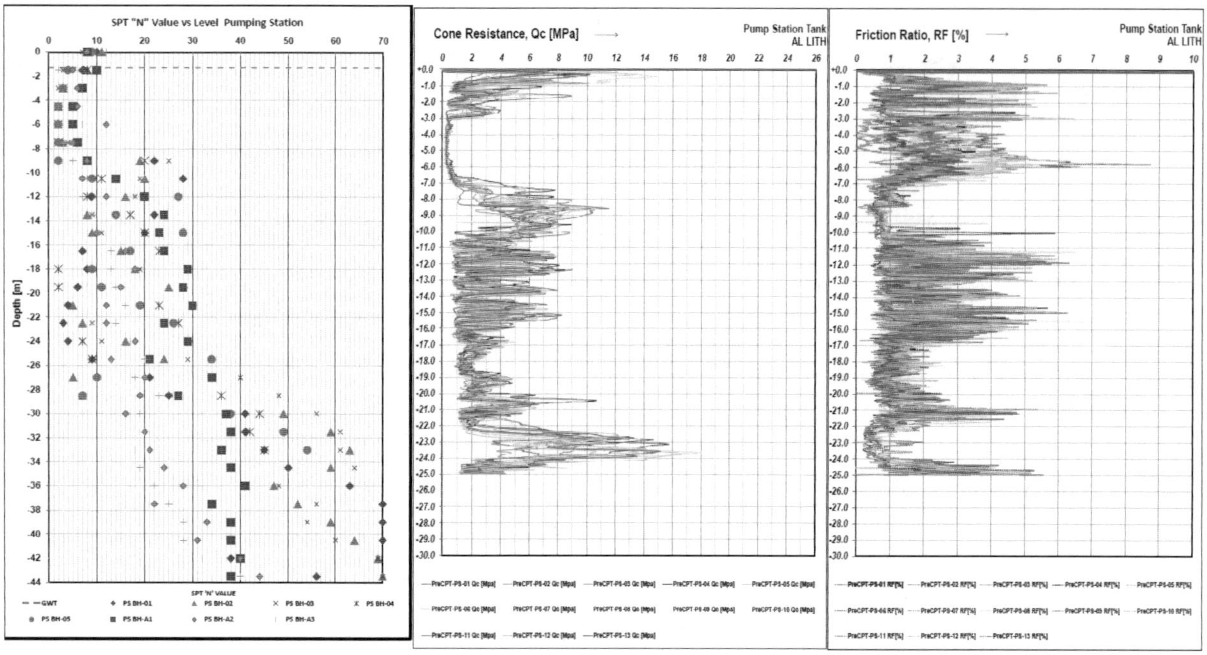

Figure 5. Sub soil profile from Borehole and CPTs at Pumping station

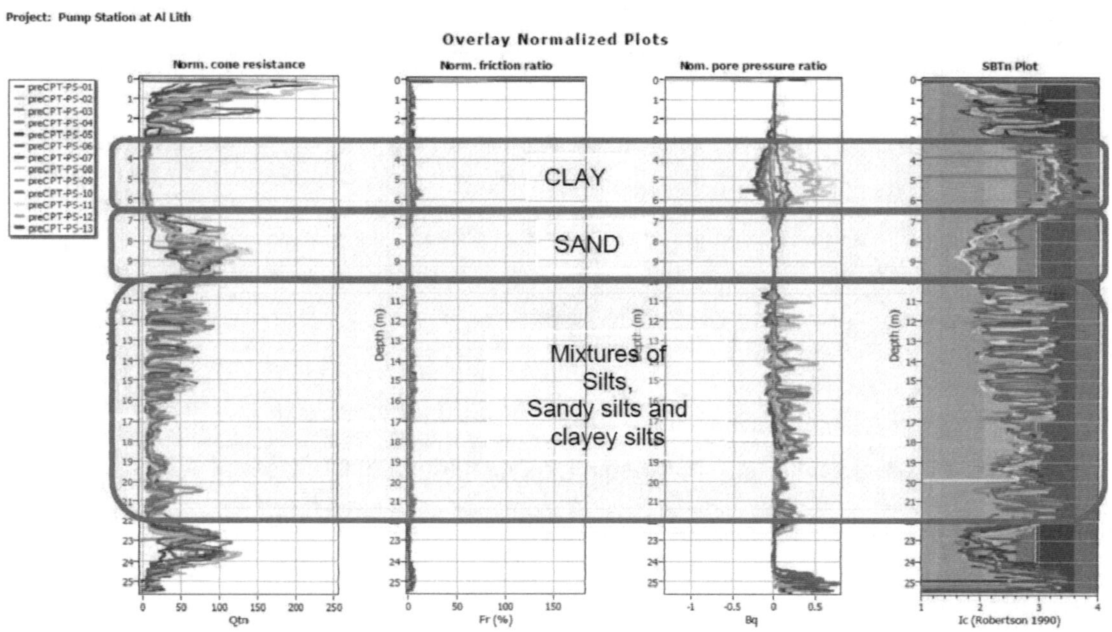

Figure 6. Robertson Soil Behavior Type Index, SBT Ic Plots at Pumping station

The soil conditions at the Pumping station were relatively weak and soft with low cone resistance, qc [MPa] values extending to deeper depths up to 30m from the ground level. The pre-improvement Cone Penetration Test, PreCPTUs were terminated prior to refusal at 25m depth due to the shortage of extension rods, but the available borehole logs showed the soil extended up to 45m depth before encountering bedrock. The weaker soils consisted of deposits of soft clay underlain by the mixtures of silts and clays falling within the depth of load influence under the 41m diameter tank and are therefore prone to very high and unacceptable total and differential settlements rendering the ground improvement solution not applicable at this location.

FOUNDATION SOLUTIONS

Terminal station - Based on the encountered soil conditions at the Terminal station for the 50 m diameter tank, ground improvement using stone columns was undertaken for 17m depth of treatment with an area replacement ratio of 19% applied under the tank area and a denser grid arrangement applied under the 1.5m wide concrete ring beam supporting the tank shell, with minimum 2 rows extending outside of the tank edge as shown in figures below.

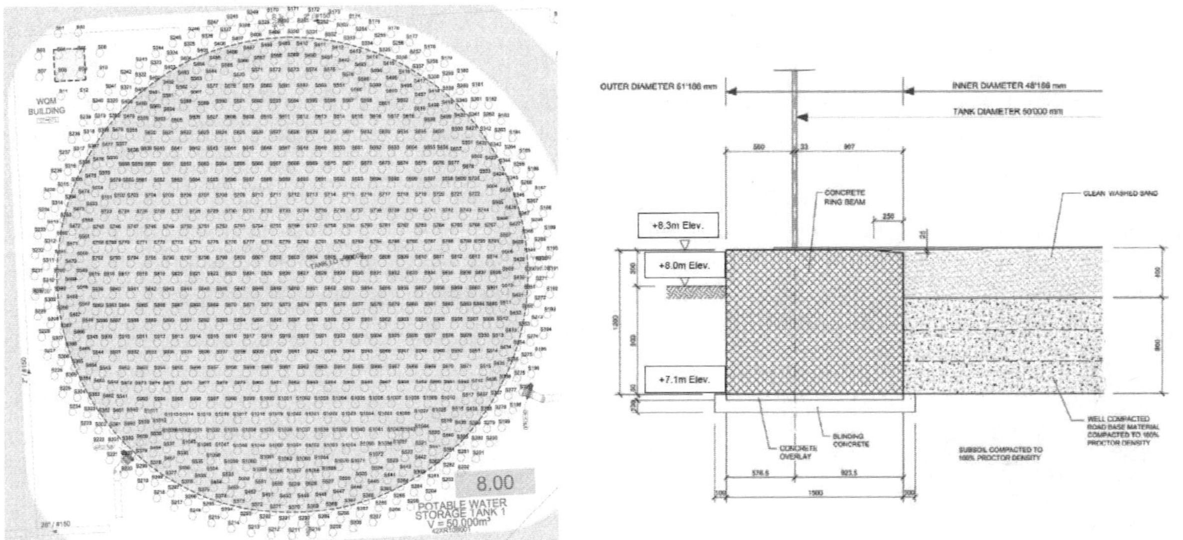

Figure 7. Details of stone column layout and concrete ring beam at 50m Diameter tank

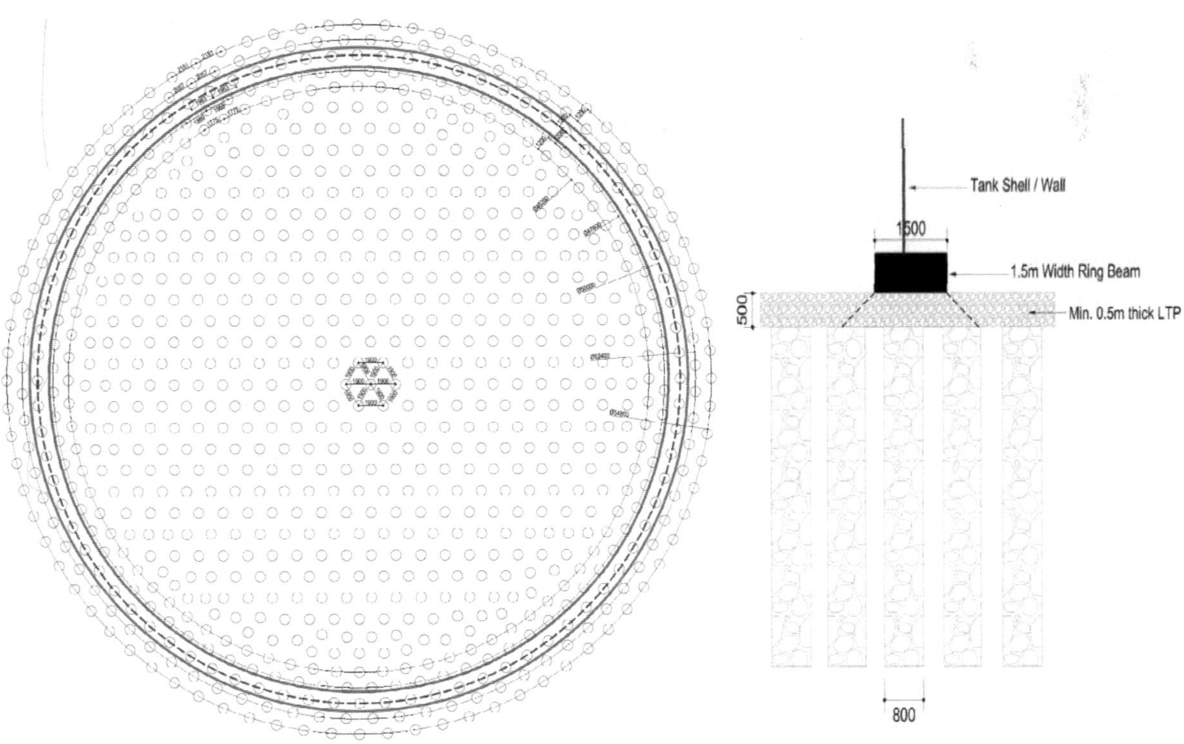

Figure 8. Arrangement of stone columns under the tank and Ring beam location

The settlements estimated for the proposed design using stone columns was 350mm (> 250 mm allowable) at the center of tank and 250 mm at the ring beam, but with controlled radial and dishing differential settlements it was agreed to proceed. Further verification tests included two (2) numbers of Zone Load Tests (ZLT) loaded up to 1.5 times of design load in the tank area (one test at tank center on 3x3m footing size and another at the location of ring beam on 1.5 x 1.5m footing size) to monitor settlement behavior, validate soil parameters and therefore verify the design assumptions. In addition to zone load tests, eight (8) numbers of Plate Load Tests (PLT) of 1m diameter have been carried out on the top of columns to check the workmanship, integrity, and stiffness of columns in accordance with German code DIN 18134:2012-04

The recorded settlement was 15.45mm against a predicted theoretical settlement of 17.10 mm in the tank, and 4.33mm for the test undertaken on the ring bean against a predicted theoretical settlement of 6.86mm. Thus, the design parameters were verified.

Pumping Station – Based on the weak and soft consistency of soil conditions presented from the soil investigation, it was considered too ambitious to attempt ground improvement in this location and this reverted to a traditional deep foundation solution. The piling solution lead to 138 numbers of 1000 mm diameter piles upto 42.5m depth from ground level.

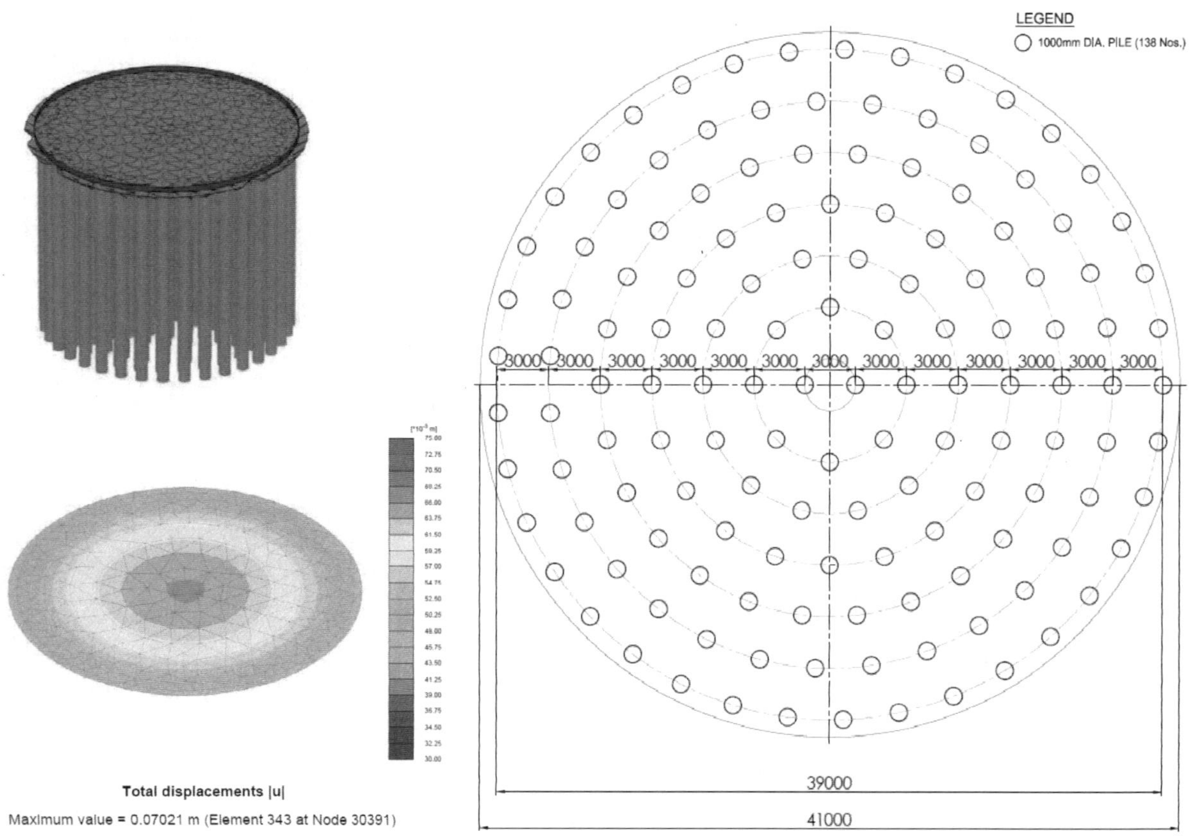

Figure 9. Pile layout and FEM 3D analysis details for 41m diameter Tank at Pumping station

The settlements estimated for the pile design was 70mm at the center of tank and 48mm at the tank edge and a base slab was provided to connect the piles with pile cut-off level at 1m below ground level. Therefore, piling solution was rendered best at the pumping station to avoid settlements.

CASE STUDY 2 – LARGE STEEL WATER TANKS 170,000 M^3 CAPACITY

The scope of work consists of construction of three (3) numbers of large steel water tanks of 106m diameter with storage capacity of 170,000 m^3 at Tank farm and Pump station as part of the development of a water transmission system between in the Eastern Region.

DESIGN CRITERIA FOR STEEL WATER TANKS

Diameter 106.00 m
Height 20.00 m
Width of Ring beam 1.50 m.
Allowable load bearing pressure 285 kPa.
Maximum allowable tank edge settlement shall be 100mm
Differential settlement 13 mm over 10.0 m along circumference
Dishing settlement from edge to center 1:120 as per API650
Factor of Safety during an earthquake FoS=1.2,
considering M=6 and PGA=0.15g at the ground surface.

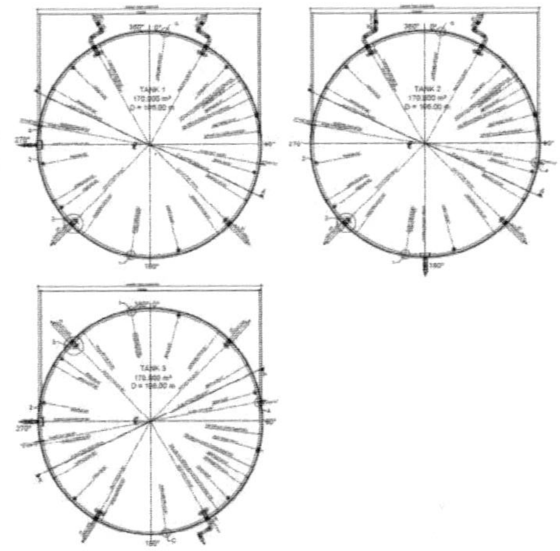

Figure 10. General layout of three numbers of 106m diameter steel tanks

SOIL CONDITIONS

The general soil profile of the site presented as typically a granular sand of variable density, some silty sand with some sandy gravels and underlain by hard silt/clay layers at lower depths. The thickness of those quaternary deposits were investigated to depths of upto 50m by boreholes. Further Cone Penetration Tests (CPTU) were undertaken in accordance with ASTM D5778 to correlate with the results from SPT "N" values gathered from the boreholes in order to confirm and verify soil profile. Based on the collected data the soil conditions appeared consistent across the site with similar soil profiles encountered under the three (3) numbers of tanks. For the presentation only Tank 2 is selected.

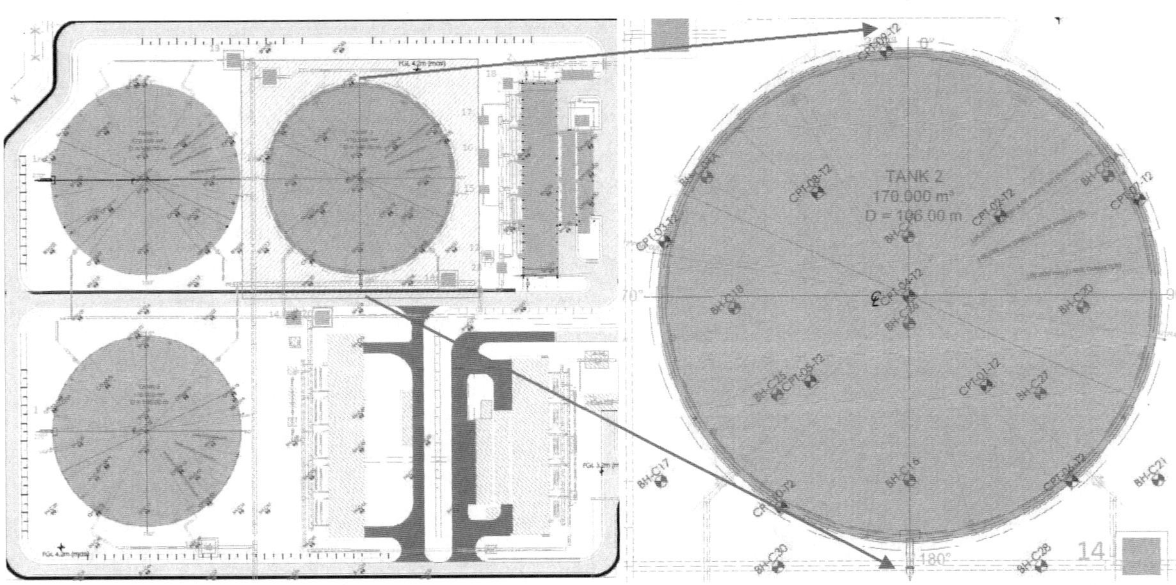

Figure 11. Soil investigation works undertaken at Tank 2 highlighted

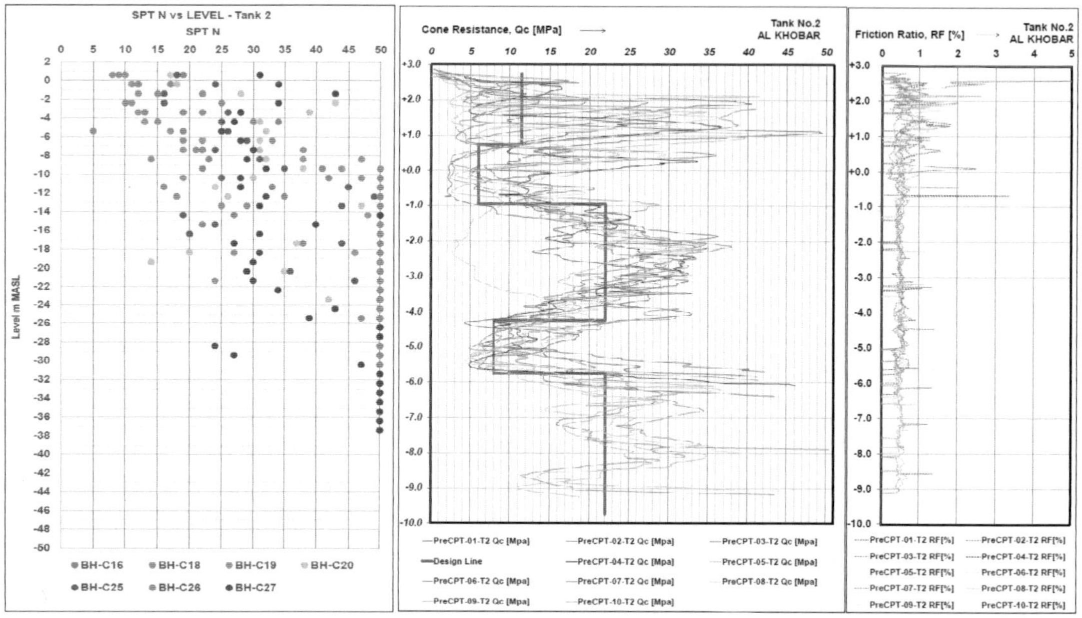

Figure 12. Sub soil profile from Boreholes and CPTs at Tank 2

Although the soil investigation suggested that in-situ vibrocompaction compaction techniques would be suitable, or even dynamic compaction, the client was insistent on vibroreplacement (stone columns) being the only applicable method to reach the requirements. As such a stone column solution based on Priebe (1995), resulted in a square grid of 2.8 x 2.8 m applied under the tank for a treatment depth of 8.5m with 800mm diameter stone columns, with four (4) radial rows applied on an equivalent dense grid of approximately 2.45 x 2.45 m c/c under the ring beam. This design satisfied the performance requirements to limit the post-construction total settlements to not exceed 100 mm at the tank edge and comply differential settlement tolerances, and ensure factor of safety, FOS > 1.2 against liquefaction risk for the design level seismic event.

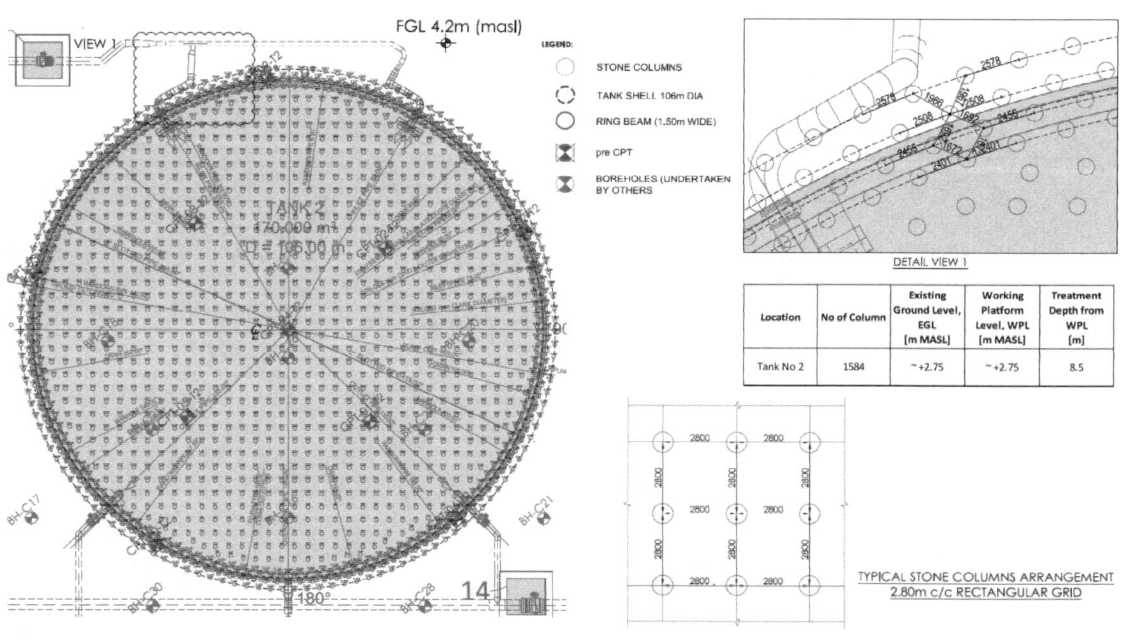

Location	No of Column	Existing Ground Level, EGL [m MASL]	Working Platform Level, WPL [m MASL]	Treatment Depth from WPL [m]
Tank No 2	1584	~+2.75	~+2.75	8.5

Figure 13. Stone Column arrangement details and drawings at Tank 2

FOUNDATION PRESSURES AT RING BEAM

The maximum foundation pressures induced under the 1.5m wide concrete ring beam were estimated based on eccentric loading conditions resulting in trapezoidal stresses. This was analyzed using both finite element FEM modelling and traditional settlement analysis based on Boussinesq (1883) and Priebe (1995). It was noted that the maximum load bearing pressures were developed on the inner edge of ring beam and these stresses were considered in settlement calculations at the tank edge.

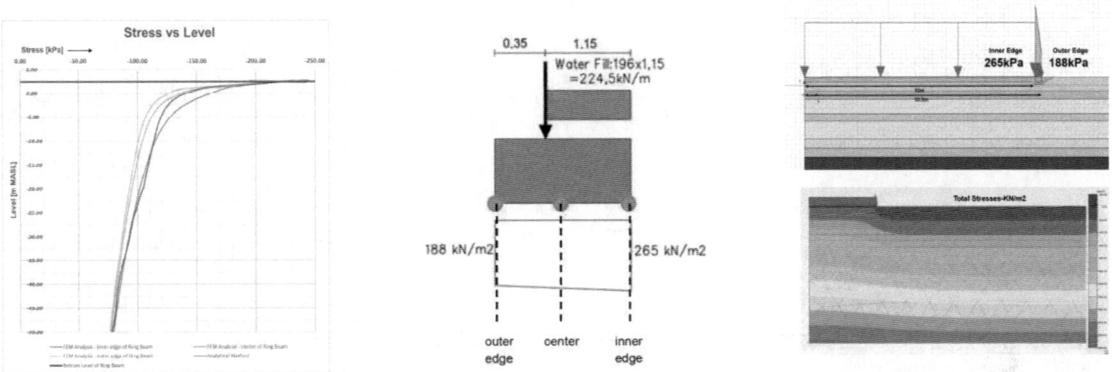

Figure 14. Calculated load bearing pressures induced under the concrete ring beam

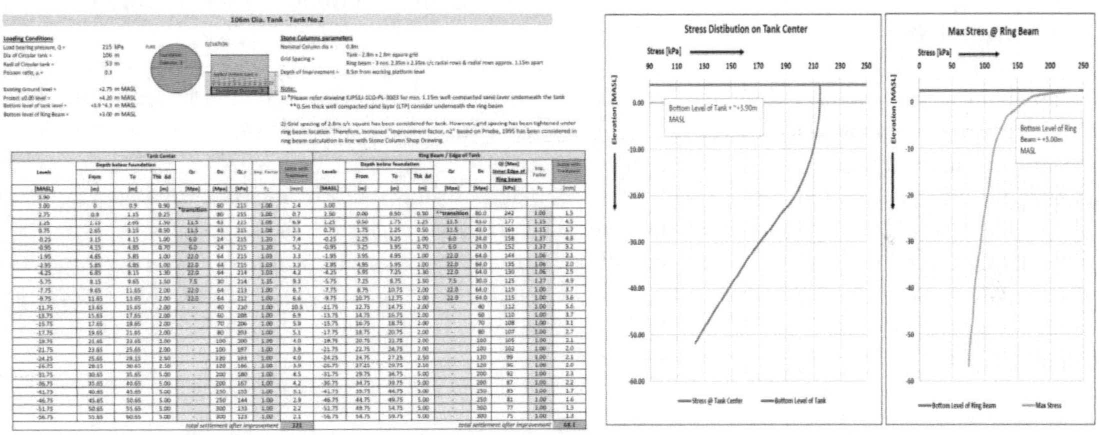

Figure 15. Settlement calculations for stresses at the center and edge of tank 2

FIELD PERFORMANCE VERIFICATION TESTS

The settlements calculated for the proposed stone column design was 121mm at the center and 68.1mm (< 100mm allowable) at the tank edge. Field performance verification tests included one (1) number Zone Load Test, ZLT undertaken at the center of tank No 2 for 165% of design load on a 3 x 3m footing size; this was to monitor the settlement behavior and verify stone column design (i.e., the assumed soil and stone column stiffness). Furthermore ten (10) numbers of Cone Penetration Tests, postCPTUs were taken in between the stone columns to assess any increase in Cone resistance, qc MPa values as a result of densification in the sands due to the vibratory compaction action during the stone column installation process. The measured settlement was 4.66 mm for ZLT at the design load of 215 kPa, which was less than predicted theoretical settlement of 8.65 mm. Furthermore, the CPTUs showed significant increase in the tip resistance, qc MPa values in comparison to CPTUs taken prior to the works indicating the relative density and bearing strength of soils has been increased within the treated depth.

The blue colored lines refer to PreCPTs and the brown colored charts refer to postCPTs plotted along with their averages as shown in the Figure 16 below.

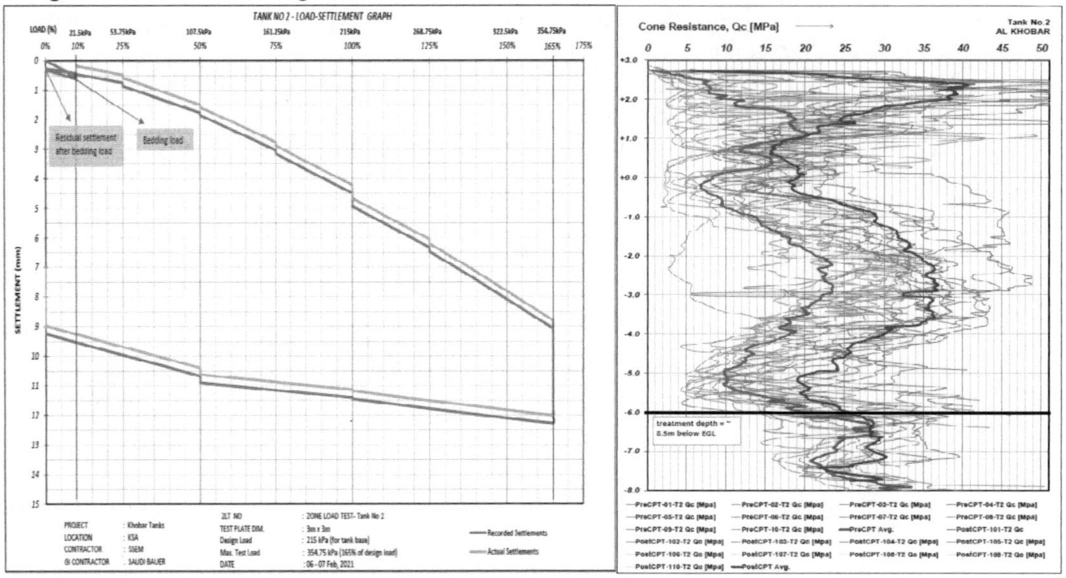

Figure 16. Zone Load Test and pre and post Cone Penetration Test Results

CONCLUSIONS

There is no universal best solution; depending on the site-specific soil conditions, shallow or deep foundations can be designed to support steel storage tanks successfully. Ground improvement including replacement and compaction techniques can be successfully applied with special additional efforts taken at the tank edge to limit the settlements to acceptable values. Careful testing, field performance tests, validation tests and suitable design would show success is possible. Thus, savings in time, money and resources could be realized where the benefits of soil improvement (ie improvement of the ground, use of ground bearing solutions) offered environmental benefits over the use of concrete, steel, suspended slabs, and arisings removal offering a sustainable option. Such figures can be seen in the DFI / EFFC Carbon Calculator.

REFERENCES

Priebe, H. J., 1995. The Design of Vibro replacement, published in Ground Engineering, December 1995

BS EN ISO 22476-12 (2009) Geotechnical investigation and testing – Field Testing: part 12: Mechanical Cone Penetration Test, British Standards Institution, London.

DIN 18134: 2001-09. Determining the deformation and strength characteristics of soil by the plate-loading test, September 2001

Lunne, T., and Christoffersen, H.P. 1983. Interpretation of cone penetrometer data for offshore sands. Proceedings, 15th Annual OTC, OTC 4464: 181-188.

Robertson, P. K. 2010. "Estimating in situ state parameter in sandy soils from the CPT." Proc., 2nd Int. Symp. of the Cone Penetration Test, CPT '10, Omnipress Publishers, Madison, Wis.

American Petroleum Institute: API Standard 650: Welded Tanks for Oil Storage, 11th edn., Washington, D.C., p. 445 (2011)

PERFORMANCE REQUIREMENTS FOR VIBROCOMPACTION ON CRUSHABLE CALCAREOUS SANDS OF THE UAE AND THEIR VERIFICATION IN COMPARISON TO SILICA/QUARTZ SANDS

Vijaya Bhushan Rao, Author, BAUER International FZE, United Arab Emirates, +971 (56) 545 0821, vijay.bhushan@baueruae.ae
Jason Redgers, Contributor, BAUER International FZE, United Arab Emirates, +971 (56) 996 0403, Jason.Redgers@baueruae.ae

ABSTRACT

The typical Vibrocompaction projects undertaken in the UAE are on reclaimed Sands. These Sands which are predominantly the crushable, calcareous, carbonate sands found locally as opposed to the more common quartz or silica sands that are featured in most design literature. The famous reclamations such as the iconic Palm Jumeirah in Dubai, often have specification for mitigation of liquefaction induced settlements, bearing capacity increases and total and differential settlement requirements. Often these performance targets are achieved by the means of achieving a particular relative density that are generally measured by a defined Cone Penetration Test, CPT targeting values of cone resistance (Qc), MPa at respective depth zones. In this paper some case studies are presented, where soil suitability for deep compaction by vibratory methods was firstly assessed for suitability using the Massarsch, 1991 compactability chart prior to the application of vibrocompaction and its subsequent evaluation. One such project named the Island Development required 17 to 23m depth of penetration only to treat the lower 5 to 6m of sand above rockhead as eventually the top 15m of sand will be further excavated to allow for the construction of basement. The paper will present pre and post CPT results, the results of initial site calibration trials and evaluation of the completed works.

Keywords: Reclamation Sands, Massarsch compactability chart, Vibrocompaction, Cone Penetration Tests

INTRODUCTION

Extensive ground improvement using vibrocompaction of reclaimed sand fills has been carried out since 2001 in the United Arab Emirates, UAE. Several artificial manmade islands on the seacoast including the iconic Palm Jumeirah islands in Dubai and other parts of the Emirates that involved dredging seabed material from the Arabian Gulf and placing the hydraulic fill on shore using different methods of fill placement. It was thought initially that the "as placed" density of the sand would be adequate to support light loads without undue settlement, whilst heavier loads would be supported on piles. It was also considered at that time that the earthquake risk was low. When the placement of sand fills for first large reclamation (The Palm Jumeirah) was well advanced this opinion changed and it was decided that ground improvement was necessary to control settlement under static load and to deal with earthquake risk which was considered to be higher than previously thought. Therefore, all the reclaimed sands had to be suitably compacted typically by deep vibrocompaction. The standard minimum imposed design criteria allowed maximum settlements of 25 mm for 3m x 3m footings with an applied bearing capacity of 150 kPa to mitigate against liquefaction for a design level seismic event with a magnitude M 6.0. with a Peak Ground Acceleration (PGA) of 0.15g at bedrock with 0.21g at the surface.

Frequently, crawler crane suspended twin vibrators (tandem system) were used for the vibrocompaction works. To prove the achievement of the required compaction, a CPT reference chart was developed. Due to the fact that the seabed material taken in the Arabian Gulf in the dredging operations largely consisted of calcium carbonate in the form of shells and shell fragments, coral detritus, and a small proportion of ooliths with accordingly lesser proportions of silicate particles (e.g., gabbro) and quartz. Such sand may therefore be classified as being "highly crushable and compressible" (Vesic 1962) and van t'Hoff (2012).

During the undertaking of the "Mega" sand reclamations in the UAE it was not fully appreciated by some clients and their advisers that the methods of analysis for the silica sands were not applicable to the calcareous materials used, i.e., the modified approach methods of analysis due the local crushability of the shells. The majority of the sand around the world is silica sand with small amounts of silicate minerals such as mica, and feldspar, so methods of analysis in the published literature relate to silica sands as this is predominantly where previous works had been carried out. Therefore modification are necessary to suit the actual materials treated and assess the degree of compressibility or crushing in relation to the testing method used.

Where there is a high carbonate content or shells, the high stresses generated at the tip of the cone during the execution of a CPT test will result locally in the crushing of carbonate particles or shells thereby recording lower-than-expected q_c values than an equivalently compacted silica sand, and hence an unrepresentative relative density (Ahmed, 2014), (Lunne 1997). Due to the highly crushable nature of these sands during CPT works, the results of CPT soundings had to be corrected in line with published works to consider the modified strength of the reclaimed sand fill in comparison with silica sand. Suitable correction factors were initially developed following the works of Al-Homoud and Wehr 2006, Belotti and Jamiolkowski, 1991 and Vesic, 1965 that were in the range of 1.3 to 2.3 applied to account for the high carbonate content where the gold standard is extensive laboratory testing in a calibration chamber. Other methods such as non-invasive shear wave measurements or back analysis of zone load tests are not yet wholly accepted as the "stigma" of low values often leads to subtle suggestions of "fixing"

Usually, quality control was executed by means of pairs of CPT every 900 m^2 of improved ground and material extractions by means of borings every 20,000 m^2. An area-wide geodetic levelling was also carried out to establish the settlements originating from the compaction processes because of voids closure resulting in increase in density. The individual compaction points were monitored by automatic logging devices attached on the crane units. Furthermore, static Zone Load Tests were carried out to monitor the load-settlement behavior and enable back analysis of design assumptions. These tests confirmed the compaction success and the theoretical assumptions of the design process.

REGIONAL GEOLOGY AND BACKGROUND

The Arabian Gulf is a shallow elongated basin of Late Pliocene to Early Pleistocene age. The basin is asymmetric, with a gentle slope on the Arabian side, a much steeper slope on the Iranian side, and water depths which rarely exceed 100 m. The basin is bound on the northwest and along the Iranian side by Zagros Mountain belt, which is the central part of the Alpine-Himalayan chain. The Gulf consists of a foreland basin with glacio-eustatically controlled mixed carbonate-clastic fill (Walkden and Williams, 1998).

As the Arabian side of the Gulf slopes gently, a major part of this sedimentation occurred in a shallow marine environment with little sediment influx from the arid mainland. Hence most of these sediments are of biogenic origin and have high carbonate content. They consist of relatively pure carbonate clay, silt and sand. According to Walkden and Williams (1998), the Gulf experienced the interaction of three equally important depositional systems throughout Pleistocene-Holocene time. These are the Arabian Gulf marine carbonate system, the Mesopotamian fluvio-deltaic system, and the Arabian continental aeolian system. Based on location, the near shore boreholes probably lie within the marine carbonate system characterized by diverse sediments, varying from high-energy oolite and bioclastic sands to low-energy marls (Walkden and Williams, 1998). The alluvial plains were a major source of aeolian siliciclastic sediment, which was later blown south-eastwards by the Pleistocene Shamal winds (Glennie, 1996). During Pleistocene glacial periods, surface sediments were exposed. This caused desiccation, oxidation, over consolidation of silts and clays, leaching and probably cementation of carbonate sands. The general geology of the UAE has been substantially influenced by the deposition of marine deposits associated with the continuous sea level changes during the relatively recent geological time. Moreover, with the existing mountainous geology across the UAE the country is considered to be of relatively low-lying area. The geological conditions in Dubai essentially consist in general of a linear coastline dissected y creeks. Superficial deposits consisting of beach dune sands with marine sands and silts. Furthermore, erosion, capillary rise phenomena as well as evaporations have led to extensive silt deposits in some areas especially near to the creeks. These superficial deposits underlined by alternating layers of calcarenite, carbonate sandstone, sands as well as cemented sand layers.

SOIL TYPES FOUND IN THE UAE

Soils are mostly inorganic in origin and have either been derived in situ or because of transportation of debris from elsewhere in the UAE. The chemistry of the soil has a rather ''special'' character, and often contains high proportions of NaCl and CaCO3. If the carbonate content is higher than 50%, the soil may be considered as carbonitic soil (Lunne et al., 1997). Soils with high calcium carbonate content occur primarily in the UAE coastal strip or in reclamations formed by dredging marine sediments derived from the Arabian Gulf. This paper concentrates of these soils. In a natural state the general soil conditions on land in the coastal strip of UAE consist either of coastal sabkhas or aeolian dune sand over bedrock, which usually consists of weak calcarenite. Aeolian dune sand is fine grained and has a very limited range of particle sizes. It falls outside the ideal range for in situ densification by vibrocompaction. In view of the huge amount of development of the coastal strip in the Emirate of Dubai, the natural geology/soil conditions have been substantially disturbed. Sand dunes have been levelled so that where crests of dunes have been pushed into intervening hollows soil may be compact, but soil in the backfilled hollows may be very loose and include "trapped" pockets of silty soil. The consequence is that there is highly variable soil density both laterally and over depth, which would result in undue total and differential settlement of shallow foundations unless some form of compaction, or ground improvement or other special foundation measures are adopted. Historically buildings in such areas, even low rise, have been supported on piled foundations.

Recent UAE reclamations consist mainly of marine dredged medium to coarse sand sourced locally from the Arabian Gulf. It consists largely of broken shells, fragments of the skeletal remains of marine organisms, coral fragments, and a small proportion of ooliths, with minor quartz and silicate-based rock particles (particularly gabbro). Calcium carbonate content is usually high, typically in the range 50% to 90%. As with any reclamation there are pockets and lenses of silty soil within the reclamations, which are usually of limited thickness (less than 1m) and of limited extent. There are sections of few reclamations where the source material has been obtained using a "cutter dredger" which has recovered broken fragments of caprock and bedrock. In these sections there may be substantial proportions of gravel, cobbles, and even boulders, which have consequences for ground improvement, methods of investigation and assessment of the effectiveness of ground improvement. In view of the hot climate, which results in high dissolved calcium carbonate content in groundwater, weak cementing of the calcareous soils may occur relatively quickly after placement. This also has consequences for ground improvement procedures and affects the interpretation of test data. Recent reclamations have been formed on existing seabed soils which consist of sandy silt, and silty sand, with thickness typically up to 3m. Beneath seabed soils there is often a layer of "caprock" typically up to 1m thick, followed by up to 3m of medium dense sand, before bedrock if reached. Caprock is absent or discontinuous in some areas, particularly close to the original shoreline. Caprock and bedrock usually consist of weak to moderately strong calcarenite. Placement of fills in older reclamations, some of which date back to the 1970's was less controlled. These consist of mainly of calcareous sand but may have a higher general silt content than recent reclamations (typically 10% to 30%), and larger and more frequent pockets and lenses of silt. The generally higher fines content exceeding > 15% necessitates the adoption of ground improvement by vibro replacement stone columns as an alternative to vibrocompaction.

EARTHQUAKE RISK

Sand that forms reclamation is mainly below water level so is saturated, and because of the methods of placement is loose to medium dense. These facts render reclamations highly susceptible to liquefaction during earthquakes. In the past it was considered that the level of seismic risk in UAE is low. However, there was a Magnitude 5.1 earthquake on 11th. March 2002 at Masafi in the Emirate of Fujairah that demonstrated that earthquakes could occur in the UAE. It is well known that there is a high level of earthquake risk in Southern Iran, but in UAE it is considered that seismic activity is generated by movements on the Dibba Fault in the Emirate of Fujairah

This is a transcurrent fault where movements are primarily horizontal. The highest risk is close to the line of this fault, and risk reduces with increasing distance from the fault. The design level of earthquake for the Emirate of Dubai is usually taken to be Magnitude M = 6, UBC Category Zone 2A/2B with peak ground acceleration (PGA) and bedrock of 0.15g and peak ground acceleration (PGA) at ground surface from 0.24 to 0.27g. At Abu Dhabi City, which is further from the Dibba Fault than Dubai, the design level of earthquake was lower than Dubai previously but now has been raised to match. The UAE is not safe from earthquake risk, and moderate earthquakes could occur.

GROUND IMPROVEMENT PERFORMANCE REQUIREMENTS

Typical requirements for performance after ground improvement in reclamation sands specified as below:

1. Provide for Allowable Bearing Pressure of 150 kPa for an isolated footing of 3x3m size placed at FGL over the full area of ground improvement using design calculations and parameters derived from testing, and reviewed and agreed by the Engineer
2. Maximum total immediate and long-term settlement (including elastic, creep, primary and secondary consolidation) due to the specified static and seismic loading conditions shall not exceed 25mm for isolated footing or 50mm for Raft foundation in (15/25/50 years)
3. Angular distortion (differential settlement) of points 5m apart shall be less than 1:500.
4. Reduce the risk of liquefaction when subjected to an earthquake of magnitude not exceeding 6 and a rock level Peak Ground Acceleration (PGA) of 0.15g after applying appropriate amplification factor, such that the factor of safety against liquefaction is 1.2 or greater. Liquefaction analyses performed using the CPT data. (Typical PGA at surface varies from 0.24 to 0.27g)
5. Medium dense to dense granular material with an angle of shear resistance not less than 35°
6. The target Relative Density to be achieved throughout the full treatment depth, except within top 1.0m from the FGL, shall be minimum 60~65% based on direct input of CPTU weighted average raw data.
7. Top 1.0m fill below FGL shall be compacted to achieve a minimum relative compaction of 95% of Maximum Dry Density (MDD) as achieved through laboratory moisture-density relationship test using 4.5kg rammer and tested by an independent laboratory.

SUITABILTY FOR VIBROCOMPACTION

As such, the critical part is the site characterization which in normal case will be undertaken by the Cone Penetration Tests, in conjunction with Boreholes and testing. The keys parts here to identify fines (passing 63 microns) and clays (passing 2 microns) and the current in-situ strength of the material and knowing this against the end requirements (specifications), with the chemical make-up of the sand.

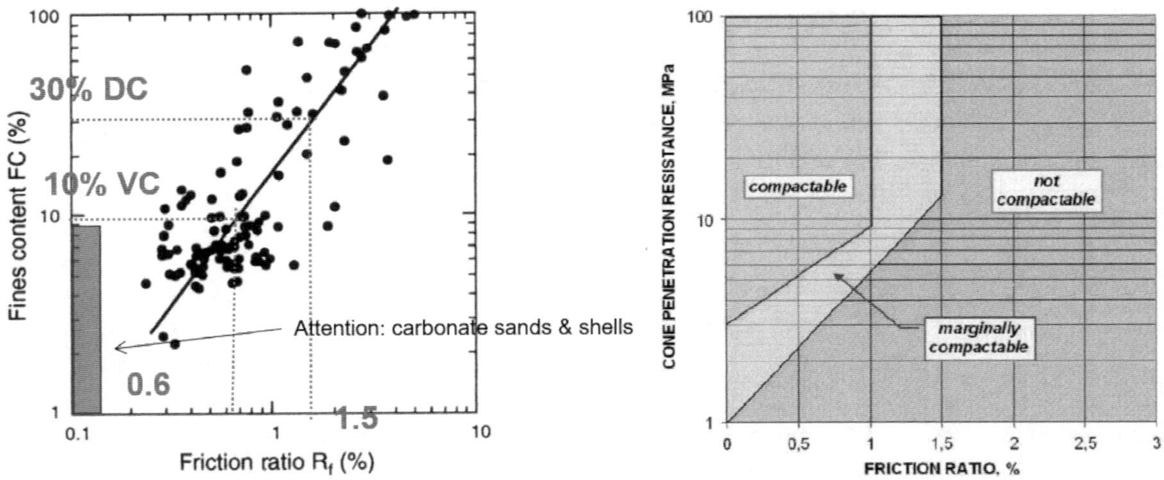

Figure 1. Soil classification for compaction based on CPT Data (Suzuki, 1991, Massarsch, 1991)

From the results of the CPT, using the software (CPTeIT) uses the published works to allow all manner of data available for assessment based on Robertson Soil Behavior Type Index, SBT Ic plots. If the soils fall within the range of SBT Ic < 2.05, it is deemed liquefiable and compactable, and greater than 2.6 that it is non-compactable, and non-liquefiable. Hence the range of 2.05 – 2.6 is where it is non compactable but liquefiable where vibro - replacement would be typically used.

Often the choice is determined by the worst soil case, and in conjunction with the suitability from classification by CPT, experience of previous works are used which have helped determine the envelopes below.

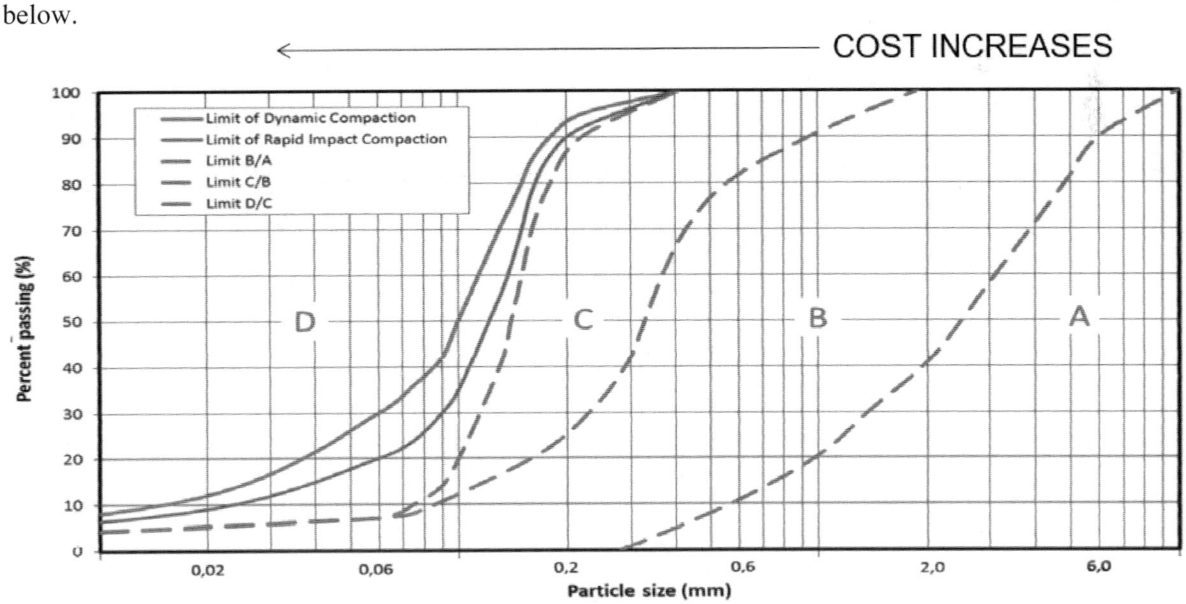

Figure 2. Soil suitability for compaction using Particle Size Distribution Curves

Zone A: The soils of this zone are very well compactable. The right borderline indicates an empirically found limit where the amount of cobbles and boulders prevents compaction because the vibroprobe cannot reach the compaction depth. Zone B: The ideal range for Vibrocompaction. They have a fines content of less than 10 %. Zone C: Compaction maybe possible with extra time and efforts, or additional filling from the surface and. Zone D: is where stone columns are a necessary solution. However due to the extra energy possible from surface compaction Rapid Impact Compaction (RIC) or Dynamic Compaction (DC) could be applied, as they also could in Zone A, B, C and up to the experience boundaries. However, these techniques are limited in depth and so other solutions may be needed.

EXAMPLE PROJECTS

The Combined Cycle Power Plant, Sharjah, UAE. The subsoil consisted of loose to medium dense Sand slightly silty to silty with shell fragments extending from existing ground levels (average ground level has been considered as +4m SMD, Sharjah Municipality Datum) to rock head. The soil profiles indicate that a consistent weak zone exists on the upper & middle part of depth which has the potential for liquefaction. The proposed structure will be supported on foundation piles with vibrocompaction being used to the mitigation of liquefaction induced settlements as well as to reduce negative skin friction on the pile design. The performance line was derived using Bellotti (1991) with an agreed shell correction factor of 1.3 to convert the silica performance line to a carbonate one.

To account for the high compressibility and grain crushability due to the common presence of carbonate content in reclamation sands, a shell correction factor of 1.3 has been applied. This Target CPT performance line if exceeded by the actual carbonate corrected CPT tip resistance, qc (MPa) after compaction; the compaction area associated with this CPT will be accepted.

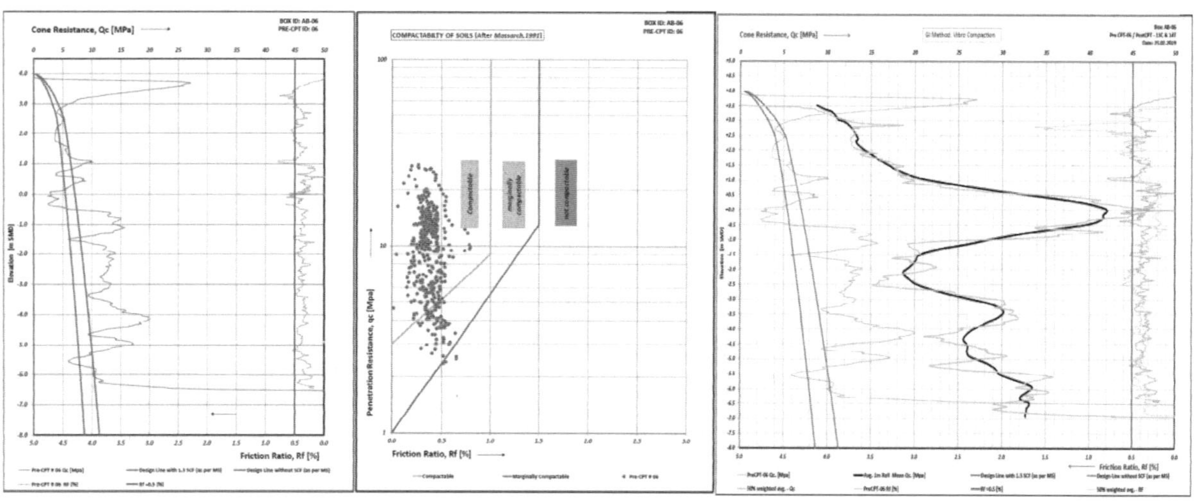

Figure 3. PreCPT soil assessment using Massarsch (1991) and PostCPT performance verification

Here we see from left to right the pre-CPT with the required (and corrected) CPT performance lines; the center graph showing the applicability of compaction after Massarsch 1991 and finally the heavy black line showing the weighted rolling mean of the post compaction CPT showing that the works exceeded the requirements.

Another project was a residential development, a circular, near shore Island. The ground surface was generally flat with the existing ground level (EGL) recorded at +6.0m DMD (Dubai Municipality Datum). However, part of the site was backfilled so that two elevated platforms at +8.0m DMD and +12.0m DMD existed. A small area of (circa 9,400 Sq.m) is occupied with existing piles at +6.0m DMD. The subsoil conditions consisted of Calcareous (Dredged) fine to medium dense SAND slightly silty with shell fragments and occasional fine gravels extending from existing ground levels to the rock head encountered at up to -11m DMD. The ground water table, (GWT) was reported at +2.0m DMD. The aim of the treatment was to mitigate for liquefaction against the design level seismic event.

A liquefaction analysis was carried out on the available CPTU's for the specified parameters of earthquake magnitude M of 6, peak ground acceleration, PGA max = 0.2g, and a factor of safety (FS) of = 1.3. This analysis indicated that a relatively loose zone -4m DMD to -9m DMD existed and therefore did not meet the specification and hence treatment was necessary to improve the ground conditions. The proposed development comprising of multistoried luxury hotels, residential towers and other ancillary building will share a combined podium that include 2 levels of basement car parking and MEP rooms. The bottom level of lower basement slab was -6.0m DMD with associated excavation to -5.0m DMD. therefore, densification was necessary for the soils below -6.0m DMD to rock head. Vibrocompaction (VC) would seem the most appropriate choice as the suitable method of ground improvement due to the soil grading and depth. The advantages here is that that VC can be applied from the existing ground level (EGL) and can effectively penetrate loose materials to densify the lower soils layers.

Table 1. Vibrocompaction areas with upper and lower working levels

Working Platform Level [mDMD]	Ground Improvement Area [Sq.m]	Soil Layers requiring Treatment	Total Depth of Penetration [m]
+6.0	50,600	-5.0m DMD to -11m DMD	17
+6.0	9,400		17
+8.0	2,100		19
+12.0	28,100		23

Further analysis of the CPTU based on Robertsons Soil Behaviour Type (SBT) (2010) showed SBT (Ic) < 2.05, and therefore compactable, a further check was undertaken by Massarsch compactabilty chart (1991) chart referenced in the clients' specifications. The available data showed the site soils are treatable by Vibrocompaction method and that it is suitable to achieve the requirements.

The performance criteria included that the tip resistance of post-CPT tests conducted at the centroid and 1/3rd point of the triangular grid shall be greater than 10 MPa for average (50% C + 50% 3rd); further, the improvement ratio (where Pre-CPT is <10 MPa) shall not be less than 25% over 1m rolling average as compared to Pre-CPT Test.

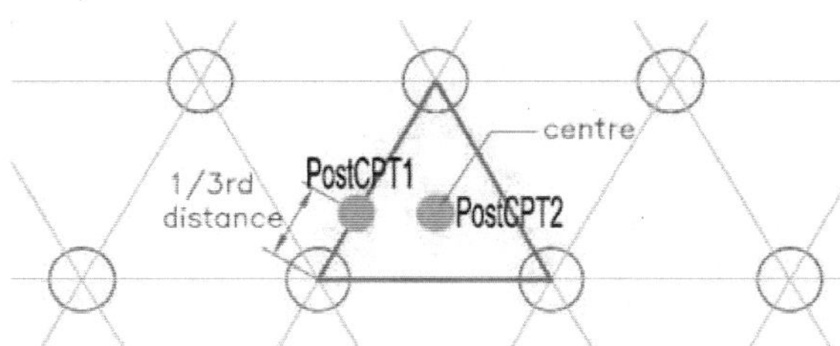

Figure 4. PostCPT Testing Sequence as per specification

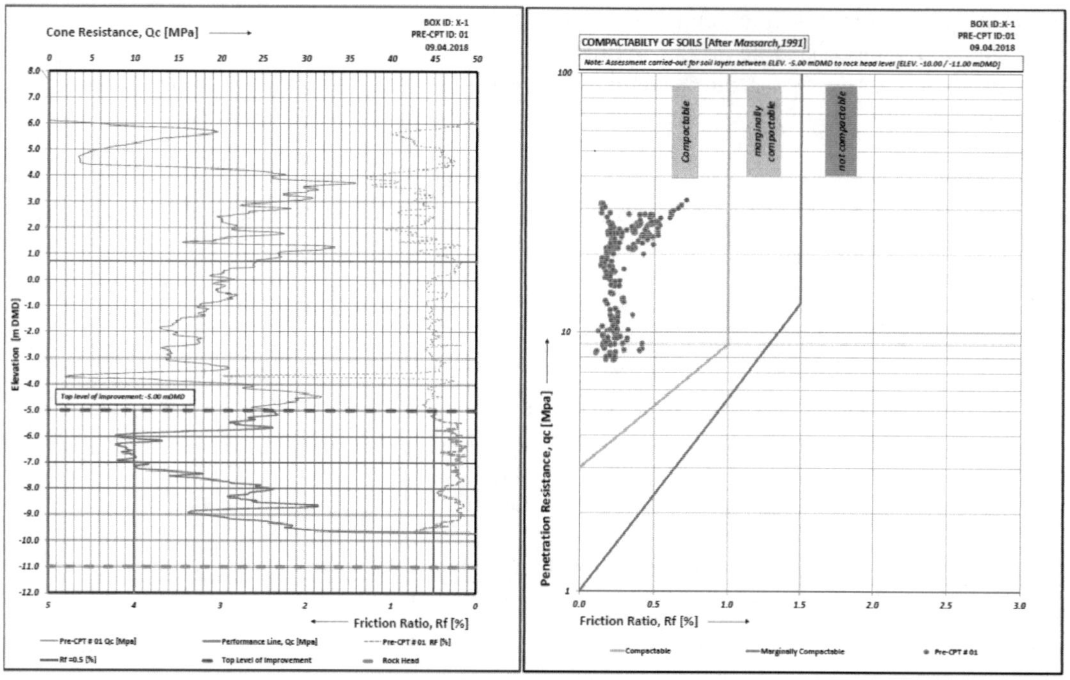

Figure 5. PreCPT soil assessment using Massarsch (1991)

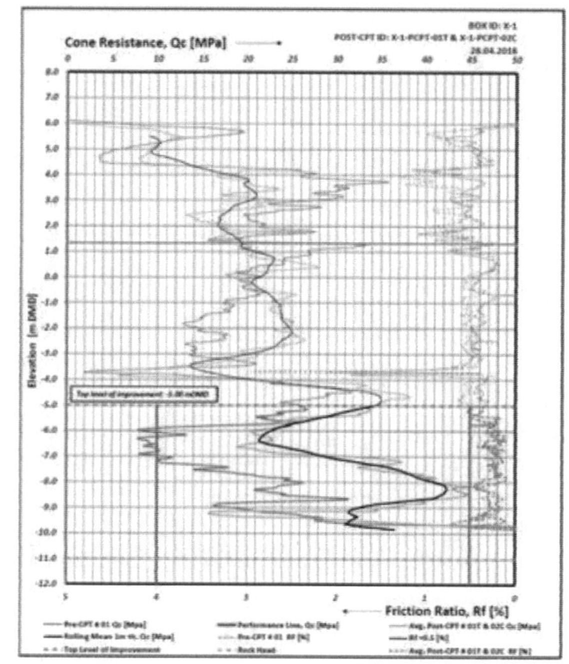

Figure 6. PostCPT Performance verification

The heavy black line shows the results of the average CPTU post treatment; far in excess of the RED Line which was the target performance line to be reached. Again, the high Ca CO3 necessitated the use of a correction factor in line with shell content and the CaCO3 and in line with project specification following the works of Homoud and Wehr to determine the relevant factor

On the iconic Palm Jumeirah Island was ground improvement to the reclamation sand up to refusal, up to 14m below EGL. The performance criteria included an analysis based on the post compaction CPTU tests to confirm that the mitigation of liquefaction induced settlement has been achieved for the design level event of PGA 0.225g and a factor of safety against liquefaction, FS = 1.3 for a magnitude M = 6. The post compaction CPTU tests were an average of (50% center + 50% 1/3rd) post CPT points of selected test triangular grid. This would ensure that overall constrained modulus of the treated soil M shall be > 50 MPa.

Table 2. CPT Performance Criteria

Depth [m]	Target CPTu Resistance, [MPa]
0.0 to 4.0	6.0
4.0 to 8.0	8.0
8.0 to Seabed	10.0

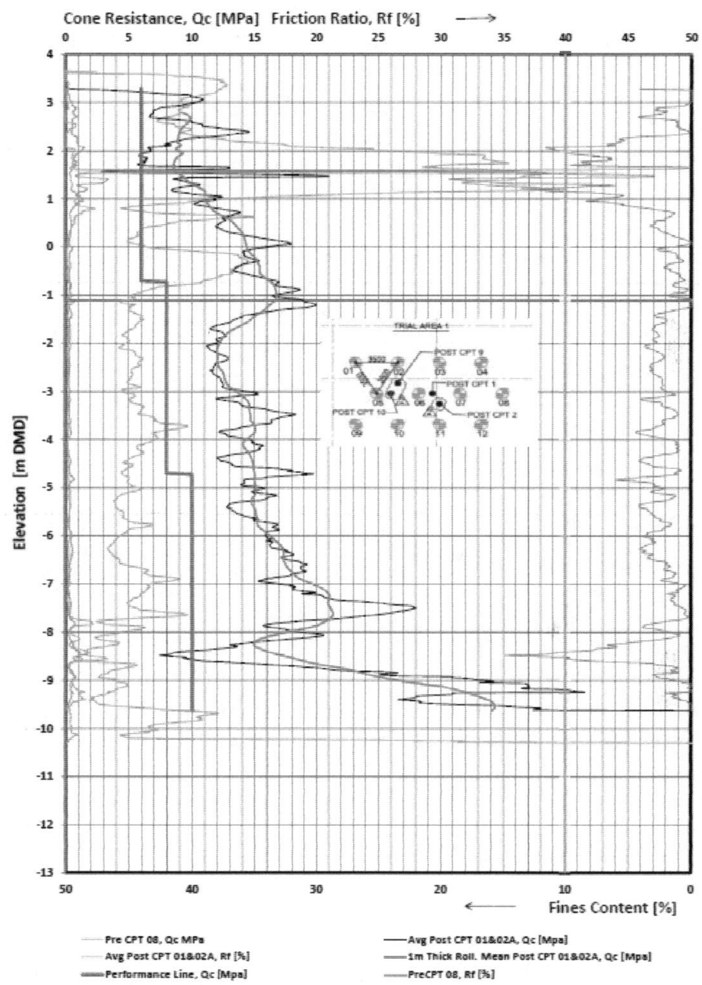

Figure 7. Pre and PostCPT vs Target CPT Performance Line Verification Chart

The graph above shows in blue the CPTU prior to work, the red line is the target performance line, and the orange line is the weighted post treatment average, showing the requirements had been met. Again, the age of the initial reclamation showed a lower shell and carbonate content, but still a suitable correction factor was agreed with the consultant based on previously published works.

CONCLUSIONS

Most of the published literature available, methods of analysis and correlations relate to silica sands and therefore to achieve the best compaction for calcareous sands, procedures must be modified slightly to suit UAE conditions and the local geology that would normally require to apply correction factors on the methos of testing and results This study would lead to more cost and time efficient works, and hence significant environmental benefits. The use of calibration chamber tests on the applied soils would lead to a greater understanding and openness in the bidding documents as opposed to the current risk of bidding with unknown contents, and hence unknown correction factors.

REFERENCES

Massarsch, K.R. 1991. Deep soil compaction using vibratory probes. In American Society for Testing and Material, ASTM, Symposium on Design, Construction, and Testing of Deep Foundation Improvement: Stone Columns and Related Techniques, Philadelphia. Edited by R. C. Bachus. ASTM Special Technical Publication, STP 1089, pp. 297–319.

Robertson, P.K., 2009a. Interpretation of cone penetration tests – a unified approach. Canadian Geotechnical Journal, 46:1337-1355

Robertson, P. K. 2010. "Estimating in situ state parameter in sandy soils from the CPT." Proc., 2nd Int. Symp. of the Cone Penetration Test, CPT '10, Omnipress Publishers, Madison, Wis.

Walkden G, Williams A (1998) Carbonate ramps and the Pleistocene-Recent depositional systems of the Arabian Gulf. In: Wright VP, Burchette TP (eds) Carbonate ramps. Geol Soc (Lond) Spec Pub 149:43–53

Glennie, K.W. 1996. The Geology of Abu Dhabi. In P.E. Osborne (Ed.), Desert Ecology of Abu Dhabi.

Published by Pisces Publications, p. 16-35.

Lunne, T., Robertson, P.K., and Powell, J.J.M, Cone Penetration Testing in Geotechnical Practice. E & FN SPON. 1997

Youd, T. L., and Idriss, I. M., eds. (1997). Proceeding of the NCEER Workshop on Evaluation of Liquefaction Resistance of Soils, National Center for Earthquake Engineering Research (NCEER), State Univ. of New York at Buffalo.

Al-Homoud AS. and Wehr W. Experience of vibrocompaction in calcareous sand of UAE. Geotechnical and Geological Engineering 2006; 24: 757-774.

Bellotti, R. and Jamiolkowski, M. (1991) Evaluation of CPT and DMT in Crushable and Silty Sands, Third interim report ENEL, C.R.I.S., Milano.

David JOHNSON., Philippe MOURRE., Stephany GALY. 2009. New Design and Quality Procedures for Ground Improvement in Middle Eastern Calcareous Sand.

Wehr W. Influence of the carbonate content of sand on vibrocompaction. 6th International conference on Ground Improvement techniques, 7.2005, Coimbra, Portugal.

Vesic, A. S. (1965) Ultimate loads and settlements of deep foundations in sand, Proceedings of the Symposium on Bearing Capacity and Settlement of Foundations in Sand, Duke University, Durham, pp. 53–68 (cited by L. Decourt, General report session 2: SPT, CPT, Pressuremeter Testing and Recent Developments in Insitu Testing, ICSMFE, Rio de Janeiro, 1989, pp. 2405–2412)

NORTHEAST BOUNDARY TUNNEL – MOUNT OLIVET ROAD DROP SHAFT

Russell H. Lutch, Brierley Associates, Brighton, MI, USA, 906-370-4265, rlutch@brierleyassociates.com

Jeremiah M. Jezerski, Brierley Associates, East Syracuse, NY, USA, 315-561-0264, jjezerski@brierleyassociates.com

ABSTRACT

The Mount Olivet Road Drop Shaft (MOR-DS) support of excavation (SOE) is a circular unreinforced concrete shaft with an inside diameter of a 30.5 ft (9.3 m) and depth of 120.5 ft (36.7 m), constructed using slurry wall methods. The SOE for the adjacent Mount Olivet Road Approach Channel (MOR-AC) and Mount Olivet Road Ventilation Vault (MOR-VV) near surface structures (NSS) consists of lean mix concrete walls and cement-bentonite walls, respectively, both constructed using slurry wall techniques and reinforced with embedded soldier piles. NSS SOE required multiple levels of preloaded internal bracing, with excavation depths of up to 75 ft (22.9 m).

The original design construction sequence for the shaft and NSS SOE required that the MOR-DS be constructed in its entirety and backfilled to ground surface prior to excavation of the adjacent NSS SOE. After installation of the drop shaft and NSS SOE an accelerated construction sequence was implemented requiring the analysis of the excavation support systems to determine if the excavations could be open concurrently without adversely impacting the partially constructed permanent MOR-DS structure.

The MOR-DS was analyzed using the finite element analysis (FEA) software, MIDAS GTS, to demonstrate that the unbalanced loads applied to the MOR-DS composite structure due to the revised sequence would not exceed the strength and serviceability requirements per ACI-350 or superseded by the more stringent project requirements.

The results of the FEA were used to set limits for geotechnical instrumentation consisting of strain gauges located on the brace levels of the NSS SOE. A comparison of the SOE design hand calculations, FEA model results and instrumentation data is presented.

Keywords: finite element modeling (FEM), geotechnical instrumentation, slurry wall construction, deep excavations, tunneling, ACI 350

INTRODUCTION

Northeast Boundary Tunnel (NEBT) Project

The Northeast Boundary Tunnel (NEBT) will relieve chronic flooding along Rhode Island Avenue NW and comprises a major component of the Consent Order driven DC Clean Rivers Program undertaken by DC Water, upon its completion. NEBT is 23 ft (7 m) inside diameter by approximately 27,700 ft (8,200 m) long and ranges in depth from about 60 to 140 ft (18 to 43 m).

The alignment passes beneath a dense urban corridor and these landmarks: a portion of the RFK Stadium Complex, Langston Golf Course, National Arboretum, Mount Olivet Cemetery, New York Avenue, Amtrak Rail Yard, WMATA Red Line, CSX rail bridge, and a large section of Rhode Island Avenue.

The project includes seven shafts ranging in depth from 76 to 154 ft (24 to 47 m) with diameters varying from about 19.5 to 54.5 ft.(5.9 to 17.1 m). Associated with each shaft are deep diversion and conveyance structures. Shaft sites encountered fill, alluvium, and Potomac Group soils. Within the Potomac Group are fine grained soils, over-consolidated clays and zones of cemented sand/gravel nodules within a clay matrix. Excavations for deep near surface structures are located in close proximity to existing sensitive structures and utilities, limiting the use of dewatering to only a few of the construction sites. The remainder of the sites utilize ground improvement to control and cut off groundwater to limit settlement and disturbance to adjacent structures.

The tunnel envelope is being excavated within the Potomac Group soils consisting of clays, sandy soils and mixed face conditions, such as a layer or layers of impermeable materials in combination with layers of water-bearing clean sands and gravels under pressurized conditions. Given the ground conditions, a Herrenknecht EPB-TBM was selected for the main tunnel. An additional smaller diameter tunnel located approximately 35 to 70 ft (10.6 to 21.3 m) beneath Mount Olivet Road was mined with an EPB-MTBM. This MTBM tunnel is approximately 700 ft (213 m) in length with a 10 ft (3 m) finished diameter.

A joint-venture comprising Salini Impregilo and Lane's S.A. Healy (SIH) is leading the Design/Build Team with Brierley Associates serving as the Prime Designer and Engineer-of-Record. Brierley was tasked with progressing the 30 percent design documents prepared during the bidding phase to final "Released for Construction". Brierley also manages the various design subconsultants including: traffic engineering/control, construction impact assessments, risk management, and concrete durability specialists. Design work began in September 2017 and construction will be completed in 2023 - two years ahead of the Consent Decree schedule.

Mount Olivet Road (MOR) Construction Site

Of the seven shafts connecting the existing near surface stormwater conveyance system to the deep NEBT CSO tunnel, one of the largest was the Mount Olivet Road Drop Shaft (MOR-DS), shown in Fig.1. Given the size of the Approach Channel (MOR-AC) and Ventilation Vault (MOR-VV) near surface structures (NSS) and their proximity to the MOR-DS unbalanced forces imparted by the NSS excavations on the MOR-DS structure was a significant contributing factor to the overall construction sequence of the MOR Shaft Construction site and the overall shaft and NSS SOE design.

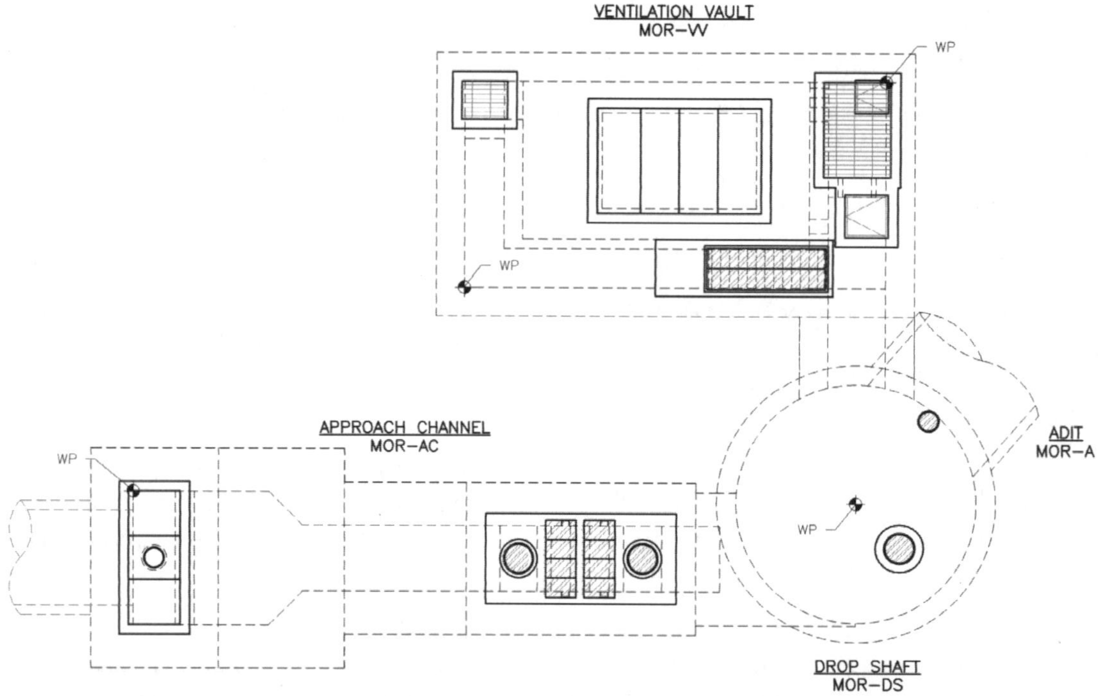

Fig. 1. MOR Shaft Construction Site Plan, NEBT Contract Drawings

The MOR-DS SOE consists of a circular unreinforced concrete shaft with an inside diameter of a 30.5 ft (9.3 m) and depth of 120.5 ft (36.7 m), constructed using slurry wall methods. The SOE for the adjacent MOR-AC and MOR-VV consists of lean mix concrete walls and cement-bentonite walls, respectively, both constructed using slurry wall techniques and reinforced with embedded soldier piles. NSS SOE required multiple levels of preloaded internal bracing, with excavation depths of up to 75 ft (22.9 m). The MOR construction site SOE plan and profile through the MOR-AC structure are shown in Figure 2-3, respectively.

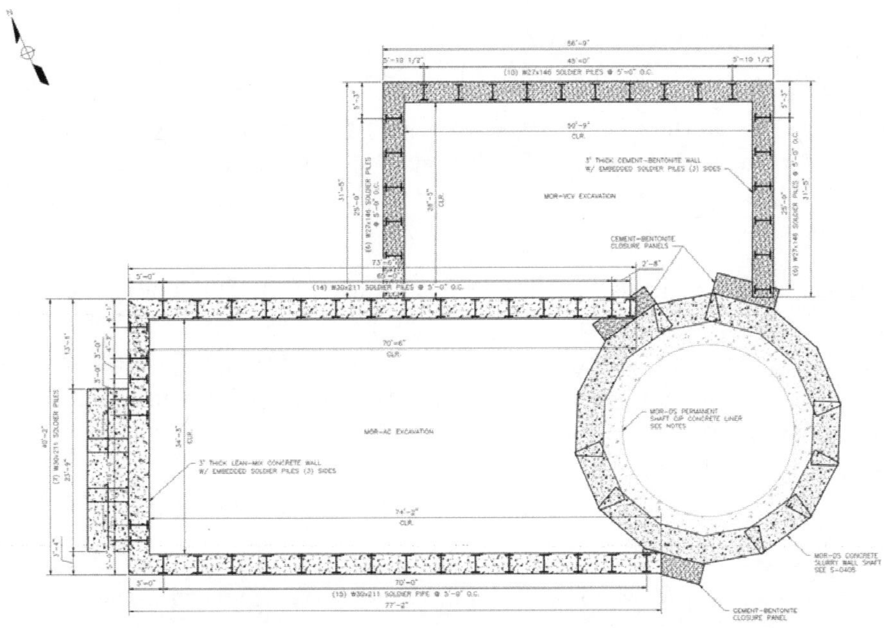

Fig. 2. MOR Shaft Construction Site SOE Plan

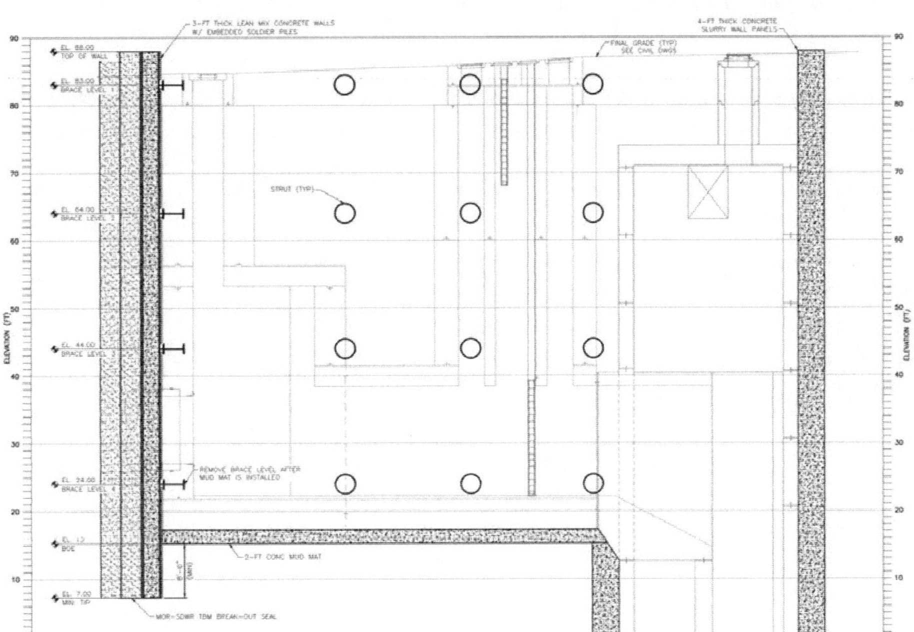

Fig. 3. MOR Approach Channel Structure and Drop Shaft Profile

In addition to facilitating the construction of the MOR-AC the NSS SOE served as the launching pit for the Mount Olivet Road Diversion Sewer (MOR-DSWR). To assist in the MTBM tunneling operations and decrease the total quantity of shoring walls to be installed, the MOR-AC excavation was enlarged to approximately the full diameter of the MOR-DS excavation and utilized a shared wall with the MOR-VV excavation. Enlarging the MOR-AC excavation to match the MOR-DS had the added benefit of balancing the loads applied to the two excavation support systems.

DESIGN

Design Criteria

While the NEBT is considered a Design-Build project, where the DB team is responsible for the design of a portion of the project permanent structures and the support of excavation for all permanent structures included in the Contract Documents, the specifications related to design of both permanent and temporary structures are generally written in a prescriptive manner with comprehensive mandatory requirements for design include the following:

- All permanent construction shall be designed and constructed with durability commensurate with the required design life of 100 years,
- Design of all underground concrete structures shall meet or exceed all provisions of ACI 350,
- SOE shall not allow groundwater forces to be exerted on any permanent structure until it has been completed in its entirety,
- Mandatory soil properties for design, and
- Consideration of both short-term and long-term conditions (drained versus undrained behavior).

Unlike many of the NEBT construction sites, dewatering/depressurization was permitted at the MOR construction site. However, minimum groundwater elevations were stipulated in the Contract Documents to limit drawdown and subsequent ground deformations associated with the construction dewatering.

Soil and Groundwater Conditions

As previously stated, the mandatory requirements for design specified the soil properties, basic soil stratigraphy and minimum groundwater levels for the MOR SOE design. Figure 4 shows the geotechnical profile at the MOR Shaft Construction Site from the project Geotechnical Baseline Report (GBR).

The MOR-AC and MOR-VV were excavated to a depth of approximately 75 ft (22.9 m) and 30 ft (9.1 m), respectively. The excavations extend through approximately 4 ft (1.2 m) of Fill, and terminate within the P/A group, which is predominantly G1 (CH) and G2 (CL) soils. The PTX group, which consists of G3A (GC), G4 (SP-SM) and G1 (CH) soils, encountered in that order, is located at great depth below the base of the NSS excavations but influences the base stability of the excavations. Table 1 provides a description of the various soil units as defined by the Contract Documents and summarizes the soil properties provided in the mandatory requirements for design.

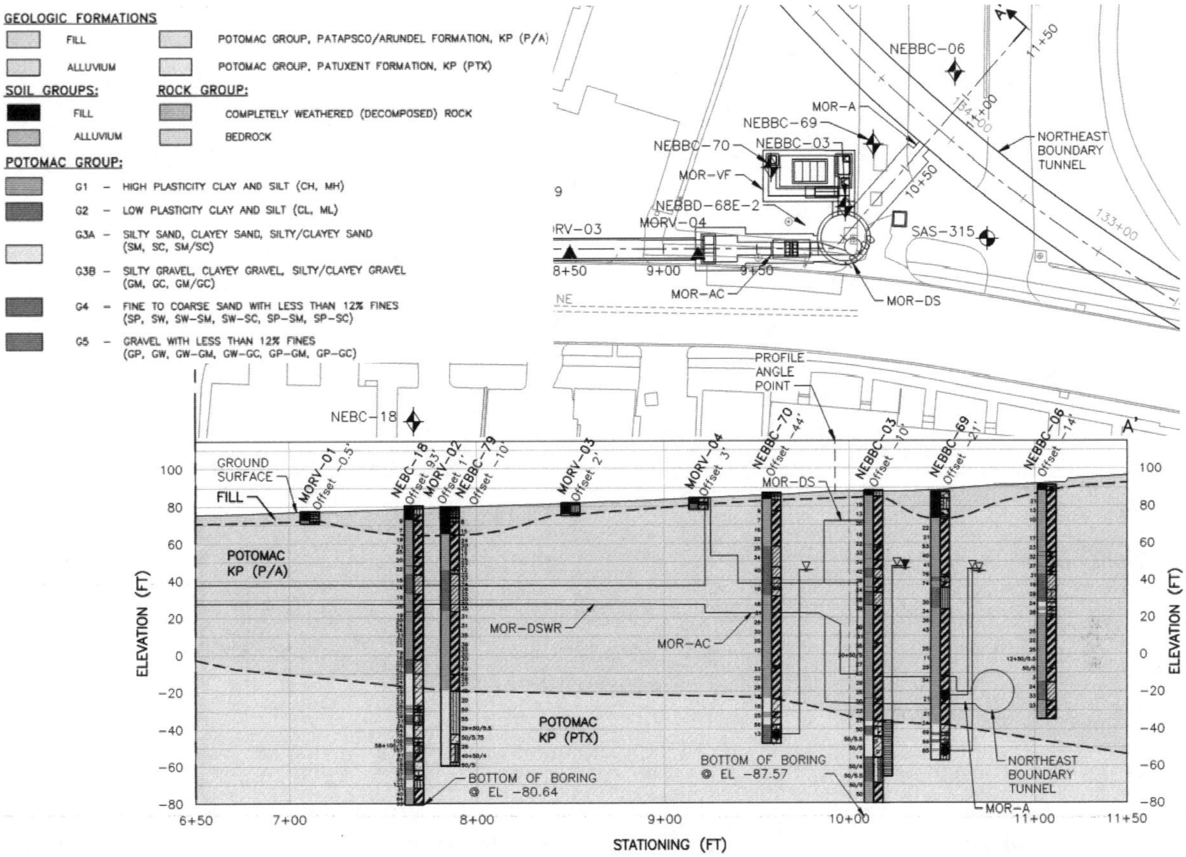

Fig. 4. MOR Plan and Geotechnical Profile, NEBT GBR

Table 1. Mandatory Requirements for Design Soil Properties, NEBT Section 01012

Parameter/Property	Fill	Alluvial Clay and Silt (Inorganic)	Alluvial Clay and Silt (Organic)	Alluvial Sands	Alluvial Gravel	G1	G2	G3A	G4
Total Unit Weight (pcf) Baseline Value/Range $^{(b)}$	110-120 $^{(c)}$	100-120	80-95	115-125 $^{(c)}$	120-130 $^{(c)}$	120-140	125-140	125-130 $^{(c)}$	130-135 $^{(c)}$
Effective riction Angle, ϕ' (dgerees), and cohesion Intercept, c' (psf)	30, 0 $^{(c)}$	28 $^{(n)}$, 0 $^{(c)}$	24 $^{(d)}$, 0 $^{(o)}$	34 $^{(c)}$, 0	36 $^{(c)}$, 0	22 $^{(e)}$, 0		36 $^{(c)}$, 0	38 $^{(c)}$, 0
Undrained Shear Strength (psf), For Design of Temprary and Permanent Structures	NA	300 + 0.27 σ_v' $^{(f)(g)}$		NA	NA	750 + 0.24 σ_v' $^{(f)(i)}$		NA	NA
At-Rest Earth pessure Coefficient, K_o	0.5 $^{(j)}$	0.53 $^{(j)(k)}$	0.59 $^{(j)}$	0.59 $^{(j)}$	0.4 $^{(j)}$	See table below		0.9 $^{(l)(m)}$	0.9 $^{(l)(m)}$

<u>Notes:</u>

(b) Based on Mean +/- 1 Standard Deviation along with other considerations such as exclusion of outliers, rounding, and engineering judgement
(c) Estimated from SPT correlations and literature
(d) The selected value of friction angle corresponding to the lower bound of the post peak strength
(e) Estimated from triaxial tests where slickensides and other defects did not play a role in the failure plane. Lower values are tyical when slickensides are exposed or participate in the failure.
(f) σ_v' = vertical effective stress
(g) Estimated from triaxial test esults
(i) Estimate is lower bound that arises when stress relief is allowed (e.g. for excavations), higher values are typically obtained otherwise
(j) Estimated from equation of 1-sin(ϕ')
(k) A high value of 0.64 should be used for desicated, slightly overconsolidated alluvium
(l) Estimated from literature correlations and OCR
(m) Valid for depth > 130 feet below ground surface. If encountered in depths < 130 feet then a K_o value of 1.1 should be used
(n) Estimated as a peak friction angle at low strains. Lower value must be used for conditions where greater soil strains may be mobilized, such as slope stability analyses
(o) An effectie cohesion intercept (c') of 160 psf may be used for temporary conditions. temporary conditions must not exceed 12 months
NA = Not Applicable

Depth Range below ground surface (feet)	Ko for G1 and G2 Soil Groups
0-20	NA
20-40	1.5
40-100	1.3
100-130	1.2
>130	1.1

Construction Sequence

The construction sequence for the MOR-DS, MOR-AC and MOR-VV required the MOR-DS permanent shaft structure to be completed in its entirety and backfilled prior to excavating the MOR-AC and MOR-VV structures. In addition, the MOR-AC and MOR-VV excavations were not allowed to be open concurrently to reduce the potential for unbalanced loading on the excavation support and partially constructed adjacent structures.

Design Approaches and Methods

The methods utilized for the design of MOR excavation support and analysis of the partially constructed permanent structures consisted of a combination of hand calculations and finite element modeling. Hand calculations were used as the basis of design for the support of excavation while FEM was used to evaluate complex three-dimensional loading conditions, develop ground deformation predictions, and during the development of a hydraulic de-pressurization scheme based on soil permeability and soil unit layering.

Empirical Hand Calculations

The design of the circular slurry wall shaft at MOR considers stability, strength and serviceability limit states at each excavation and dewatering stage. Applicable stability limit states involve structural integrity of the temporary shaft structure and basal stability of the shaft soil plug until the partial permanent structure is installed, and after the depressurization systems are turned off.

The design of the circular shaft was performed by calculating the lateral earth pressure based on the at-rest earth pressure coefficients, Ko, presented in Table 1 and the groundwater pressure based on the maximum groundwater design elevation per the mandatory design requirements. Evaluation of shaft excavation basal stability was conducted utilizing hand calculations and to develop depressurization requirements as part of the dewatering system performance criteria.

The near surface structure (NSS) support of excavation (SOE) geometric shape, excavation sequence, bracing installation, system stiffness and wall deflection result in earth pressures that differ from a triangular at-rest earth pressure distribution in terms of distribution shape and magnitude with depth. During the excavation and bracing process, a redistribution of the lateral earth pressure occurs which concentrates loads into the bracing levels. This pressure redistribution has been found to be trapezoidal in shape, based on back-calculations of field measurements of bracing loads, and is the generally accepted standard of practice in the industry for design of multi-level braced excavations. FHWA GEC4 was used to develop the apparent earth pressures (AEP) used in design of the NSS SOE.

Groundwater was controlled at the Mount Olivet Road site using depressurization well located within the confined aquifer at great depth for base stability of the MOR-DS shaft.

Surcharge loadings are considered in addition to the soil and groundwater loads. These surcharge pressures consist of the larger of a general lateral surcharge pressure as specified in the mandatory design requirements or surcharge loads associated with the large crane used to support the diversion sewer tunneling operations completed out of the MOR-AC excavation.

To satisfy the short-term and long-term (i.e. undrained versus drained soil behavior) design requirements for the project, Brierley analyzed the NSS SOE for the undrained (short-term) condition during intermediate excavation stages and the drained (long-term) condition after excavation was completed.

Finite Element Modeling

Brierley Associates employs a commercially available finite element software, Midas Geotechnical Tunnel System (GTS) NX, to simulate a variety of soil-structure interactions. A hardening soil constitutive model was utilized to predict maximum ground deformations induced during shaft construction. A linear elastic-perfectly plastic constitutive model, with a Mohr-Coulomb shear strength envelope definition, was utilized in some cases when predicted ground deformations are not critical.

When developing soil parameters for the finite element model (FEM) Brierley follows a rational approach which relies heavily on the project GDR. As part of this process, the GDR laboratory test data was interpreted by our geotechnical engineers and modelers. Brierley relied on our interpretation of the laboratory test data, test specimen quality, adjustments for sample seating effects, and related approaches to develop constitutive models representative of field conditions to supplement those shown in the GBR as input parameters in our numerical models.

CONSTRUCTION

Construction at the Mount Olivet Road site started with slurry wall installation for the MOR-DS, followed by jet grout ground improvement for the MOR-A adit between MOR-DS and NEBT. Once the shaft excavation had been completed, installation of the near surface structure excavation support walls consisting of lean mix and cement-bentonite walls with embedded soldier piles was completed. To facilitate break-out of the MOR-A excavation from the MOR-DS, the permanent shaft collar was installed along with a portion of the MOR-DS shaft permanent lining.

After completion of the MOR-A excavation between the MOR-DS and NEBT, the installation of the MOR-AC and MOR-VV excavation support systems was completed. Concurrently, the final lining of the MOR-DS permanent shaft structure was installed to the underside of the shaft roof.

ACCELERATED CONSTRUCTION SEQUENCE REDESIGN

To reduce the overall construction schedule at the Mount Olivet Road site, the Contractor requested the excavation support design and analysis for the permanent structures be revised to allow for excavation of the MOR-AC and MOR-VV to occur concurrently and begin prior to the completion of the MOR-DS permanent structure. To analyze the more complex three-dimensional load effects of the irregularly shaped excavations and their impacts on the partially constructed permanent structures, a revised and more complex finite element model (FEM) of the MOR excavations and structures was required.

Revised Finite Element Modeling

The MOR-DS was reanalyzed using the finite element analysis software, MIDAS GTS, to demonstrate that the forces and moments induced on the MOR-DS permanent structure due to the revised sequence would not exceed the strength and serviceability requirements per ACI-350.

The shaft slurry wall and permanent liner were both modeled using shell elements using an offset function in the finite element software. The slurry wall shaft was modeled at the SOE centerline with a diameter of 34.5 ft (10.5 m) with a wall thickness of 2.5 ft (0.8 m) to simulate the effective wall thickness achieved due to installation tolerances. The permanent shaft was modeled as shell elements offset from the slurry wall. The modeled offset is equal to half the slurry wall thickness plus half the permanent liner thickness to replicate the design thickness of the permanent shaft. The modeled geometry gives the permanent shaft liner a diameter of 30 ft (9.1 m), which is conservatively larger than the MOR-DS design centerline diameter of 27 ft (8.2 m). The choice to model the system by maintaining the actual slurry wall diameter was made to maximize the predicted loads carried by the system.

The SOE walls for the MOR-AC and MOR-VV were also modeled using shell elements. The stiffness of the walls was based on the design wall thicknesses and design properties for the lean mix concrete and cement-bentonite. The bracing was installed using truss elements for the struts and beam elements for the wale beams. The excavation depths of the MOR-AC and MOR-VV are approximately 75 ft (22.9 m) and 30 ft (9.1 m), respectively. Figure 6 presents a section cut through the MOR-AC and MOR-DS to illustrate the modeled elements, penetrations and portions of the MOR-DS slurry wall removed during construction.

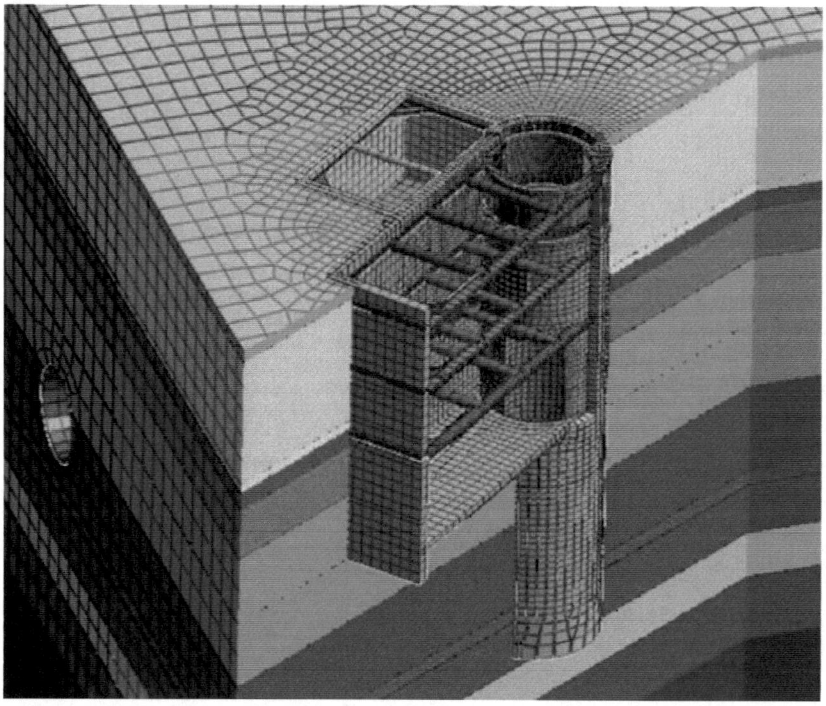

Fig. 5 Mount Olivet Road Three-Dimensional Finite Element Model

Given the anticipated MOR-AC and MOR-VV construction durations and the retained soils primarily consisting of cohesive G1 and G2 soils, an undrained (short-term) analysis is most appropriate when analyzing the excavations. However, to satisfy contract requirements three analyses were completed to analyze both the short-term (undrained) and long-term (drained) conditions. All three models were developed to demonstrate that the excavation of the MOR-AC and MOR-VV may be open concurrently and to develop contingency measures if excessive ground or structural deformations were observed during construction.

The first FEM analysis considered the undrained properties of the G1 and G2 soil units during excavation and accounts for long term behavior by reducing the strength and stiffness to the drained parameters after the MOR-AC and MOR-VV excavations are complete and base slabs are installed. Based on the relatively short duration of the excavation and bracing installation sequence as compared to the time rate of consolidation of these materials, this approach is extraordinarily conservative.

The second FEM analysis conservatively considers only drained soil properties throughout the excavation sequence. This conservative approach was required by the project Owner to serve as a worst-case scenario or upper bound to the construction loading to be applied to applied to the partially completed MOR-DS structure.

The third FEM analysis also conservatively considers drained soil properties throughout the excavation sequence but includes the installation of contingency bracing to demonstrate the effect of stiffening the shaft and to provide potential contingency measures if the soil does not behave as expected. The construction stages simulated in the models differ slightly for each model due to the slight variations to account for changes in the soil properties and the installation of the contingency bracing.

As previously indicated soil strengths were characterized using a Modified Mohr-Coulomb Failure Criterion. The strength properties used for the three models are in accordance with the drained parameters presented in Table 1.

Based on the ground conditions presented in the Contract Documents and observed in the field during jet grout, slurry panel installation, and excavation of the shaft and adit, it was Brierley's opinion that the G1 and G2 material will act undrained throughout the duration of the MOR-AC excavation and construction. However, the first FEM analysis considered the effect of reducing the strength and deformation properties to the drained condition immediately following the base slab installation and conservatively prior to the sequential removal of the internal bracing and construction of the permanent MOR-AC structure. That analysis demonstrated that the partially constructed permanent MOR-DS shaft was not overstressed.

To support that the first FEM analysis appropriately considers the short-term (undrained) and long-term (drained) soil behavior as it pertains to the actual construction sequence and duration, the rate of consolidation and dissipation of pore water pressure of the cohesive G1 and G2 soil units was assessed. The data provided in the project Geotechnical Data Report indicates that the coefficient of consolidation would be on the order of 0.03 square feet per day. Per our discussions with the Contractor, it is understood that the excavation and bracing installation would be completed in approximately 3 months. However, Brierley conservatively assumed an excavation duration of 6 months in the consolidation analysis.

Given the thickness of the P/A soil group containing the G1 and G2 soils is approximately 100 ft (30.5 m) at the location of the MOR-DS shaft, as shown in Fig. 4, the time factor, T, is approximately 0.002. Most industry references do not provide correlations for time factors less than 0.05, and therefore a time factor of 0.05 was conservatively used to estimate the percent consolidation at various levels below the bottom of MOR-AC excavation. Note that the use of a time factor of 0.05 corresponds to the use of a coefficient of consolidation of 0.69 square feet per day which is significantly higher than any test data collected in the G1 or G2 soil units.

Using the conservative time factors described above, the consolidation analysis indicated approximately 3 percent consolidation at the MOR-AC bottom of excavation (BOE), 11 percent at 10 ft (3 m) below the MOR-AC BOE and 32 percent at a depth of 20 ft (6.1 m) below the MOR-AC BOE. Given that the soils at depth below the final BOE will not be "loaded" due to the excavation process until near the end of the excavation sequence, the analysis is extraordinarily conservative, and it is demonstrated the G1 and G2 soils will behave undrained and will likely not reach drained conditions over the full duration of construction.

The second FEM analysis was developed to serve as an upper bound for structure loads and ground deformations and to conservatively analyze the case where the G1 and G2 soil units behave in a drained condition throughout the full construction sequence. That analysis indicated substantial overstress of the partially constructed MOR-DS permanent structure associated with the MOR-AC excavation, as the soil strength properties shown in Table 1 for the drained condition are significantly less than those for the undrained condition resulting much larger lateral earth pressures applied to the excavation support system and ultimately transferred to the adjacent MOR-DS structure.

The third FEM analysis is a modified version of the second model which included five cross struts installed within the MOR-DS structure and a 4 ft (1.2 m) thick diaphragm slab installed just below the MOR-AC opening in the MOR-DS prior to the MOR-AC excavation. The FEM analysis was developed to present a design that considered stiffening of the MOR-DS shaft with bracing as a contingency in the event the observed shaft deformations during the MOR-AC excavation indicate ground behavior and subsequent shaft loading representative of the drained condition. The results of the third FEM analysis indicated that such contingency measures could be implemented to protect the MOR-DS structure and prevent the shaft internal forces from exceeding the limits of ACI 350 for the conservative drained condition.

INSTRUMENTATION AND MONITORING

To develop the deformation limits to trigger installation of contingency measures or completion of construction of the shaft internal drop structure prior to advancing the MOR-AC excavation, the results of the first two FEM analyses were compared. The maximum estimated shaft deformations are approximately 0.1 in (0.027 m) for the first FEM analysis (undrained case) and 0.25 in (0.068 m) for the second FEM analysis (drained case). Based on these estimated deformations Brierley developed action and maximum limits of 0.1 (0.027 m) and 0.17 in (0.046 m), respectively. To monitor the shaft deformations during the MOR-AC excavation shape arrays were installed on the inside face of the MOR-DS permanent shaft liner at various elevations.

To provide an additional means of monitoring the soil behavior and loads applied to the shoring during excavation, strain gauges were installed on the MOR-AC cross struts. Figure 6 shows the load in one of the MOR-AC cross struts during excavation and the corresponding air temperature over time.

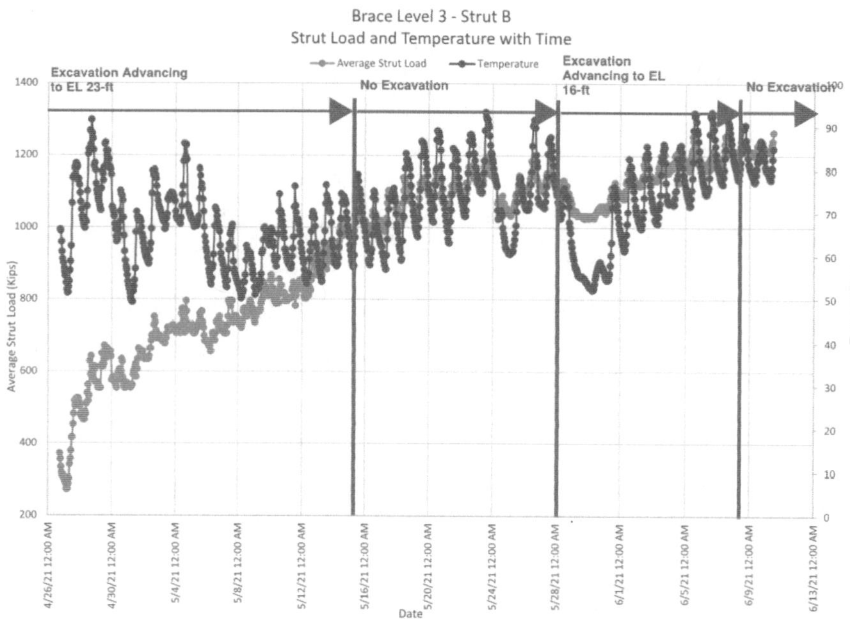

Fig. 6 MOR-AC strut load and temperature readings during excavation

The fluctuation in the strut axial loads are related to temperature changes and not direct indications increases in lateral earth and groundwater loads. Considering the temperature effects, the load in the struts due to earth and groundwater pressures are generally closer to the lower bound of the load ranges provided by the first FEM analysis based on undrained soil behavior.

SUMMARY AND CONCLUSIONS

In conclusion, the FEM analysis which utilized undrained soil parameters most appropriately modeled the excavation support and permanent structure behavior and provided estimates of both excavation support loading and permanent structure deformation most in agreement with the instrumentation data.

The FEM analyses which considered drained soil parameters for the short-term condition were presented for completeness and are not considered representative of the actual loading the excavation support and partially constructed permanent structures experienced. However, modeling of the drained condition allowed for the development of contingency measures in the unlikely event such ground conditions were observed during construction.

With the implementation of monitoring and setting of the Action and Maximum levels the risk associated with the excavation of the MOR-AC was substantially reduced. If at any point the monitoring had indicated that movements or bracing loads had exceeded the action levels, contingency measures were ready to be implemented.

ACKNOWLEDGEMENTS

The authors would like to thank Salini Impregilo, The Lane Construction Corporation, and DC Water for opportunity to work with them on the NEBT Project.

REFERENCES

DC Clean Rivers Project Division J - Northeast Boundary Tunnel (NEBT) Contract Drawings.

DC Clean Rivers Project Division J - Northeast Boundary Tunnel (NEBT) Geotechnical Baseline Report.

DC Clean Rivers Project Division J - Northeast Boundary Tunnel (NEBT) Section 01012 Mandatory Requirements for Design.

Geotechnical Engineering Circular No. 4 Ground Anchors and Anchored Systems, Federal Highway Administration, 1999.

PERFORMANCE BASED EXCAVATION SUPPORT DESIGN FOR CHALLENGING URBAN SITE GEOMETRY IN TORONTO, CANADA

Erika Acton, Isherwood Geostructural Engineers, Mississauga, ON, Canada, 1-905-820-3480, erika@isherwood.to
Shawna Munn, Isherwood Geostructural Engineers, Mississauga, ON, Canada, shawna@isherwood.to
Carol Domitric, Isherwood Geostructural Engineers, Mississauga, ON, Canada, carol@isherwood.to

ABSTRACT

The EQ Bank Tower, a 24-storey office tower with 4 levels of underground parking, is currently under construction at 25 Ontario Street in the downtown core of Toronto, Canada. The tower geometry required an 18 m (60 ft) deep excavation inside an existing 3-storey 100-year-old façade, to be retained in place throughout construction. This presented a variety of design challenges that required several innovations and a customized monitoring program. In this paper, we will discuss two major challenges, and highlight the innovative and adaptable solutions carried out using performance-based design leveraging the observational method. The first was installation of a shoring system in a 500 mm (20 in) gap between the historical façade and a live gas main. The second involved excavating 10 m (33 ft) into expansive shale bedrock adjacent to a sensitive neighbouring building.

Key words: heritage façade, micropile, performance based design, monitoring, excavation shoring, support of excavation, SOE, expansive shale bedrock, glacial formations.

BACKGROUND & PROJECT DESCRIPTION

The EQ Bank Tower is a 24-storey office tower currently under construction in downtown Toronto at 25 Ontario Street, located at the southeast corner of Adelaide Street and Ontario Street. It will add 466,000 sq ft of office space just east of the historical St. Lawrence Market neighbourhood. The building has 4 underground levels used for a combination of parking, storage, and mechanical/electrical purposes. The below-grade excavation footprint measures 50 by 45 m (164 by 147 ft) and ranges from 16 to 18 m in depth (50 to 60 ft).

Heritage Façade

The majority of the site was occupied by the Drug Trading Company office building, built in 1941 and designated by the City of Toronto to have "architectural, associative and contextual value". Since 2015, the City of Toronto has required historical façades to be supported in-place to retain the original craftmanship, rather than dismantling buildings and re-building brick-by-brick. Based on this policy, the existing 2.5 storey, 11 m (36 ft) tall, brick façades along the North and East walls were required to be supported in place throughout the duration of construction and later incorporated into the façade of the new office tower (City of Toronto, 2012).

Adjacent Utilities

This project is located in a historical Toronto neighbourhood, blocks from the historical Distillery District to the south and St. Lawrence Market to the west. As a result, the underground city right-of-way is crowded with municipal services, several in close proximity to the project footprint. Most critically, a 150 mm (6 in) gas main was located adjacent to the North and East walls where the existing historical façades already limited the available shoring footprint.

Neighboring Buildings

The project site is adjacent to two high-rise buildings founded on the shale bedrock. The Toronto Community Housing Corporation (TCHC) building is located east of the under-construction development. The TCHC building is a 12-storey steel and glass building supported by drilled shafts founded within the first few metres (3 to 8 ft) of shale bedrock, directly adjacent to the 25 Ontario property line. The excavation plan for the new development involved excavating an additional 7 metres (23 ft) beneath the underside of the existing foundations. As a city-owned property, the managing team was risk-averse and very concerned about damage due to movements and construction vibrations.

The SAS Institute, located south of the proposed development, is an 8-storey steel and glass building supported by conventional spread footings founded on shale bedrock. The proposed excavation was located 1 to 2 m (3 to 6 ft) away from the footing footprint and 2 to 4 m (6 to 12 ft) beneath the underside of the existing footings.

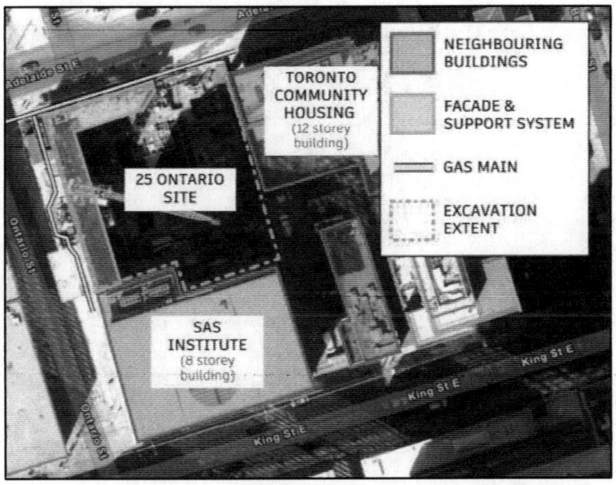

Figure 1 25 Ontario Location Plan

Figure 2 Site Renderings of the EQ Bank Building Looking Southeast

GEOLOGY & GEOTECHNICAL CHALLENGES

Previous site use and existing structures on the site resulted in various depths of man-made fill and existing foundations up to 2.3 m (8 ft) in depth. The natural overburden is a native clayey silt glacial till to a depth up to 11 m (35 ft), underlain by the Georgian Bay Formation, a shale containing inter-beds of calcareous shale, limestone, and calcareous sandstone (Isherwood, 2012). The surface of the bedrock is relatively flat at about 11 m (43 ft) below ground level. It exhibits pronounced horizontal jointing typical of shale, but is well-studied and is known to provide a reliable foundation with allowable bearing values of 2.4 to 7.2 MPa (25 to 75 tsf) (GHD, 2016).

The shale bedrock under Toronto is also known to contain high locked-in horizontal stresses on the order of 4 MPa (40 tsf), induced by periods of glaciation up to 2 km thick (Dyke, 2002). These locked-in stresses are much higher than can be resisted by shoring techniques, resulting in inevitable wall movement when excavating deep into the rock. Vertical joints in the rock are held in place by these stresses but open as the excavation proceeds, with the highest stresses locked in the east-west direction (Trow & Lo, 1988).

PROXIMITY OF GAS MAIN AND FAÇADE

As the design developed, it became apparent the vertical shoring elements would have to be squeezed between the spatial requirements of the utility owners, and the city mandate for the façade to be supported in place.

Early in the project, two possible shoring design scenarios were developed based on the assumed gas main location from City record drawings, which showed a 700 mm (28 in) gap between the gas main and property line. A typical full-depth secant pile wall was initially designed, which would have been directly adjacent to the assumed gas main location would have required relocating the gas main to allow for vertical drilling. The alternate option developed was an anchored micropile shoring design allowing for approximately 250 mm (10 in) clearance to the gas main (See Figure 3). These were presented for coordination and signoff to utility owners

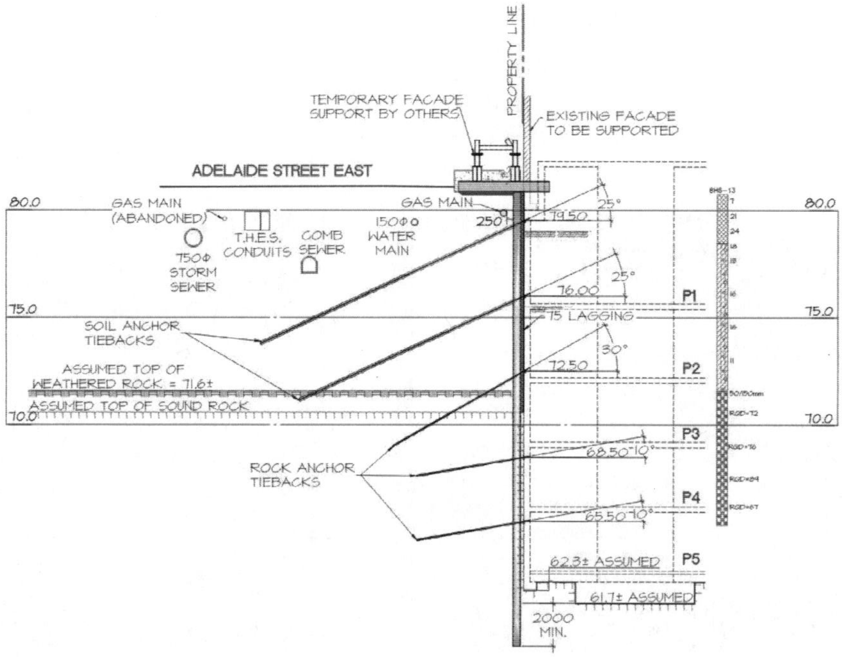

Figure 3 Design section through north wall based on City of Toronto record gas main location

As a result of our consultation, the gas main owner, Enbridge, communicated two key design requirements:

a) Unimpeded access from grade to all gas main infrastructure must be maintained throughout construction, including the ability to excavate and expose any parts of the gas main with a trench box at any time.
b) Shoring would not be allowed within 300 mm (12 in) of the gas main due to minimum clearances required around the gas main pipe for gas line tooling spatial needs.

These requirements eliminated the possibility of an externally supported façade support system (including counterweight blocks) above the Enbridge gas main on the sidewalk or street. As a result, the historic façade and support system were re-designed to be suspended on the partially constructed ground floor slab inside the excavation. This slab was supported by the shoring wall and temporary king piles inside the excavation.

Due to city boulevard improvement works, daylighting and surveying the as-built location of the gas main was significantly delayed. Unfortunately, the as-built location was another 250 mm (10 in) closer to the property line, reducing the clearance between the façade and the edge of the gas main to 450 mm (18 in). The reduced clearance envelope for the SOE vertical members presented new challenges:

- 450 mm (18 in) would allow insufficient clearance for micropile shoring based on the drilling clearances required to the façade and the 300 mm (12 in) clearance required for the gas main.
- A conventional anchored shotcrete solution would not be possible at the top of the shoring wall due to adjacent utility geometries (specifically the gas main).
- Encroaching into the underground footprint with shoring on the inside of the façade was not acceptable at the P1 level due to minimum electrical room footprint requirements. Any encroachment into the P1 level would trigger a major building redesign.
- Gas main relocation would take 2 years to complete based on Enbridge availability and was no longer feasible due to the advanced stage of the project.

These limitations prompted the consultant team to develop an innovative solution to safely support the façade and the utilities in the street while respecting the space limitations of the underground garage. The first underground level (P1) was left as-is to avoid impacting the critical mechanical space, while levels P2 and downward were shifted into site to make room for micropile shoring. Although these changes resulted in a more complex construction sequence, constructability and economic viability were not compromised. Through this design process, the shoring system was broken into the two components described below.

Spile Shoring Outside the Historical Façade

Isherwood designed an upper shoring system comprised of W150 (W6") "spiles" installed using hydrovac methods to a depth of 7.5m (24 feet). The depth was determined to be a reasonable and constructable target depth, based on consultation with hydrovac contractors, and the method was tested prior to construction inside the site to confirm constructability. The depth of 7.5m (24 ft) allowed excavation to the underside of the structural grade beam used to step the perimeter wall in between P1 and P2 levels.

Enbridge approved the hydrovac procedure in close proximity to their gas main, as the design maintained the minimum 300 mm (12 in) clearance required. Steel plate lagging was installed between the spiles using hydrovac methods in the top 1.5 m (5 feet) from grade to protect the gas main during tieback installation prevent soil face collapse as the historical underground foundation wall was removed in panels during excavation. This spile system was supported with five levels of hollow-core-bar anchors, and allowed further façade support and shoring work to proceed inside the historical façade.

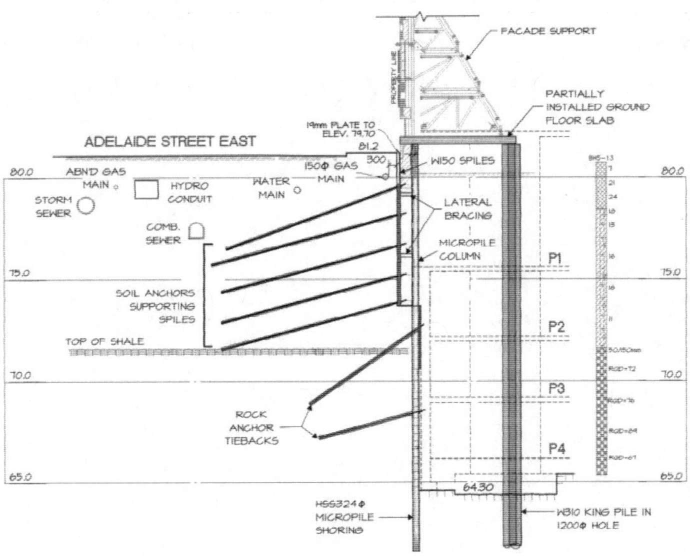

Figure 4 Design section through north wall based on as-built gas main location information

Lower Micropile Shoring Inside the Historical Façade

Once the spiles were installed, the interior of the historical building was partially demolished, leaving the structure in place between the perimeter wall and the first interior existing column bay to temporarily support the heritage façade. A low headroom micropile drill rig was then used to install 20 m (60 ft) deep micropiles from the basement of the partially demolished building as shown in Figure 5.

Once the micropiles were installed, they were extended upwards to the new ground floor slab elevation (visible in Figure 5) to act as columns from the ground floor to the underside of the P1 grade beam. From the underside of the grade beam downward, the micropile column transitioned to a traditional micropile shoring pile application. Next, the king piles were installed within the site to support the site-facing edge of the partial ground floor slab. The partial ground floor was then installed, the façade support system installed, and remainder of building demolition completed. As excavation proceeded, the micropile columns were braced laterally at two levels to the W150 spile wall.

Figure 5 View of micropile drilling (left photo) and first row of anchor drilling (right photo) inside the historical facade

Figure 6 View of site looking northeast at final excavation grade

EXPANSIVE SHALE AT NEIGHBOURING BUILDING FOUNDATIONS

Along the east side of the proposed development, available space for the shoring system was limited to a 360 mm (14 in) gap between the property line and the proposed foundation wall. In addition, the neighbouring 12-storey apartment building was located directly on the property line with two levels of underground parking, and with structural caissons bearing on rock as shown in Figure 7.

Isherwood designed a shoring system to suit this gap and allow excavation 10 m (33 ft) below the underside of the neighbouring building. This included an 8 m (26 feet) excavation into the expansive shale bedrock, below the assumed founding level of the existing structural caissons. As-built information of the structural foundation design of the 2010-era building was minimal, so conservative estimates on the depths of the existing structural caissons were used for design.

The design solution was developed to address the following key requirements:

a) Design a slender shoring system to fit in the narrow gap.
b) Protect the existing structural caissons and allow them to be safely exposed during excavation.
c) Control movement of the shoring and adjacent building, in particular in relation to the potential locked-in stresses in the Georgian Bay Shale formation.
d) Adequately demonstrate the safety and robustness of the shoring system to a neighbour who had a very low risk tolerance.

As shown in Figure 7, an anchored shotcrete shoring system was designed to support the soil and rock face and the neighbouring building loads. The shotcrete anchors were designed with additional pre-load to limit movement as the locked-in lateral stresses were gradually released. Line drilling was also employed along this wall to allow a cleaner line of excavation, reducing the risk of accidental over-excavation of shale pieces directly below the existing structural caissons, and reduce movement due to locked in stresses. Anchored channels were used where excavating directly adjacent to the structural caissons at the property line as shown in Figure 8 to laterally support the caissons during excavation.

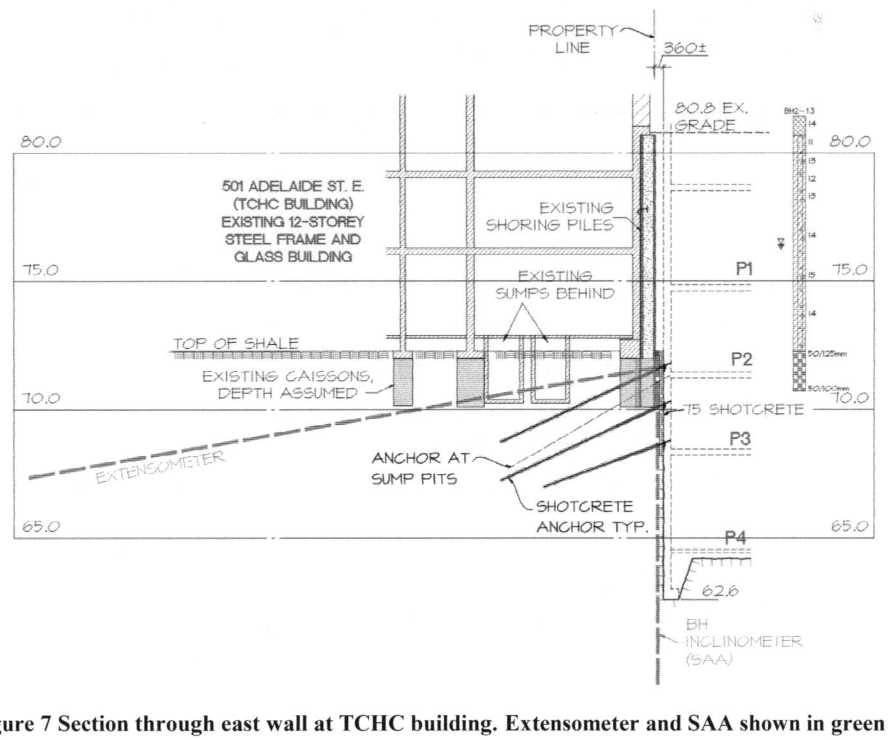

Figure 7 Section through east wall at TCHC building. Extensometer and SAA shown in green

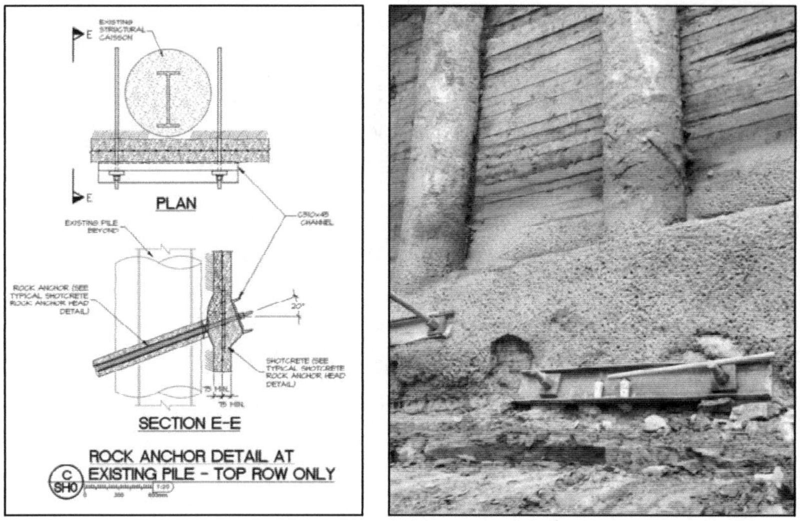

Figure 8 Anchored channel detail used at existing caissons on East wall (left diagram) and as-built detail (right photo)

CONSTRUCTABILITY CHALLENGES

Spile Installation Challenges

When installing the north wall spiles, boulders were encountered at several locations, limiting the installation depth of over half of the 19 spiles. 10 of the spiles were installed up to 2 m (6.5 ft) shallower than target depth. Where spiles didn't reach target depth, anchored shotcrete was installed in a panel sequence to control movement and prevent surface collapse/sloughing.

Threaded Casing Defects

During micropile installation, the shoring contractor (Keller) observed defects in the manufactured micropile casing threads due to shearing of the threaded casing joints while drilling.

Since the defect was determined while drilling, Keller attempted to extract the casing but was unsuccessful. Isherwood and Keller developed a solution of installing a smaller steel pipe inside the micropile prior to grouting as shown in Figure 9. This internal casing was designed to provide continuity through the damaged zone, with compressive resistance to take the façade loading on its own. Two additional monitoring targets were also installed above and below the defective joint to more closely monitor this micropile's performance.

The defective micropile material not yet installed was replaced with new material. This was not possible for the three casings already installed using the defective material. For these, Isherwood considered that:

a) A "torque test" had been carried out on each of the micropiles during drilling, and the driller hadn't observed any loss in resistance while drilling.
b) The grout could adequately transfer the load, even if the defective joints were not capable of taking 100% of the load themselves.
c) Excavation would proceed at a controlled pace for the entire length of the exposed micropiles, since the shotcrete spiles behind had anchors at 1.5 m (5 ft) vertical centres. At each lift of excavation, Isherwood was able to visually review the micropile casing joints as they were exposed and take monitoring readings to confirm movement levels remained acceptable.

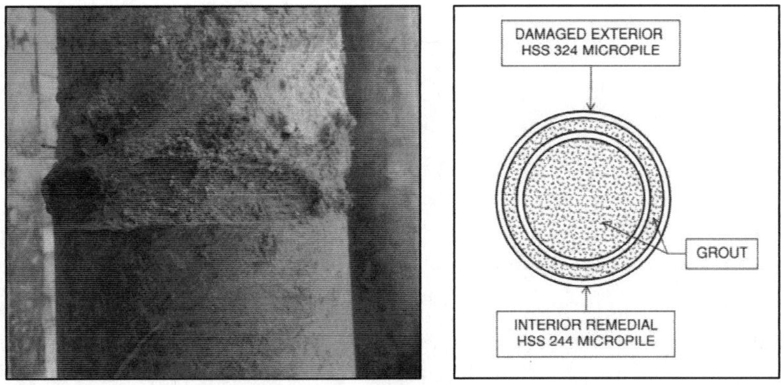

Figure 9 Micropile casing thread defects (left photo) and diagram of solution (right figure)

MONITORING & PERFORMANCE

Monitoring Program

Isherwood designed and carried out a monitoring program, with instrumentation focused in the areas the design team identified as the greatest risks (refer to Figure 11 for partial monitoring plan). The following instrumentation was employed:

- An array of targets installed on the elevation of the spile shoring system.
- Inclinometers in two of the micropiles along the north wall.
- Extensometers (MPBX) and Shape Accel Arrays (SAAs) at two locations each along the east wall.

Caution was required while excavating near the micropiles, since accidental over-excavation would increase the unbraced length of the micropiles (i.e., decreasing the compressive capacity), and equipment bumping into the micropiles could cause excessive lateral forces on the micropiles. Monitoring this closely allowed the chance to add additional lateral bracing if unacceptable deflections were observed.

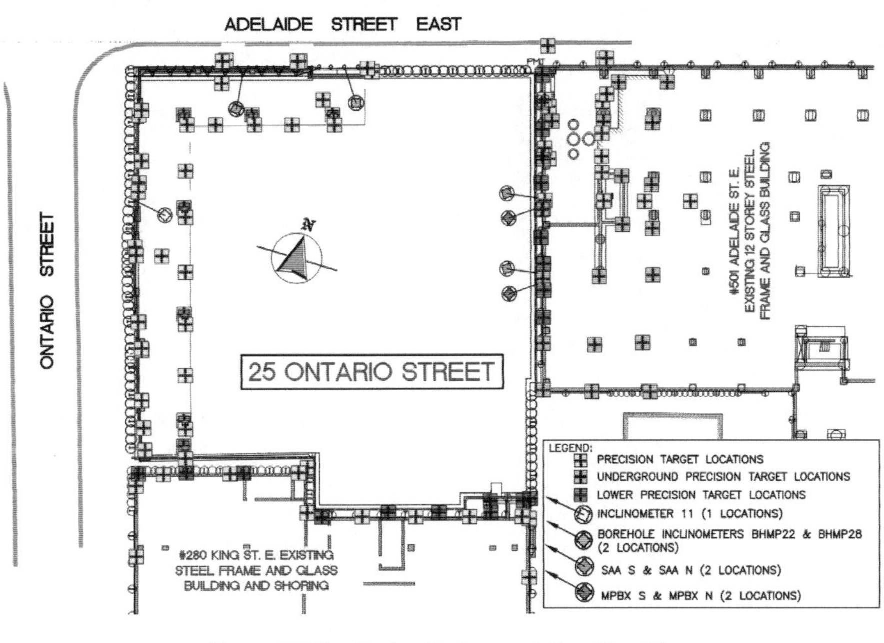

Figure 10 Monitoring Instrumentation Plan View

Expansion of the shale bedrock was a known risk based on Isherwood's experience on nearby projects and previous research done to determine the forces and typical expansion of shale when excavating into it. The SAA and MPBX instruments allowed opportunities to install additional tieback bracing or revisions to excavation sequencing if unacceptable movement levels were observed.

Performance of Spile Shoring Wall

Of the five rows of anchors supporting the spiles, four of the levels performed as anticipated. The third row from grade underperformed, with half of the anchors locked in between 35 to 80% of design load. Because of the array of monitoring targets on the façade, we were able to avoid installing an extra row of remedial anchors. Instead, the loads on subsequent anchor rows were slightly increased, while relying on monitoring to confirm the revised system performed as expected. In the end, the spiles performed well with a maximum of 14 mm (0.56 in) of into-site movement.

Performance of Micropiles at the Façade

The micropiles performed well throughout construction, with up to 10 mm (0.4 mm) of movement observed by the time the site was excavated to the final grade, and 12 mm (0.5 in) total movement observed at the time of the final reading – after lateral bracing was removed during the build-out process.

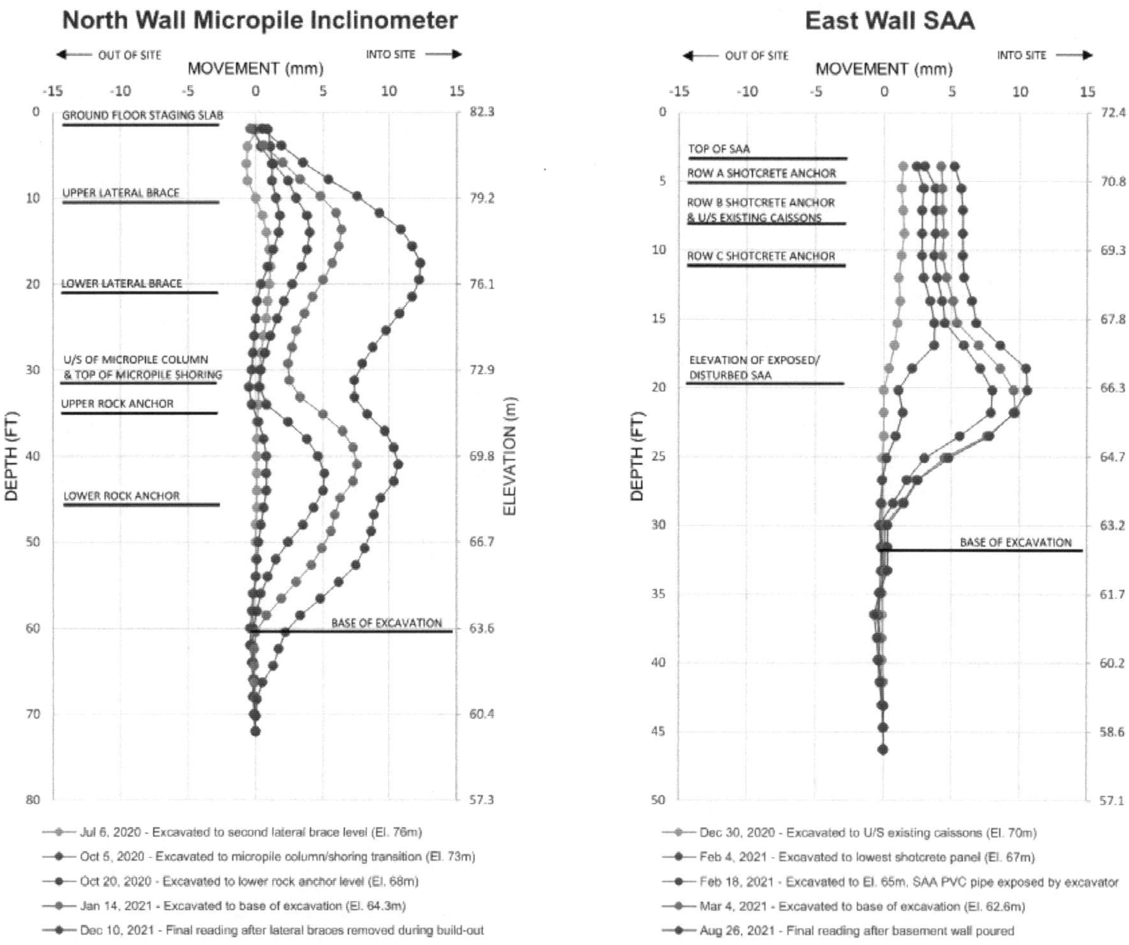

Figure 11 Shoring performance at north wall (left) and east wall (right)

Performance of East Rock Face and Adjacent Neighbouring Building

Both the SAAs and the MPBX instrumentation indicated lower than anticipated movement during construction and excavation. The two MPBX instruments indicated a maximum of 7 mm (0.3 in) of movement at the excavation face (i.e., the first line of structural columns of the neighbouring building), reducing to 4 mm (0.2 in) of movement 6 m (20 ft) back from the face (i.e., directly below the first interior line of neighbouring footings). The SAAs corroborated this data, showing a maximum of 7 mm (0.3 in) of movement, with spikes of movement up to 14 mm (0.6 in) at one elevation where the wall was over-excavated and the SAA PVC casings were directly exposed in error.

The precision survey monitoring inside the neighbouring building indicated negligible movement throughout construction, with less than 5 mm (0.2 in) of cumulative movement throughout construction.

CONCLUSIONS AND LESSONS LEARNED

The shoring system proved to be a success in controlling the risks that came with excavating through expansive shale bedrock beneath a sensitive historical façade as well as directly adjacent to a sensitive neighbouring 12-storey apartment building. Overall movements of the shoring around the entire site were on the order of 0.06% of the excavation depth at each wall).

In addition to these successes, we learned several lessons throughout the course of the project.

1. Repeated and clear communication is essential on project sites when executing a complex solution in the field. Many of the mistakes on site happened during transition times. Regular site meetings helped alleviate this and reduced miscommunications.
2. Reliable monitoring was critical to the success of the project. Because of the aggressive nature of the design, we actively relied on monitoring to confirm our conclusions and allowed us to adjust the design as changed conditions were encountered on site.
3. Accurate utility information is vital and should be considered early in the process. If test pits of the gas main were able to be carried out earlier in the process, the project team would have had the flexibility to explore more efficient design solutions.

ACKNOWLEDGEMENTS

The authors would like to express their appreciation to:

1. First Gulf, the developer and construction manager, who were critical to the success of the project.
2. RJC Structural Engineers and Facet Group, for the collaborative approach on the complex design.
3. Keller Canada, who carried out the installation of the intricate shoring system.
4. Monir Precision Monitoring, who carried out the monitoring work.

REFERENCES

1. "Deep Excavation Support – Shangri-La – Toronto", Isherwood, Brown, Earle & Kreycir, 2012.
2. "Horizontal Displacement Induced by Rock Excavation: Scotia Plaza, Toronto, Ontario", Trow & Lo, 1988.
3. "The Laurentide and Innuitian ice sheets during the Last Glacial Maximum", Dyke, Clark, England, Miller, Shaw, & Veillette, 2001.
4. "Geotechnical Investigation – Proposed Multi-Storey Building, 25 Ontario Street, Toronto, Ontario", GHD, 2016.
5. "Alterations to a Heritage Property, Intention to Designate under Part IV, Section 29 of the Ontario Heritage Act –25 Ontario Street", City of Toronto, 2012.

SAN DIEGO COURTHOUSE COMMONS SHAFT SHORING AND TUNNEL CANOPY TUBES

Ali Nasirian, Ph.D., Project Engineer, Condon-Johnson & Associates, Inc., 3434 Grove St, Lemon Grove, CA; 310-971-3177; ANasirian@condon-johnson.com

Gabriel Carvajal, Project Manager, Condon-Johnson & Associates, Inc., 3434 Grove St, Lemon Grove, CA; 858-530-9165; GCarvajal@condon-johnson.com

ABSTRACT

To facilitate the safer and more secure transportation of inmates from San Diego Central Jail to the new San Diego Courthouse, the County of San Diego committed to building a new transfer facility as part of the downtown developments. The transfer facility includes a three-story underground structure adjacent and connected to the Central Jail and a 60-ft deep, 23-ft diameter, 320-ft long tunnel. The tunnel starts at County Jail, passes underneath Western Regional Detention Facility (Old Jail) and, ultimately, connects to the State Courthouse. Condon-Johnson & Associates, Inc. (CJA) was selected to install the temporary shoring shaft for the underground structure as well as three arrays of canopy pipes to support the tunnel's Sequential Excavation Method (SEM) construction.

CJA installed the rectangular 56-ft deep access shaft shoring with three levels of bracing, and a combination of secant piles, soldier piles, and timber lagging. The shoring also included a sliding bridge at top to provide access for trucks to enter and exit from county jail during the duration of construction.

Following shoring and shaft excavation, CJA installed and grouted 7-5/8" dia. canopy pipes in 3 reaches, parallel to the longitudinal design line of tunnel with a slight lookout angle. CJA installed 24 each canopy pipes, 99-ft from face of shaft at reach 1, 21 each, 80-ft length canopy pipes at reach 2, and 21 each, 75-ft length pipes at reach 3. CJA confirmed canopy pipe alignment via Inertial Sensing Gyro.

To mitigate the possibility of settlements caused by the tunneling operation underneath Old Jail strip footing and grade beams, CJA also performed void fill grouting. Sleeve port grout pipes (SPGP's) were installed inside the Old Jail basement and grouted with the use of packers as the tunneling operation passed underneath the Old Jail.

In detailing various aspects of the construction, this paper aims to magnify the innovative solutions for shoring, canopy pipe installation and void fill grouting challenges as well as the state of the art quality control measures undertaken by CJA.

KEYWORDS: Shaft Shoring, Secant Piles, Soldier Beam, Canopy Pipe, Void Filling Grouting Technique

INTRODUCTION

Completed in 2017, the new San Diego Courthouse building replaced the previous County Courthouse, the Family Courthouse, and the Madge Bradley Courthouse in downtown San Diego. The new building provides a safe and adequate area for modern court operations for criminal, probate, family court, and small claims services. To eliminate the transportation of detainees via the public streets, the City of San Diego City Council confirmed construction of the underground facility, which included a transfer shaft and tunnel between the County Jail and Courthouse.

Condon-Johnson and Associates, Inc. (CJA) was subcontracted by the first-tier contractor, Atkinson Construction, to install temporary shoring and three reaches of canopy pipes to support tunnel excavation.

This paper focuses on the construction aspects of temporary shoring, canopy pipes and void fill grouting.

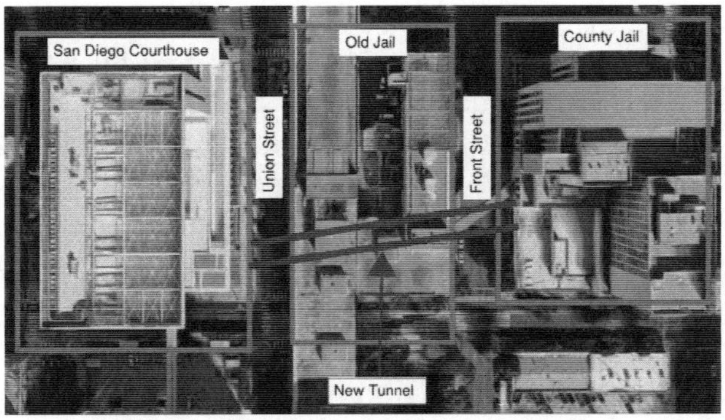

Fig. 1: Project plan view (by CJA)

GEOTECHNICAL CONDITIONS

Borings and cone penetration tests (CPTs) showed soil conditions consisting of artificial fill over old paralic deposits at shaft shoring. The tunnel was located entirely within the old paralic layer which was composed of marine and non-marine terrace units of poorly to well-consolidated medium-grained sandstone and siltstone (Kleinfelder, 2019).

Two groundwater monitoring wells showed groundwater elevation at +1 to +2 feet MSL. Multiple dewatering wells were installed to lower the groundwater to facilitate construction (Kleinfelder, 2019).

A Fault Hazard Investigation revealed an approximate 4.5 to 7 ft wide zone of active traces of the San Diego fault cross the tunnel at approximate station + 45 ft (see Fig. 2 with soil conditions and active fault zone location) (Kleinfelder, 2019).

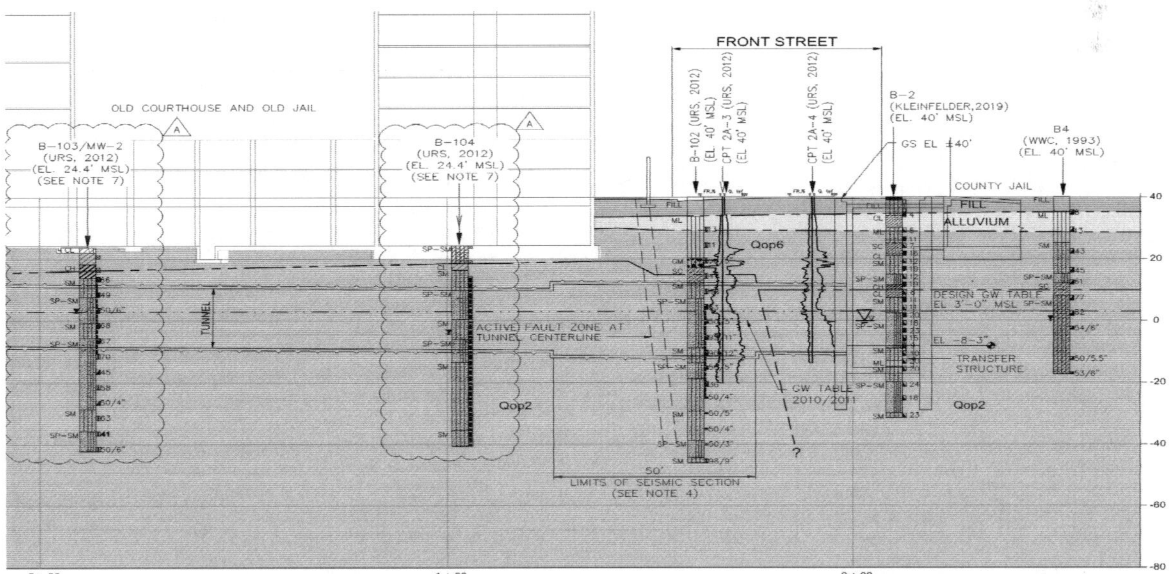

Fig. 2: Soil condition and boring logs (MJA, 2020)

TEMPORARY SHORING AND EXCAVATION

The rectangular shaft was designed by McMillen Jacobs & Associates (MJA) with 41 each, 36-inch diameter, 69-ft long piles and 3 levels of internal bracing. To reduce the possibility of settlement of the County Jail, the east side of the shoring included 23 each secant piles and 18 each soldier beam and lagging were utilized at the north, west and east sides of temporary shaft (see Fig. 3 for plan and sectional view of the temporary shaft).

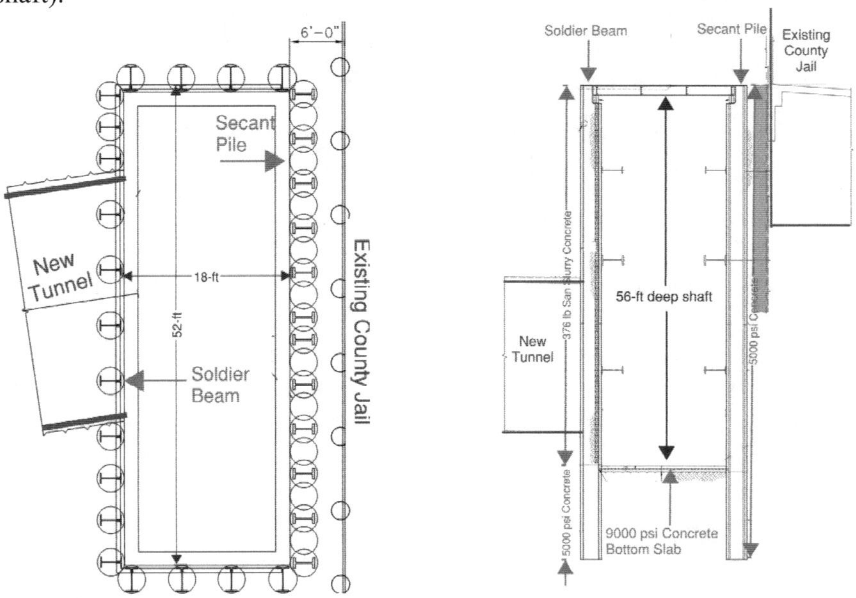

Fig. 3: Left: plan view of shaft, Right: sectional view of shaft

Prior to drilling secant piles, a reinforced guide wall to assist in maintaining secant pile location was constructed along the east perimeter of the shaft. For secant pile installation, first a series of primary piles are drilled and concreted, then secondary piles are constructed by cutting into already concreted primary piles. It is important that secondary piles are always constructed between two primary piles; otherwise, the secondary piles will have a tendency to deviate from their proposed location.

All piles were constructed using 6 to 20 feet sections of 36-inch outer diameter (OD) double wall casing connected to the drill rig's drive adapter and advanced with an oscillator.

The first double-walled casing section was set into the formed opening in the guide wall and leveled. After the casing had been rotated a several feet into the ground, the verticality of the casing (in both axis) was checked again using a level to ensure the pile would be plumb within 1%.

Excavation with an auger or drilling bucket was performed to within 2 feet of casing tip elevation during casing advancement. Once the casing reached the designed tip elevation, the soils inside the casing were excavated to casing tip elevation. At this point, the pile base was cleaned using a cleanout bucket.

The verticality of the mast of the drilling rig and depth of drilled hole were constantly monitored and displayed to the operator, allowing for adjustments as required.

Figure 4 shows pile installation sequence.

Fig. 4: Left: set casing at top of pile. Center: drill and remove soil inside casing. Right: clean the bottom of pile using a cleanout bucket (photo by CJA)

Prior to concreting, the depth of the hole was rechecked with a weighted measuring tape. A tremie pipe was then installed to pile tip. The tremie pipe was internally free of hardened concrete to allow for a smooth concreting procedure. Just prior to the concreting procedure, a plug was inserted into the tremie funnel to separate the fresh concrete from the water.

As the concrete was pumped and rose within the hole, the tremie pipe was withdrawn progressively. The lower end of the tremie pipe was kept approximately 5 feet into fresh concrete at all times. The depth of concrete in the hole was measured using a measuring tape with a weighted end. The tape was lowered into the hole until it met resistance caused by the concrete and a measurement was taken. Water that was displaced during concreting was discharged to the storage tank at the site.

Once the concrete level in the pile had reached approximately 5 feet into the temporary casing, the casing was withdrawn with the oscillator or the pile boring rig in sections. This kept the concrete level at least 5 feet above the casing shoe at all times. Figure 5 shows pile concreting method.

Fig. 5: pumping concrete using tremie method (photo by CJA)

Once the temporary casing had been removed from the hole and the concrete was flush with the top of guide wall, the steel beams were installed. The reinforcement was set using a 100-ton lattice boom crane fitted with a vibratory hammer. The vibratory hammer was used only occasionally when the steel pile did not reach the design tip elevation under its own weight. As the steel pile was lowered into the hole, the guide wall (on secant piles holes) was used as a template to set the beam in the correct location. For soldier piles, batter boards with string lines to face of beams were utilized to ensure accurate placement. The verticality of the pile was checked with a hand level. As necessary once the steel pile reached its tip, the steel pile was suspended from the concrete guide wall to ensure that it would not settle. Figure 6 shows wet beam setting method.

Fig. 6: wet set beams (photo by CJA)

A sliding bridge, designed for maximum load of 80,000 pounds, was installed at top of shaft to provide access for trucks to the County Jail during construction. Soldier and secant piles were designed to account for this additional surcharge load.

Fig. 7: Sliding bridge at top of shaft (photo by CJA)

To keep storm water out of the shaft, a concrete curb was installed around the shaft. During excavation, ground water table was lowered to at least 5 feet below bottom of excavation for each lift. A Komatsu PC490 was utilized to excavate the top approximately 25 feet. A skid steer was lowered inside shaft to support the excavation/lagging activity. CJA backfilled behind lagging utilizing a flowable 3-sack lean concrete. After 25 feet, a mini-excavator and skid steer loaded the soil into a bucket inside the shaft, and a crane lifted the bucket out for loading into trucks.

Fig. 8: Excavating top 25 feet of shaft (photo by CJA)

Fig. 9: Excavating lower portion of shaft (photo by CJA)

The design required three levels of internal bracing and a temporary truss to transfer the loads after removal of soldier beams located at tunnel opening.

CJA installed the third level of bracing, completed excavation, and installed a 10-in thick reinforced concrete slab at the bottom of the shaft. After the structural slab was installed and set, the shaft was backfilled, and the third level of bracing was removed to provide access for the tunneling operation.

Fig. 10: Three levels of internal bracing (photo by CJA)

Fig. 11: Temporary truss (photo by CJA)

CANOPY PIPES

In order to support tunnel sequential excavation method (SEM), canopy pipes were installed. Cored holes were provided at the face of the tunnel to ensure each canopy pipe was installed at its designed location. The 7-5/8" dia. canopy pipes were installed in three reaches. Each reach overlapped with the subsequent reach by 20 feet. Pipes were drilled parallel to the longitudinal design line of the tunnel with a slight lookout angle. 24 each, canopy pipes installed 99-ft from face of the shaft starting at station 0+1.5. After tunneling reached station 0+74, 21 each, 80-ft length canopy pipes were installed from within the tunnel, outside of the tunnel face. At station 1+34, 21 each, 75-ft length pipes were similarly installed.

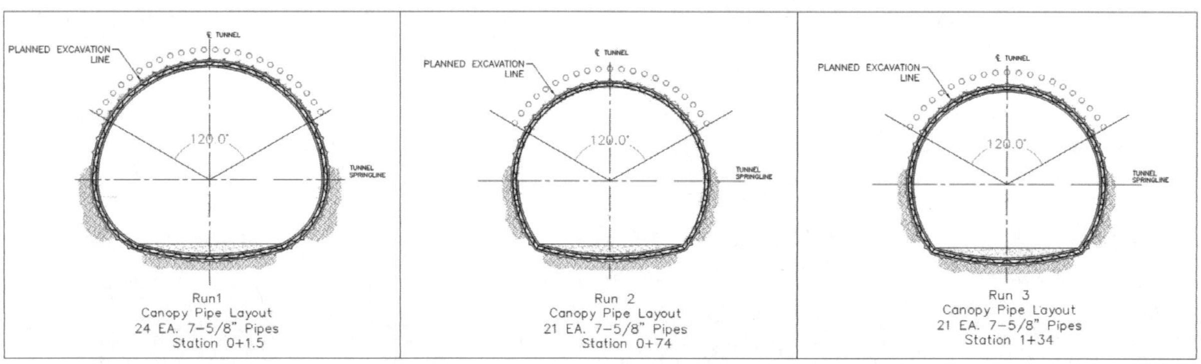

Fig. 12: Canopy pipe layout

Prior to installation, a test program was conducted to determine the bending strength of the fully grouted canopy pipes planned for use. 4-each, 6-ft long grout-filled pipes, 5-1/2" diameter x 0.562" wall casing and 7-5/8" diameter x 0.500" wall casing were grouted and shipped to be tested. The bend testing was performed on a 500 kip Universal Loading Machine. All 4 pipes were loaded to failure. At failure, the threads broke or striped with a vertical displacement between 2" to 3". Bend testing confirmed the adequacy of 7-5/8 inch pipes.

Fig. 13: Canopy test program and typical failure at joint (CEL Consulting, 2020)

Drilling was performed with a Hutte 202 for the first run and with a Commachio MC-8 for the second and third runs. Rotary duplex drilling technique were used at all runs, with the canopy pipe acting as the outer casing.

Fig. 14: Canopy pipe installation; Left: reach 1; Center: reach 2; Right: reach 3 (photo by CJA)

Pipe alignment was surveyed using Inertial Sensing Gyro. The Gyro was able to produce rapid results with an accuracy of 0.2° inclination, 0.5° azimuth, and less than 5% position.

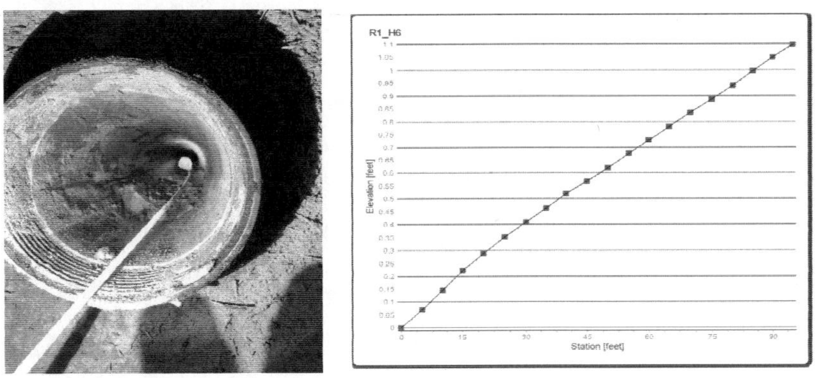

Fig. 15: Canopy pipe survey using Gyro (photo by CJA)

Grouting of the inside and outside of the pipes, by way of ports in the pipe wall , occurred after installation. Grout, consisting of Portland cement Type I/II with water/cement ratio of no greater than 0.5, achieved strength of 700 psi at 1 day and 4500 psi at 28 days.

The grout operation consisted of an Obermann batch plant, flow measuring devices, pressure gauges, hoses, packers, and miscellaneous valves and fittings. The pump was manually controlled with visual monitoring of a digital flow meter. Two different methods of grouting were used. At reach 1, packers spaced 15 ft apart were used to grout the pipe to a maximum of 50 psi. Use of double packer allowed isolation of port zones to be independently grouted. Grouting was considered completed when there was no grout take over a 2 minute period at the maximum pressure or when volume reached 1.0 cubic foot per linear foot of pipe. Reaches 2 and 3 were grouted full length with a pressure cap with a vent pipe. Grout logs were maintained for each pipe, with depth, time, volume, and pressure.

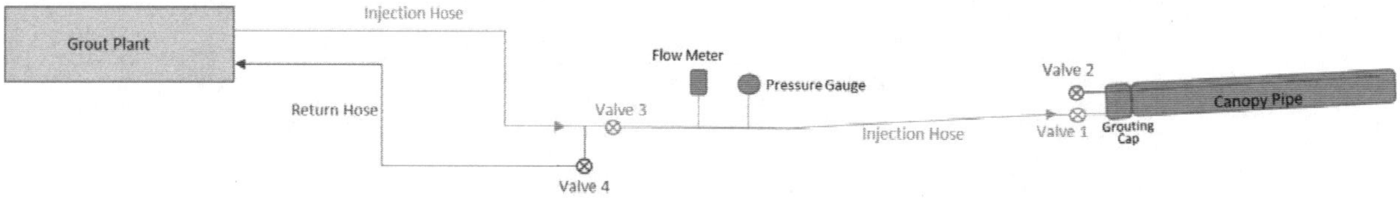

Fig. 16: Schematic drawing for canopy pipe grouting method 2

Fig. 17: canopy pipe grouting (photo by CJA)

OLD JAIL BASEMENT VOID FILLING GROUTING

To both fill possible voids created during SEM tunneling underneath old jail footings and to avoid settlement, void fill grouting was conducted from within the basement of the Old Jail. Grout locations provided by designer are shown below (Figure 18).

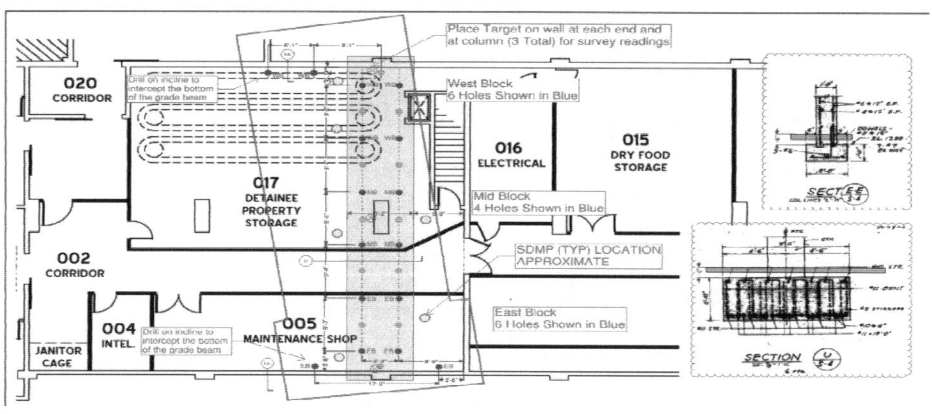

Fig. 18: void filling grout hole location inside basement of Old Jail (KPFF, 2020)

CJA cored 6" holes through the slab, and then drilled with a 4" auger to depth at the angled locations and a 2" core barrel at the vertical locations within the footing. 1.5" diameter sleeve port grout pipes (SPGP's) were installed and grouted in place.

Fig. 19: Left: Core concrete footing inside jail basement, Right: grouting underneath footing (photo by CJA)

Prior to the grout injection, a momentary high pressure (low volume) injection of grout was used to provide a passageway for the grout to travel in the angled holes that have SPGP's. Each hole had grout injected within the treatment zone, immediately below the footing.

CJA recorded pressure, flow, and volume for each hole grouted. SPGP holes were cleaned after each shift to ensure access to grout zones, which required re-grouting.

SUMMARY AND CONCLUSIONS

Secant pile walls have been used widely as a cost effective method to mitigate movements, especially when shafts are located close to existing heavily loaded buildings. CJA installed the temporary shaft shoring, and the survey readings confirmed movements were within design tolerance.

The canopy pipes assisted Atkinson Construction with their tunneling operations. Conducted to backfill potential voids during the tunneling operations; CJA's void filling operations under the old jail footings were a success. This project proves that a combination of canopy pipe and void filling techniques could effectively work together and minimize the possibility of settlements and collapses during SEM tunneling operations.

Teamwork, technical knowledge, work force, and quality of equipment were key factors contributing to the success of this project. Collaboration between all parties during design and construction played an important role to help the project meet schedule and produce a high quality results, meeting all design requirements.

REFERENCES

1. *Courthouse Commons Tunnel, Complete Building Permit, Volume 2, Foundation and Frame Drawings, prepared by McMillen Jacobs Associates dated February 18, 2020.*
2. *Courthouse Commons Tunnel, Grouting Plan for Old Jail Strip Footing and Grade Beam, prepared by McMillen Jacobs Associates dated August 8, 2020.*
3. *Courthouse Commons Tunnel, Old Jail Settlement Mitigation, Grout Injection Plan, prepared by KPFF Consulting Engineers dated August 7, 2020*
4. *Condon Johnson Courthouse Tunnel Project, Canopy Pipe Testing Program, prepared by CEL Consulting dated Jan 21, 2020.*
5. *Geotechnical Design Summary Report, Inmate Tunnel, Courthouse Commons Tunnel, prepared by Kleinfelder dated April 11, 2019.*

"HORSESHOE" SECANT COMPRESSION SHORING SYSTEM FOR TUNNEL REPAIR

Rob Jameson, Malcolm Drilling, San Francisco, CA, USA, 650.333.9057, rjameson@malcolmdrilling.com
Noah Miner PE, Malcolm Drilling, Milwaukee, WI, USA, 414.242.7860, nminer@malcolmdrilling.com
Giuseppe Gaspari PE MBA, AECOM, Mississauga, Ontario, Canada, 905.206.8578, giuseppe.gaspari@aecom.com
Kaveh Talebi, PhD, Jay Dee Contractors, Livonia, MI, USA, 734.793.8918, KTalebi@jaydee.us

ABSTRACT

A storage and drainage tunnel advancing below I-75 in Detroit, Michigan experienced sudden subsidence. The Tunnel Boring Machine (TBM) was damaged to the extent that its shield, cutting head and extensive interior components needed to be extracted and replaced along with several completed tunnel segments. The tunnel was 16 feet (4.9m) in diameter with crown over 79 feet (24m) deep at the time.

The tunneling contractor immediately engaged their designers and specialty subcontractors to develop a scheme to access the machine and repair the tunnel. They required a shaft 99 feet (30m) deep, with unobstructed plan dimension of 50 feet (15m) to accommodate the large components that needed to be removed and replaced. This precluded the application of an internally braced excavation and instead earth support solutions focused on use of a compression ring. Since the completed tunnel section behind the TBM obstructed closure of the compression ring below its invert, this paper describes the innovative access shaft solution employed.

A conventional compression ring was constructed using 5 feet (1.5m) diameter drilled shafts down to the tunnel crown. In front of, and to the sides of the tunnel heading, the secant shafts were extended to over 150 feet (46m) depth, creating a horseshoe of support extending 50 feet (15m) below the planned excavation invert. Larger diameter "King Pile" shafts were drilled immediately to either side of the tunnel, overlapped into each end of the secant "horseshoe". The king pile shafts, together with a heavily reinforced ring beam above the crown elevation, acted as a door frame to transfer hoop loads from the incomplete secant arc into reaction supports above and below the tunnel. These oversize king pile shafts were extended and socketed into rock over 70 feet (21m) below tunnel invert.

Very tight installation tolerances were necessary to maintain secant shaft alignment to depths exceeding 150 feet (46m). The project operational, geotechnical, and quality data was managed and visualized using a custom GIS interface. This paper will present the design approach employed and review the construction and quality program selected to repair the tunnel.

Keywords: Cased Drilled Shafts, Secant Pile Wall, Compression Ring, Support of Excavation, FEM, Tunnel, TBM, Infrastructure

INTRODUTION

The overall I-75 Modernization project consists of the design and reconstruction of approximately 18 miles (29 km) of the freeway, north of Detroit, Michigan. This stretch of freeway has been divided into three segments for reconstruction. As part of the I-75 reconstruction, a proposed drainage/storage tunnel from north of 8 Mile Road to north of 12 Mile Road is planned to be constructed within southernmost Segment 3. The current drainage system is almost 50 years old and in need of improvements. The proposed 14.5 foot (4.4 m) diameter drainage tunnel will provide storage and conveyance to help avoid future flooding in the depressed portion of the I-75 freeway. The proposed tunnel is approximately 22,370 feet (6820 m) long and is generally located along the northbound I-75 service drive, 100 feet (30m) below ground.

Figure 1 - I-75 Modernization Project Map With Segment ID's

On July 30, 2021 the tunnel excavation experienced mine subsidence resulting in earth sinking and shifting, casing settling, cracking and disarrangement of the constructed tunnel lining. The Tunnel Boring Machine (TBM) was damaged to the extent that required its shield, cutting head and extensive interior components to be extracted and replaced. The tunnel was 16 feet (4.9m) in diameter with the crown over 79 feet (24m) below ground surface at the time.

The tunneling contractor immediately engaged their designers and specialty subcontractors in order to develop a scheme to access and repair the machine and tunnel liner. To accommodate the large components that needed to be replaced, a 50 foot (15m) unobstructed excavation was required. The invert of the tunnel was 96 feet (29m) below surface, so the excavation would need to be supported beyond 100 feet (30m). Damage to the tunnel lining meant that the excavation support would have to encapsulate the machine, without the ability to construct directly below it, so an innovative solution was required.

SUBSURFACE CONDITIONS

Based on the completed borings for the Contractor's supplemental geotechnical exploration, subsurface conditions were encountered as follows: Surficial pavements, bedding, and predominantly cohesionless fill were encountered to depths of approximately 3 to 5 feet (0.9 to 1.5m). The fill materials were underlain by native gray silty clays. Clay consistency was hard to stiff immediately below the surficial fills, then transitioned to become medium to stiff with increasing depth to 108 feet (33m) below the ground surface, as measured at the closest boring to the shaft location. Granular soils, consisting of saturated dense silts and sands were encountered beneath the silty clay, alternating with layers of hard gray silty clay to the explored depths of the borings, or to bedrock in those borings in which bedrock was encountered. Highly to completely weathered shale, in combination with sand and gravel, was encountered within the upper 3 to 10 feet (0.9 to 3m) of the bedrock unit.

It should be noted that, due to the mine subsidence experienced in the tunnel, construction dewatering was on-going and drew the groundwater table below tunnel invert during the performance of the supplemental geotechnical exploration and the repair shaft construction. Prior to the start of this construction dewatering effort, the groundwater piezometric level ranged between approximately Elevation 603 and 615 feet, or approximately 21 to 33 feet (6.4 to 10m) below the ground surface. See Figure 2 below for a geologic profile of the tunnel alignment.

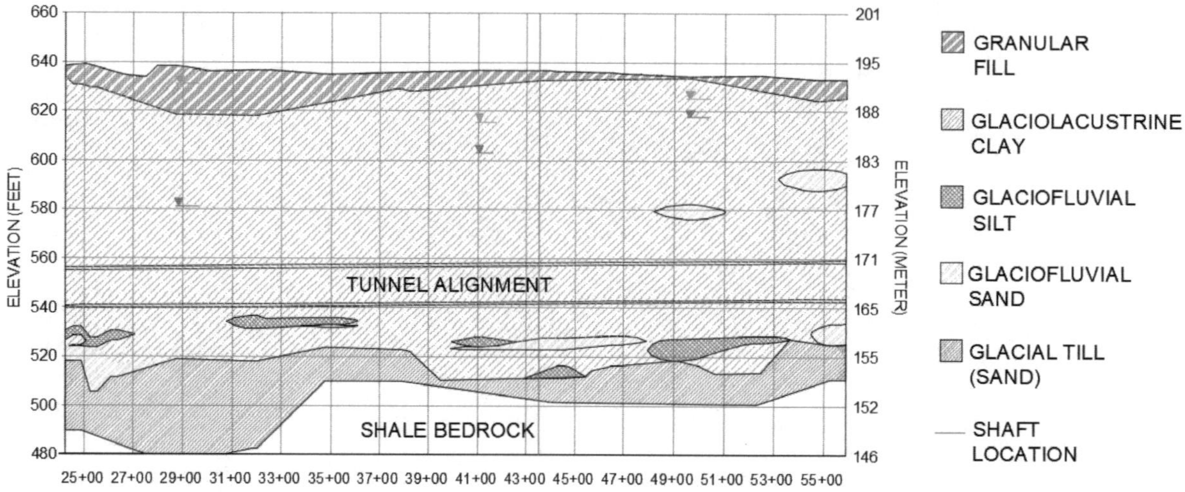

Figure 2 - Subsurface Profile

SITE CONDITIONS

The TBM experienced the damage at approximately ST 43+50 of the tunnel alignment. This placed the TBM directly beneath N Stephenson Highway, which is a local access road running parallel, and just to the east of I-75. The centerline of the tunnel alignment is roughly 75 feet (23m) from the edge of I-75, which has two lanes in each direction at this location. There was a 15 feet (4.6m) tall brick sound barrier wall just to the west of N Stephenson Highway, and a 40 foot (12m) wide grass shoulder between the wall and I-75. To further facilitate construction, a portion of the sound barrier wall was removed, and K-rail was placed at the edge of the I-75 shoulder.

DESIGN

Soon after the subsidence incident, the tunnel contractor started working with their repair team members to develop the most expedient and efficient solution to the challenging problem. The design team developed preliminary concepts. The tunnel and drilling contractor would provide operational constraints and constructability inputs, and the designer would revise schemes accordingly. The coordination between the three partners rapidly converged on the unique solution that created the required access to repair the tunnel and TBM.

Given that the TBM and tunnel final liner damage occurred in a densely populated area, the impacts on engineered structures and other urban edifices was a major risk for the design of a repair shaft. The ground subsidence above the TBM caused a depression which extended to the interstate and prompted a temporary closure of the north bound traffic. In addition, the nearby apartment building, was also within the impacted zone with one of the entrances to the apartment complex and its parking lot affected. To limit damage to surrounding existing buildings and infrastructure, a robust approach to both geo-structural analyses and reciprocal interaction of the ground and structural elements was undertaken.

Description of the Shaft Structural System

The selected support of excavation system comprised a circular shaft with internal diameter of 55 feet. (16.8m). A ring beam with dimensions of 60 inches x 60 inches (1.5m x 1.5m) was introduced at 67.2 feet below grade (20.5m), 4 feet (1.2m) above the TBM and the existing precast concrete tunnel liner system (PCTL). The circular shaft wall consisted of interlocking primary and secondary secant drilled shafts. The shafts extended close to bedrock or were terminated above the tunnel where layout conflicted with the existing liner segments. Primary secant piles were unreinforced and were drilled first. Secondary secant piles were reinforced with circular steel cage and were installed keyed into two previously installed primary piles. The secant piles had a minimum outside diameter of 59 inches (1.5m) and were installed at 41 inch (1.1m), center-to-center intervals, resulting in an overlap of 18 inches (0.46m). On either side of the tunnel liner, composite cast-in-place oversized secant piles (King piles) embedded in bedrock were required. King piles had a minimum outside diameter of 79 inches (2.0m) and were reinforced with steel circular reinforcing steel cages and a steel HP 14x117 beam (metric HP360x174). After excavation to shaft subgrade, a 20-inch (0.51m) thick cast-in-place structural concrete slab was required to provide support and bracing while mining through the soft-eye. Figure 3 below shows plan and isometric views of the shaft design.

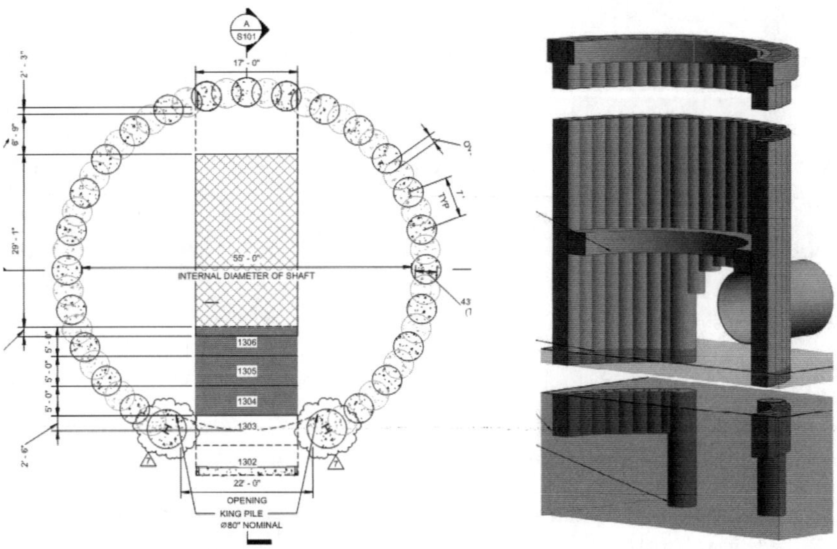

Figure 3 - Shaft Plan and Isometric Views

Two main analysis regimes were instituted to design the complex shaft. Through coordination across geotechnical and structural teams of the designer, it was agreed to develop an approach that could utilize the soil-structure numerical modeling results provided by PLAXIS while allowing full structural detailing of the various members via a rational load distribution and load combinations in a SAP2000 structural model. Both numerical models are based on the finite element method and the geometry of the structural elements was studied to be fully consistent. Finally, the design of structural members was performed using S-Concrete software and was based on outputs from the SAP2000 model.

Applied lateral earth pressure values were based on the output of the PLAXIS numerical model created by the designer's geotechnical and tunnel teams. Lateral soil sub-grade springs were introduced to the shaft with horizontal subgrade modulus values based on empirical equations and soil properties provided by the geotechnical team. The PLAXIS model does not consider the viability of shell structural elements to distribute horizontal moments across the unreinforced primary piles after the tensile concrete stresses have been exceeded, but it properly describes the soil/structure interaction and addresses the load transfer from the ground into the structural elements for a variety of construction phases. PLAXIS readily provides net force output which is used for design of the shaft members. However, it does not perform a design check to determine the resulting net stresses on the inside and outside faces of the wall section. Stress output is critical in determining the performance of the primary piles. The limitation of PLAXIS to model the complexity of redistribution of stresses among multiple structural elements, such as between the walls and the ring beam, is addressed with the SAP2000 structural model. See Figure 4 below for selected outputs from PLAXIS and a 3D model view from SAP2000 which will be discussed in subsequent paragraphs.

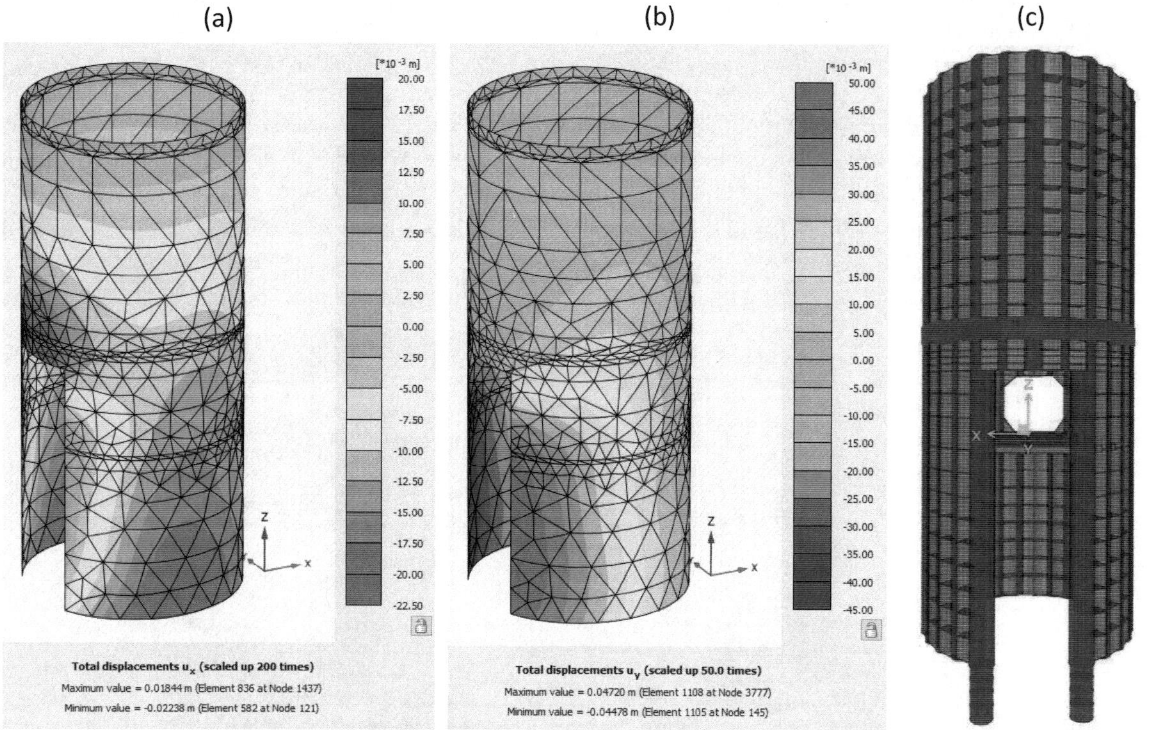

Figure 4 – PLAXIS 3D Model X-displacements(a) and Y-Displacements(b) Output (c) SAP2000 Model Frame

Analysis of the structural system was carried out by creating a non-linear elastic frame analysis using SAP2000 software with rational load distribution between structural framing members. Staged construction steps were created in the model. Proposed structural member sizes for the secant piles, ring beam, king piles, and structural slab, as well as their properties, were introduced into the SAP2000 model. King piles and secondary secant piles were introduced as frame elements. The ring beam consisted of continuous frame elements and was located 67.2 feet (20.5m) below ground surface. Shell elements simulating the unreinforced primary secant piles were modeled as 42.5 inch (1.08m) thick concrete walls with a compressive strength of 3600 psi (25MPa) and were set up to distribute the soil lateral pressure onto the secondary secant piles in one-way action. A reinforced structural slab at the bottom of the excavation was modeled as a shell element with a thickness of 20 inches (0.51m).

The reinforcing steel design for the king pile, ring beam, secondary secant piles, and structural slab were derived from the force envelope diagrams from the SAP2000 and PLAXIS models. In most cases, SAP2000 results were equal to or higher than the results from the PLAXIS model. The reinforcing for structural elements was designed using the reaction forces from SAP2000. For king piles, higher shear results from the PLAXIS model were used for design of structural reinforcing between El 480' and El. 467'. Figure 4.c above contains a 3D view from SAP2000 of the shaft structural system, including the secant piles, king piles, ring beam, and structural slab.

Structural Design Using S-Concrete

The structural resistance of the structural concrete members was calculated in accordance with ACI-318-14 with a load factor of 1.6 applied to the lateral earth pressure (H) and 1.2 for dead load (DL) according to ASCE-7-16. Ring beam, secant piles, and king piles were subjected to combined moment, shear, and axial loads. The structural slab-on-grade, which provided lateral support to the shaft at invert level, was subjected to only axial load. Verification of the adequacy of the concrete compressive strength against hoop stress at the location of interlock between primary and secondary secant piles was completed. Finally, S-Concrete was used to verify the stability of the secant piles above the opening, and check to ensure the concrete bearing in the rock sockets was not compromised.

CONSTRUCTION

Working Platform

The I-75 access shaft required 4.90 feet (1.5m) diameter secants and 6.56 feet (2.0m) diameter king piles to extend over 150 feet depth (46m). The Bauer BG50 drilling rig is uniquely capable of handling the size and weight of drill tooling necessary to construct these elements. However, the size and weight of this equipment required careful evaluation to ensure the drill working platform was suitable and stable to manage the imposed operational loading. Further, it was important to verify that any ground disturbance or voiding that could undermine the working platform had been repaired prior to mobilization. The drilling contractor performed a series of 21 each Cone Penetration Tests (CPTs) at the site to verify the consistency of subgrade materials and probe for potential soft spots which could have remained after the subsidence event and subsequent void grouting undertaken by the tunnel contractor. This testing did not identify any data of concern, but since ground investigation is only performed at discrete points, work still needed to proceed with caution.

The ground surface above the TBM subsided in a "crater" fashion as much as 5.5 feet (1.7m). The work platform was constructed by backfilling this area with 9 inch (14cm) thick, compacted lifts of crushed concrete and MDOT 21AA stone, to bring grade to one foot (30.5cm) below top of pile elevation. The working platform was then capped with hardwood crane mats for load spread, and a thin layer of compacted sand to provide a clean wear surface. This system was evaluated according to the Temporary Works Forum "Design of Granular Working Platforms for Construction Plant" and determined to have sufficient bearing support for the expected operating pressures imposed by the drilling and concreting operations. See Figure 5 below for depictions of the working pad being constructed.

Figure 5 - Working Platform Construction (photo courtesy of Drilling Contractor)

Verticality Control

The innovative access shaft design relied on maintaining drilled secant shaft verticality tolerances of less than 0.5% at depths exceeding 150 feet (46m). Secant shafts must be drilled straight and true from ground surface and cannot be steered down-hole during the advance. While secant piling methods have been regularly employed for support of excavations reaching 60 to 80 feet (18 to 24m) depth, it is unusual for this technique to be employed for elements with the depth required at I-75.

All the secant shafts were to be advanced with temporary sectional casing. The stiffness and rigidity of the casing maintains drilling alignment during advance, and the casing cuts into the primary shafts to create required overlap as the secondary shafts are installed. The drilling contractor was engaged to provide guidance during the interactive design process. Based on the drilling contractor's experience, 59 inch (1.5m) diameter secant piles were selected as the optimum size to minimize the number of elements and construction joints in the secant ring. The size and weight of this casing size helps to maintain alignment during advance, but due to the weight and diameter of the full casing string, a Bauer BG50 drilling rig, the largest production rig currently operating in North America, was required for the operation.

The initial step in verticality control was construction of a reinforced concrete guide wall. As the working pad was constructed, a leveling mat was poured and formwork was braced forming inner and outer edges of the guide wall ring. Slightly oversized sonotubes were used to form each pile location. The forms were cut, fitted together, surveyed to ensure accuracy, and then fixed in place. Reinforcing steel was placed between the sonotubes and the formwork and then the trench backfilled with structural concrete. The guide trench enables secant shafts to be located within 1 inch (2.5cm) of design location at ground surface. Figure 6 below shows guide wall construction.

Figure 6 - Guidewall Construction Details

The key to successful secant installation is to ensure the shaft is started straight, since the rigid steel casing effectively maintains its direction. The rig mast can be plumbed using built in inclinometers, however the most critical measure is actual casing alignment. The digital level is the tool of choice for field control of casing alignment. The operators ensured that casing was within 0.1 degrees of vertical in all directions at start of advance. The casing alignment was repeatedly checked and adjusted at grade using the drill rotary attachment during advance. After casing was advanced to between 60 and 80 feet (18 to 24m), a downhole verticality measurement using sonic caliper or SHAPE tool was performed to verify verticality was within tolerance, and once confirmed, casing advance was continued to designed depth. The as-built alignment was surveyed after completion of drilling using a sonic-caliper to confirm construction met tolerance requirements. Figure 7 below shows a sonic caliper report from a 6.6 foot (2.0m) king pile, confirming that verticality of less than half a percent is achievable.

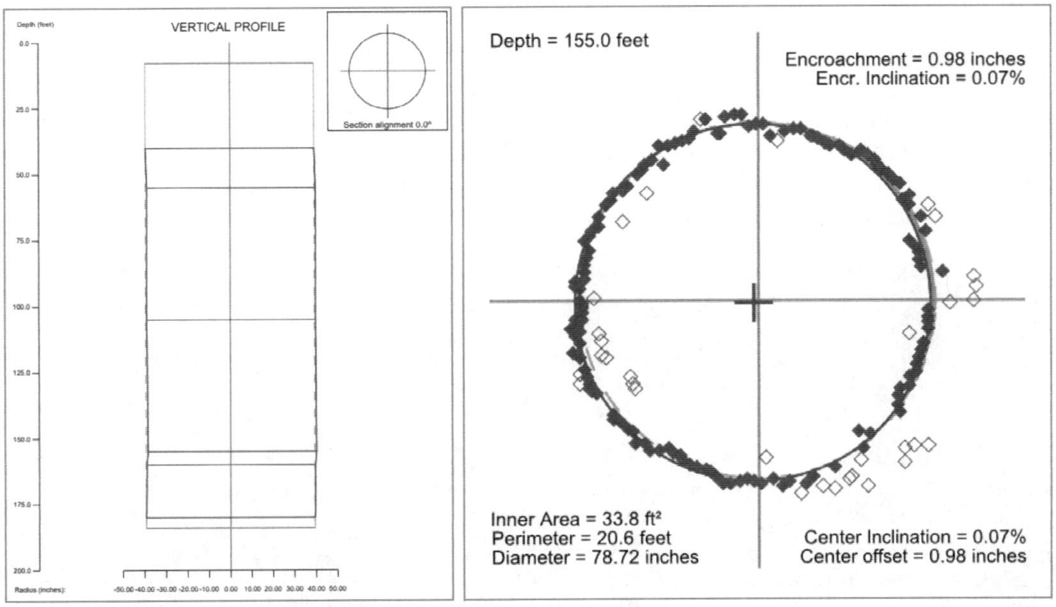

Figure 7 - King Pile Sonic Caliper Verticality Report

Concrete Strength Control

Primary secant shafts are constructed in native ground which generally has uniform radial characteristics within the drilled diameter. In contrast, the secondary piles cut into and overlap the previously placed primary elements. It is more challenging to maintain verticality in the secondary elements since there is a non-uniform drilling condition with concrete on two opposite sides of the shaft and softer soil on the remaining perimeter. The secant operations for this project were planned to ensure that secondary piles were cut into two adjacent elements with similar strengths to avoid the casing being pushed preferentially in one direction by harder concrete. Specific to this project, the king piles required construction cutting into concrete on only one side of the shaft. Exceptional care was required to ensure these shafts could be constructed without excessive deviation due to the unbalanced drilling condition.

To assist with secant construction, the drilling contractor aimed to limit early strength gain in the concrete and sequence work such that secondary shafts could be constructed within the period of 3-5 days after the adjacent primary elements were placed. A series of trial batches were performed with variable additive doses in order to limit strength gain during this target construction window.

Excavation, Reinforcing Cages, and Concrete Placement

The secant shafts were fully cased to mitigate concerns associated with verticality, potential voids, and damage to surrounding structures. Casing was advanced with the rotary head of the BG-50, and overburden was removed with various augers and buckets. For the two king piles, bedrock sockets were completed using progressive rock augers and roller bits. See Figure 8 below for a picture of a drilling method used on site.

The size, complexity, and variability of the cage designs required careful attention. Large shear forces required heavy lateral reinforcement which went through multiple iterations of design to meet the loading demand, while ensuring that there were large enough windows to allow free flow of the tremie concrete. Splicing cages over the hole was chosen as the method to place cages with lengths of over 100 feet (30.5m). In order to allow the TBM to exit the shaft after completion of repairs, composite cages were constructed with fiberglass reinforcement for partial lengths across the soft eye. See Figure 8 below for an example of a composite cage being lifted on site.

The most critical operation in the construction of a drilled shaft is the placement of concrete. During concreting on a cased drilled shaft, many moving parts need to be coordinated with precision to ensure a smooth operation. Pulling casing, monitoring fluid and concrete levels, hanging cages, breaking lines and more need to be synchronized. On this project, the drilling contractor elected to utilize a boom truck to deliver fluid concrete to the tremie and to extract casing with Leffer Oscillators. Figure 8 below shows a picture of the concreting operation on site.

Figure 8 - Excavation, Composite Cage Placement, and Concreting (Photo courtesy of Drilling Contractor)

GIS Database

There were numerous data streams that were relevant to the project teams. With this in mind, the drilling contractor teamed with an instrumentation specialist, Geosyntec, to create an interactive project map. All relevant pile production information, pile quality control information, preproduction CPT data, and more was built into a real-time project map. Using the information from the verticality surveys, a minimum pile overlap tool was developed to instantly devise if two proximal piles had met the verticality and minimum overlap requirements. Screenshots of the map and tool can be seen in Figure 9 below.

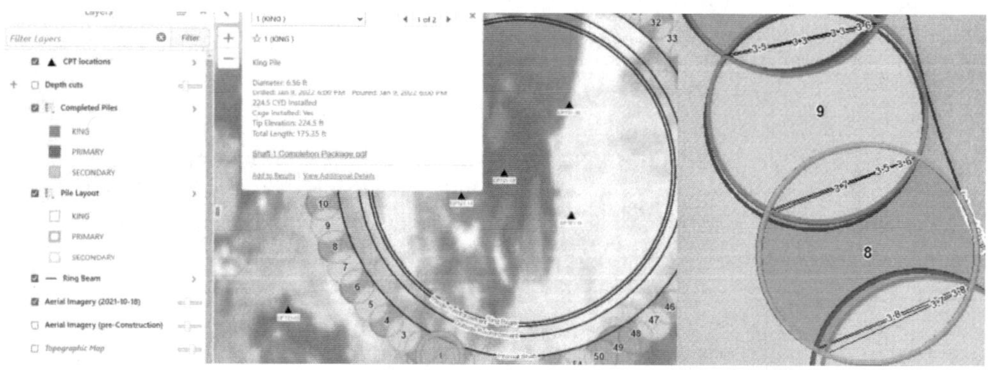

Figure 9 - GIS Screenshots

INSTRUMENTATION AND PERFORMANCE

There were very real concerns that movements of the shaft structure could lead to unacceptable external impacts. Various instrumentation methodologies were employed to monitor the shaft and surrounding structures. The most compelling data on the performance of the shaft design and construction is the proximal inclinometer data which showed a maximum lateral deflection of just under one quarter inch (6.3mm), at the top of the shaft after being fully excavated. This appears to closely match the structural analysis using SAP2000 model as seen in Figure 10 below.

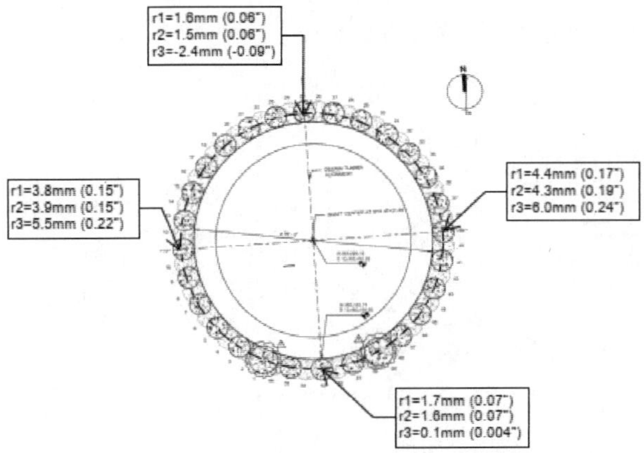

Radial movement at top of shaft from SAP2000 analysis

Note: r1, r2 and r3 denotes radial movement in Stage 1, Stage 2 and Stage 3 respectively. At stage 3 (final stage) the new TBM has bored through the soft-eye.

Figure 10 – Radial movement at top of shaft

CONCLUSION

At the time of writing, the once stranded TBM has been fully exhumed (Figure 11 below) and replaced with a new unit set to finish the tunnel. The timely execution of planning, design, construction, and commission speaks to the effective collaboration of the tunnel contractor, designer, and drilling subcontractor. The efficient accomplishment of this novel challenge could only have been achieved through inclusion of the full project team throughout the planning, design, and execution. From the moment it was stranded to the day the final element was installed in the shaft system, the TBM had an escape route in less than six months.

Figure 11 - Retrieval of TBM After Shaft Completion (Photo Courtesy of Tunneling Contractor)

REFERENCE DESIGN CODES AND SOFTWARES

The design team have optimized an efficient design for the shaft using relevant ACI codes and sophisticated 3D FEA analysis software as noted below.

-ACI 318-14
-ACI 360-16
-ASCE 7-16
-PLAXIS 3D Geotechnical Analysis
-SAP 2000 3D Structural Analysis
-S-Concrete

NICHOLSON IN THE NEIGHBORHOOD: VIRGINIA'S NORTH HILL PROJECT

Dominic Ammon, Nicholson Construction Company, Pittsburgh, PA (412) 260-1110
dominic.ammon@nicholsonconstruction.com
Mark Rothbauer, Nicholson Construction Company, Pittsburgh, PA (412) 334-1479
mark.rothbauer@nicholsonconstruction.com
Paul Landers, P.E., GEI Consultants, Inc., Washington, DC (202) 828-9510
planders@geiconsultants.com

ABSTRACT

Across the Potomac River from the location of the DFI47 Annual Conference (October 2022), in Alexandria, VA, two large permanent anchored walls with a decorative shotcrete face were recently constructed for the North Hill project, a new multi-family and townhome development. The project was executed with a design-build approach for the Fairfax County Redevelopment and Housing Authority with review by the Fairfax County Geotechnical Review Board. The project was constructed in highly variable Potomac group sediments that include high plasticity clay/silt terrace deposits overlying coarser grained materials. The soils were over consolidated and exhibited potential for slickensides from relief of horizontal stresses from past erosion. The wall system was designed to include deep soldier piles and permanent long strand anchors considering past slope instability in the area. As the construction progressed, problematic soil conditions and water issues required modifications to drilling methods. The two permanent walls spanned 800 and 1,000 feet long with a maximum height reaching 40 feet.

Keywords: anchors, soldier piles, difficult soils, construction, retaining wall

A DEVELOPMENT DECADES IN THE MAKING

The new North Hill residential community will open in Fairfax County, VA in 2022. The community is situated on a 33-acre parcel located along US Route 1 approximately 10 miles from the heart of Washington, DC. When completed, the development will include 175 townhomes on the east side of the site and five multi-family housing units with 279 apartments on the west side (Fig. 1).

In 2017, the Fairfax County Redevelopment and Housing Authority (the County) issued a Request for Proposal to develop approximately 20-acres of the 33-acre parcel for the housing units. The remaining acreage will become a park maintained by the County. The accepted proposal was provided by K. Hovnanian and CHPPEN (a joint venture of Community Housing Partners (CHP) and PennRose, a private developer). The County engaged the Fairfax County Geotechnical Review Board (the Review Board) for engineering review of the design and construction.

A mobile home park previously located on the site was abandoned nearly 30 years ago after slope instability caused movement in the mobile home pads. The site terrain slopes from the northeast to southwest with an elevation change across the site of approximately 106 feet. Development grading required two retaining walls: one adjacent to the townhomes, and the other along the multi-family development parallel with Route 1.

Nicholson teamed with GEI Consultants, Inc. (GEI) to develop a design-build solution including two permanent anchored soldier pile walls. This project was unique and challenging because of the size of the required walls, past slope instability, variable soil conditions, and demanding project constraints.

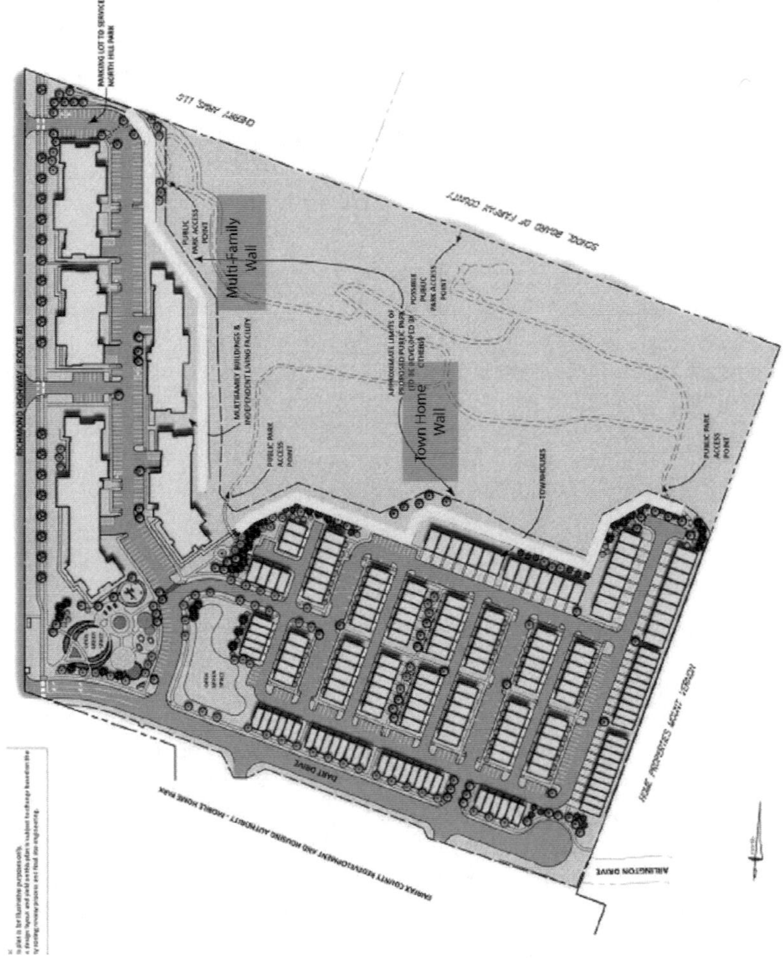

Fig. 1. Overview of North Hill Community Development

SOIL CONDITIONS AND PROJECT CONSTRAINTS

The local geology comprises coastal plain sediments consisting of Terrace deposits and the Potomac formation. The site soils are primarily fined grained highly plastic clays and silts interbedded with coarser grained sands. These sediments are overly consolidated and exhibit potential for slickensides from relief of horizontal stresses from past erosion. The Review Board was concerned about potential slip surfaces based on previous movement at the site. The soil conditions vary widely across the alignment of both walls, creating challenges for establishing soil parameters and subsurface profiles for design

The County defined the following project requirements for the design and construction of the wall:

- Disturbance adjacent to the wooded park area was limited to 10 feet from the wall face.
- The top of wall needed to match the existing contours of the site.
- A block façade was required to meet the aesthetic requirements for the new neighborhood.
- A minimum factor of safety of 1.25 was defined for global stability for both walls.
- Top of wall movement was limited to 2 inches.
- Anchors stressing was required to be in accordance with Post Tensioning Institute (PTI, 2004). requirements to a maximum test load of 1.33 times design load.

DESIGN

The GEI design team analyzed the wall system in an iterative process using three sets of shear strength parameters: short-term values for construction stages, intermediate values for serviceability, and long-term values for global stability. SLOPE/W was used to assess the global stability (Fig. 2) using residual shear strength values. In this analysis the wall shear strength values were adjusted until the minimum factor safety was achieved. LPILE was then used to model the pile stresses below the excavation considering the zone of soil movement above the anticipated slip surface determined through the slope stability analysis. Limit equilibrium analyses were performed utilizing fully softened soil parameters to design the internal stability of the wall which determined the stresses in the piles and in the tiebacks. Finite element analyses were run using Plaxis 2D to demonstrate the serviceability of the retaining wall and as a confirmation of the internal forces and movements calculated from the limit equilibrium models. Piles and tiebacks were designed for the maximum stresses determined from all the analyses performed. Residual shear strength parameters ultimately controlled the design requiring large diameter drilled piles with multiple rows of high-capacity strand anchors bonded beyond deep failure surfaces.

The results of the analysis confirmed the need for deep soldier piles and long anchors. The multi-family wall is 800-feet long, with approximately 8,000 sf of exposed face with a maximum height of 15.5 feet. The multi-family wall is cantilevered along most of the wall, with one row of 31 anchors in the middle section. The townhome wall is approximately 1,000 feet long, with approximately 13,000 sf of exposed face and a maximum height of 40 feet. The townhome wall is cantilevered in sections, but most of the wall is anchored with a total of 127 anchors. At its peak height, three rows of anchors are installed in the townhome wall.

As the wall height and soil conditions varied along the wall, the soldier pile diameters were designed to be either 30 or 36 in depending on the required flexural strength of the beam or to provide additional passive resistance. The beams were generally spaced on 5- to 7-foot centers. The soldier pile lengths ranged from 10 to 80 feet, and the bottom of the piles extended to a maximum depth of 45 feet, with a ratio of the exposed height of roughly 3:1. Both 5- and 7-strand anchors were used, with the bond zones ranging from 20 to 40 feet. The longest anchor was 120 feet long with a 40-foot bond zone. Design loads ranged from 88 to 210 k. After discussion, the County approved an easement for the permanent anchors drilled below the park property once it was ensured that the anchors would be installed below expected depths for future utility or infrastructure development in the park.

Temporary wood lagging was designed to be placed between the beam webs during the excavation. A permanent architectural shotcrete facing was designed and approved to meet the aesthetic requirements.

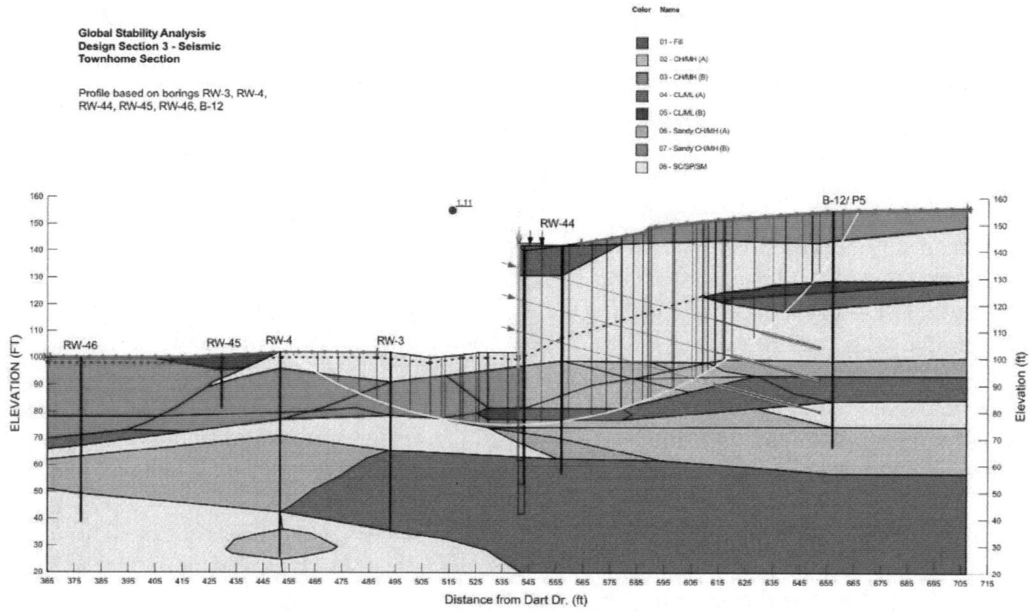

Fig. 2. Cross section of global stability analysis.

PANDEMIC CHALLENGES

Construction at the North Hill project occurred during the Covid-19 pandemic, adding another layer of logistical challenges to an already complicated job. These challenges included a 3-week operations shutdown, extended lead times for the fabrication of anchor materials, and limited availability of timber lagging boards.

SOLDIER PILE INSTALLATION

The work sequence started at the multi-family wall and then progressed to the townhome wall. The work on the multi-family wall started at the south end of the wall with the shallowest piles (10 feet long) and progressed northward with deeper piles. Work benches were constructed along the front face of the walls to minimize disturbance on the park side of project. The piles were constructed by drilling 30- or 36-inch-diameter holes using auger methods (Fig. 3), and then H-beams were set in the open holes (Fig. 4). The holes were then either tremie or gravity filled with concrete to the elevation of the bottom of the exposed wall face. Piles that would have anchors installed were backfilled from the concrete level to the top of the H-beam with 200 psi flowfill mix. The flowfill was used to provide additional support and minimize deflection during anchor stressing. The beams were placed with the tops of the beams a minimum of 2 feet higher than the planned top of the wall to allow for backfill behind the wall prior to excavation and setting of the lagging. Piles were cut to the planned top of wall elevation after anchor stressing.

At the height of site activity, both walls were under construction with up to 30 employees and three drill rigs performing soldier pile excavation and anchor installation (Fig. 5). The work started steadily with stable ground conditions at the shallow drilling depths at the south end of the multi-family wall, and the initial holes were drilled without any casing.

Fig. 3. Soldier pile drilling using auger methods at the townhome wall.

Fig. 4. Soldier pile drilling and setting H-beams on the townhome wall.

Fig. 5. Overview of project during construction

As work progressed along the multi-family wall and the entry point elevation got higher and the pile tip elevation got lower, variable soil and groundwater conditions were encountered that required modifications to the drilling techniques used. Multiple perched water layers were encountered on top of the finer, highly plastic clay deposits, causing hole collapse during deeper pile installations. Sandy old root layers were also encountered. In one location, an abandoned capped storm sewer was found at depth. The sewer structure was draining into sandy root layers that were interbedded with the clays, providing no path for the water to drain. These saturated layers caused the drill holes to collapse at varying depths, causing concern about the stability of the construction bench. Temporary top casing pipe (42-in-diameter) was installed to a depth of 10 feet to seal off these perched water layers and stabilize the holes, thereby keeping the bench area stable and allowing the remainder of the hole to be drilled without casing. As construction progressed along the multi-family wall, the depth to groundwater increased, and casing lengths were increased to 15 feet.

After modified drilling methods on the multi-family wall were established, work began on the townhome wall east of the multi-family wall. More water issues were encountered during drilling of the first piles of the townhome wall. A prominent running sand layer was found at approximately 20 to 30 feet below grade at the furthest east point of the project. The 15-ft temporary top casing pipe and open hole drilling were no longer effective. Segmental casing was used to keep the holes open while drilling through this layer to depths of 80 feet. Using segmental casing helped alleviate the difficulties of the water issues; however, it prevented the use of a template beam for setting piles. Without the template, the pile orientation and location was difficult to maintain during concrete placement and removal of the casing.

The following steps were used to resolve this problem:

1. Stop the hole excavation 1 foot above the planned pile tip elevation.
2. Confirm pile position by GPS, then drive the pile to the proper tip elevation using the drill rig. The 1-foot embedment locked the piles positioning and orientation.
3. Weld a piece of rebar to the top of the pile and to the top piece of the segmental casing.
4. Pour the theoretical amount of concrete to fill the embedment length.
5. Remove the temporary rebar.
6. Pull the segmental casing, completing the pile installation.

As the work progressed along the wall, it was realized that the as-designed bench would require major earthwork. The current grade elevation would need to be increased 15 feet with a 30-foot-wide platform and a 2:1 slope from that elevation. Nicholson and the GC agreed to reduce the bench elevation to be closer to the current grade elevation, resulting in a section of piles being drilled with approximately 15 feet of beam cantilevered to the top of the wall. This section was later backfilled behind the wall prior to anchor drilling.

The soldier piles were intended to be drilled to within 2 feet of the top of pile elevation. In this area, the timber lagging was installed and backfill was placed behind the face of the wall up to the top of pile elevation. While this was very helpful for the soldier pile drilling, this led to a future issue down the road when stressing the anchors.

ANCHOR INSTALLATION

A total of 158 anchors were installed within the two walls (Figs. 6 and 7). Anchor holes (7-5/8 in diameter) were drilled using a rotary duplex system with water as the flushing medium. The anchors were installed at a 15 degree angle measured downward from horizontal and tremie grouted while extracting the temporary casing. Free lengths ranged from 20 to 80 feet, and bond lengths ranged from 20 to 40 feet. The long free lengths were needed to place the bond zones beyond the failure planes identified behind the wall. Post grout tubes were installed in all the anchors.

Due to the variability in site soils, anchor bond zone material was expected to vary across the walls. Two pre-production anchors were installed to verify the bond capacity within the finer grained material which was determined to be the weakest expected bond zone material. The two pre-production sacrificial anchors were installed vertically in front of the location of the multi-family wall. These sacrificial anchors were performance tested to a maximum test load of 1.5 x design load, and then stressed to 2.67 x the design load until geotechnical failure. All production anchors on both walls were proof tested, and 8 anchors (5%) were subjected to performance tests. All tests were conducted with a maximum test load of 1.33 times the design load. Creep tests were conducted at a hold load of 1.5 x the design load and run in accordance with PTI recommendations.

During stressing of the top row of the townhome wall anchors in the region where the soldier piles were drilled with the 15 feet cantilever, the beams experienced significant displacement towards the fill soils. This excessive beam movement was attributed to the weaker fill soils. The anchors were eventually tested and locked off at the design load after accounting for the beam movement.

Fig. 6: Anchor drilling on the townhome wall.

Fig. 7. Anchor drilling and grouting on the townhome wall.

WALL PERFORMANCE

The movement of the top of the wall movement was estimated in design to be up to 2 in. As beams were placed, survey points were installed at the top of the wall and surveyed weekly by GEI to monitor the movement of the top of the wall throughout excavation. Based on the monitoring, the tops of the walls have moved on average less than 0.5 in. The wall monitoring continued for 5 months after the completion of the walls, and no signs of long-term creep were evident.

SHOTCRETE APPLICATION

A specialized shotcrete subcontractor applied an architectural shotcrete facing after the Review Board accepted the wall construction. The blown shotcrete wall facing was constructed over reinforcing steel and drainage fabric (Fig. 8). A photo of the completed townhome wall is shown in Fig. 9.

Fig. 8. Preparing wall facing for architectural shotcrete finish on multi-family wall

Fig. 9. Architectural shotcrete finish of completed townhome wall

FINAL REMARKS

When completed, the North Hill development will include 216 multifamily apartments, 63 senior/independent living apartments, 175 townhomes and a 12-acre public park. The top of wall monitoring has shown movement well within estimated values, indicating that appropriate design values were assigned for the geology present. The excavation for the wall was completed in September 2021, and the finishing architectural touches were in December 2021. The remaining construction on the North Hill project continued through the summer of 2022. For projects of this large size with expected soil and water variations, quickly adjusting drilling methods is critical to success.

Although there were many challenges and variables encountered, the wall is high quality in both its construction and its appearance. The finish surface blends into the background and adds value to the neighborhood. The permanent wall will serve the residents and support the park for many years.
North Hill will be open for new residents in the Fall of 2022.

ACKNOWLEDGEMENTS
Nicholson Construction Company and GEI Consultants gratefully acknowledges the project owners and developers, K. Hovnanian and CHPPEN.

REFERENCE

Post Tensioning Institute. (2014). Recommendations for Prestressed Soil and Rock Anchors, Phoenix, AZ., 109 p.

DESIGN OF GROUND IMPROVEMENT BY INCLINED DSM COLUMNS FOR RAILWAY EMBANKMENT (TÜRKİYE)

Candan Gokceoglu[1], Hacettepe University, Geological Engineering Department, Ankara, Türkiye
+905324730851, cgokce@hacettepe.edu.tr
Ahmet Sirin[2], General Directorate of Turkish Railway System, Ankara, Türkiye
Asli Can[3], Iksa Engineering Company, Ankara, Türkiye, +905300419149, a.can@iksamuhendislik.com.tr
Emre Turk[4], Akademi Zemin Company, Ankara, Türkiye
Suat Gullu[5], General Directorate of Turkish Railway System, Ankara, Türkiye

ABSTRACT

In this study, the ground improvement application applied under the live railway bridge's approach embankments in Sivas, Türkiye was investigated as a case study. The soil profile is mainly composed of orangish conglomerate, clay, silty sandy clay, and under claystone, and mudstone units in the study area. As a soil improvement method, the Deep Soil Mixing (DSM) technique based on the principle of simultaneous mixing the ground while injecting with the water-cement mix, was selected. The application of DSM columns is designed to be 45 degrees horizontal, in order to reach the deepest point of the soft clay units for better improvement against settlement. The core samples were taken from the DSM columns and the compressive strength tests were performed in the laboratory. The results were higher than the minimum designed values. It means that inclined DSM columns were applied successfully, and the excessive settlement problem was solved.

Keywords: ground improvement, inclined deep soil mixing, railway embankment

INTRODUCTION

In this study, the problems encountered in railway approach embankments are described and the solution to the embankments with the improved ground is presented. The existing live railway approach embankment over a length of some 310 meters and up to 13 meters high, founded on soft clay, and weak claystone units, is investigated as a case study. In geotechnical engineering, soft clays are a well-known category of problematic soils, moreover, taking engineering measures to the problematic deep soft clay layers under approach embankments can be a challenging problem with an existing live railway. Thereby, a ground improvement technique with the advantages of ease and speed of application, the application's depth, and the variety of soil conditions must be chosen. Ground improvement using soil mixed columns is an effective technique for reducing settlement and stability of soft clay under embankments for transport infrastructure such as major highways and rail transport routes (Ishikura et al 2016). In many studies, the DSM has been shown to be a successful method for consolidation settlement on soft ground, bearing capacity, and slope stabilization (Arulrajah et al. 2008, Barrett et al. 2012, Maghsoudloo and Can 2014, Jamsawang et al. 2015, Gundogdu et al, 2020). In this method, improving the strength of the soft ground limits the settlement and provides slope stability. Constitutively, the DSM is a mechanical mixing method of soft soil by adding water/cement injection. Therefore, considering the settlement and stability problems of the approach embankments, the DSM method was chosen as a ground improvement.

In order to investigate the settlement performance of the DSM improvement technique, consolidation settlement calculations for both unimproved and improved ground conditions were performed, and the uniaxial compressive strength of the deep mixing columns was investigated. This study presents the challenges in predicting the settlements of the embankments and the analytical calculations to assess the performance of ground treatment by DSM columns. Moreover, this study is important as it shows that the DSM columns can be successfully applied as inclined structures within an existing live railway.

PROJECT DESCRIPTION AND SITE INVESTIGATION

The ground improvement method was applied to the approach embankment of a bridge within the scope of the Ankara - Sivas Railway Project in Sivas, Türkiye. The bridge is located between Km:358+200 - 359+100. Figure 1 shows the location of the study area. Earth concrete wall constructions in the approach fillings were completed as of September 2017. 90 cm of weak ground excavation was made, and no additional ground improvement was applied. During the field studies carried out by Fugro Sial in June 2021, the morphology and geological conditions of the area where the bridge is located were studied, and the deformations were observed in the reinforced earth concrete walls on the approach embankments.

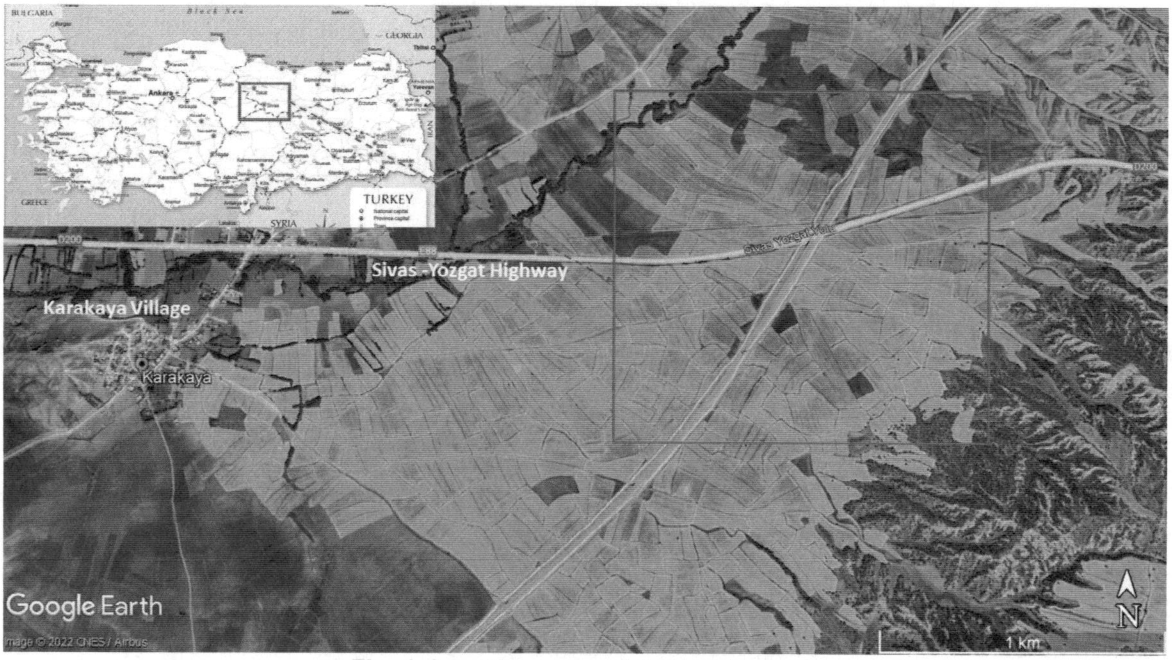

Fig. 1. Location map of the study area

GEOLOGICAL CONDITIONS

The study area is located on the Derindere member (Tid), which is a member of the İncesu formation (Ti) and is represented by Upper Miocene-Pliocene aged river and lake formations. While the upper levels of the unit are showing more soil characteristics, composed of mudstone-sandstone levels belonging to the İncesu formation, completely - highly weathered, relatively more resistant rocks are observed below them. 6 boreholes with a total depth of 185 m were carried out in the study area (Fig. 2). The unit generally consists of orangish conglomerate, clay, silty sandy clay, claystone, and mudstone. The standard laboratory tests were carried out on the specimens obtained from the boreholes, and the units are determined as CL, CH, and MH according to the Unified Soil Classification System. Clay, sandy clay, and claystone levels were observed in borehole SK-01, and the SPT-N values of the boreholes are given in Fig.3a and b respectively (Fugro Sial, 2021). According to the borehole data, it is seen that the groundwater level varies between 4.3 and 18.0 m. The undrained shear strength used in the short-term analysis of clay units is calculated with the help of the correlation based on the plasticity index, cu and the SPT-N value (Stroud, 1974). The effective internal friction angle (Φ) is assumed 0 as the sample is fully saturated in the undrained condition. In the long-term analysis of clay units, the correlation between the Effective Shear Resistance Angle and the Plasticity Index (Terzaghi, 1996) was used to determine Φ. The plasticity index values of the boreholes are given in Fig.3c, and the generalized soil parameters are given in Table 1.

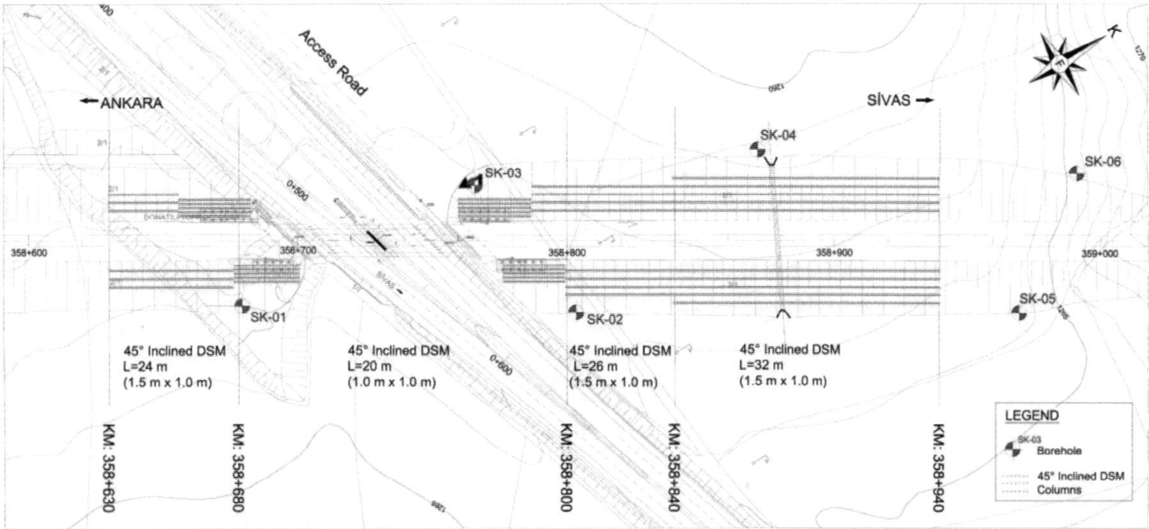

Fig. 2. General layout plan and borehole location (Fugro Sial, 2021)

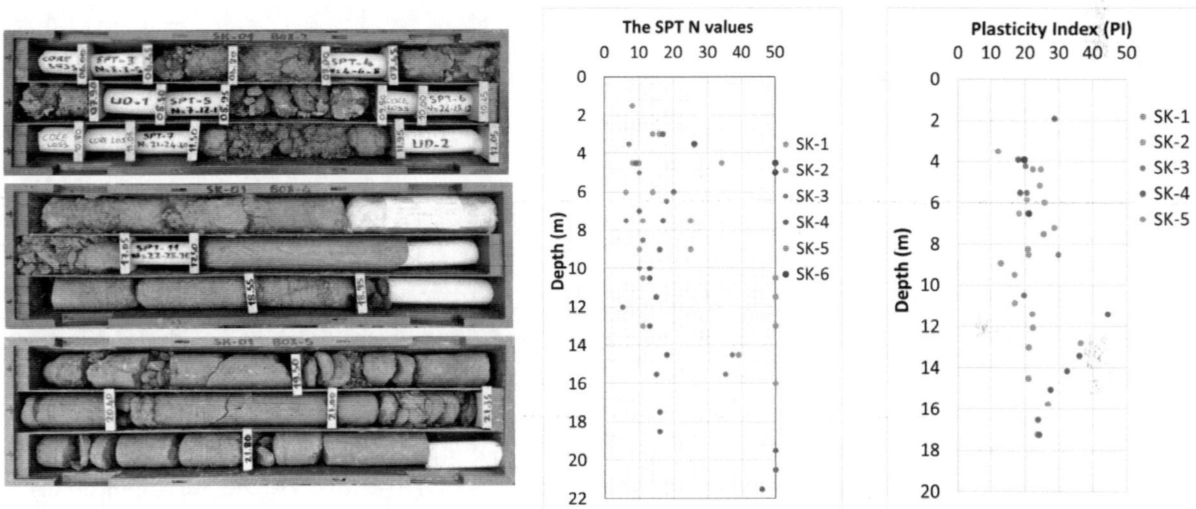

(a) (b) (c)

Fig. 3. (a) Clay, sandy clay, and claystone levels observed in borehole No: SK-01, (b) the SPT-N values, and (c) Plasticity index of the boreholes (Fugro Sial, 2021)

Table 1. Generalized soil parameters

Soil Formation	Short-term			Long-term		
	γ (kN/m^3)	c_u (kN/m^2)	Φ (^{o})	γ (kN/m^3)	c (kN/m^2)	Φ (^{o})
Filling Material (Rock)	20	3	42	20	3	42
Clay-1 (SK-1)	18	62	0	18	6	31
Clay-2 (SK-2/SK-3)	18	51	0	18	5	30
Clay-3 (SK-4)	18	54	0	18	5	29
Claystone (SK-1/SK-2/SK-3/SK-4)	18	250	0	18	15	31

SETTLEMENT ASSESSMENT

When stresses increase in saturated cohesive soils, the pore water pressure of the soil increases, and due to the low permeability of the soil, settlements occur in a long time and these settlements are called consolidation settlements in the literature (Bowles, 1997, Das, 2010). During the field studies carried out on May 2021 (Fugro Sial, 2021), the morphology and geological conditions of the area where the bridge is located were examined, and settlement problems were observed on the bridge approach embankments. Therefore, the amount of consolidation settlement and settlement times are calculated based on the data obtained from the one-dimensional consolidation test for the study area (Fugro Sial, 2021). As a result, it was seen that engineering measures were necessary for the approach embankments.

The settlements of the clays that occur under the applied load first progress rapidly but are damped by decreasing speed over time. It was aimed to determine the progress rate of settlements and how long the final settlement takes place. The consolidation settlements of the cohesive units are calculated with Eq.1. According to the consolidation theory, the unitless time factor (T_v) and consolidation percentage (U_{av}) are calculated as in Eq.2 and Eq.3, respectively.

$$s_c = m_v * H * \Delta q * \mu \qquad [1]$$

$$T_v = (c_v * t) / (H / N)^2 \qquad [2]$$

$$U_{av} = s_t / s_c \qquad [3]$$

Where s_c = total consolidation settlement, Δq = effective stress increase in the middle of the layer due to loading, m_v = coefficient of volume compressibility, μ = Skempton and Bjerrum coefficient (1957), H = layer thickness, c_v = consolidation coefficient, t = time, N = drainage coefficient, s_t = consolidation settlement amount due to construction period. According to Craig (1974), μ takes a value between 0.7 and 1.0 for normally consolidated clays. In the settlement analysis, μ=1.0 was taken. N takes the value of 1 or 2, depending on whether the drainage is unidirectional or bidirectional, respectively. Considering the sand interlayers observed in the clay unit where settlement is expected, it is assumed that the compressible layer will drain during consolidation, and it is a bidirectional drainage situation. The consolidation coefficient, c_v can be determined by the graph in NAVFAC Engineering Design Manual (1988) depending on the liquid limit (LL), c_v=1.5 m^2/year considering the results of the liquid limit tests (LL_{ave}=51) on clay samples. The consolidation percentage U_{av} value is obtained by using the consolidation percentage and the time factor (T_v) relationship graph by Olson (1977). Table 2 shows the total consolidation settlements calculated by the depths of the samples used in the oedometer tests for critical sections.

Table 2. Consolidation settlement analysis for the sections Km: 358+800 - 358+880

PARAMETERS	Km: 358+800			Km: 358+880	
	1. Layer	2. Layer	3. Layer	1. Layer	2. Layer
Layer Thickness (H)	2.1m	3.0m	4.3m	6.0 m	7.0 m
m_v (1/kPa)	0.00019	0.00008	0.00011	0.00018	0.00006
Pressure Increase Sum (σ_{vo}+$\Delta\sigma$)	241 kPa	298 kPa	367 kPa	366 kPa	490 kPa
Skempton-Bjerrum Coefficient (μ)	1.0	1.0	1.0	1.0	1.0
Consolidation Settlements (s)	9.6 cm	7.1 cm	17.7 cm	39.3 cm	20.2 cm

Total Consolidation Settlement (s_c)	**34.4 cm**	**59.5 cm**

The timeline prepared for the different km sections, in order to define the calculated consolidation settlements and the settlements that occurred before the ground improvement application, is presented in Fig.4. Considering the 15-months of embankments filling construction process, the time-dependent variation of the expected settlement amount has been examined.

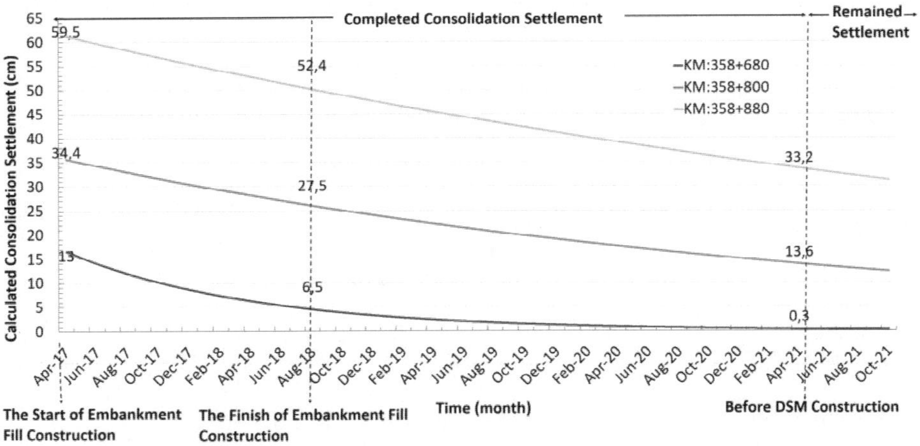

Fig. 4. Timeline of consolidation settlement (Fugro Sial, 2021)

GROUND IMPROVEMENT BY DSM COLUMNS

According to the graph of consolidation settlement versus time (Fig. 4), the remaining settlement amounts for critical sections Km:358+800 and Km:358+880 are 33.2 cm and 13.6 cm, respectively, well above the allowable limits. In the consolidation settlement calculations for the railway embankments, the upper limit of settlement is targeted as 5 cm (TCDD, 2020). In order to bring the remained consolidation settlements to permissible limits, an improvement with inclined deep mixing applications was proposed, taking into account the thickness of the compressible clay layer and applicability. To achieve the best results in the approach embankments, the application of inclined DSM columns is designed to be 45 degrees with horizontal. Schematic representations of the inclined DSM application under the embankments are given in Fig 5.

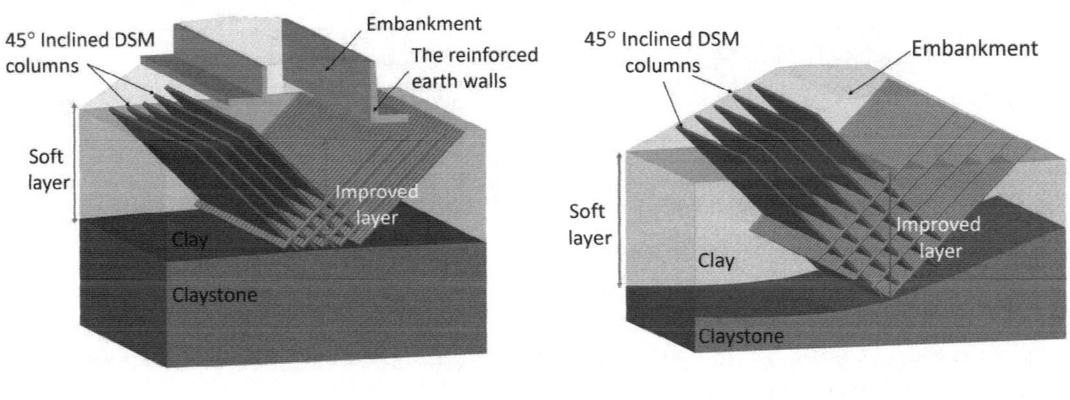

(a) (b)

Fig. 5. Schematic representations of ground improvement by the inclined DSM column method (a) Km:358+800, (b) Km:358+880

The DSM columns were designed with a diameter of 0.6 m, 3 to 4 rows in cross-section, 1.5 m vertically, 1 m apart horizontally, and with various lengths, considering the soil conditions. In DSM construction with a vertical spacing of 1.5 m and an angle of 45°, the distance of the vertical plane is taken as 3 m. Settlement calculations after ground improvement with the deep mixing method were carried out in accordance with FHWA-HRT-13-046 (2013). To obtain the composite soil modulus of elasticity (M_{comp}) which is given in Eq. 7, the Eq. 4-5-6 were used respectively. The settlement calculations of the improved ground are then calculated by Eq.8, considering the length of the column to be formed along with the compressible layer.

$E_{dm} = 300 * q_{dm,spec}$ $\quad$ $q_{dm,spec} = 2.50$ MPa, $E_{dm} = 750{,}000.00$ kPa [4]

$M_{soil} = 1/m_v$ $\quad$ $M_{soil} = 9.091$ kPa [5]

$a_{(s,center)} = \pi * d^2 / 4 * s_x * s_y$ $\quad$ $a_{s,center} = 0.089$ [6]

$M_{comp} = a_{(s,center)}\ E_{dm} + (1 - a_{(s,center)}) * M_{soil}$ $\quad$ $M_{comp} = 75.000$ kPa [7]

$\Delta H_{dm} = H_{dm} * \frac{q}{M_{comp}}$ $\quad$ $\Delta H_{dm} = 4.97\ cm$ [8]

Where E_{dm} = modulus of elasticity of the DSM columns and $q_{dm,spec}$ = uniaxial compressive strength of the DSM columns at the 28th day, M_{soil} = the constrained modulus of the untreated soil, $a_{s,center}$ = center columns area ratio, s_{center} = distance between columns (m), M_{comp} = modulus of elasticity of the composite soil (kPa), ΔH_{dm} = settlement (m), H_{dm} = length of the column (m), and q = fill load (kPa).

Finally, the expected consolidation settlement of the embankments after ground improvement by the DSM columns was calculated with the Settle 3D software according to the calculated improved soil composite modulus of elasticity (M_{comp}) of the DSM. The results of the Settle 3D are 1.85 cm and 3.71 cm for the critical sections which are below the desired settlement limit, given in Fig. 6 a and b, respectively.

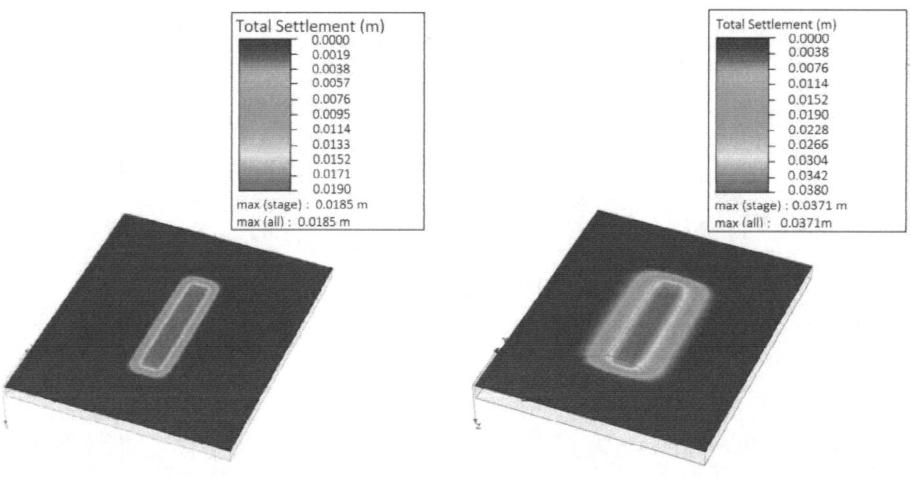

(a) (b)

Fig. 6. Consolidation settlement analysis by Settle 3D (a) section KM:358+800 (b) section KM:358+880

The water-cement mixture ratio of the DSM columns was from 0.8 to 1.0 and injected by 50-100 bar pressure from 14 nozzles on the mixing equipment. The rotation rate of the mixing equipment was 80 rpm. To control the mechanical properties of the DSM columns, a total of 12 core samples were taken, and uniaxial compressive strength tests were performed on the samples. The average obtained compressive strength of the samples was 11.24 MPa which is higher than the minimum designed value. The test results of the core samples are given in Table 3.

Table 3. Uniaxial compressive strength test results of the constructed deep mixing columns

Sample No.	Compressive Strength (MPa) DSM 763	DSM 1712	DSM 1857	Average
1	6.04	12.08	13.51	11.24
2	6.81	12.31	12.99	
3	7.85	12.74	13.32	
4	8,19	13.06	12.96	

STABILITY ASSESSMENT

For the critical sections, slope stability analyzes were performed by using Slide2 Limit Equilibrium Analysis (LEA) software. Considering the presence of clay units under the embankment the stability analyzes were carried out for short-term and long-term both in static and dynamic situations, given in Fig. 7a and b, and Fig. 8a and b respectively. The superstructure loads were applied to the system as q = 30.0 kPa. The horizontal earthquake acceleration was used as k_h = 0.124 g in the seismic analyzes carried out on the fill slopes, according to Türkiye's Earthquake Hazard Map which is published by the Turkish Disaster and Emergency Management Authority (AFAD). The calculated safety coefficients are given in Table 4 and are within the allowed limits (TCDD, 2020). This shows that the designed ground improvement has successfully fulfilled its task.

Table 4. Stability analysis for the sections KM: 358+800 and KM: 358+880

Section KM	H_{max}	Short-term (static)	Short-term (dynamic)	Long-term (static)	Long-term (dynamic)
358+680	13.8 m	1.707	1.241	2.118	1.629
358+800	10.6 m	1.723	1.342	2.010	1.746
358+880	11.7 m	1.605	1.118	2.508	1.919

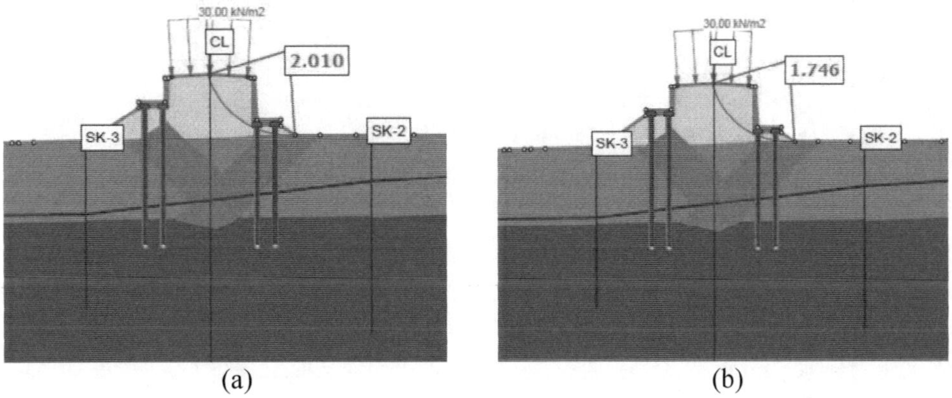

Fig. 7. Stability analysis of section Km:358+800 in long-term (a) static state (b) dynamic state

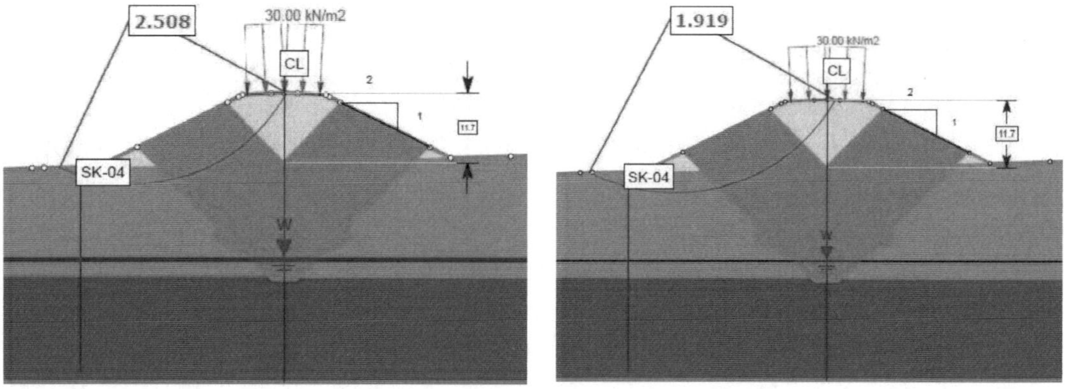

(a) (b)

Fig. 8. Stability analysis of section Km:358+880 in long-term (a) static state (b) dynamic state

CONCLUSION

In this study, the settlement problem that occurred on bridge approach embankments was evaluated, and the ground improvement methodology with DSM columns was chosen for the project in terms of applicability, time, and economy. The DSM design in the soft ground was evaluated through consolidation calculations and presented.

As a result of the calculations, it was determined that the settlement and stability problems are prevented with the DSM application for the embankments. The DSM columns were designed with a diameter of 0.6 m, 3 to 4 rows in cross-section, 1.5 m vertically, 1 m apart horizontally, and with various lengths, according to the soil conditions. To achieve the best results in the rail embankments, the application of DSM columns is designed to be 45 degrees horizontal.

In the study area, the DSM columns application was completed successfully as of November 2021. The strength of DSM columns was confirmed through the core samples taken from the DSM columns and the compressive strength tests were performed in the laboratory. The results were higher than the desired values. After DSM construction, settlement measurements were done continuously on the railway, and no settlement changes were observed during six months, and the railway system runs safely. It is concluded that the suggested ground improvement methodology with inclined DSM columns is found satisfactory.

(a) (b) (c)

Fig. 9. (a) View of the approach embankments (b,c) the construction of inclined DSM from the site.

ACKNOWLEDGMENTS

The authors thank the General Directorate of Turkish Railway System (TCDD) for the support during the project studies.

REFERENCES

Akademi, 2021. Ankara – Sivas Demiryolu Projesi KM:358+630 – KM: 358+940 Arası Geoteknik Değerlendirme Raporu. 381 p. (in Turkish)

Bowles, J.E. 1996. Foundation Analysis and Design. 5th Edition, McGraw-Hill Companies, Inc., Singapore.

Bruce, M.E., Berg, R.R., Collin, J.G., Filz, G.M., Terashi, M., & Yang, D.S. 2013. Federal Highway Administration Design Manual: Deep Mixing for Embankment and Foundation Support.

Craig, R.F. 1974. Craig's Soil Mechanics (7th ed.). CRC Press. https://doi.org/10.4324/9780203494103

Das, Braja M., 2010. Principles of Geotechnical Engineering (Ed. 7th). Stamford: Cengage Learning.

Fugro Sial, 2021. Ankara – Sivas Demiryolu Projesi KM:358+180 – KM: 359+100 Arası Jeolojik – Jeoteknik Değerlendirme Raporu. 354 p. (in Turkish)

Gundogdu, A. E. K., Can, A., Acar, K., Gokceoglu, C., 2021. "A Soil Improvement Methodology for a Shallow Tunnel Excavated in Extremely Weak Clay: A case study from Türkiye" EURENGEO2020, Athens, Greece, Oct 7, 2021

Ishikura, R., Yasufukua, N., Brown M. J., 2015. An estimation method for predicting final consolidation settlement of ground improved by floating soil cement columns. The Japanese Geotechnical Society. Soils and Foundations 2016; 56 (2), 213–227.

Jamsawang, P., Voottipruex, P., Boathong, P., Mairaing, W., Horpibulsuk, S., 2015. Three-dimensional numerical investigation on lateral movement and factor of safety of slopes stabilized with deep cement mixing column rows. Engineering Geology Vol. 188, 7 April 2015, 159-167.

Jamsawang, P., Yoobanpot, N., Thanasisathit, N., Voottipruex, P., Jongpradist, P., 2015. Three-dimensional numerical analysis of a DCM column-supported highway embankment. Computers and Geotechnics 72 (2016) 42–56.

Maghsoudloo, A. and Can, A., 2014 "A Case Study of Wet Soil Mixing for Bearing Capacity Improvement in Türkiye" International Conference on Piling and Deep Foundations, Stockholm, Sweden, 21-23 May 2014

NAVFAC DM-7.2, 1986. Foundations and Earth Structures. Design Manual 7.2, Department of the Navy, Naval Facilities Engineering Command, Alexandria, Va., 253 pp.

Olson, R. E., 1977. Consolidation under time-dependent loading, J. Geotech. Eng. Div., ASCE, 103(GT1), 55–60.

Skempton, A.W. and Bjerrum, L. (1957) A Contribution to the Settlement Analysis of Foundations on Clay. Géotechnique, 7, 168-178.

TCDD Soil Mechanics and Tunnels Project Branch Office, Considerations, 2020

A STUDY OF ASPECTS OF THE HISTORICAL DEVELOPMENT OF GEOTECHNICAL ENGINEERING

Michael D. Bennett, E.I.T., A.G.E.S., Inc., King of Prussia, PA, USA, 610-354-0333, mbennett@agesinc.com
Thomas J. Kennedy, Geopier Foundation Co., Davidson, NC, USA, 704-439-1790, tkennedy@geopier.com

ABSTRACT

The study of geotechnical history is well-established. Most geotechnical practitioners, teachers, and graduate students have some grasp of the origins of the field. However, this understanding often consists of a jumble of names and anecdotes. This paper discusses how a deeper comprehension of geotechnical history can benefit academia and practice alike and describes how the authors have begun a wide-ranging study of the subject. The paper reviews the dual-track methodology the authors have generated for the study, which consists of written articles about different works in geotechnical history and a database tracking how often pieces have been cited in geotechnical literature. The first technique has shed light on aspects of geotechnical history such as the connection of the Darwin family to the origins of geotechnical engineering, the St. Francis Dam failure of 1928, and the rivalry between theoreticians and practitioners in the early days of the field. The second has helped quantify subjects such as the outstanding contributions of Karl Terzaghi, the disagreements between Terzaghi and Donald Taylor, and the specialization of the field over time. The paper concludes by discussing the authors' planned future work on geotechnical history.

Keywords: geotechnical history, literature review, citation database, Karl Terzaghi, George Darwin, St. Francis Dam, theory, practice, high-impact pieces, super-high-impact pieces, Donald Taylor

INTRODUCTION

Studying the history of geotechnical engineering is not a new practice. Every thesis and paper in the discipline incorporates a review of historical literature on the topic under consideration. In addition, stand-alone literature reviews, case histories, and even purely historical pieces are already part of geotechnical discourse. Yet the systematic thinking that is a key part of geotechnical engineering as a technical discipline has largely been absent from studies of the history of the field. As a result, the history of geotechnical engineering is often related as a series of stories about a few founders (Coduto et al. 2011). The current state of geotechnical history is similar to the state of the field overall in the early 1920s – many key facts are there, but their significance has, often, not yet been grasped.

This narrative may be worth changing for several reasons. First, a study of the history of geotechnical engineering could provide opportunities for critical reevaluation and revision of extant theories. Second, redefining how the history of geotechnical engineering is told might give the profession insight into potential new avenues of research to be explored. Third, as geo-professionals review the works of and learn lessons from their predecessors, they will better appreciate the shared history that binds them.

This paper describes an ongoing study of geotechnical history consisting of web articles and a citation database. First, the origins of the study will be discussed. Next, the early phases of work and the findings to date will be recapitulated. Finally, ongoing and potential future work will be summarized.

BACKGROUND

The authors met during their graduate studies of geotechnical engineering. After graduating, they continued their technical educations informally by reviewing geotechnical literature. Each piece reviewed included details about the historical development of the discipline, which eventually led to the decision to further explore this history. Some preliminary investigation into the subject revealed that geotechnical history had mainly been used to introduce the field and had primarily been shared as a collection of anecdotes about Karl Terzaghi and other pioneers. Thus, while the technical aspects of geotechnical engineering had been explored systematically and documented extensively, the historical development of those technical aspects had not been given similar treatment.

Ultimately, it was decided to study geotechnical history in greater depth using a dual-track approach. First, review articles on select historical pieces would be posted on a publicly accessible web page. The articles would attempt to identify the origins of geotechnical principles and to indicate unexplored avenues of inquiry in the field. Then, once the web page and articles were well-established, a data-driven methodology would be developed to help identify trends in geotechnical history which had previously been obscured.

FIRST PHASE – INITIAL ARTICLES

The geotechnical pieces reviewed in the first phase of the study included one complete book (Hool and Kinne 1923), part of a second book (Wheeler 1877), and seven technical papers: Gregory (1844), Rankine (1857), Baker (1881), Darwin (1883), Reynolds (1886), Bell (1915), and Terzaghi (1925). The nine technical sources were selected based on their age and their significance to the field as assessed by previous geotechnical groups and authors, notably BSCE (1940), Cooling et al. (1969), and Skempton (1985).

The review of these works generated numerous findings of interest. First, it became clear that few, if any, technical lessons remained to be gleaned from early literature in the field. This reflected that the study of geotechnical history to date has been robust, albeit not always presented in an organized manner. Second, the pieces reviewed showcased the origins of a number of geotechnical principles, including Rankine lateral earth pressure theory (Rankine 1857), the dilatancy of sands and negative pore pressures (Reynolds 1886), and the drained and undrained shear strengths of soils (Bell 1915). Reading these works represented a powerful experience and brought the authors a strong sense of connection to the roots of the profession.

Third, writing the articles provided valuable background to the pieces and allowed them to be placed in their appropriate historical context. For instance, the influence of relative density and stress history on the behavior of sands behind a retaining wall was first noted in an 1883 study performed by George Darwin, a son of Charles Darwin (Cooling et al. 1969; Caldwell 2013). Over thirty years later, Karl Terzaghi cited the younger Darwin's work in his own studies of the lateral earth pressures of sands, which were among his first works in soil mechanics (Terzaghi 1920). It seems fitting that George Darwin, whose father first articulated the modern theory of evolution, made an invaluable contribution to the evolution of geotechnical engineering.

The study also contextualized geotechnical advances in the case of the St. Francis Dam, which failed in 1928 due in part to uplift forces. Some scholars have stated that the failure could not have been prevented by the contemporary standard of care, claiming that uplift was not a common consideration for dam design at the time. However, Hundley and Jackson (2015) cited numerous technical sources from the 1910s and 1920s to argue that civil engineers in the early 20th century did in fact recognize that, as one of them stated in 1913, "It is a crime to design a dam without considering upward pressure" (Hundley and Jackson 2015). To provide further detail on this point of historical contention, the study consulted the 1923 book *Foundations, Abutments, and Footings* by Hool and Kinne, a source not reviewed by Hundley and Jackson (2015). The guidance provided on dam foundations in *Foundations, Abutments, and Footings* confirmed that civil engineers already widely knew of the threat uplift posed to dams at the time the St. Francis Dam was designed and constructed in the mid-1920s, and the book even referenced a 1912 article cited by Hundley and Jackson (2015). Thus, the study also reiterated the importance of continuously updating the standard of care for geotechnical projects and of strictly adhering to the latest best practices.

Finally, the first phase of the study suggested that, prior to the emergence of Karl Terzaghi in the mid-1920s, an intense rivalry existed between engineers who approached geotechnical problems mainly from a theoretical perspective and others who tackled these problems primarily using common field practices and rules of thumb. For example, Rankine (1857) used pages of elaborate equations to derive his namesake theory of lateral earth pressure. A quarter-century later, Baker, in discussing case studies of earthwork failures, sharply critiqued theoretical approaches like Rankine's: "The numerous published tables giving the calculated required thicknesses of retaining walls to three places of decimals stand really on exactly the same scientific basis, and have the same practical value, as the weather forecasts for the year in Old Moore's [Almanac]" (Baker 1881).

The theory versus practice debate raged on in geotechnical circles until the mid-1920s, when Terzaghi became prominent in part by realizing that there was considerable interplay between theory and practice within the field. It was Terzaghi who first and most effectively advocated uniting the theoretical and practical camps using a series of what he called "semi-empirical procedures" (Terzaghi 1944) whereby data from the field are used together with theories to generate designs. An example is the equation generated by Terzaghi and his mentee Ralph Peck for predicting the settlement of sands from SPT N-values. The observational method pioneered by Peck, whereby designs are checked against field data during construction and modified as necessary, complements the semi-empirical procedures (Coduto et al. 2011). Today, semi-empirical procedures and the observational method remain critical tools for geotechnical engineers, even with the advent of newer techniques such as electronic in-situ testing and numerical methods.

SECOND PHASE – CITATION DATABASE

The opening phase of the study made clear that the semi-empirical procedures and observational method used on the technical side of geotechnical engineering could also be used to study the history of the field. Per the initial goals of the study, the decision was made to incorporate an element of data-driven analysis into the study prior to writing more entries. The simplest and most readily apparent empirical method available for the study involved creating a database of how many times each of a set of geotechnical pieces was cited in a set of sources. This database of citations in geotechnical literature could then be used to guide the selection of pieces for review in future articles.

A total of 30 geotechnical sources (29 books and one course syllabus) dating from 1937 to 2020 were consulted to create the database, and bibliographic information was recorded for select pieces cited by each source. To keep the database manageable in scale, the pieces selected were primarily works in English published before 1970 to which the authors had ready access. The year 1970 was chosen for the first iteration of the database because most geotechnical fundamentals were well-established by then, although much remained to be developed. While the selection criteria excluded many significant pieces from the database, they also allowed both for its completion in a reasonable timeframe and for its easy future expansion. Select additional pieces were also included on the basis of historical judgment.

The final database included nearly 1,200 pieces. The structure of the database allowed the data to be interpreted in several ways, such as how many times each piece was cited and which pieces from certain time periods were cited most. The roughly 100 pieces which had been cited by five or more sources were referred to as high-impact pieces, meaning that they appeared to have significantly influenced the geotechnical field judging by how often they had been cited. Similarly, the 25 pieces with 10 or more citations were labeled super-high-impact pieces. The numbers five and 10 were arbitrarily chosen to simplify the process of selecting significant works. In future iterations of the database, as the number of sources in the database increases, the required numbers for high-impact and super-high-impact pieces may be increased as well.

The database allows for the use of a semi-empirical historical procedure to select historic pieces of geotechnical literature for future review. This procedure offers several powerful advantages over the haphazard method of literature selection used earlier in the study. It provides a first-order approximation of how impactful a piece has been in geotechnical history by quantifying how often it has been cited. It also allows students of geotechnical history to avoid examining only pieces deemed significant by one or two previous anthologies in the field. Lastly, it represents a systematic way of thinking about geotechnical history which may help bring this sub-discipline in line with the rest of the field.

The semi-empirical historical procedure described above has distinct limitations in addition to the criteria for selecting pieces. The number of times a piece has been cited by geotechnical sources does not necessarily indicate its value to the profession. For example, although Baker (1881) is a high-impact piece, having been cited eight times in the database, Skempton (1985) noted that Baker's critique of overly precise calculations based on earth pressure theory "is based more on a lack of understanding than shortcomings in the theory" – i.e., that Baker's paper does not represent a great geotechnical advance. In addition, since the procedure depends on how many times a piece has been cited, it is biased toward pieces that have extensively been covered and are thus less likely to yield new geotechnical ideas. However, these limitations may be addressed at least in part by utilizing an element of historical judgment in selecting pieces. This element of judgment makes the procedure semi-empirical, thus tying it to the rest of geotechnical engineering.

To date, the database has led to three major observations, all of which suggest that the number of citations for each piece in the database does reflect, to some extent, broader contemporary trends in geotechnical engineering. The first observation relates to how often certain authors have been cited. The database supports the reputation of Karl Terzaghi as the father of geotechnical engineering. During his career, Terzaghi authored or co-authored at least 19 high-impact pieces, including eight super-high-impact pieces. For comparison, the database also indicates that two of Terzaghi's most accomplished mentees, Arthur Casagrande and Ralph Peck, combined for a total of 14 high-impact pieces, including six super-high-impact pieces, during their very illustrious careers. Such a comparison quantifies how outstanding Terzaghi's contributions to geotechnical engineering were.

The second trend noted relates to the disagreements between Karl Terzaghi and his fellow geotechnical pioneer Donald Taylor. Christian and Baecher (2015) report that Terzaghi and several of his mentees felt that Taylor took an overly theoretical approach to soil mechanics. Taylor's use of Terzaghi's MIT lecture notes in writing the 1948 textbook *Fundamentals of Soil Mechanics*, while appropriately acknowledged, may also have fueled animosity between the two. The database partially quantifies the tension between Terzaghi and Taylor, showing that, within *Fundamentals of Soil Mechanics*, Taylor cited 12 pieces authored or co-authored by Terzaghi (Taylor 1948). By contrast, Terzaghi cited just one paper by Taylor in his 1943 book *Theoretical Soil Mechanics* (Terzaghi 1943), while Terzaghi and Peck cited none of Taylor's work in their 1948 textbook *Soil Mechanics in Engineering Practice* (Terzaghi and Peck 1948). Notably, the disparity between the number of pieces Terzaghi and Taylor had each authored as of 1948 – 184 for Terzaghi versus 11 for Taylor – may account in part for the discrepancy between how often they cited each other's work. Yet the database suggests that, while Taylor held Terzaghi's work in high regard, the opposite was not necessarily the case.

The third aspect of geotechnical history highlighted by the database is how often pieces from different time spans were cited. The 20-year periods from 1930 to 1949 and from 1950 to 1969 had relatively similar numbers of high-impact pieces included in the database; there were 44 from the 1930s and 1940s, while there were 41 from the 1950s and 1960s. However, the number of super-high-impact pieces differed sharply between the two eras; there were 20 from the 1930s and 1940s, compared with four from the 1950s and 1960s. This disparity is partly because seven of the sources reviewed for citations were written before 1950. Yet it also reflects how the highest impact pieces from the 1930s and 1940s were more broadly applicable in their technical content than the highest impact pieces from the 1950s and 1960s. Clearly, by 1970, geotechnical engineering had separated into multiple distinct sub-disciplines, even as practitioners continued to work across several of them.

The trends observed within the database to date are mainly reiterations of established points about the field of geotechnical history. However, every historical investigation involves assessing the findings of previous studies. The parallels between the numbers from the database and certain phenomena in geotechnical history indicate that the database serves as a reasonably robust method of studying this history. Furthermore, while the initial findings from the database are familiar, the semi-empirical historical procedure supporting the database is new to geotechnical history. The evidence offered by the procedure to reaffirm well-observed trends in the history of the profession suggests that it will likely reveal additional, previously unrecognized historical trends as the study progresses.

FUTURE PLANS

Now that a semi-empirical procedure has been applied to the study of geotechnical history, the observational method of revising plans based on field monitoring and data may be applied similarly. Moving forward, the current study will be enhanced in several ways to improve its prospects for long-term growth. First, more web articles on historical geotechnical works are being written and posted. The works covered in these articles are being selected using the semi-empirical historical procedure described. While the first phase of the study focused primarily on geotechnical pieces written before the rise to prominence of Terzaghi in the mid-1920s, these newer articles mainly concentrate on works written between approximately 1920 and 1970. These pieces from the more recent past may yield more ideas for new areas of geotechnical study and research than their older counterparts.

The database will also be expanded to include additional sources and pieces. Over time, the criteria for selecting pieces for the database will gradually be expanded to allow for a more thorough analysis of geotechnical history. As the database grows, the semi-empirical historical procedure for examining the impact of vintage geotechnical pieces will be continuously evaluated and refined. To maximize the resources available for expanding the database, collaborations will be pursued both with private collectors and with corporate and university libraries.

A third enhancement relates to broadening the study beyond its anglophone origins. It is worthy of note, especially in a day and age where English is the *lingua franca* of the scientific and engineering world, that many pioneering works in what is now geotechnical engineering were written in other languages. Several steps are being taken to give the study a more international perspective. Guest entries by multilingual geotechnical engineers from around the world who can draw upon non-English sources are being commissioned. In addition, English translations of several prominent non-English geotechnical works, such as those by Charles-Augustin Coulomb, Alexandre Collin, and Henri Darcy, have been obtained and will be reviewed.

Finally, as the study grows in size and scope, it may eventually require bringing in more members of the geotechnical community. Accordingly, partnerships will be established with larger institutions such as the Deep Foundations Institute and the Geo-Institute of the American Society of Civil Engineers domestically and the Institution of Civil Engineers and the International Society for Soil Mechanics and Geotechnical Engineering abroad. These partnerships will be used to encourage exploration of geotechnical history among small groups of academics, professionals, and retirees in the field. The partnerships will also solicit feedback from geo-professionals on what topics in the history of the profession are of interest. Over time, the partnerships will ensure that the study of the history of geotechnical engineering will become robust and self-sustaining.

SUMMARY AND CONCLUSIONS

A study of aspects of geotechnical history has been undertaken which uses a dual-track approach. Literature review articles are written and posted on a webpage while a citation database is used to generate findings using a consistent, semi-empirical historical procedure. Although the initial phases of the study have located few new avenues for correcting or expanding geotechnical theory, they have highlighted the origins of geotechnical principles, placed vintage geotechnical literature in its historical context, and illustrated how theory and practice in the field went from competing to complementing each other. The early portion of the study has also partially quantified several established trends in early geotechnical history, including the considerable contributions of Terzaghi, the disagreements between Terzaghi and Donald Taylor, and the increased specialization of the field over time. Future phases of work will include additional articles, the expansion of the database, and institutional partnerships. Ultimately, the goal of the study is to generate findings on geotechnical history that can positively influence both how the past of the field is told and where its future may lie.

ACKNOWLEDGMENTS

Professor Michael McGuire, P.E. of Lafayette College provided initial guidance on organizing the study. His suggestion to map the connections between pieces of historical geotechnical literature eventually inspired the creation of the database. Travis Shoemaker, E.I.T. of the University of Illinois at Urbana-

Champaign and Samuel Consolvo, E.I.T. of CDM Smith reviewed the manuscript prior to submittal and gave invaluable suggestions on improving it.

REFERENCES

Baker, B., 1881. The actual lateral pressure of earthwork. Minutes and Proceedings of the Institution of Civil Engineers, 65, 140-186. Reprinted in Cooling et al. (1969).

Bell, A. L., 1915. The lateral pressure and resistance of clay, and the supporting power of clay foundations. Minutes and Proceedings of the Institution of Civil Engineers, 199, 233-272. Reprinted in Cooling et al. (1969).

Bjerrum, L., Casagrande, A., Peck, R.B., and Skempton, A.W. (Eds.), 1960. From theory to practice in soil mechanics: selections from the writings of Karl Terzaghi. John Wiley and Sons, Inc., New York, 425 p.

Boston Society of Civil Engineers (Eds.), 1940 (reprinted 1963). Contributions to soil mechanics, 1925-1940. Boston Society of Civil Engineers, Boston, 413 p.

Caldwell, E., 2013. Darwin's children represent highs and lows of famous scientist's personal life. https://news.osu.edu/darwins-children-represent-highs-and-lows-of-famous-scientists-personal-life/. Accessed 28-02-2021.

Christian, J.T., and Baecher, G.B., 2015. D. W. Taylor and the foundations of modern soil mechanics. American Society of Civil Engineers, ASCE Journal of Geotechnical and Geoenvironmental Engineering, 141 (2), DOI 10.1061/(ASCE)GT.1943-5606.0001249.

Coduto, D.P., Yeung, M.R., and Kitch, W.A., 2011. Geotechnical engineering: principles and practices, 2nd edition. Prentice Hall, Upper Saddle River, 814 p.

Cooling, L.F., Skempton, A.W., and Little, A.L. (Eds.), 1969. A century of soil mechanics. Institution of Civil Engineers, London, 482 p.

Darwin, G.H., 1883. On the horizontal thrust of a mass of sand. Minutes and Proceedings of the Institution of Civil Engineers, 71, 350-378. Reprinted in Cooling et al. (1969).

Gregory, C.H., 1844. On railway cuttings and embankments; with an account of some slips in the London clay, on the line of the London and Croydon Railway. Minutes and Proceedings of the Institution of Civil Engineers, 3, 135-145. Reprinted in Cooling et al. (1969).

Hool, G.A., and Kinne, W.S. (Eds.), 1923. Foundations, abutments, and footings. McGraw-Hill Book Company, New York, 414 p.

Hundley, N., and Jackson, D.C., 2015. Heavy ground: William Mulholland and the St. Francis Dam disaster. University of California Press, Oakland, 439 p.

Rankine, W.J.M., 1857. On the stability of loose earth. Philosophical Transactions of the Royal Society of London, 147, 9-27.

Reynolds, O., 1886. Experiments showing dilatancy, a property of granular material, possibly connected with gravitation. Proceedings of the Royal Institution of Great Britain, 11, 217-227.

Skempton, A.W., 1985. A history of soil properties, 1717-1927. Proceedings, 11th ICSMFE, San Francisco, August 12-16, Golden Jubilee Vol., pp. 95-121.

Taylor, D.W., 1948. Fundamentals of soil mechanics. John Wiley and Sons, Inc., New York, 700 p.

Terzaghi, C., 1920. Old earth-pressure theories and new test results. McGraw-Hill Publishing Company, Engineering News-Record, 85 (14), 632-637. Reprinted in Bjerrum et al. (1960), 108-113.

Terzaghi, C., 1925. Modern conceptions concerning foundation engineering. Boston Society of Civil Engineers, Journal of the Boston Society of Civil Engineers, 12 (10), 397-423. Reprinted in BSCE (1940), 1-27.

Terzaghi, K., 1943. Theoretical soil mechanics. John Wiley and Sons, Inc., New York, 510 p.

Terzaghi, K., 1944. Ends and means in soil mechanics. Engineering Institute of Canada, Engineering Journal, 27, 608-615.

Terzaghi, K., and Peck, R.B., 1948. Soil mechanics in engineering practice. John Wiley and Sons, Inc., New York, 566 p.

Wheeler, J.B., 1877. An elementary course of civil engineering, for the use of cadets of the United States Military Academy. John Wiley and Sons, Inc., New York, 472 p.

ESTIMATING KINEMATIC DEMANDS FROM LATERAL GROUND DEFORMATIONS ON PILE-SUPPORTED WHARVES

Arash Khosravifar, Ph.D., P.E.[1], Thomas Torkelson, P.E.[2], Milad Souri, Ph.D., P.E.[3]

[1]Assistant Professor, Portland State University; 1825 SW Broadway, Portland, Oregon 97201; Tel 503-725-4280; email: akhosravifar@pdx.edu
[2]Project Engineer, GeoPacific Engineering, Inc.; 14835 SW 72nd Avenue, Portland, Oregon 97224; Tel 503-961-5160; email: ttorkelson@geopacificeng.com
[3]Engineering Staff, Geotechnical Resources, Inc.; 16520 SW Upper Boones Ferry Road, Suite 100, Tigard, OR 97224; Tel 971-336-3228; e-mail: msouri@gri.com

ABSTRACT

Kinematic demands from lateral ground deformations can be a major cause of damage to maritime and highway transportation structures such as wharves, piers, and bridges. These kinematic demands are different from the ground deformations in the free field in that the resisting forces from piles will reduce the ground deformations to some extent (i.e., pile pinning effect). Current design codes for wharf structures allow the beneficial resisting forces from pile pinning effects to be incorporated in pseudo-static slope stability analysis (e.g., ASCE/COPRI 61-14). While design guidelines that are specifically developed for bridge foundations provide recommendations on how to incorporate the pile pinning effects (e.g., Caltrans 2012), their applicability to wharf structures need to be evaluated. This study uses the results of a large-scale centrifuge test on a pile-supported wharf structure to evaluate the accuracy of available Newmark sliding block analysis in estimating the lateral ground deformations. Pile pinning effects are incorporated based on the approach outlined by Caltrans. The results of this study support the use of existing pile pinning methods that were primarily developed for bridge foundations in maritime structures (e.g., wharves and piers). It was found that median Newmark displacements provide a good estimate of the permanent end-of-shaking displacements and that the median plus 1 standard deviation Newmark displacements provide a good estimate of the peak transient displacements. Design considerations for use in the Newmark sliding block analysis to estimate the kinematic demands on piles are discussed.

Keywords: Piles, liquefaction, lateral spreading

1.0 INTRODUCTION

The damage to pile-supported bridges and wharves due to liquefaction-induced lateral spreading has been documented in many earthquake case histories where soil displacements exceeding 1 m were reported as the likely cause of damage (e.g., Hamada et al. 1986, Turner et al. 2016, and Cubrinovski et al. 2017). Lateral ground deformations can result from liquefaction and cyclic softening as well as the degradation of foundation soils due to earthquake loads. While most design standards require piles to be designed for the effects of liquefaction-induced lateral spreading, the existing guidelines vary significantly on how to estimate the soil displacements (i.e., kinematic demands). First, there are large uncertainties in estimating seismic slope displacements when employing the commonly used Newmark sliding block analysis. Second, there are practical challenges on how and when to apply the restraining effects of piles (also known as pile pinning effects). While the design guidelines developed by highway bridge agencies provide some guidance on how to account for pile pinning effects in estimating lateral ground deformations (e.g., Caltrans 2016), the applicability of these guidelines for maritime transportation structures—such as wharves and piers, which are commonly supported on a large number of closely spaced, relatively flexible piles (steel or concrete piles up to 2 feet in diameter)—needs to be assessed.

The kinematic demands on piles can be estimated using different methods with varying levels of complexity, ranging from the simplified Newmark sliding block analysis (Newmark 1965) to a more detailed two- or three-dimensional dynamic analysis that incorporates soil–structure interactions. The simplified approach includes estimating free-field soil displacement profiles with depth, adjusting the free-field displacements to account for the restraining effects of pile foundations (i.e., pile pinning effects), and applying the pile-restrained soil displacements to the end-nodes of the p-y springs to estimate stresses and strains in piles. The objective of this study is to evaluate the accuracy of using the Newmark sliding block analysis including pile pinning effects (as presented in Caltrans 2012) in estimating the lateral ground deformations on piles supporting a wharf structure. Results from a centrifuge test on a pile-supported wharf are used in this study. The details of the experiment are described in McCullough et al. (2001). The piles in the centrifuge tests were subjected to lateral ground deformations due to the liquefaction of foundation soils. Pile pinning effects were included in the analysis by incorporating the lateral pile resistance in the limit equilibrium slope stability analysis.

The next section of this paper describes the characteristics of the centrifuge test. It is followed by a step-by-step description of the approach that accounts for pile pinning effects, which is consistent with the procedures outlined in Caltrans (2012, 2016). The results are then compared with the experimental measurements, and some conclusions are made regarding design considerations that should be made when estimating the liquefaction-induced kinematic demands on the piles.

2.0 CENTRIFUGE EXPERIMENT

The results of a centrifuge test on a pile-supported wharf were used in this study to evaluate the accuracy of the current methods for estimating lateral ground deformation (kinematic) demands on piles. Figure 1 presents the cross section and plan view of the centrifuge model. The piles in this centrifuge tests were subjected to liquefaction-induced kinematic loads. The structure consisted of a wharf deck, supported by 21 piles configured in a 7-by-3 matrix. The soil profile consisted of loose sand, dense sand, and a multi-layer rock dike. The piles consisted of aluminum pipe piles that represent reinforced concrete piles with a diameter of 0.64 m and a flexural stiffness (EI) of 2.1e5 kPa-m^4 (all units are in prototype scale.) The piles were 27.2 m long and were driven using a vibratory hammer and a template such that their tips sat on the base of the centrifuge container. The wharf deck was attached to the top of the piles with connections that constrained rotation at the pile head.

The model was subjected to a sequence of shaking with different amplitudes at a centrifugal acceleration of 40.1 g. The results in this study are presented for the first shaking (Event 11) with the input motion corresponding to the 1989 Loma Prieta earthquake recorded at Outer Harbor Station scaled to a peak ground acceleration (PGA) of 0.15 g. More details on the centrifuge test can be found in McCullough et al. (2000, 2001) and McCullough and Dickenson (2004).

3.0 ESTIMATION OF LATERAL GROUND DEFORMATIONS

The lateral ground deformations are estimated using an iterative process to incorporate pile pinning effects. This process is divided into three main steps that are described in the following sections.

3.1 Estimating Pile Resisting Forces (Step 1)

Pile shear forces along the ground failure surface provide resisting forces that act as reinforcements and result in a reduction of cyclically induced ground displacements. In Step 1, these resisting forces are estimated by performing a lateral load analysis on the piles using p-y solutions. A soil displacement profile is imposed to the end nodes of p-y springs, and the internal shear forces are extracted from the piles at the approximate depth of the shear failure surface. (The shear failure surface is determined using slope stability analysis in Step 2, and the soil displacement profile is updated in each iteration based on the results of the Newmark sliding block analysis in Step 3).

The p-y analyses were performed using LPILE (Ensoft 2014). The p-y spring properties were generally consistent with those recommended by API (1991). However, the initial stiffness of the p-y springs was back-calculated from a series of centrifuge tests that were softer than the API curves. The back-calculation of experimental p-y springs from centrifuge tests is presented in Souri et al. (2020). The p-y spring input properties are listed in Table 1.

An idealized soil displacement profile was used to impose soil displacements to the end nodes of the p-y springs. Figure 2a shows the soil displacements that were imposed to the end nodes of the p-y springs in LPILE for a case where the ground surface displacement was 0.07 m, as an example. These displacements were only applied for piles that intersected the failure surface. As seen in Figure 2a, it was assumed that the rockfill moved monolithically with equal displacements throughout. The shear strains were assumed to be uniform within the liquefying loose sand. It was assumed that the underlying dense sand did not accumulate significant shear strains.

The p-y springs were modified to account for the first-order softening effect of pore water pressure and liquefaction using p-multipliers (P_m) that were linearly proportional to the pore water pressure ratio R_u generated during the ground motion: $P_m = 1.2 - 1.1*Ru$ for $Ru > 0.2$ and $P_m = 1.0$ for $Ru \leq 0.2$. The effect of pore pressure generation was assumed to be negligible when Ru was below 0.2. This relationship is generally consistent with the nonlinear relationship proposed by Liu and Dobry (1995) as presented in FHWA (2011). Derivation of the simple linear relationship for P_m used in this study is described in Souri et al. (2020). The recorded excess pore pressure ratios in centrifuge tests were used in this study. In practice, these values can be estimated from an effective stress site response analysis or from simplified correlations with the factor of safety against liquefaction.

3.2 Pseudo-static Limit Equilibrium Analysis (Step 2)

In the second step, pseudo-static limit equilibrium analysis is performed with the primary goal of calculating the yield acceleration (*ky*), defined as the horizontal downslope acceleration that results in a factor of safety against slope failure that is equal to 1. The limit equilibrium analysis was performed using the program SlopeW (Geo-Slope 2021). The Spencer method of slices was used in this analysis. Both non-circular and circular failure surfaces were analyzed, and the failure surface that generated the lowest yield acceleration under seismic loading was considered the critical failure surface and was used in the analysis. Figure 3 shows the soil profile modeled in SlopeW and the calculated failure surface. In this study, the centrifuge container walls were used as the boundaries for the failure surface. When modeling field conditions, the boundaries should be extended far enough to prevent boundary effects on the failure surface.

The beneficial resistance of piles against the laterally moving ground (pile pinning effect) was considered by including the piles as reinforcement elements in the limit equilibrium analysis. The pile shear resistance values were extracted from the p-y analysis performed using LPILE in Step 1. Pile spacing is an input parameter in the limit equilibrium analysis that defines the tributary out-of-plane width for the pile resisting forces (i.e., foundation influence zone). The extent of the foundation's influence zone on the laterally spreading ground depends on the width and length of the laterally spreading ground, as explained in Brandenberg and Turner (2017). For bridge structures, Caltrans (2016) recommends using an effective width based on the width and height of the embankment or 1.5 times the foundation width in a case with no embankments, where the width of the failure mass is much larger than the foundation width. In the centrifuge test analyzed in this study, the effective width was assumed to be equal to the width of the centrifuge container. This assumption is supported by the dissected data measured after the test, which showed fairly uniform deformations across the width of the centrifuge container. Therefore, the pile spacing in the limit equilibrium analysis was set to 9.15 m; this value was calculated by dividing the effective width (27.45 m) by the number of rows of piles (n = 3). This analysis resulted in a reasonable agreement between the estimated and measured ground deformations, as described later. Sensitivity analysis showed that setting the pile spacing in the limit equilibrium analysis to be equal to the actual center-to-center spacing of the piles in the out-of-plane direction (i.e., 6.1 m) resulted in an underestimation of the ground deformations.

The shear strength in the liquefiable sand layer in the limit equilibrium analysis was determined based on the equivalent friction angle determined by Ebeling and Morrison (1992). In this method, the equivalent friction angle is calculated by reducing the drained friction angle with a factor proportional to the average pore water pressure ratio within the soil unit, as recorded in the centrifuge test (R_u = 80% and 24% in loose and dense sand, respectively, in the test analyzed in this study). The drained friction angles for Nevada sand with different relative densities were obtained from McCullough et al. (2000) and are listed in Table 1. For comparison, the shear strength profile calculated using the Ebeling and Morrison (1992) method is plotted against the residual undrained shear strength of liquefied soil (*Sr*) calculated using the methods of Stark and Mesri (1992), Seed and Harder (1990), Idriss (1999), Olson and Stark (2002), Kramer and Wang (2007), and Idriss and Boulanger (2008), as indicated in Figure 4. This figure shows that for the average R_u values experienced in this test, the equivalent friction angle method results in higher shear strengths compared to all other *Sr* methods. This conclusion is generally valid for R_u values smaller than 90%. For the test analyzed here, the effective friction angle method resulted in a reasonable estimate of ground deformations when compared to the value measured in the centrifuge test.

3.3 Newmark Sliding Block Analysis (Step 3)

In the third step, permanent soil displacements were calculated using Newmark sliding block analysis, which was conducted using the program SLAMMER (Jibson et al. 2013). The yield acceleration (*ky*) was determined from the pseudo-static limit equilibrium analysis (Step 2). The uncoupled, rigid block methods were utilized for performing the Newmark analysis. Newmark analyses are typically performed in practical applications using accelerations that are obtained from site response modeling; however, in this study, the recorded accelerations from the centrifuge tests were used as input in the Newmark analysis. As a result, the uncertainties in ground motion estimation associated with site response analysis are minimized. Figure 5 shows the input acceleration and accumulated soil displacement time histories using a yield acceleration of *ky* = 0.053 as an example. The acceleration used in this example is recorded within the failure mass (labeled as "Acc 3164"), as indicated in the schematic inside the figure. As can be noticed from this figure, the Newmark displacements incrementally accumulate at every large bayward cycle that exceeds the yield acceleration. The soil displacement estimated using the Newmark method in Step 3 is used as the input soil displacement imposed on the p-y analysis in Step 1. The iterative process continues until convergence in soil displacements is achieved.

4.0 RESULTS

4.1 Pile Pinning Curves

The iterative analysis described in the previous section can be performed by developing the pile pinning curve shown in Figure 6 for the test analyzed in this study. The figure includes the following relationships: (a) pile shear resistance shows the summation of pile shear forces for all piles that pass through the failure surface developed from the p-y analyses in LPILE for a range of imposed soil displacements as described in Step 1, and (b) sliding mass shear resistance used in the limit equilibrium analysis (Step 2) versus estimated slope displacement using Newmark block analysis (Step 3). The point where the two curves intersect corresponds to the converging solution from the iterative process described earlier.

To account for the flexibility of the failure mass and the spatial distribution of accelerations, the Newmark analysis was performed using all acceleration time histories that were located inside the failure mass. The failure surface was determined from the limit equilibrium analysis and was found to be generally in agreement with the inferred failure surface from the experimental data. Figure 6a shows the failure surface inferred from the centrifuge measurements (shown as a red dashed line) and the failure surface determined from the limit equilibrium analysis (shown as a black dashed line). The inferred failure surface from the centrifuge test was determined from the peak transient soil displacement profiles that were back-calculated by double integration of the acceleration time histories at the locations of the accelerometers. The location of the seven accelerometers within the failure mass are shown in Figure 6a. At each iterative step, the Newmark displacements were calculated based on the mean as well as the mean plus one standard deviation (σ) of the seven accelerations in order to evaluate the uncertainties in the predicted displacements. These two estimates (mean and mean + 1σ) are shown as two different curves for the sliding mass in Figure 6b. In the test analyzed here, the predicted slope displacements are 7.3 cm and 10 cm for the median and the median + 1σ, respectively. It is useful to note that the measured permanent slope displacement in the centrifuge test was 9 cm. However, the measured transient slope displacement was larger, as will be discussed in the following section.

It is worth noting that the displacement corresponding to a resisting force of zero (i.e., the intersection of the sliding mass shear resistance curve with the horizontal axis) indicates the free field displacements without considering the pile pinning effects (i.e., 19 cm and 32 cm corresponding to the median and the median + 1σ estimate, respectively). The significant difference between the free-field slope displacements and the pile-restraint displacements highlights the benefits of including the pile pinning effects in the analysis.

4.2 Residual versus Permanent Soil Displacements

The time history of measured soil displacement in the centrifuge test is plotted in Figure 7 for a point located upslope and behind the deck. Therefore, this displacement represents the pile-restrained slope displacements. Also plotted in this figure are the soil displacements calculated from the Newmark analysis using the yield acceleration of $ky = 0.053$ corresponding to the iteration after which convergence is achieved. The Newmark displacement time histories are plotted for the seven accelerometers within the failure mass. The permanent measured slope displacement is 9 cm, while the measured peak displacement during ground motion (i.e., the transient displacement) is 12 cm (a factor of 33% larger). The estimated Newmark displacements ranged between 1 cm and 13 cm with a median value of 7.3 cm and a median + 1σ value of 10 cm. This comparison suggests that the permanent soil displacements are better estimated using the median displacements from the Newmark analysis and that the peak transient displacements are better predicted using the median +1σ value from the Newmark analysis. This conclusion is generally consistent for four other centrifuge tests on pile-supported wharves and when using multiple ground motions applied in each centrifuge test (a total of 10 experiments). Details for these comparisons can be found in Torkelson (2021). For comparison, the median +1σ displacements estimated from the prediction models of Bray and Travasarou (2007) are approximately 1.9 times larger than the median values.

4.3 Spatial Distribution of Estimated Soil Displacements

The selection of acceleration time histories within the failure mass is further examined by comparing the Newmark displacements calculated using the data from all the accelerometers in the centrifuge test. Figure 8 shows the spatial distribution of the estimated Newmark displacements. Considering that the measured permanent displacement was approximately 0.1 m, the contour corresponding to 0.1 m delineates the accelerometers that over- or under- predict the permanent soil displacements. It is observed that the failure surface determined from the limit equilibrium analysis (shown as a red dashed line) reasonably agrees with the experimental data (indicated by the contours that show a rapid reduction in soil displacements). It is noted that using the accelerations at the base of the model (i.e., those within the dense sand) will result in an underestimation of the measured slope displacements and that using the accelerations at the ground surface will result in an overestimation of the measured slope displacements. Therefore, it is recommended to use the accelerations estimated within the failure mass as a basis for performing the Newmark analysis.

5.0 CONCLUDING REMARKS

The results of a centrifuge test on a pile-supported wharf subjected to liquefaction-induced lateral ground deformation were used to evaluate the accuracy of a Newmark sliding block analysis in estimating the kinematic demands on piles. The shear resistance of the piles in constraining the ground deformations (i.e., the pile pinning effects) were incorporated in the limit equilibrium analysis using the shear forces in piles determined from the p-y analysis in LPILE. The analyses and evaluations performed in this study support the use of guidelines that have been primarily developed by highway bridge agencies (i.e., Caltrans 2012) in estimating kinematic demands on maritime structures (e.g., piers and wharves). The following observations should be considered in the design:

- Neglecting the beneficial resistance of pile pinning effects results in grossly overestimated slope displacements. For the centrifuge test examined here, where the average R_u in the loose and dense sand layers were smaller than 90%, the shear strengths in sand layers were calculated using the equivalent friction angle (Ebeling and Morrison 1992) and were used in the pseudo-static limit equilibrium analysis, which provided a reasonable estimate of the displacement measured in the centrifuge test.
- The results of the analyses showed that the permanent end-of-shaking displacements in the centrifuge test were better estimated using the median Newmark displacements, while the peak transient displacements were better estimated using the median + 1σ Newmark displacements. The measured peak transient displacement was 1.3 times larger than the measured permanent displacement.
- It was found that using the accelerations at the base of the model (i.e., within the dense sand) in Newmark analysis resulted in an underestimation of the measured displacements and that using the accelerations at the ground surface resulted in an overestimation of the measured displacements. Therefore, it is recommended for design to use the accelerations estimated within the failure mass as a basis for performing Newmark analysis.

Uncertainty associated with estimation of soil accelerations and cyclically induced pore water pressures were not considered in this study. The results presented in this paper are based on the properties of the analyzed centrifuge test.

6.0 ACKNOWLEDGEMENTS

The authors would like to thank Dr. Steve Dickenson, Dr. Nason McCollough, and Mr. Scott Schlechter for sharing their centrifuge test data. Support for the recent analyses of the centrifuge tests was provided by Grant No. CMMI-1761712 from NSF and Grant No. CPF-2021-SSMC-1 from the Deep Foundations Institute. Any opinions, findings, and conclusions or recommendations expressed in this article are those of the author(s) and do not necessarily reflect the views of the funding agencies.

7.0 REFERENCES

AASHTO, (2014). "Guide Specifications for LRFD Seismic Bridge Design." Second Edition with 2014 Interim, AASHTO, Washington, D.C.

Armstrong, R.J., Boulanger, R.W., (2014). "Equivalent Static Analysis of Piled Bridge Abutments Affected by Earthquake-Induced Liquefaction.", Journal of Geotechnical and Geoenvironmental Engineering, Vol. 140 (1): 04014046.

American Petroleum Institute, API, 1993. Recommended practice for planning, design, and constructing fixed offshore platforms. API RP 2A–WSD, 20th Ed., API, Washington, D.C.
ASCE/COPRI 61-14, (2014). "Seismic Design of Piers and Wharves.", prepared by the ASCE Standards Comm. on Seismic Design of Piers and Wharves." ASCE, Reston, VA.
Bray, J.D., Travasarou, T., (2007). "Simplified Procedure for Estimating Earthquake-Induced Deviatoric Slope Displacements." Journal of Geotechnical and Geoenvironmental Engineering, Vol. 113, (4): 381–392.
Caltrans (California Department of Transportation). 2012. Guidelines for Foundation Loading and Deformation Due to Liquefaction Induced Lateral Spreading. Sacramento, CA: Caltrans.
Caltrans (California Department of Transportation). 2016. Bridge Memo to Designer (MTD) 20-4: Seismic Retrofit Guidelines for Bridges in California. Sacramento, CA: Caltrans.
Cubrinovski, M., Bray, J. D., de la Torre, C., Olsen, M. J., Bradley, B. A., Chiaro, G., Stocks, E. and L. Wotherspoon. 2017. "Liquefaction effects and associated damages observed at the Wellington Centreport from the 2016 Kaikoura earthquake." Bull. N. Z. Soc. Earthq. Eng., 50 (2): 152–173.
Ebeling, R. M., and E. E. Morrison, Jr. 1992. The Seismic Design of Waterfront Retaining Structures. Tech. Rep. No. ITL-92-11. Port Hueueme, CA: U.S. Naval Civil Engineering Lab.
Ensoft, (2014). "User's Technical Manual for LPILE".
Federal Highway Administration (FHWA), 2011. LRFD seismic analysis and design of transportation geotechnical features and structural foundations. Publication No. FHWA-NHI-11-032, GEC No. 3. Washington, D.C.: U.S. Department of Transportation.
Geo-Slope International Ltd, (2021) "Stability Modeling with GeoStudio".
Hamada, M., Yasuda, S., Isoyama, R., and Emoto, K. 1986. ''Study on liquefaction induced permanent ground displacements.'' Research Rep., Association for Development of Earthquake Prediction, Japan, November, 87.
Idriss, I. M., 1999. An update to the Seed-Idriss simplified procedure for evaluating liquefaction potential, in Proceedings, TRB Workshop on New Approaches to Liquefaction, Publication No. FHWA-RD-99-165, Federal Highway Administration, January.
Idriss, I. M., & Boulanger, R. W. (2008). "Soil Liquefaction During Earthquakes". Earthquake Engineering Research Institute.
Jibson, R. W., Rathje, E. M., Jibson, M. W., and Lee, Y. W., (2013). "SLAMMER – Seismic Land Movement Modeled using Earthquake Records", U.S. Geological Survey Techniques and Methods, book 12, chap. B1, unpaged.
Kramer, S. Wang, C.H., (2007). "Estimation of the Residual Strength of Liquefied Soil", in preparation.
Liu, L., and Dobry, R., 1995. Effect of liquefaction on lateral response of piles by centrifuge model tests. NCEER Bulletin, 91, 7–11.
Makdisi, F. I., & Seed, H. B. (1978). "Simplified procedure for estimating dam and embankment earthquake-induced deformations". Journal of the Geotechnical Engineering Division, 104(7), 849–867.
McCullough, N.J., Dickenson, S.E., Kutter, B.L. and Wilson, D.W., (2000). "Pile-Supported Wharf — Centrifuge Model NJM01." Report No. GEG01-2000, OSU-Geotechnical Engr.
Newmark, N.M. 1965. "Effects of Earthquakes on Dams and Embankments." Geotechnique 15 (2): 139–160.
Olson, S. M., & Stark, T. D. (2002). Liquefied strength ratio from liquefaction flow failure case histories. Canadian Geotechnical Journal, 39(3), 629–647.
Seed, R. B, and Harder, L. F., 1990. SPT-based analysis of cyclic pore pressure generation and undrained residual strength, in Proceedings, Seed Memorial Symposium, J. M. Duncan, ed., BiTech Publishers, Vancouver, British Columbia, pp. 351–76.
Souri, M., Khosravifar, A., Schlechter, S., McCullough, N. & Dickenson, S. E. (2020), "Development of Experimental P-Y Curves from Centrifuge Tests for Piles Subjected to Static Loading and Liquefaction-Induced Lateral Spreading", DFI Journal The Journal of the Deep Foundations Institute: Vol 14, Issue 1.

Stark, T. D., & Mesri, G. (1992). Undrained shear strength of liquefied sands for stability analysis. Journal of Geotechnical Engineering, 118(11), 1727–1747.
Torkelson, Thomas John II, (2021), "Kinematic Demands on Pile-Supported Wharves Due to Liquefaction-Induced Lateral Ground Deformations" M.S. Dissertation, Portland State University.
Turner, B., Brandenberg, S. J., and Stewart, J. P. (2014). "Evaluation of collapse and non-collapse of parallel bridges affected by liquefaction and lateral spreading." PEER Rep. 2014/10, Pacific Earthquake Engineering Research Center, UC, Berkeley, CA, 122.

Table 1. P-Y relationship properties and multipliers

Soil unit	Modulus of subgrade reaction	Depth-dependent ultimate strength (P_{ult}) based on	Excess PP Ratio, Ru (%)	P-multiplier
Loose sand (D_R = 39%)	3500 kN/m^3	API Sand with $\phi = 33°$	80	0.32
Dense sand (D_R = 82%)	3500 kN/m^3	API Sand with $\phi = 37°$	24	0.94
Rockfill	5200 kN/m^3	API Sand with $\phi = 45°$ and pseudo-cohesion = 15 kPa	0	1.0

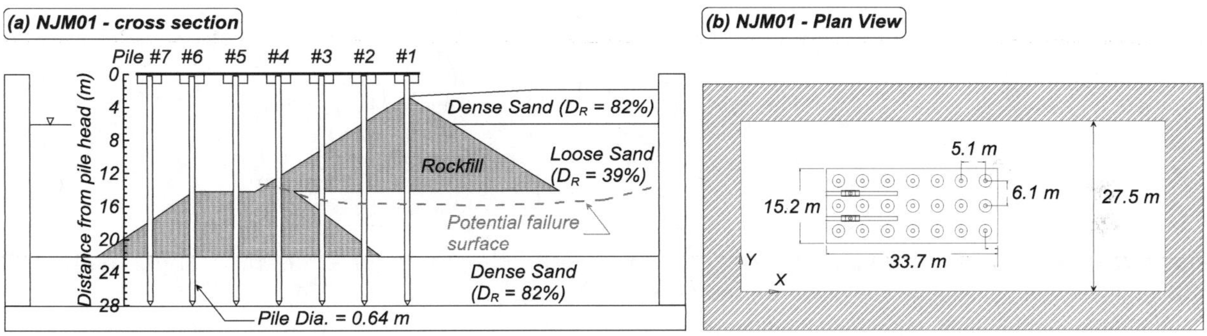

Fig. 1. Centrifuge test NJM01 layout properties: (a) Cross section, and (b) plan view

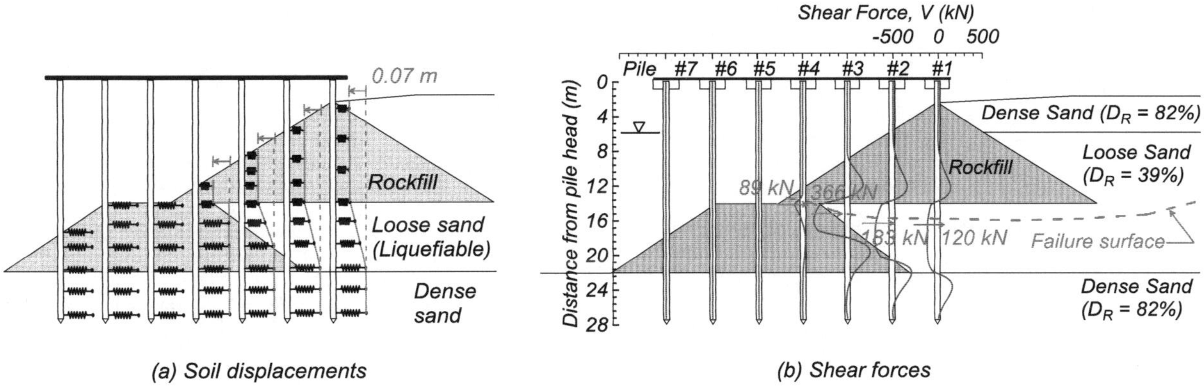

Fig. 2. (a) soil displacements applied to the end node of p-y springs, (b) shear forces along the piles and at intersections with the failure surface

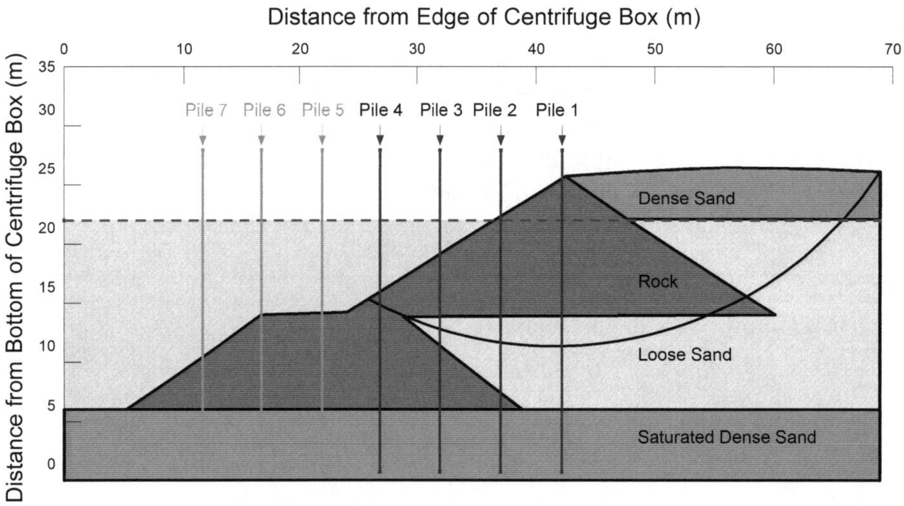

Fig. 3. Pseudo-static limit equilibrium analysis model in SlopeW incorporating shear resistances in piles

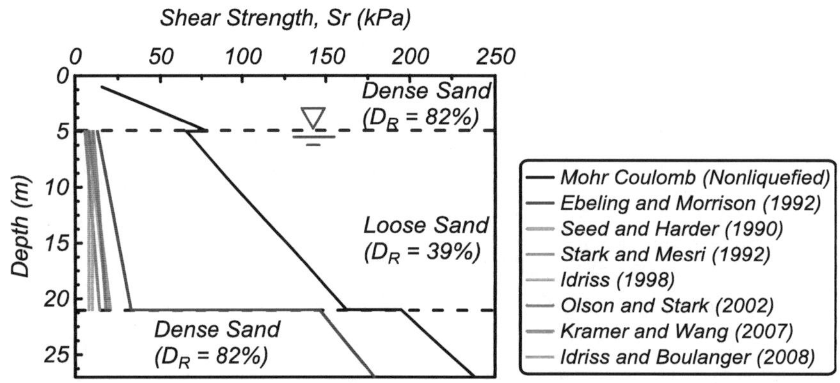

Fig. 4. Comparison of the soil shear strength with depth using the effective friction angle method and residual undrained shear strength methods

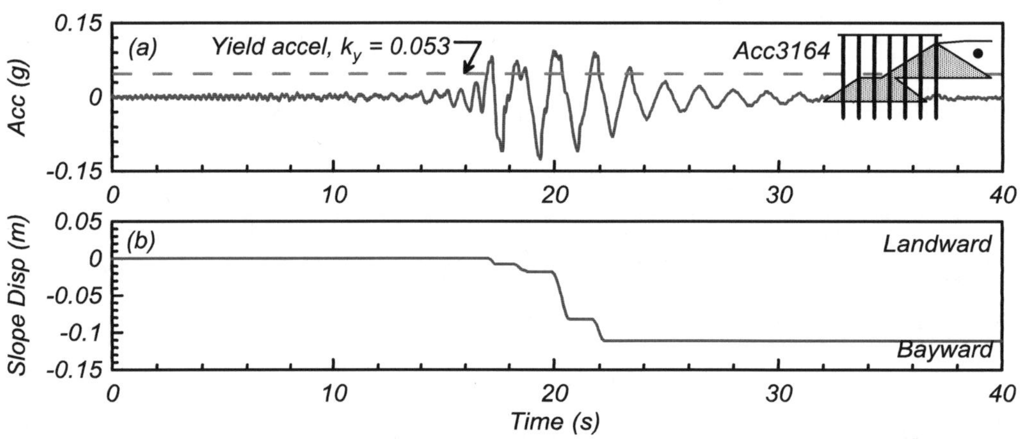

Fig. 5. Time histories of (a) recorded acceleration within the failure mass and (b) calculated displacement using Newmark sliding block analysis

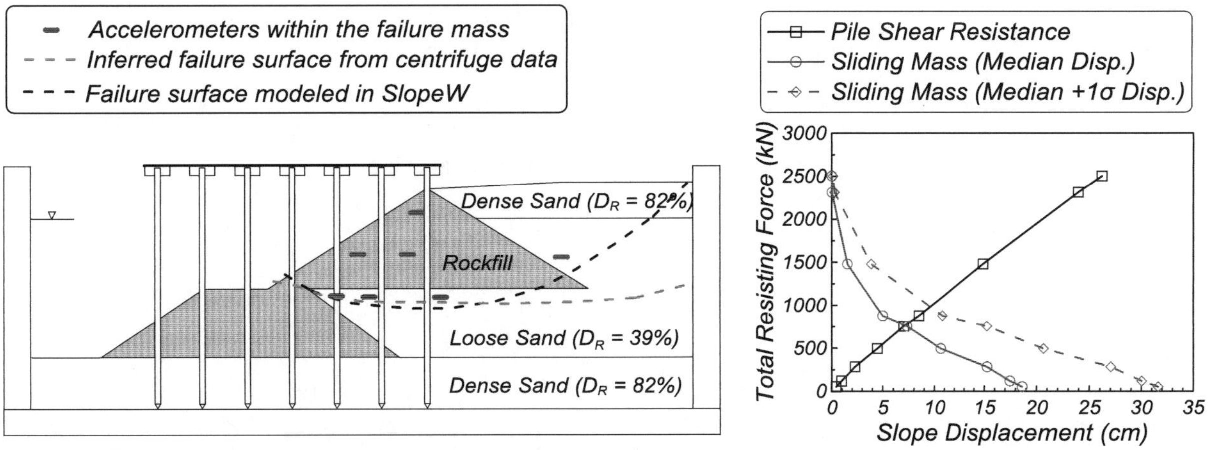

Fig. 6. (Left) The location of accelerometers within the failure mass, (Right) Pile pinning curves from sliding mass and pile shear resistance

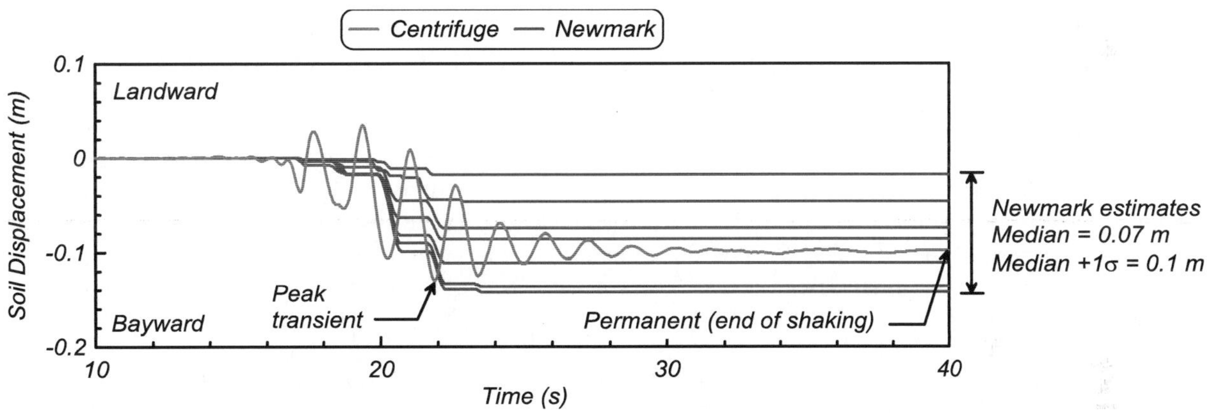

Fig. 7. Displacement time histories of recorded soil displacements and estimated Newmark displacements

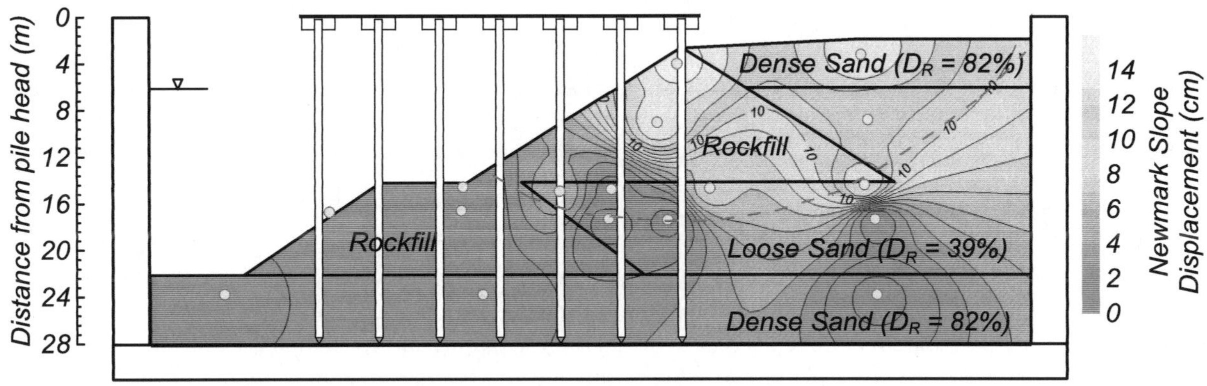

Fig. 8. Newmark displacements from all the accelerometers in the centrifuge model

ESTIMATION OF WIND TURBINE FOUNDATION ROTATIONAL STIFFNESS BASED ON FULLY NON-LINEAR DYNAMIC ANALYSIS

Spencer E. Ness, P.E. University of Notre Dame, Notre Dame, IN, USA, sness@nd.edu Yazen Khasawneh, Ph.D. P.E., University of Notre Dame, Notre Dame, USA, IN, ykhasawn@nd.edu Will Tjaden, University of Notre Dame, Notre Dame, IN, USA wtjaden@nd.edu Mohammed Nasim, Ph.D. P.E., Terracon Consultants, Germantown, MD, USA, mohammad.nassim@terracon.com

ABSTRACT

Wind turbine components undergo many loading cycles during their operational life. Additionally, wind turbines may undergo extreme loading conditions from natural hazards, mechanical failures, or emergency stops. Current design practice for the foundations of wind turbines accounts for these complex loading conditions through a simplified pseudo-static approach based structural loading combinations. A rotational stiffness for the foundation is estimated from the pseudo-static analysis and then compared to an acceptable minimum rotational stiffness provided by the turbine manufacturer. These minimum rotational stiffnesses are based on dynamic analysis of the tower and nacelle system with the foundation represented as linear springs. This approach provides a reasonable approach to the foundation design but does not accurately capture the true dynamic performance of the entire soil-foundation-tower-nacelle system. This becomes especially important for pier-type foundations, which are being more commonly used to support modern and future generations of larger turbines. Additionally, the dynamic performance of the foundation will vary as the adjacent soil undergoes a large number of loading cycles of varying strain magnitudes. In this study, the full turbine-tower-foundation-soil system was modeled using finite element methods. An advanced non-linear soil constitutive model and dynamic loading time histories were used to investigate changes in stiffness of the soil, and the possible impact on the dynamic response of the system. Results were compared to conventional design methodologies with the goal of reducing foundation capital costs while avoiding increases in O&M costs and which can lower the LCOE of wind energy.

Keywords: Wind turbine foundation, wind energy, numerical modeling, dynamic analysis, 3D FEM, pier foundation, intermediate foundation, cyclic degradation,

INTRODUCTION

Wind energy is a key technology in the effort to decarbonize energy generation. While the environmental impacts of energy sources is a key consideration, utility scale power generation is often driven by economics. The key economic metric in comparison of energy sources is the levelized cost of energy (LCOE). LCOE includes the capital costs, operations and maintenance (O&M) costs, fuel costs, utilization rates and other factors to assess the net-present value of the energy that a generation source will produce over its lifetime. The LCOE of wind energy is driven primarily by capital costs as there are no fuel costs and O&M costs are low unless there are premature issues with components of the turbine-tower-foundation system. Some of these premature issues have arisen with foundations and foundation soils which have caused original equipment manufacturers (OEMs) of the turbines, wind energy developers, and wind energy operators to take conservative approaches to foundation design.

Many of these foundation issues are driven by the dynamic nature of the loads in the system. As discussed in other literature (Khasawneh et al., 2021), the nature of the mechanical dynamic loads, typically identified as the 1P (rotor revolution frequency) and 3P (blade shadowing frequency, equal to three times the 1P frequency) frequencies. The key challenge of wind turbine design is that the natural frequency of the overall system is typically very close to these frequencies leading to significant concerns of resonant motion and amplification of the mechanical loading. Further complicating the issue is that the 1P and 3P frequencies are not singular frequencies that can be targeted but instead a function of the operating speed of the turbine which can vary from 10-20 RPM in utility scale turbines. This creates bands of 1P and 3P frequencies that must be considered in the design of the system. Figure 1 below highlights these operating frequencies along with the wind and wave frequency spectrums often used in the design of offshore wind structures.

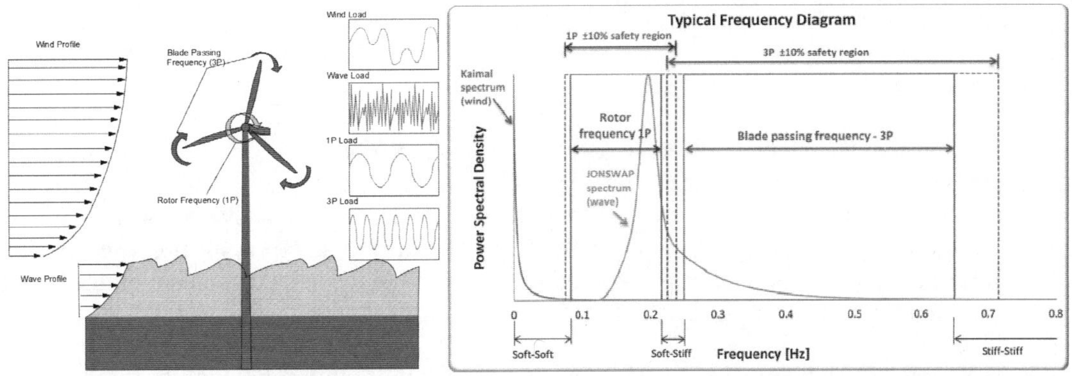

Fig. 1. Typical Loads and Load Frequencies for an offshore wind turbine. Taken from Halici (2016, left) and Arany et al. (2014, right).

It is not feasible to design and construct a turbine-tower-foundation system that is stiff enough to have a natural frequency above the 3P frequency range and a system that was flexible enough to have a natural frequency below the 1P frequency range would have significant deflections which would impact the mechanical elements of the power generation components.

These dynamic problems have been assessed and studied by many researchers and members of the wind energy community (Khasawneh et al., 2021, Halici, 2016, Arany et al., 2014). Typically to address this issue the problem has been simplified to address computational and analytical complexity. It is well documented that machine foundations will lead to a portion of the adjacent soil acting as an effective mass with the foundation (Richart et al., 1970). This is also the case with the shallow/gravity and intermediate foundations often used to support wind turbines. This creates an overall system consisting of the effective adjacent soil body, foundation, tower, and turbine. The turbine includes the blades, hub, rotor, nacelle, and mechanical energy generation equipment. This creates a system of highly complex interactions of structural and mechanical components that is then founded on often complex soil profiles. The turbine OEMs often simplify this complex and dynamic system with a simple beam-mass system supported on a series of translational and rotational springs to represent the soil and foundation systems. While this simplified approach can capture the general behavior of the turbine and tower, it has significant shortcomings in addressing the non-linearity of the soil behavior which can be sensitive to frequency contents, strain accumulations, and stiffness degradation. Figure 2 below illustrates the simplified approach in comparison with the actual system.

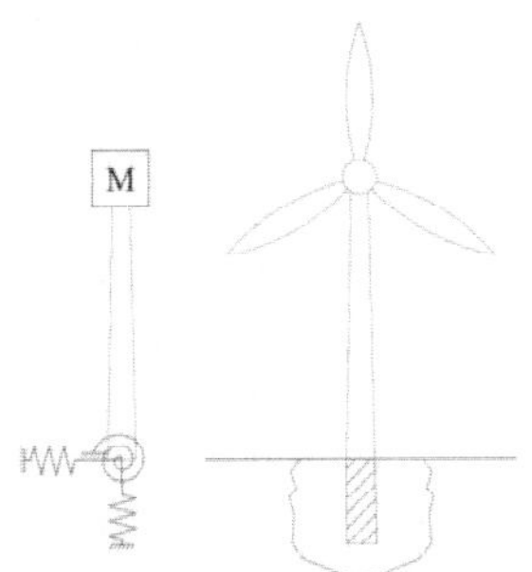

Fig. 2. Illustration of typical beam mass system used in frequency analysis of wind turbines compared to theoretical system.

A key component of these simplified models is the stiffness of the rotational spring which has a significant role in the determination of the natural frequency of the beam-mass system. OEMs tend to target rotational stiffnesses very close to a fixed condition. This leads to significant design requirements for the foundation designer. These complex motions and demands are typically reduced to pseudo-static applied moment applied to a finite element model of the foundation. Additionally, it is common for elastic-perfectly plastic soil models to be used in this modeling which leads to further issues with capturing the behavior of the soil. Each of these simplifications requires conservative assumptions which can compound and lead to an overall system that can be significantly overdesigned.

To evaluate the long-term foundation stiffness degradation, a simplified methods were proposed by Vitali, et al. (2021 and 2022) for both fine-grained and coarse-grained foundation soils. The proposed simplified method utilizes the mobilized cyclic loads and number of cycles for cohesive soils and utilizes the mobilized shear strain and number of cycles for granular soils. In the Vetali et al. proposed method the cyclic loads were applied monotonically using modified Cam-Clay model for the fine-grained soils and the UBCSAND model for the coarse-grained case.

This study evaluates the use of the pseudo-static approach and the rotational stiffness as design methodology as a means to avoid resonance and amplification conditions.

MODEL DESIGN AND PROCEDURE

To better assess the performance of the soil-foundation-tower-turbine system, a more advanced approach was taken to model the entire soil-foundation-tower-turbine system and use dynamic acceleration time histories to capture more accurately the performance of the system. To simplify the geometry of the system the blades, hub, rotor, and nacelle were reduced to a mass applied to a volume representing the nacelle dimensions. The foundation and tower dimensions, nacelle masses, and acceleration time histories were based on a 2.0 MW turbines, with the overall model composited from several turbines at different sites. The commercially available Plaxis3D® was used to perform the finite element modelling and analysis. As the primary focus of this work is the performance of the foundation and adjacent soil, Plaxis3D® was selected given its optimization for soil-structure interaction problems while its relative limited libraries for steel or concrete materials will be addressed further below.

An acceleration time history taken from a similar in-service tower was then applied to the system as an acceleration to the nacelle mass. This time history was applied for 60 seconds of simulated time and the system was left in free vibration to assess its natural frequency. Both prior to the application of the acceleration time history and after the free vibration period, the foundation's rotational stiffness was measured in a static calculation phase. Dynamic rotational stiffnesses were measured based on the performance of the model during the acceleration application. The foundation dimensions were then modified, and the process repeated to assess how changes in the rotational stiffness of the foundation impacted the dynamic performance of the entire system.

Foundation Description

With the growth of turbine sizes, many modern turbines are founded on intermediate-type foundations which leverage more capacity of the soil compared to shallow foundations but do not have the costs of a deep foundation. A pier type foundation was used for this analysis. The dimensions and material properties are described in Table 1. An elastic material model was used for the concrete used in the pier. It was assumed that for the strain level the foundation would undergo, an elastic model would effectively estimate the material performance. Reinforcing steel was neglected for this analysis and the foundation material was based purely on the properties of concrete. As discussed above, the depth of the foundation was varied to assess how accurately rotational stiffness measurements captured dynamic performance.

Table 1. Foundation dimensions and material properties

Parameter	Value
Pier Inner Diameter	16 ft
Pier Outer Diameter	12 ft
Pier Wall Thickness	4 ft
Pier Length (variable)	32 ft, 28 ft, 24 ft, and 20 ft
Material Model	Linear Elastic
Unit Weight, γ	150 pcf (0.15 kcf)
f'_c of Concrete	6000 psi (6 ksf)
Elastic Modulus	4,415,280 psi (635,800 ksf)
Poisson's Ratio	0.15

Ground Profile

A ground profile was selected of a single granular material for simplicity in modeling. Additionally, no groundwater was considered in this analysis (dry condition). The Hardening Soil with Small Strain (HSSS) model was selected for use with the model. While the original Hardening Soil model has shortcomings in capturing the response under cyclic loading, the HSSS performs better under cyclic loading by capturing the hysteresis in the loading cycles. The properties adopted for the HSSS model are presented in Table 2.

Table 2. Soil properties for HSSS Model

Parameter	Value
Unit Weight	125 pcf
Initial Void Ratio, e_{init}	0.50
E_{50}^{ref}	350 ksf
E_{oed}^{ref}	350 ksf
E_{ur}^{ref}	1050 ksf
Power, m	0.50
Effective Cohesion c'_{ref}	0 ksf (granular)
Effective Friction Angle, φ'	35°
Dilatancy Angle, ψ	5°
Small Strain, $\gamma_{0.7}$	5×10^{-5}
Small Strain Shear Modulus, G_0^{ref}	440 ksf
Poisson's Ratio	0.20
K_0	0.4264

Tower and Nacelle Mass Modeling

As previously discussed, the blades, hub, rotor, nacelle, and all mechanical elements present at the top of the turbine were modelled as a mass applied to a volume with the general dimensions of a typical 2.0 MW nacelle. This simplification reduced the complexity of the geometry of the system and related meshing issues while also reducing the number of materials that would need to be considered which aided in model development and computational effort. This lumped mass was modeled as a linear elastic material with a very high stiffness to avoid any excess deformations of this mass under the acceleration which may have impacted the results. While these simplifications do modify the mass distribution of the overall system, the simplifications were deemed appropriate given the diminishing returns of modeling the entire system including mechanical elements.

The tower was modeled as a 2-dimensional shell element. The shell thickness was determined based on available information on typical tower wall thicknesses. A single shell thickness was used for the full height of the tower when this would typically vary to better optimize the tower's material consumption. Additionally, no internal stiffening of the tower was considered. Typical values for steel density and stiffness were used in a linear elastic material model. Fully tied conditions were used at the tower-nacelle and foundation-tower connections as these connections are well-designed and for typical operating conditions and would not be anticipated to be overstressed or have plastic deformations under the loading conditions.

Model Dimensions, Boundary Conditions, and Meshing

Given the dynamic nature of portions of this analysis, model parameters were of particular importance. The overall model dimensions were 100 ft x 100 ft x 100 ft. This provided several foundation diameters between the any edge of the foundation and the model boundary. Additionally viscous boundary conditions were applied on all lateral boundaries of the model to ensure that energy propagating from the foundation through the soil was not reflected into the model. Inspection of the model results shows little reflection or impacts from the boundaries.

Meshing was performed using Plaxis 3D's standard mesh generation of 10-node tetrahedral elements for volume elements and 12-node interface elements for modeling soil-structure interaction. The tower shell element was modeled as a 6-node plate (Brinkgreve, R. B. J., et al., 2012). Mesh fineness is varied throughout the model to optimize calculation time. An overall connectivity plot of the model is shown in Figure 3. along with additional images showing meshing details.

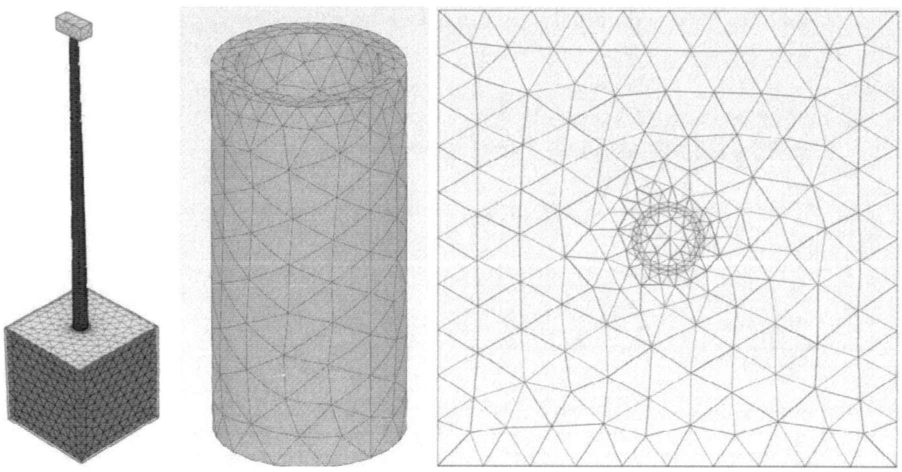

Fig. 3. Mesh Connectivity Plot of Entire Model, Pier Foundation, and Cross Section through Model at Top of Foundation

Model Phases and Loading

Phased loading and element activation were utilized to simulate typical portions of the life cycle of the turbine. The initial phase consists of only the soil body followed by foundation installation. Due to the short duration required to excavate for and install the foundation, an intermediate open-excavation phase was not considered. Prestressing is then applied to the pier foundation as surface loads on the top and bottom faces of the pier. This prestressing is typical for the pier-type foundations to avoid tension developing in the concrete pier walls. Following prestressing of the pier, the tower and nacelle mass are installed.

Following these construction phases, the acceleration time history is applied to center of the nacelle mass. This time history is applied for 60 seconds, approximately fifteen revolutions of the turbine. The acceleration has both and fore-and-aft and side-to-side component to represent the complex motion of the system. This allows for straining of the adjacent soil body and impacts to the effective soil mass's stiffness. The acceleration is then removed, and the model is allowed to free vibrate for 60 seconds.

Frequency Domain Analysis

During the analysis, various points on the system are monitored and displacement, velocity, and acceleration time histories are recorded. These include the bottom of foundation, top of foundation, midpoint of tower, and top of tower, among others. A particular focus is the top of foundation level as this is the level often used for rotational stiffness calculations in the pseudo-static approach. Time histories from this level are then passed through a Fast Fourier transform (FFT) to generate the response in the frequency domain. MATLAB software including the FFT function and Signal Analyzer app are utilized in this analysis.

Pseudo-Static Rotational Stiffness Calculation and Measurement

To assess the rotational stiffness, a typical operating moment, based on OEM guidance for 2.0MW turbine, is applied to the foundation. This applied moment already included P-Δ effects from the turbine mass however as the tower and turbine mass were present in this calculation, the P-Δ impacts were likely duplicated. Given the small magnitudes of displacement it was assumed that these impacts were negligible. Rotational stiffness was then calculated based on this applied moment and the rotation of the top of the foundation.

Dynamic Rotational Stiffness Calculation

In order to assess the dynamic rotational stiffness of the model, the forces being transferred from the nacelle motion into the tower and ultimately in to the foundation were extracted from the model. Forces in the tower shell were extracted for at the tower-foundation level and used to determine an applied moment. Deflection results from the dynamic loading phase were then used to calculate rotation at the top of the foundation.

RESULTS

As noted above, the procedure was repeated on pier foundations with varying lengths and constant diameters. Results were then compared for to assess changes in natural frequencies of the system with variation of dimensions.

Results for 60 seconds of Acceleration Time History

Figure 4 below displays the displacement time history for the downwind edge of the top of the foundation. The first 60 seconds correspond to the application of the load time history while 60-120 seconds correspond to free vibration.

Given the complex, multi-axis nature of the movement it is difficult to assess magnitudes of the motion, particularly in the 20 ft and 24 ft piers where there is a considerable increase in the vertical displacement when compared to the longer piers. However, in the free vibration period it is apparent that the amplitude of the vibration decreases with an increase in pier depth.

While the magnitudes of these displacement time histories are important in assessing impacts to the system, particularly the mechanical components, the designer must also focus on the frequency content of the motion to avoid the 1P/3P ranges. Figure 5 illustrates the frequency content of the motion of frequencies of interest, less than 1 Hz.

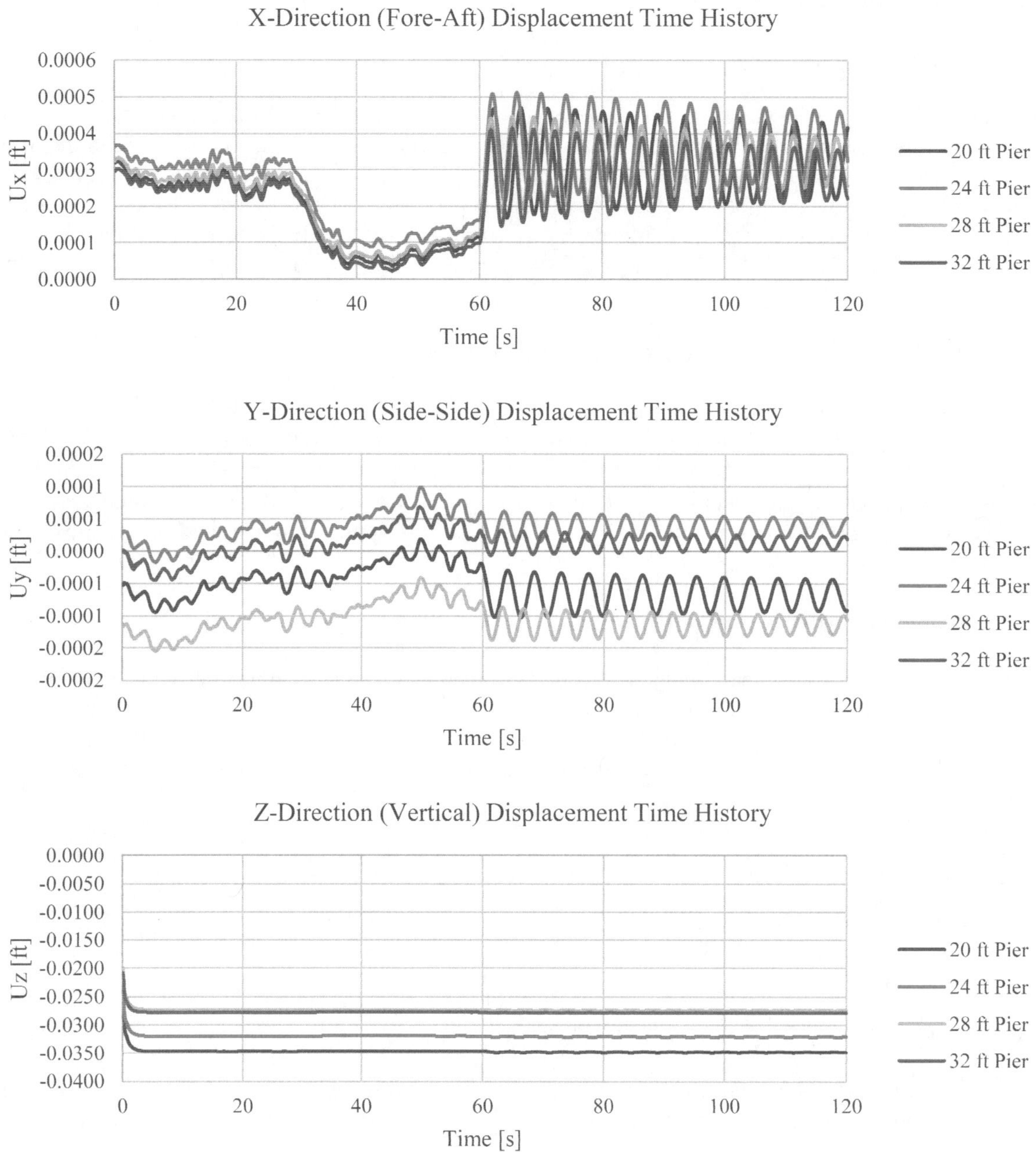

Fig. 4. Displacement Time Histories of Downwind Edge of Top of Foundation

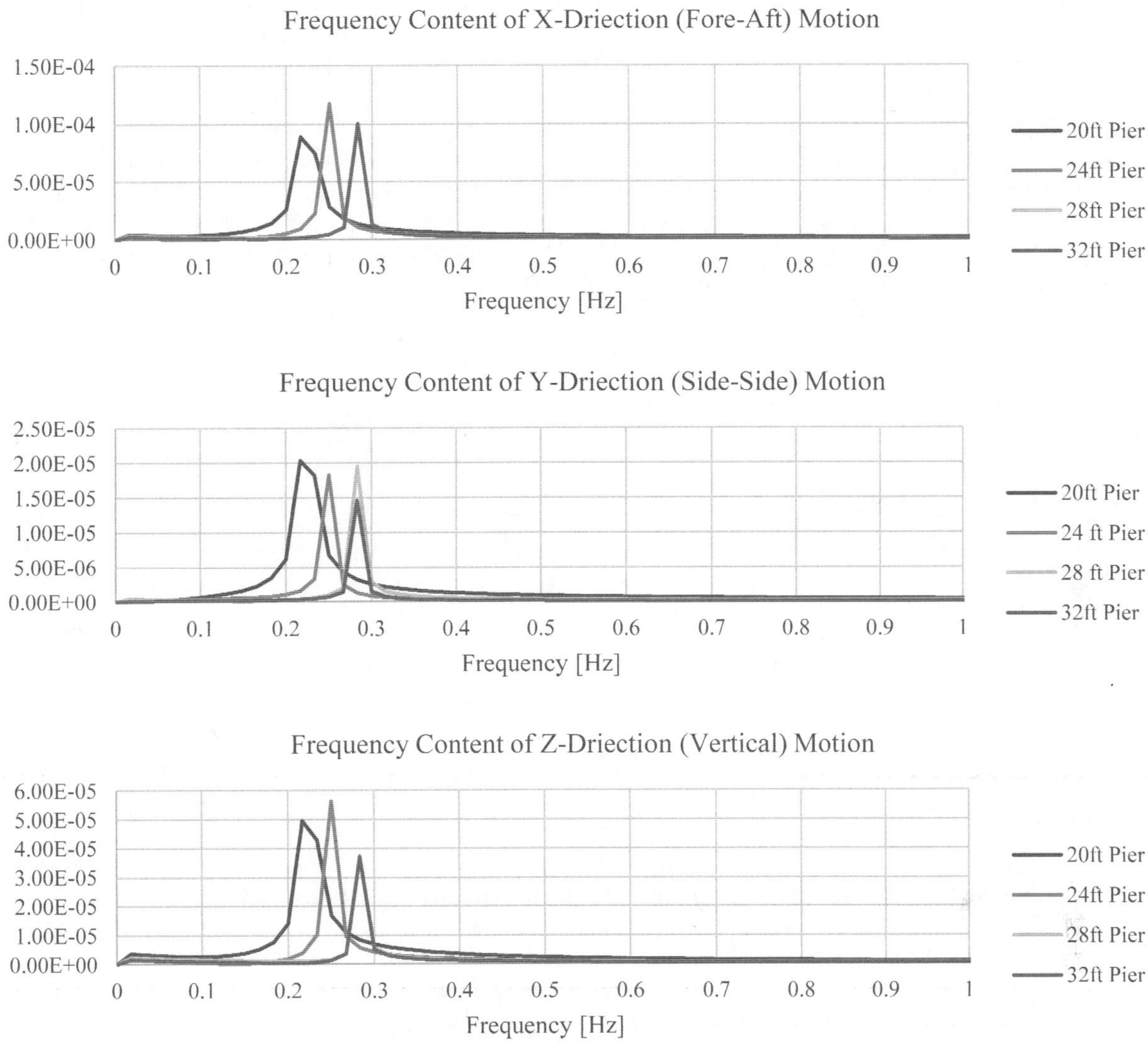

Fig. 5. Fourier Transforms of Frequency Content of Free Vibration Period

As Figure 5 displays, there is very little difference in the frequency content of the soil-foundation-tower-turbine system when comparing the 28 ft pier and the 32 ft pier. While the magnitude of motion is an important criterion given the sensitive nature of the electrical generation equipment, the results show that for this ground condition the foundation's length could be reduced a significant amount without real changes in resonance. The shorter piers (20 ft, 24 ft) behaved as expected with a decrease of the fundamental frequency of the system reflecting a softer foundation.

Pseudo-Static Rotational Stiffness Impacts

Rotational stiffness of the foundation was assessed prior to the application of the acceleration time history and then after the completion of free vibration. This allows for a comparison of the impacts of the applied cyclic loading to the foundation stiffness. Figure 6 below highlights the rotational stiffness results.

As expected, the foundation stiffness generally increases with an increase in the pier length. However, there is a reduction in stiffness between the 28 ft and 32 ft piers. It is expected that this is a result of the 32 ft pier's additional depth reaching a partial fixity condition. As the pier becomes more slender and reachs greater depths, partial fixities may develop changing the nature of the pier displacement from a rigid body to a cantilevered condition.

The rotational stiffness also increases from the pre-acceleration application condition to the post-acceleration condition. The soil profile is cohesionless and not very dense in it's initial state. As the motion from the nacelle acceleration propagates to the soil, the soil densifies and increases in stiffness. As this analysis was limited to a 60 second time history it is possible that further densification and increase in soil stiffness is possible. It should be anticipated that the cohesionless soil densification reaches steady state after about 5 to 10 cycles.

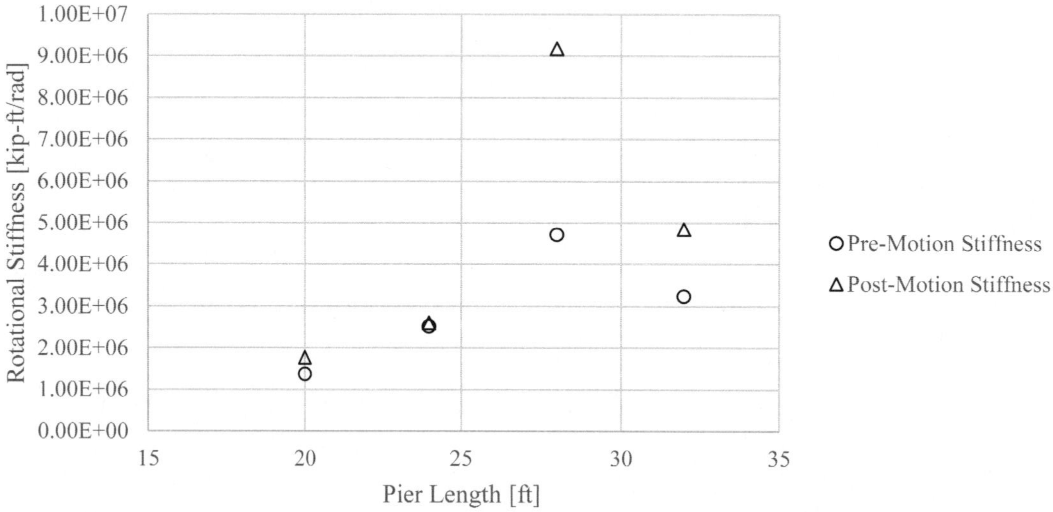

Fig. 6. Rotational Stiffness vs. Pier Length for Investigated Foundation Sizes

Dynamic Rotational Stiffness Results

To determine the rotational stiffness, it was first necessary to determine the dynamic applied moment applied to the foundation. As described above, this was accomplished by examining the forces in the tower at the tower base. The time history of these moments is shown in Figure 7 below. These values follow the expected trend that as the foundation length and therefore stiffness increases, the foundation attracts more moment demand. It also further highlights the similar performance of the 28ft and 32ft long foundations.

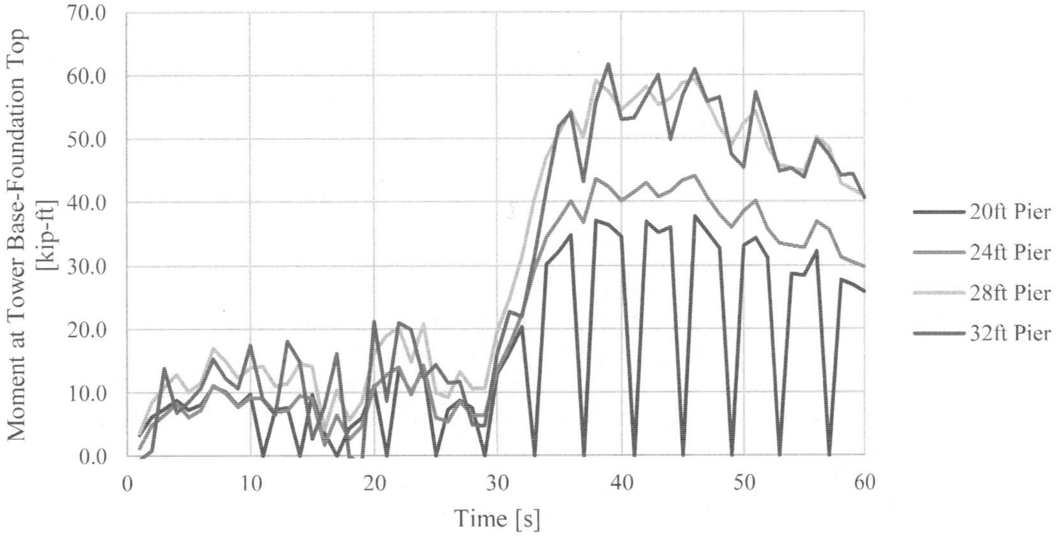

Fig. 7. Moment at Tower Base vs. Simulated Time

From these moments and the previously presented displacement data it is possible to calculate a dynamic rotational stiffnesses for the foundations. These are presented in Figure 8 below.

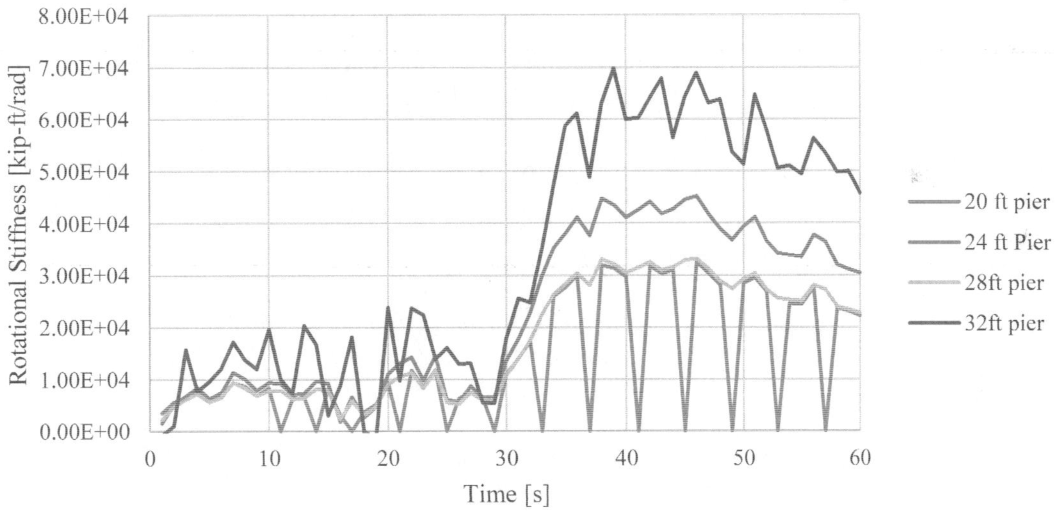

Fig. 8. Dynamic Rotational Stiffness Time Histories

From these results it is apparent that the reported dynamic rotational stiffness is quite sensitive to the moment demand being placed on the structure as the rotational stiffnesses vary closely with the applied moment. This reinforces that a linear spring does not capture the performance of the soil-foundation system. It should be noted that both the reported moments and rotational stiffnesses in the dynamic analysis are several orders of magnitude less than that of the pseudo-static analysis. This is not expected as the low applied moment would have been anticipated to return a larger rotational stiffness as the soil model should be stiffest at low strain levels.

Additionally, it is not apparent why the 28ft pier performed so poorly on the dynamic rotational stiffness analysis especially as it was a standout in the pseudo static analysis. It may be driven by the development of partial fixities as discussed in the previous section as this foundation develops nearly the same moment demand as the longer 32ft pier but has larger deflections. Further investigation into this phenomenon is warranted.

SUMMARY AND APPLICATION TO DESIGN PROCESS

While a simple linear rotational spring can be useful as a simplified or preliminary approach to the design of wind turbine foundations, it fails to capture many essential behaviors. The presented analysis showed that changes in pier foundation length could have limited impact on the system natural frequency. Additionally, the measurement and calculation of rotational stiffness will be impacted based on the underlying behavior of the foundation such as the development of fixity and the dominance of foundation deformation or rigid body motion in the overall behavior. Further, true dynamic behavior can vary in ways that are not well captured in pseudo-static approaches.

The use of non-linear constitutive models and dynamic loading analysis better capture the true behavior of the structure. While these approaches lead to greater computational effort and complex results, they can provide significant savings to energy developers and drive down the LCOE of wind energy. Future and ongoing work by the authors will focus on developing more simplified design guidelines to gain the benefits of these advanced analyses without requiring intensive computing for each project or turbine location.

ACKNOWLEDGEMENTS

The authors acknowledge the support of the Department of Civil and Environmental Engineering and Earth Sciences at the University of Notre Dame; the Center for Sustainable Energy at Notre Dame; General Electric Company; NextEra Energy Resources, LLC; and Terracon Consultants, Inc.

REFERENCES

Aasen, S., Page, A., Skau, K., & Nygaard, A. (2016). Effect of the Foundation Modelling on the Fatigue Lifetime of a Monopile-based Offshore Wind Turbine.

Arany, L., Bhattacharya, S. , Macdonald, J., & Hogan, J, 2015. Simplified critical mudline bending moment spectra of offshore wind turbine support structures. Wind Energy, 18, 2171-2197. https://doi.org/10.1002/we.1812

Arany, L., Bhattacharya, S. , Macdonald, J., & Hogan, J. (2017). Design of monopiles for offshore wind turbines in 10 steps. *Soil Dynamics and Earthquake Engineering*, 92, 126–152. https://doi.org/10.1016/j.soildyn.2016.09.024

Brinkgreve, R., Kappert, M., Bonnier, P. (2007). Hysteretic damping in a small strain stiffness model, Plaxis Manual

Brinkgreve, R. B. J., et al. (2012) "Plaxis 3D V22. Manual", AA. Balkema.
Halici, O. F. (2016). *Assessment of simulation codes for offshore wind turbine foundations*. [Unpublished master of science thesis]. Istanbul Technical University.

Khasawneh, Y., Nasim, M., Nafisi, A., Javadi, S., and Duan, L. (2021). "Wind Turbine Intermediate Foundation Performance through Frequency Analysis". In: Proceedings of the International Foundations Congress and Equipment Expo (IFCEE 2021), Dallas, Tx, May 10-14, 2021, pp. 384–393.

Richart, F.E. Jr., Hall, J.R. Jr., and Woods, R.D. (1970) "*Vibrations of Soils and Foundations*". Prentice-Hall, Inc., Englewood Cliffs, New Jersey.

Vitali, Osvaldo & Nassim, Mohammad & Khasawneh, Yazen. (2021). "Cyclic Degradation of Wind Turbine Foundations in Cohesive Soils". in Proc. DFI 2021, Las Vegas, NV, USA.

Vitali, Osvaldo & Nassim, Mohammad & Khasawneh, Yazen. (2022). "Cyclic and Monotonic Behavior of Onshore Wind Turbine Foundation Systems in Cohesionless Ground". in Proc. Geo-Congress 2022, Charlotte, NC, USA.

LONG-TERM PERFORMANCE OF WIND TURBINE FOUNDATIONS THROUGH NUMERICAL MODELING

William Tjaden, Department of Civil and Environmental Engineering & Earth Sciences,University of Notre Dame, Notre Dame, IN, USA. wtjaden@nd.edu

Spencer Ness, P.E., Department of Civil and Environmental Engineering & Earth Sciences,University of Notre Dame, Notre Dame, IN, USA. sness@nd.edu

Yazen Khasawneh, Ph.D. P.E., Department of Civil and Environmental Engineering & Earth Sciences,University of Notre Dame, Notre Dame, IN, USA. ykhasawn@nd.edu

Mohammad Nassim, Ph.D. P.E., Terracon Consultants Inc, GDI Sector, Germantown, MD, USA. Mohammad.nassim@terracon.com

ABSTRACT

The evaluation of long-term performance of the wind turbine structures is essential in lowering the Levelized Cost of Electricity (LCOE) from wind sources. The required maintenance of the wind turbine foundation during the design-life time of the wind turbine may result in escalated cost, if not properly evaluated during the planning and designing phases. The degradation of the foundation soils' stiffness due to the accumulation small plastic strains resulting from normal and/or extreme operating cyclic loading may result in escalating the O&M cost. The foundation's stiffness degradation will result in a shift of the wind turbine system natural frequency, which may result in amplifying the vibrations generated at the hub. The anticipated shift of the wind turbine system natural frequency during the design-life time should be investigated through numerical simulations of the anticipated dynamic loading utilizing soil constitutive models that are capable of capturing the foundation stiffness degradation due to the accumulation of such small plastic strains. This study presents the validation of an element model that will be used to evaluate the performance of existing and/or proposed constitutive model.

Keywords: Wind Turbine Foundation, Wind Energy, Numerical Modeling, 3D FEM, Cyclic Degradation, Long-term Performance, Constitutive Models.

INTRODUCTION

Throughout the life cycle of a wind turbine (WT), the foundation will experience $\sim 10^8$ of loading and unloading cycles that may result in the stiffness degradation of the foundation soils. The current design guidelines do not clearly cover the small plastic strain accumulation in the foundation soils and their effect on the degradation of the foundation stiffness, which may lead to an adverse effect on the long-term dynamic response of the wind turbines. There were attempts to evaluate stiffness degradation on the long-term dynamic response of the wind turbines, by using degradation coefficients for large number of loading cycles for both granular and fine grained soils. The evaluations were conducted using numerical simulations using cyclic loads that were applied monotonically using the hardening soil model and the cam clay models for granular and fine grained soils, respectively (Vitali et al, 2021 and 2022).

In general, there are several factors that control the undrained cyclic response of cohesive soils. The most significant factors are the shear strain amplitude, void ratio or density, confining pressure, plasticity index, OCR, number of cycles and frequency presented by Ishihara (1996). The stiffness degradation is traditionally presented in the form of a modulus reduction curve that is in the shape of an "S" (Eg. Idriss, I et al, 1979).

The typical loads acting on a wind turbine during normal operations are wind loads, dynamic loads due to blades rotation, and wave loads (in case of Offshore wind farms) as can be seen in Figure 1. The dynamic response fundamental frequency of wind turbine systems will be within three possible ranges as shown in Figure 2. The ranges are namely, soft-soft, soft-stiff, and stiff-stiff and are based on the rotor frequency (1P) and the blade passing frequency (3P). The wind turbine foundation provides the required degree of fixity to ensure that the dynamic response is within the acceptable range. If the foundations provide a soft-soft response, then large deformations are anticipated, which may include significant differential settlement. On the other hand, it will be cost prohibited to design the foundations to provide a stiff-stiff response. The most viable option is to design the foundations to provide soft-stiff response, however, this range is sensitive to any shift in the fundamental frequency, because any shift brings the fundamental frequency closer to the 1P & 3P loads which may lead to dynamic motions amplifications. As such, it is essential to quantify the cyclic degradation of the foundation soil's stiffness and the associated shift of the fundamental frequency during the design-life of the wind turbine.

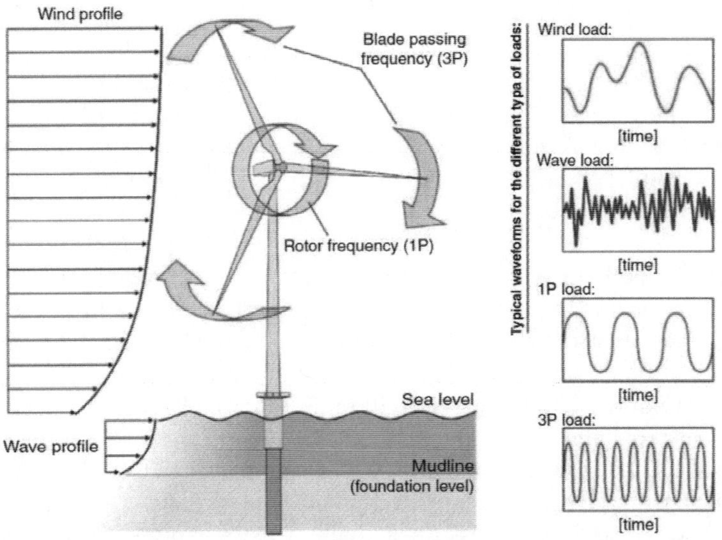

Figure 1: Representative Loads acting on WT (Bhattacharya, 2019)

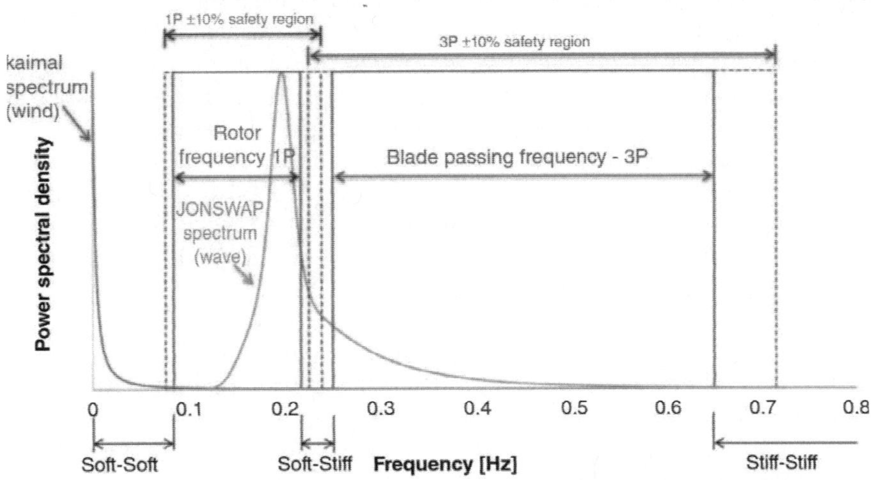

Figure 2: Typical Frequency Diagram (Bhattacharya, 2019)

This study outlines the initial steps to adopt/ develop a soil's constitutive model that will reasonably capture the response of undrained cyclic loading of cohesive soils, which will then be utilized to evaluate the long-term foundation soils stiffness degradation and associated shift of the wind turbine natural frequency.

CONSTITUTIVE MODELING

In order to accurately model the mechanical/ hydromechanical response of soils in numerical simulations, a constitutive model is required to capture the stress-strain relationship/ volumetric change or pore water pressure response. A few constitutive models were studied to assess their prediction capabilities of spoils response to undrained cyclic triaxial tests. Namely, the Hardening soil model with small strain stiffness, NGI-ADP, and UDCAM-S constitutive models. In an effort to decide which constitutive model is more capable to capture the long-term degradation of the foundation soils stiffness. It is important to understand the fundamental complexity of the problem trying to be solve, which will eventually predict wind turbine fundamental frequency changes under normal operating loads during the design-life time.

The Linear Elastic model, in the case of soils, can be useful in predicting the small stain behavior (strains $\sim 10^{-5}$). It is based on Hooke's law to capture the three-dimensional stress strain. The linear elastic model doesn't capture yield of the evolution of plastic strains. As such, this model is not suitable to simulate soils response because of the highly nonlinearity in their response when subjected to stresses. However, we use the linear elastic model to check the different aspects of the model, such as the validity of the boundary conditions and the applied loads.

The Mohr-Coulomb (MC) model is a linear elastic, perfectly plastic constitutive model that adopts the Mohr-Coulomb yield criteria. It is based on Hooke's law to capture the elastic portion of the stress strain curve. Once the failure criterion is reached, plastic strains evolve, which will show increase in the strain at constant stress unless dilation angle is introduced. This model is suitable only to provide a first order approximation of the behavior and should not be used to provide final conclusions or recommendations.

Another constitutive model that was investigated for its applicability to capture the soil's response to undrained cyclic loading is the Hardening Soil Model with and without small strain Schanz et al (1999) and Benz (2006). This model exhibits a stress dependent stiffness expressed by a hyperbolic stress strain relationship. The principal stress state in this model is not fixed and expands due to strain hardening. Two types of hardening are present within the Hardening Soil Model: shear and compression hardening. The shear hardening captures the strain hardening due to deviatoric stresses while the compression hardening captures the strain hardening due to isotropic loading or 1D compression like odometer condition. The Mohr-Coulomb yield criteria is adopted in this model. The hardening soil model overcomes the limitations of previous hyperbolic models by its formulation in the plasticity framework, while the previously developed hyperbolic models are formulated non-linear elasticity framework. The original model was incapable of capturing the hysteresis response of soils under dynamic loading. The introduction of small strain to the model allowed capturing the hysteresis response, however, the model does not allow for strain accumulation with increasing number of cycles and it incapable of capturing the evolution of the shearing induced excess pore water pressure.

The NGI-ADP model was developed by the Norwegian Geotechnical Institute (NGI) to capture the clays response to undrained loading. It was developed in the critical soil mechanics framework and is capable of capturing clay's response to undrained loading under various shearing mechanisms. It captures the response under active shearing (triaxial compression), direct simple shear (DSS), and passive shearing (triaxial extension). The model was formulated in anisotropic framework and is capable on capturing anisotropic stress response (Grimstad et al., 2011). Currently, the model is incapable of capturing cyclic response.

The UDCAM-S constitutive model has a similar formulation as the NGI-ADP model, with cyclic capabilities added to capture modulus degradation with number of cycles. The model utilizes plastic strain accumulations to capture stiffness degradation with the number of cycles by establishing the loading path on contours of equal shear strain level imposed on the normalized cyclic shear and number of cycles space (Anderson, k. et al., 1998 and Rasch, C., 2016).

MONOTONIC ELEMENT TEST CALIBRATION

The selection/ development of the appropriate soil's constitutive model with the required capabilities, for evaluating the long-term performance of wind turbine foundations, will require verification and calibration of the selected model through simulating element tests. The element tests should cover different shearing modes, hydraulic boundary conditions, type of loading, and rate of loading. The commercially available software provides a tool box that will aid in calibrating constitutive models. PLAXIS® has a built-in element test called SoilTest that allows users to simulate the response to several laboratory tests (shearing mechanism) with different hydraulic boundaries. The tests include triaxial and direct simple shear, with both monotonic and cyclic capabilities. In the SoilTes, tt is assumed that the element test is a $1m^3$ cube. However, the SoilTest tool is a "black box", which doesn't allow the user to inspect mechanical and hydraulic boundaries as well as the validity of the loading steps. As such, element tests were simulated in PLAXIS® using cubical and cylindrical shapes and were compared to the results from the SoilTest toolbox. The cubical shape is dimension is 1m, while the geometry of the cylinder is consistent with the size of a typical tri-axial soil specimen (144.1 mm x 71.1mm). The PLAXIS® element test was set up to simulate a monotonic triaxial test and a cyclic undrained tri-axial test (the cyclic test is discussed later in the paper). The discretization for the two geometries utilized 2nd order tetrahedral volume elements as presented in Figure 3. The stages are an "initial phase" that calculates the in-situ stresses of the sample based on its unit weight and height of the specimen. The next phase is all-around confinement to simulate the consolidation stage using consolidation calculation type in PLAXIS®. The subsequent phases simulates the shearing stage with the desired hydraulic boundary condition (drained or undrained) with compression (loading) or extension loading (unloading).

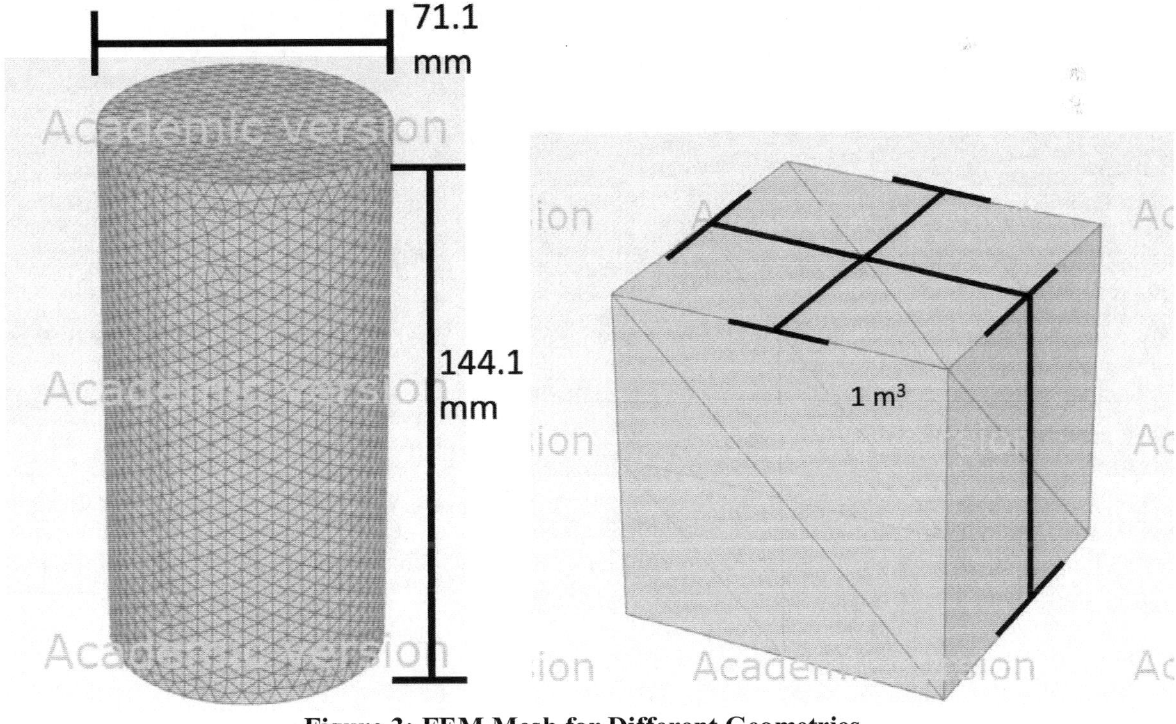

Figure 3: FEM Mesh for Different Geometries

Elastic-Perfectly Plastic

To ensure the accuracy of the cylindrical element test formulation and consistency with the built-in SoilTest feature, the cylindrical model was verified by simulating Consolidated Undrained test (CU). The selection of the elastic perfectly plastic model was for a first order approximation and is used to verify boundary conditions and loading sequence. The results from the simulation of cubical element, cylindrical element, and from SoilTest tool kit are presented in Figure 4 for the elastic perfectly plastic constitutive model (MC model). The material parameters used on the simulations are presented in Table 1.

Table 1 – Soil parameters for the Mohr-Coulomb soil model.

Stiffness		Strength		
$E'(kPa)$	v'	$c'_{ref}(kPa)$	φ' (°)	ψ (°)
$50e^3$	0.35	1.0	35.0	0

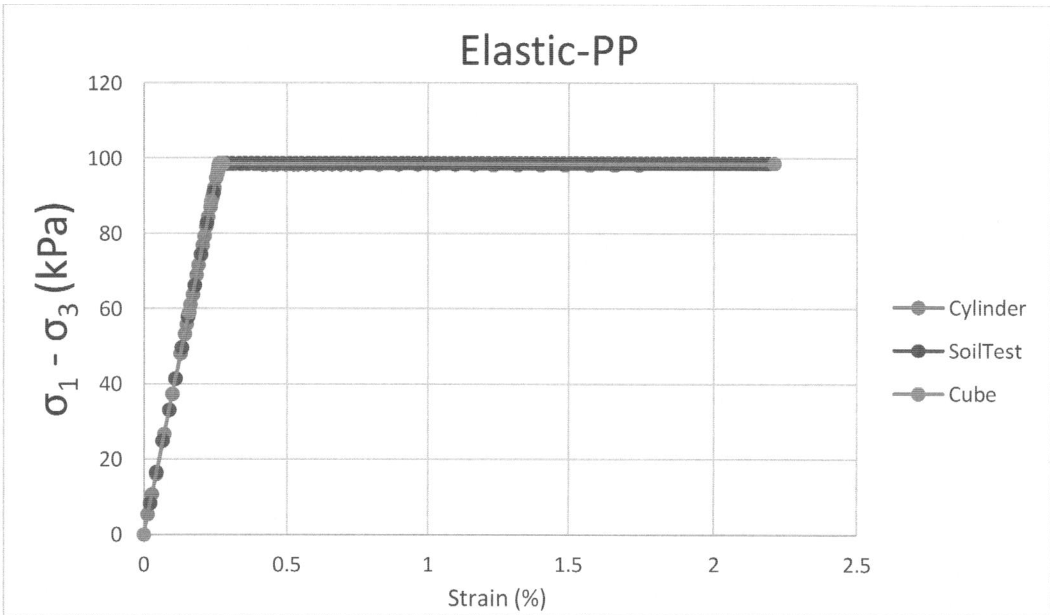

Figure 4: Elastic-Perfectly Plastic Verification

As it can be seen in Figure 4, the proposed simulation with cylindrical model agrees will with the SoilTest and the cubical element test.

Hardening Soil Model

The performance of the element test using relatively advanced soil constitutive model. The parameters used in the element test are presented in Table 2. They were chosen based off an available triaxial verification listed on Bentley's website. Figure 5 shows the stress-strain comparison. The calibration was completed with three different confining pressures to create the stress path and the critical state line shown in Figure 6.

Table 2 – Soil parameters for the Hardening soil model.

Stiffness				Strength		
E_{50}^{ref} (kPa)	E_{oed}^{ref} (kPa)	E_{ur}^{ref} (kPa)	Power (m)	c'_{ref} (kPa)	φ' (°)	ψ (°)
$20e^3$	$20e^3$	$60e^3$	0.5	0.0	35.0	0

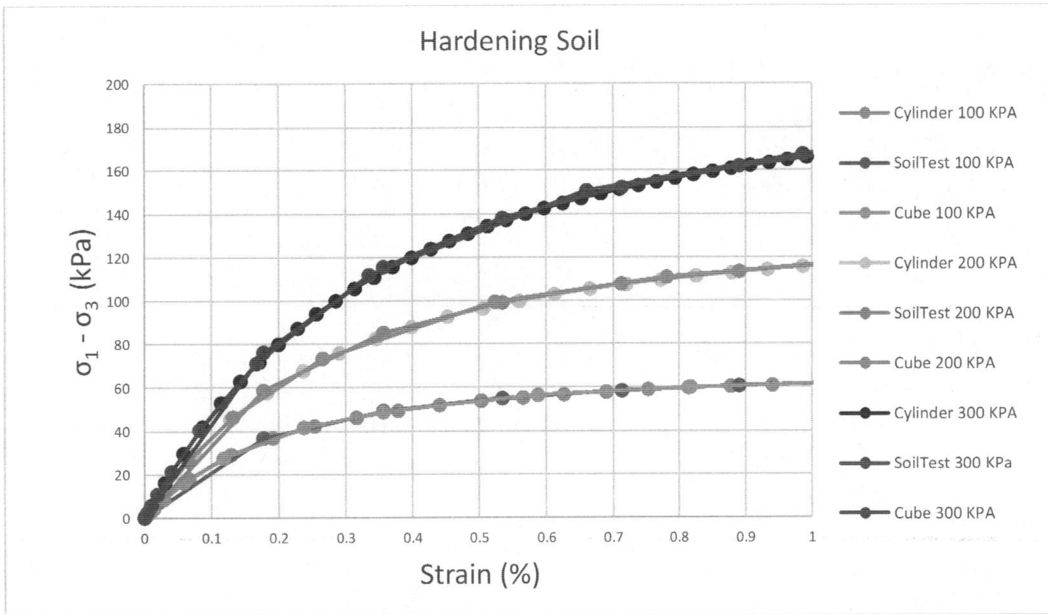

Figure 5: Hardening Soil Verification

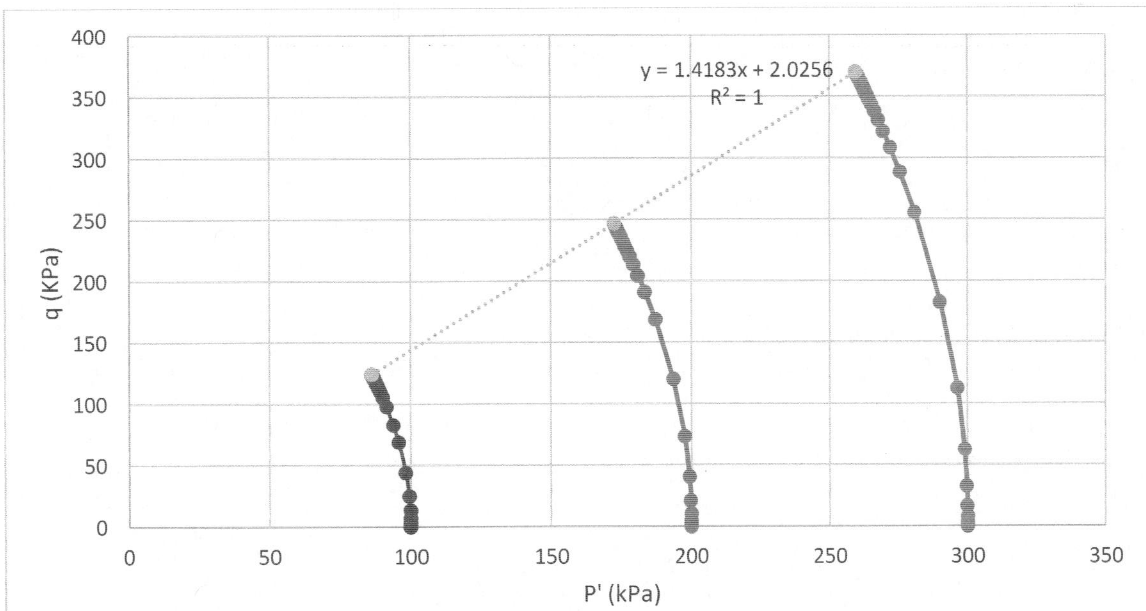

Figure 6: Stress Path

As it can be seen from Figures 5 and 6, the element test performed well and

NGI-ADP

The NGI-ADP constitutive model was simulated with the element test. The parameter used in the element test are presented in Table 3 and are based on Grimstad et al (2010). The comparison between the element tests and the SoilTest toolbox is presented in Figure 7, which shows a good agreement between the element test and the toolbox.

Table 3 – Soil parameters for the NGI-ADP soil model.

Stiffness				Strength						Advanced
G_{ur}/s_u^A	γ_f^C (%)	γ_f^E (%)	γ_f^{DSS} (%)	$s_{u,ref}^a$ (kPa)	z_{ref} (m)	τ_{inc}^C (kPa/m)	τ^E/τ^c	τ_0/τ^c	τ^{DSS}/τ^c	ν_u
420.4	6.43	7.87	11.97	149.7	0.0	0.0	0.5389	$2.32e^{-3}$	0.786	0.4950

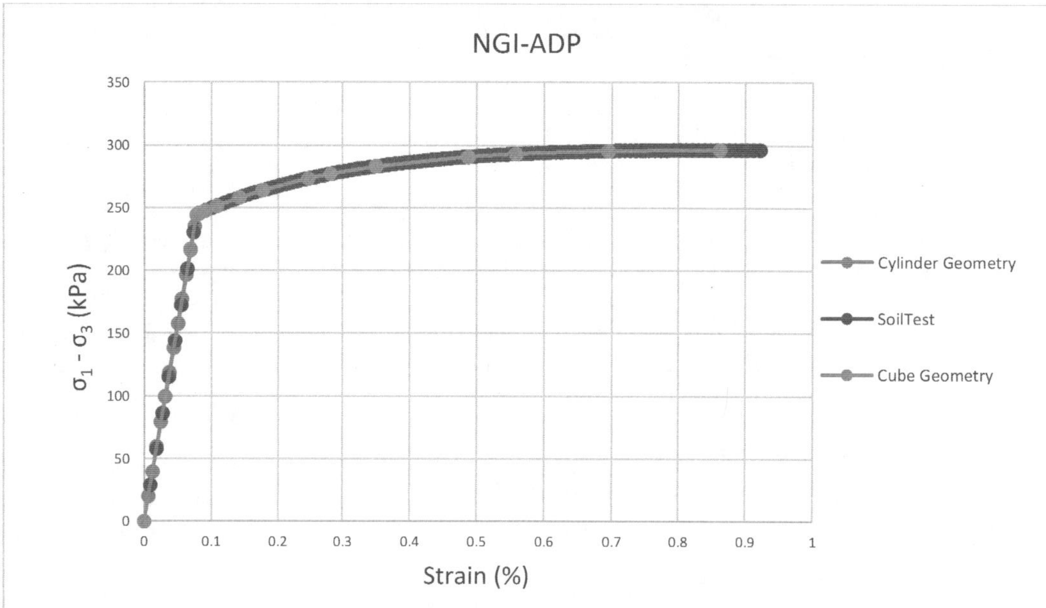

Figure 7: NGI-ADP Verification

UDCAM-S

Additional inspection was performed using the element test for the UDCAM-S model, with the parameters presented in Table 4. The parameters were based on Rasch (2016). The comparison between the element tests and the toolbox is presented in Figure 8.

Table 4 – Soil parameters for the UDCAM-S soil model.

Stiffness				Strength						Advanced
G_{ur}/s_u^a	γ_f^C (%)	γ_f^E (%)	γ_f^{DSS} (%)	s_{ref}^a (kPa)	z_{ref} (m)	τ_{inc}^C (kPa/m)	s_u^p/s_u^a	τ_0/s_u^a	s_u^{DSS}/s_u^a	ν_u
700	1.5	6.0	3.0	149.7	0.0	0.0	0.4	0.285	0.7	0.490

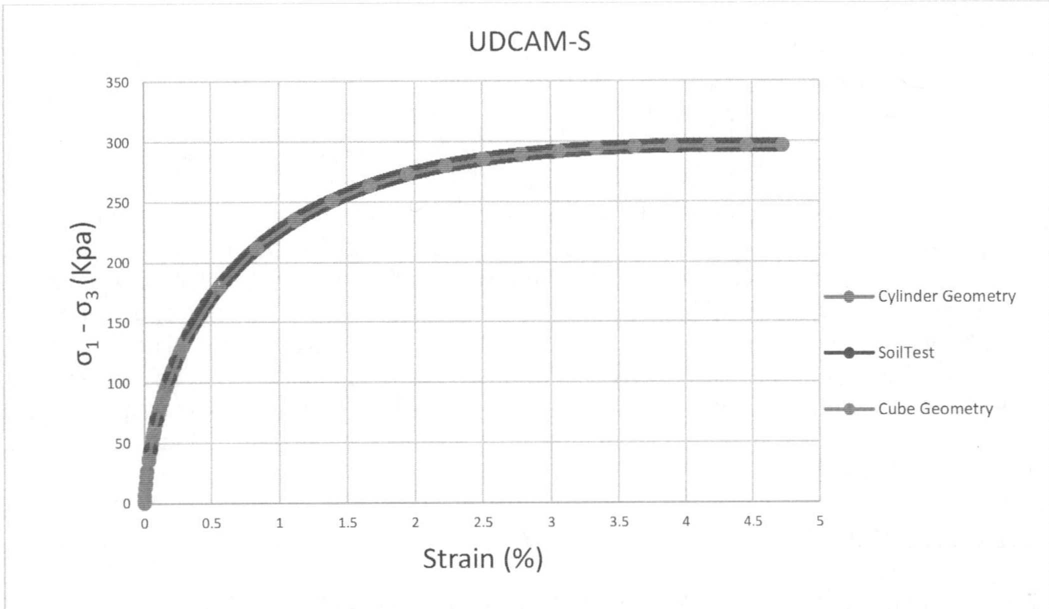

Figure 8: UDCAM-S Verification

CYCLIC ELEMENT TEST CALIBRATION

The element tests under cyclic loading using the dynamic type of analysis showed in consistency with the toolbox. The simulation of the cyclic tests used a one-way sinusoid of 0.1 Hz in the vertical direction. The comparison between the element tests and the toolbox is presented in Figure 9.

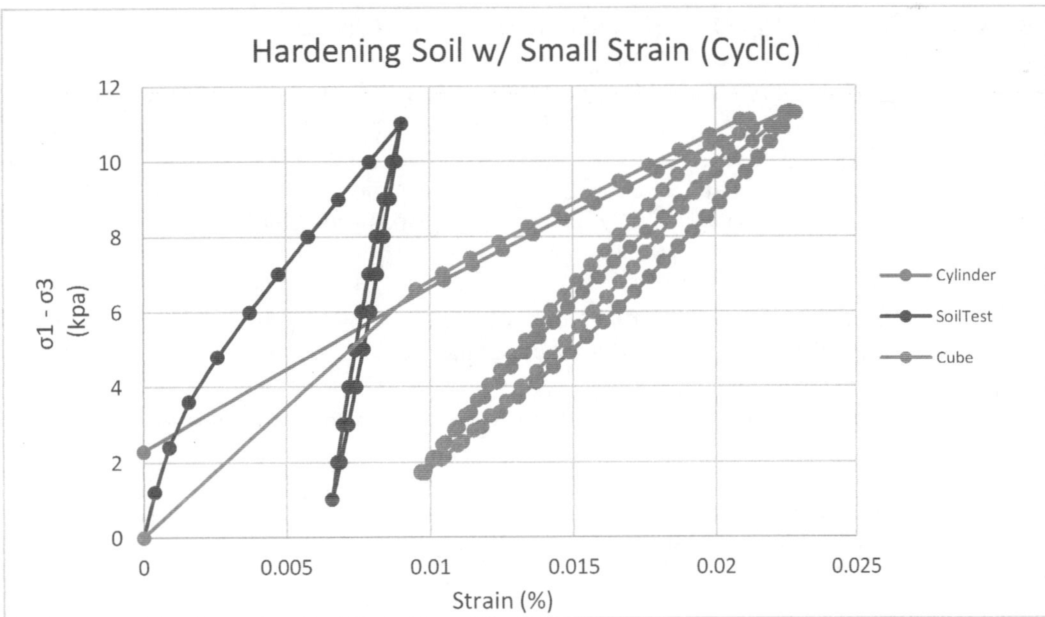

Figure 9: Element Test Performance Under Cyclic Loading

It can be seen from Figure 9 that there is a discrepancy between the element tests and the toolbox, the most important finding from this simulation that the deviatoric stress at the beginning of cyclic loading is not zero, although the toolbox showed zero deviatoric stress at the beginning of cyclic loading. In addition, the element cubical test showed zero deviatoric stress at the beginning of cyclic loading, but it showed closer response to the cylindrical element test.

CONCLUSION

Studying the effect of many cycles on WT foundations has paramount importance on gaining insight into the long-term behavior of the structural system. This paper highlights the calibration process of an element test to determine the appropriate soil's constitutive model. The monotonic loading scheme shows agreement between the built-in SoilTest toolbox, while the simulation of the cyclic loading showed geometrical dependency of the results. Furthermore, the issue of having a deviatoric stress at the beginning of cyclic loading suggests that the condition during consolidation was not entirely isotropic with shearing in the consolidation stage in the case of cylindrical geometry. As such, a more careful evaluation should be considered when using dynamic loadings with some geometries when using commercially available toolboxes.

ACKNOWLEDGEMENTS

The authors acknowledge the support of the Department of Civil and Environmental Engineering and Earth Sciences at the University of Notre Dame, the Center for Sustainable Energy at Notre Dame, General Electric Company, NextEra Energy Resources, LLC, Terracon Consultants, Inc., and Prof. Muhannad Sulimen's research group from Lehigh University.

REFERENCES

Benz, T. (2006) Small-Strain Stiffness of Soils and its Numerical Consequences. Ph.D. dissertation submitted to the University of Stuttgart.

Brinkgreve, R. B. J., et al. (2012) "Plaxis 3D V22. Manual", AA. Balkema.

Bhattacharya S. (2019) "*Design of Foundations for Offshore Wind Turbines*". John Wiley & Sons Ltd 1st edition.

Grimstad, G., Andresen, L., Jostad, H.P. (2010). "Anisotropic shear strength model for clay". International Journal for Numerical and Analytical Methods in Geomechanics, (Accepted for publication).

Idriss, I.M., Dobry, R., and Sing, R.D., (1978). "Nonlinear behavior of soft clays during cyclic loading". Journal of Geotechnical and Geo-Environmental Engineering, 104, 1427–1447. ASCE.

Ishihara, K. (1996). "*Soil Behaviour in Earthquake Geotechnics*". Oxford: Oxford Science Publication. Clarendon Press.

Kok, Sien & Ti, & Huat, Bujang & Noorzaei, Jamaloddin & Jaafar, Saleh. (2009). "A Review of Basic Soil Constitutive Models for Geotechnical Application". Electronic Journal of Geotechnical Engineering. 14.

Rasch, C. (2016). "Modelling of cyclic soil degradation: Development of a cyclic accumulation model and the application to a gravity based foundation", Ms. dissertation submitted to the University of Delft

Schanz T, Vermeer PA, Bonnier PG (1999). "The hardening soil model: Formulation and verification". Beyond 2000 in Computational Geotechnics

Vitali, Osvaldo & Nassim, Mohammad & Khasawneh, Yazen. (2021). "Cyclic Degradation of Wind Turbine Foundations in Cohesive Soils". in Proc. DFI 2021, Las Vegas, NV, USA.

Vitali, Osvaldo & Nassim, Mohammad & Khasawneh, Yazen. (2022). "Cyclic and Monotonic Behavior of Onshore Wind Turbine Foundation Systems in Cohesionless Ground". in Proc. Geo-Congress 2022, Charlotte, NC, USA.

